D0850192

ANNUAL REVIEW OF MICROBIOLOGY

ANNUAL REVIEW OF MICROBIOLOGY

Volume 36, 1982

MORTIMER P. STARR, *Editor*
University of California, Davis

ALBERT BALOWS, *Associate Editor*
Centers for Disease Control, Atlanta, Georgia

JEAN M. SCHMIDT, *Associate Editor*
Arizona State University, Tempe

ANNUAL REVIEWS INC. 4139 EL CAMINO WAY PALO ALTO, CALIFORNIA 94306 USA

ANNUAL REVIEWS INC.
Palo Alto, California, USA

International Standard Serial Number: 0066-4227
International Standard Book Number: 0-8243-1136-1
Library of Congress Catalog Card Number: 49-432

PRINTED AND BOUND IN THE UNITED STATES OF AMERICA

PREFACE

With the publication of Volume 36 of the Annual Review of Microbiology an era comes to an end, and we extend our deep thanks to Dr. Mortimer P. Starr, who is retiring as Editor. Dr. Starr has given 24 years of unstinting and enthusiastic service to the Annual Review of Microbiology, as an Associate Editor starting with Volume 12 and as Editor since Volume 27. His editorial expertise and exacting academic standards have helped place the Annual Review of Microbiology at the forefront of scholarly publications in the field. He will be greatly missed, and we wish him equal success with whatever publications may next benefit from his keen editorial attention. We also thank Dr. Jean M. Schmidt, who took over as an Associate Editor in 1981 on the departure of Dr. John L. Ingraham and most ably completed his term in that position. Dr. Ingraham, who became an Associate Editor with Volume 27, served the Annual Review of Microbiology with great dedication and patience, and we thank him warmly (if belatedly) for his rich contribution to the high repute of the series.

Alister Brass, Editor-in-Chief
Annual Reviews Inc.

Annual Review of Microbiology
Volume 36, 1982

CONTENTS

INDEXES

OTHER REVIEWS OF MICROBIOLOGICAL INTEREST

Special Announcement: Volume 1 of the *Annual Review of Immunology* (Editors: William E. Paul, C. Garrison Fathman, and Henry Metzger) will be published in April, 1983.

From the *Annual Review of Biochemistry,* Volume 51 (1982):

Wittmann, H. G. *Components of Bacterial Ribosomes*

Hershko, A., Ciechanover, A. *Mechanisms of Intracellular Protein Breakdown*

Stoeckenius, W., Bogomolni, R. A. *Bacteriorhodopsin and Related Pigments of Halobacteria*

Englund, P. T., Hajduk, S. L., Marini, J. C. *The Molecular Biology of Trypanosomes*

Cabib, E., Roberts, R., Bowers, B. *Synthesis of the Yeast Cell Wall and Its Regulation*

Jelinek, W. R., Schmid, C. W. *Repetitive Sequences in Eukaryotic DNA and Their Expression*

Challberg, M. D., Kelly, T. J. *Eukaryotic DNA Replication: Viral and Plasmid Model Systems*

From the *Annual Review of Biophysics and Bioengineering,* Volume 11 (1982):

Parson, W. W. *Photosynthetic Bacterial Reaction Centers: Interactions Among the Bacteriochlorophylls and Bacteriophenophytins*

From the *Annual Review of Genetics,* Volume 16 (1982):

Bentvelzen, P. *Interaction Between Host and Viral Genomes in Mouse Mammary Tumors*

Botstein, D. *Genetic Analysis of Microbial Development*

Chilton, M.-D. Tumor-Inducing Plasmids of *Agrobacterium tumefaciens*

Magasanik, B. *Genetic Control of Nitrogen Assimilation in Bacteria*

Nasmyth, K. *Molecular Genetics of Mating Type in Yeast*

Sussman, M. *Development Genetics of Dictyostelium*

Yanofsky, C. *Attenuation and Mutations that Alter It*

From the *Annual Review of Medicine,* Volume 33 (1982):

Downs, W. G. *The Rockefeller Foundation Virus Program: 1951–1971 with Update to 1981*

Tesh, R. B. *Arthritides Caused by Mosquito-borne Viruses*

Skinsnes, O. K. *Infectious Granulomas: Exposit from the Leprosy Model*

Zaleznik, D. F., Kasper, D. L. *The Role of Anaerobic Bacteria in Abscess Formation*

Armstrong, D., Wong, B. *Central Nervous System Infections in Immunocompromised Hosts*

Melish, M. E. *Kawasaki Syndrome (The Mucocutaneous Lymph Node Syndrome)*

From the *Annual Review of Pharmacology and Toxicology,* Volume 22 (1982):

Neu, H. C. *The In Vitro Activity, Human Pharmacology, and Clinical Effectiveness of New -Lactam Antibiotics*

From the *Annual Review of Phytopathology,* Volume 20 (1982)

Anderson, N. A. *The Genetics and Pathology of Rhizoctonia solani*

Blakeman, J. P., Fokkema, M. J. *Potential for Biological Control of Plant Diseases on the Phylloplanes*

Hildebrand, D. C., Schroth, M. N., Huisman, O. C. *The DNA Homology Matrix and Nonrandom Variation Concepts as the Basis for the Taxonomic Treatment of Plant Pathogenic and Other Bacteria*

Huisman, O. *Interrelations of Root Growth Dynamics to Epidemiology of Root Invading Fungi*

Murant, A. F., Mayo, M. A. *Satellites of Plant Viruses*

Purcell, A. H. *Insect Vector Relationships with Procaryotic Plant Pathogens*

Thresh, J. M. *Cropping Practices and Virus Spread*

Van Alfen, N. K. *Biology and Parasitology for Disease Control of Hypovirulence of Endothia parasitica*

Dahlberg, K. R., Van Etten, J. L. *Physiology and Biochemistry of Fungal Sporulation*

Zeyen, R. J. *Application of In Situ Microanalysis in Understanding Disease: X-Ray Microanalysis*

From the *Annual Review of Public Health,* Volume 3 (1982)

Berger, N. N. *Water and Wastewater Quality Control and the Public Health*

Chang, S.-L. *The Safety of Water Disinfection*

Ongerth, H. J., Ongerth, J. E. *Health Consequences of Wastewater Reuse*

ANNUAL REVIEWS INC. is a nonprofit scientific publisher established to promote the advancement of the sciences. Beginning in 1932 with the *Annual Review of Biochemistry,* the Company has pursued as its principal function the publication of high quality, reasonably priced *Annual Review* volumes. The volumes are organized by Editors and Editorial Committees who invite qualified authors to contribute critical articles reviewing significant developments within each major discipline. The Editor-in-Chief invites those interested in serving as future Editorial Committee members to communicate directly with him. Annual Reviews Inc. is administered by a Board of Directors, whose members serve without compensation.

Sidney Raffel

Ann. Rev. Microbiol. 1982. 36:1–26

FIFTY YEARS OF IMMUNOLOGY

Sidney Raffel

Stanford University School of Medicine, Stanford, California 94305

INTRODUCTION

From time to time when talking to young physicians or scientists, I try to recover in my thoughts what biomedical science was like in my formative days about a half a century ago. It is fairly apparent from the facial and verbal expressions of my dialogists—though they are usually discreet—that they regard the state of science of that period as belonging to a geologically distinct epoch. At such times, the same picture flashes through my mind: a conversation with a young clinical investigator in Baltimore, in about 1931. He was interested, as was I, in infectious mononucleosis and in the nature of and reason for its antibody response. His brisk, informed manner and his cognizance of the laboratory findings of the patient we were studying at the time were so akin in tenor to what such a conversation might bring forth today—with some important details added, such as the role of the Epstein-Barr virus in this disease—that I thought I would try more broadly to recall the state of knowledge at that time about those areas of immunology and related topics with which I have had experience.

This essay records personal observations about developments in the science that has been my main interest for about 50 years—immunology—and something about my experiences with students of various categories: medical, graduate, and undergraduate. Of medical practice I had only a taste during the years of World War II, and of full-time administrative duties I had a year as Acting Dean of the Stanford Medical School. Otherwise, with the exceptions—customary for midcentury scientists—of travel, committees, and the like, research and teaching have accounted for my life away from the hearth.

I came to research and teaching at a rather tender age, hence the length of exposure. My interest in immunology was sparked by a simple enough

0066-4227/82/1001-0001$02.00

stimulus, but memory of it has lasted for over half a century. I was 18 years old, a senior undergraduate at Johns Hopkins University, enrolled in a course in general biology taught by Ethan Allen Andrews. Professor Andrews was not one to pander to the interests of individual students, but he and assistants provided lectures and laboratory periods that, over the course of a year, pretty thoroughly covered the natural history of living objects, from the habits of the hydra to the anatomy of the frog and the development of the chick embryo, including, in the latter case, cutting sections of the prenatal fowl at various points in their 3 weeks of development, mounting them on slides, and executing drawings of their component parts. Of the then extant chemical understanding concerned in the workings of these and other beasts and plants, there was little word. But one day, reading one of our textbooks, I came across a few paragraphs describing antigens and their ability to induce the formation of antibodies specifically reactive with them. I thought this information was fascinating, a response sharpened considerably by a book called *Microbe Hunters,* by Paul de Kruif, which appeared at about that time. I could conceive of no more exciting way to spend my remaining years than to learn more about infections and immunity. So, at the end of that year, in 1930, I enrolled as a graduate student in the Department of Immunology at the Johns Hopkins University School of Hygiene and Public Health.

The fact that such a department existed suggests that the science had already attained a level of sophistication considerably beyond the preparation of vaccines and antisera. The Department was then small in personnel but spacious in size; it occupied an entire floor in the fairly new nine-story structure that had been erected, with the support of the Rockfeller Foundation, for research with emphasis on the training of health personnel from this and other countries. The Chairman of Immunology was Roscoe Hyde, a geneticist turned viral immunologist. Viruses were known chiefly by their effects in plants and animals, and by their filterability through porous membranes. Dr. Hyde was given to pithy aphorisms; what I recall best about his scientific teachings was his description of viruses as "genes gone wild." Another of the three faculty members of the Department was G. Howard Bailey, whose main interest at the time was the "heterophile," or Forssman antigen, a substance he and others tracked through a wide range of fauna and flora, from bacteria to man. Dr. Bailey became my mentor for 3 years, and subsequently I became his assistant for 2 years, working under a grant from the National Research Council that yielded for me $400 per annum. This was pretty scanty pay even for that period of the Great Depression, but the NIH had not yet been invented.

In those days, one leading textbook of immunology was a weighty tome by John Kolmer of the University of Pennsylvania. A more compact one by Hans Zinsser called *Infection and Resistance* was widely used. Dr. Kolmer's book described everything then known about immunology and how to do it, from bleeding a sheep to setting up serological tests. Zinsser's book was much more concise: It dealt with immunologic events and their explanations, and only little with experimental manipulations. There were other useful texts, one especially helpful by the British author W. Topley, and a small volume by H. G. Wells of the University of Chicago on *The Chemical Aspects of Immunity.*

The tendency for inhabitants of every generation to consider that developments in their area evolved simultaneously with the kindling of their own interest has been alluded to. This is perhaps particularly so now in the era of the biologic molecule. In fact, 50 years ago, the questions that interested immunologic investigators were surprisingly similar to those that engage attention now, but of course at a much more phenomenological level.

The concept of antigenicity was well defined in chemical terms and it had been known for years that polysaccharide as well as protein could function as antigen. Landsteiner and some predecessors had provided much detailed information about the influence of simple chemical groupings in determining antigenic specificity and had devised the term "hapten" for such chemical determinants.

The antigens of many infectious agents and their subgroups were becoming known, mainly by serologic distinctions [e.g. the salmonellae, classified by Kauffmann & White (1936),] the dysentary and cholera bacilli, and others. The lipopolysaccharide antigens of Gram-negative bacilli, and the Vi (virulence) antigen of the typhoid bacillus had been characterized. The capsular polysaccharide of the pneumococcus had been isolated and purified by Avery and Heidelberger in the 1920s, and here again antigenic distinctions within the species were known and were being exploited by Felton and others to produce antibodies in horses and rabbits for the treatment of lobar pneumonia in those pre–antibiotic times (although the sulfonamides and, not long after, penicillin were about to appear).

Loss and gain of antigens by bacteria was recognized. Smooth-rough dissociation had been described by Arkwright back in 1921, and the H↔O (flagellated to non-flagellated) fluctuation in Gram-negative bacilli had been established years before by Smith & Reagh (1903) and Weil & Felix (1917). Antigenic variation was also well established. Thus, phase variation —the shift between two expressions of antigenicity in *Salmonella* flagella —had been described (Andrewes, 1925). A harbinger of things to come, premonitory of the molecular biologic era, was reported by Griffith in 1928. He manipulated the transformation of one type of pneumococcus to another by injecting into animals a noncapsulated derivative of one antigenic type with killed capsulated cells of another type, and he isolated viable capsulated organisms of the latter kind. Some years after (1944) came the apocalyptic demonstration by Avery and colleagues that this could be done more directly by the transfer of nucleic acid.

A half century ago, polymorphism of antigens of animal cells was also well known. Karl Landsteiner and colleagues had defined the major human blood groups at the turn of the century, and the second human system of erythrocyte antigens (MN) was uncovered in 1922. The fact that some infectious agents could undergo changes in a major antigen during the course of infection, thus evading the specific immunity that had been acquired, had been established for *Borrelia* and for trypanosomes.

The influenza virus, which was to become the most notorious of infectious agents in this regard, was known by 1938 to show more than one antigenic form in the predominant type.

In this milieu of information concerning the diversity of antigens and their expressions in infectious agents, I undertook as a neophyte in graduate work to study *Trypanosoma equiperdum,* a flagellated blood protozoan related to the agents of African sleeping sickness. This parasite multiplies vigorously in rodents, so that a few days after infection a drop of fresh blood swarms with them, lashing about and kicking aside blood cells with their flagella. Giemsa stain of a blood film shows beautifully colored crescent-shaped organisms, each with a pale membrane along its border terminating in a single flagellum. Aside from the beauty of the beast, I was intrigued by the immunologic caprice of the infection it produces. Mounting numbers in the blood over a period of several days suddenly disappear completely, simultaneously with the appearance of antibodies against them. Then a few organisms reappear and again rapidly proliferate, unaffected by the coexisting antibodies. An individual animal might experience several such crises, i.e. the sudden disappearance of trypanosomes and their subsequent reimergence. With each crisis an antibody of new specificity appears, and each relapse strain evades the series of preformed antibodies. Since the trypanosomes could not be cultivated, strains of each antigenic specificity had to be carried in mice, with transfer to a fresh animal before the first crisis occurred.

This interplay of host and parasite is a fascinating compacted version, at least in principle, of what was later found to be the case with the influenza virus: a game between host and parasite, the former responding with an almost completely protective immunity, the latter in turn responding by discarding one antigenic guise and adopting another. In the case of the trypanosome, the game is played in an individual host; in the case of the virus, it is played in a community of hosts, worldwide, so that after a particular antigenic type has run its course, i.e. has exhausted the reservoir of nonimmune subjects, a new antigenic version appears.

In the 1930s, the nature of the trypanosomal antigen participating in this game was unknown, as was the process of antigenic alteration. I found, as had others, that the number of trypanosomal variants was finite for, if relapse strains were collected from a series of infected hosts, about a dozen could be accumulated. But any individual animal was eventually fatally overcome by a relapse strain, usually after the third or fourth recurrence.

The membrane antigen is now recognized as a glycoprotein, and the variant antigens appear to result not from mutations, but from rearrangements of the DNA, as might be anticipated from the nonrandomness of the number of variations.

HETEROGENETIC ANTIGENS

The diversity of antigens that provides distinctions among members of a species—of bacteria or of cells of higher organisms—has its obverse image in the existence of components of biologically diverse "containers" antigenically the same, or closely related. This was recognized, long before my neophytic years, as residing in a sharing of chemical structures reactive with the same antibodies. The best studied of these heterogenetic or so-called heterophile antigens was the Forssman substance, characterized back in 1911 by the investigator whose name it was given. This substance was found to be extractable by organic solvents from thc tissues of various animals and other forms of life. Its lipoidal extractability did not reflect the basis for its antigencity; it contains also carbohydrate and protein, and therein resides its antibody-inducing ability.

Because of the ubiquity in nature of this Forssman antigen, antibodies to it are also widespread in those species that do not themselves carry it. For example, in man, a non-Forssman-containing creature (except those with Type A erythrocytes, whose carbohydrate moiety is similar to that of Forssman antigen), low titers of anti-Forssman antibodies occur universally.

In the early 1930s, a wide range of bacteria, plants, and animal tissues had been surveyed for possession of this antigen; my preceptor, Professor G. H. Bailey, with Mary Shorb, had carried out a considerable part of these surveys. Its presence in infectious agents was especially enticing for the immunologists interested in disease, for it evoked such visions as possible cross-immunization by environmental entities (e.g. foods containing the antigen) that might protect against pathogenic organisms, such as certain of the pneumococci, which also contain it.

During this period, infectious mononucleosis came to be recognized as a problem for young people, though it must have existed for some millenia before and is now fairly commonly seen. A pair of investigators, Paul and

Bunnell, reported that patients with this disease had increased titers of antibodies against sheep erythrocytes—cells that had become the prototype for study of Forssman-bearing antigens. The disease affects lymphoid organs and lymphocytes, inter alia, and it became of much interest to us to determine the nature of this antibody response as a possible lead to a more specific diagnostic test for the disease, as well as for providing a possible clue to its etiology.

Hence, we tested sera from cases of the disease against a wide variety of Forssman-containing tissues and bacteria and found that the antibody concerned was in fact not against antigen of the Forssman type, but rather against an entirely independent heterogenetic antigen of the sheep red blood cell that occurs also in the erythrocytes of cattle. This finding did provide a more specific diagnostic serologic approach to the disease, but at the time it gave us no lead to etiology. The question of a virus as a causative agent certainly occurred to us, and we tried vainly to use tissues or tissue debris from diseased areas to absorb the antibodies, but these efforts bore no fruit. Methods for viral cultivation were not yet available; only many years later was it found that the Epstein-Barr virus is the causative agent. This virus infects the B class of lymphocytes (the antibody-producing type), but whether the cattle cell antigen is shared by this virus is still not clear. Since the potential antibody-synthesizing lymphocytes are infected, and since a potpourri of other antibodies (such as the Wassermann antibody) as well as excessive immunoglobulins often accompany this disease, it may be that the characteristic antibody is one of the many non-specifically stimulated by the resident virus, though the fact that this particular specificity is so constant an accompaniment of the disease makes this unlikely.

ANTIBODIES

Fifty years ago antibodies were being enthusiastically exploited, if not too well characterized chemically. They were respected for their ability to modify the courses of certain diseases, either by administration of serum from animals (usually rabbits or horses)—as in the treatment of diphtheria, lobar pneumonia, and tetanus—or through active induction by vaccination, as was the case for smallpox, typhoid fever, whooping cough, diphtheria, yellow fever, scarlet fever, plague, leptospiral disease, and some infections of domestic animals. Serologic tests for antibodies (or antigens) had been in use for decades, including the classical ones of precipitation of soluble antigens, agglutination of particulate substances such as bacteria, lysis (with complement) of certain bacteria and animal cells, particularly erythrocytes, complement fixation as in the famous Wassermann test for syphilis, opsonization for phagocytosis of particles by leucocytes and macrophages, and neutralization of bacterial and other toxins and of viral infectivity.

The insertion of markers into antibodies was being developed by Albert Coons with fluorescein isothiocyanate—a major prelude to the later localization of antibody synthesis in lymph nodes and spleen, and soon after in cells, the plasmacytes.

Exemplary studies of the protective mechanisms of antibodies in bacterial infections were carried out, during this period, by Cannon at the University of Chicago, and by Arnold Rich at Johns Hopkins University. Rich found that in rabbits infected with pneumococci and provided with antibodies, but deprived of most of their leucocytes by a bone marrow poison, the bacteria remained clumped (agglutinated) at the injection site for a time; but, in the absence of polymorphonuclear cells, they eventually broke loose to spread through the body. With these cells present, the bacteria were phagocytosed and destroyed: a beautiful demonstration of what had been inferred from more piecemeal evidence in the past.

There was also an appreciation adduced experimentally of the influence of heredity on specific antibody formation, and on the abilities of certain mice to resist particular infectious agents.

Little was known about the molecular nature of antibodies beyond the fact that they were "globular proteins" separable from other proteins of plasma by salting out or dialysis. Nothing was known about their cellular source. If anyone surmised that the lazy-looking lymphocytes might be the culprits, this was well hidden in the literature. I recall that, in a lecture in pathology in 1931, the lecturer, one of the sharpest existing exponents of his specialty, remarked that it was a constant rebuke to the intelligence of members of his profession that a function could not be assigned to those cells so ubiquitous in their distribution and in their intrusion into pathologic areas.

The brilliant quantitative methods and deductions by Heidelberger in this country and Marrack in England had demonstrated the mutual multivalence of antibodies and antigens in their reactions in varying proportions. But the precise number of combining sites per antibody molecule was not known, and some that precipitated poorly with soluble antigen were considered univalent, hence incapable of forming the lattice of precipitate resulting from mutual multivalence. Only some years later did studies with simple antigens (those possessing only one determinant group) establish in the ultracentrifuge and by equilibrium dialysis that antibodies have two combining sites, unless polymerized.

There were indications that all the antibodies made against a particular antigen were not the same in a number of respects: in their "avidity" for antigen, their capacity to activate complement, their ability to aggregate antigen, their protective capacities against infection, and their location among the globulins precipated by different concentrations of salt.

I found intriguing the notion that the association of antibody with different globulins might be related to their variable activities, and that this association might change with progressive immunization. Charles Pait, then a recent medical school graduate, was interested in these questions; an instructorship was available for him, so we collaborated in a study that eventually answered some of our questions.

The entities we used as antigens were in fact complex mosaics of antigens: sheep erythocytes, horse serum, and the bacterium *Proteus vulgaris.* The antisera obtained from rabbits at different stages in immunization were dialyzed against distilled water for prolonged times, with periodic removals of globulin precipitates. These were quantified as to nitrogen content and then were tested for antibody activity by various methods intended to indicate the changing quality of antibody per unit of nitrogen: for example, the ability of anti-erythocyte serum to promote lysis of the cells as compared with agglutination.

At this time, Arne Tiselius in Sweden had devised an apparatus for characterizing proteins by their migration in an electrical field, and a colleague, Eloise Jameson of the Stanford University Department of Chemistry, had journeyed to his laboratory to learn something of the new technology. She brought back one of the earliest models of the electrophoresis apparatus; this employed the schlieren mode of photographing moving protein boundaries. She and her young co-worker, Claudio Alvarez-Tostado, kindly turned their time and expertise to making photos of our antibody-containing fractions; the areas under the peaks I then laboriously measured by planimetry. But the basic knowledge required for interpretation, or perhaps the precision of the measuring techniques, were not adequate for informing our project beyond the point obtainable with Kjeldahl determinations and serologic assays.

The structures of antibody molecules and their globulin associations and functional differences had to await the work in the late 1940s of Porter, and subsequently Edelman and many other immunochemists, for precise definition. But, at the time, it was exciting to add to the evidence that all antibodies are by no means functionally the same, and that their activities and globulin associations shift with repetition of an antigenic stimulus and with time.

IMMUNOLOGIC HYPERSENSITIVITY

Fifty years ago immunologic hypersensitivity was a well-advanced concept among researchers, in both the laboratory and the clinic. The recognition of anaphylaxis was several decades along, dating from Portier and Richet's discovery back in 1902, when Prince Albert of Monaco, weary of being stung by Portugese Men-of-War while bathing in his coastal waters, importuned these scientists to find out what could be done about these poisonous hydrozoans. They injected extracts of these creatures into dogs and found them to be not only toxic, but also sensitizing to a shock syndrome on reinjection. At about the same time, in the US, Theobald Smith saw a form of respiratory shock on re-exposure of guinea pigs to diphtheria toxoid. Other manifestations of antibody-associated reactivities had also been described in the early years of the century: between 1903 and 1906 the Arthus reaction, human allergic reactivity or atopy, and the serum sickness syndrome in man. Furthermore, efforts were being made to do something about these; the concept and application of methods of desensitization were being discussed by Weil and others by 1913. In the mid-1930s, Cooke and Loveless reported the appearance in people undergoing desensitization of antibodies of the same specificity for antigen as the sensitizing ones, but differing from these in that they did not cause reactions; instead, they blocked the sensitizing antibodies from combining with antigen. Again, this was evidence for populations of functionally diverse antibodies, long before knowledge existed of their different molecular classes.

Histamine had been known for some years as a probably important mediator of hypersensitive reactions. Its origin from the basophil granule was to be uncovered later. In this period another, and what eventuated as the most important mediator of blood-vessel dilatation and smooth muscle constriction in hyperactive man, was found by Kellaway and associates. It was called slow reacting substance (SRS), in current parlance a leucotriene; like prostaglandin, it is a derivative of arachidonic acid.

Antibody-mediated immunologic hypersensitivity, or "immediate hypersensitivity," has been of recurring interest to me beginning in the mid-1930s, when Bailey and I worked with the pneumococcal polyaccharide capsular antigen. In those times, horses were commonly used by pharmaceutical companies for producing antibodies against pneumococci of different antigenic types for the specific treatment of pneumonias. Hypersensitive reactions to equine serum were frequent, and sometimes fatal, so that enlightenment about the conditions that might modify these occurrences was eagerly sought. Horse antibodies were known to be incapable of sensitizing guinea pigs for anaphylactic shock, whereas antibodies made by rabbits were very adept in this regard.

Our hope was to get some insight into these functional differences. Since molecular structures were unknown, our efforts were directed at phenomenological characterizations. If we used very small quantities of horse serum we found we could sensitize guinea pigs fairly regularly. Without knowledge of the structure of antibody molecules, or of the cells to which such molecules need to be anchored to combine with antigen and release the mediators of anaphylaxis, we could go no farther at the time.

Years later, in the mid-1960s, and in a different context, I collaborated with Alfred Amkraut, Leon Rosenberg, Oscar Frick, and others to learn something about the requirements for eliciting anaphylaxis with antigens possessing single reactive sites for antibody. By now, the knowledge that only some classes of antibody could attach to appropriate cell membranes was well established, and it had been found that—for the cell to be provoked into excreting anaphylactic mediators—it was necessary that the antigen link two adjacent antibody molecules at the cell surface. For this purpose, antigens with single combining sites would obviously not do. But we found that anaphylactic reactions could be induced with certain single-determinant compounds.

I believe that a thread links these findings separated by thirty years: the inability of horse antibody to sensitize subjects unless provided in very small amounts, and the ability of certain monovalent antigens to provoke the anaphylactic reaction despite the general requirement for antibody linkage by antigen. I think both point to an alternate path by which hypersensitive reactions can come about. Antibody-antigen complexes formed in the circulation may activate the formation of substances, e.g. bradykinin, that like the products of the mast cell constrict smooth muscle and dilate small vessels. This view was reinforced by the later finding with my postdoctoral fellow Baldwin Tom that isolated rabbit antibody of a molecular subclass known to be non-membrane attaching, when combined with antigen in vitro, would induce hypersensitive reactions upon injection into guinea pig skin, and with Frick and Liacoupoulis that very excessive doses of antigen injected into sensitized animals or added to lung strips from such animals would not elicit reactions. Both findings can be ascribed to the occurrence of non-membrane-associated activities by complexes of antigen with antibody in appropriate ratio, so that complement is activated, bradykinin is released, or both.

AUTOIMMUNITY

Autoimmunity was a well-developed concept a half century ago in all its aspects—theoretical, experimental, and clinical. The fertile mind of Paul Ehrlich envisioned the body reacting immunologically against itself and, with his aptness for the Latin bon mot, had dubbed this "horror autotoxicus." By 1904, Landsteiner and Donath had found a clear-cut clinical instance of this: Patients with paroxysmal hemoglobinuria destroyed their own red cells by means of antibodies and excreted hemoglobin in the urine. These rather unusual antibodies attach themselves to erythocytes at low temperatures, as when the subject immerses a limb in cold water. On return to the usual body temperature, the attached antibodies activate complement, and the cells are lysed. All of these conditions were revealed by the investigators seventy-five years ago, and no clearer example of autoimmunity has since been described, although many instances are now known or suspected.

In the 1930s, there was considerable speculation about an autoimmune basis for sympathetic ophthalmia, in which injury to the uveal tract of one eye sometimes results in destruction of this tissue in the other. In 1933, came the initial experimental demonstrations that injections into monkeys of nerve tissue—even that derived from the same species—would eventually induce an immunologic response with damage to their own myelin sheaths, a syndrome called "experimental allergic encephalomyelitis" or EAE.

The fact of such occurrences, spontaneously or experimentally induced (and, in the case of EAE, this may perhaps occur in humans after repeated doses of rabies vaccine prepared from animal nerve tissue), suggested a number of interesting points regarding the nature of an immunologic responding system that could countenance such gaffes as reacting against one's own tissues, and the nature of the tissue antigens that would permit such responses. Obviously, these must be organ-specific antigenic determinants, a thought not precisely apocalyptic even in those days before knowledge of the DNA code. A number of examples of such tissue specificities were known.

My entrée into these considerations started during research with antipneumoccoccal antibodies in the early 1930s, with G. H. Bailey. The bacteria were cultivated in beef muscle infusion broth and, in our experiments on anaphylaxis alluded to earlier, the fluids from the culture medium were sometimes used for eliciting shock in animals passively sensitized with antibodies against the bacteria. We were surprised to find that when we used sterile broth as an experimental control, anaphylaxis occurred, as with the bacterial soup. It was evident that the bacteria used to vaccinate rabbits for the production of antibacterial antibodies were simultaneously inducing antibodies to constituents of the broth itself, and we questioned whether or not this might be specific for the tissue from which the broth was prepared. Indeed, these antibodies were specific, directed against a thermostable antigen in the muscle. Broths made from other organs had their own specificities, as did tissues from a variety of species, including man.

We were unable to induce autoimmune disease, e.g. by injecting rabbits with vaccines cultured in rabbit muscle broth, or by administering antitissue antibodies produced in another species. This inability is of course a frequent finding in studies of autoimmunity: In many instances the presence of circulating antibodies cannot be correlated with pathologic effects, possibly because of the inaccessibility of the antigens involved.

Allergic encephalomyelitis has been of recurrent interest to me and colleagues, one of whom, Elizabeth Roboz-Einstein, was among the initial discoverers that a basic protein isolated from myelin carries the antigenic determinant that induces the disease. Antibodies were produced in rabbits against this protein and were labeled with a fluorescent marker; Helene Rauch in our group was able with this procedure to localize the antigen in the myelin sheath and, interestingly, to show that the antibodies induced by bovine myelin protein reacted as well with the nerve sheaths of humans and other species. This underscored the well-known fact that encephalomyelitis could be induced as well by heterologous as by homologous nerve tissue or its protein.

The central role of lymphocytes in cellular immune responses was still novel at this time, and we wanted to find whether or not cells from the immunized animal might combine with the protein antigen. Attachment of the labeled protein to some cells was evident, but at the time there was no distinction of antibody-producing (B) from cellular-reactive (T) lymphocytes; hence we were unable to make a final assessment of this central question. However, we tentatively interpreted our observations to suggest a direct lymphocyte-antigen combination.

INFECTION AND VACCINATION

A half century ago, the art of prophylactic vaccination was well advanced; after all, it began, almost simultaneously with the origin of the United States, with Jenner's vaccine for smallpox. It is difficult to pinpoint the true beginning of almost anything, but it seems that earlier in that century Lady Mary Wortley Montagu brought back from Turkey to England the idea that a crust of smallpox lesion itself could be given in "small" doses to induce a milder-than-usual case of the disease which would be protective. If one wished to pursue the origin of prophylaxis further, it is recorded that Mithridates of Pontus, a king whose thirst for power led him to kill every conceivable pretender to his throne, including his mother and children, dosed himself with small amounts of the regicidal poisons then in vogue by drinking the blood of ducks that had imbibed these in sublethal quantities. This was about 2100 years ago. It is unlikely he thought in terms of acquired immunity, but he certainly had in mind adapting himself to these noxious substances.

By the mid-1930s, the vaccines—as all active immunizing agents came to be called—included toxins or toxoids of diphtheria, scarlatina, clostridia and staphylococci, not all yet suitable for application, but one of them, diphtheria toxoid, had altered the visage of childhood. Although viruses had not yet been seen, vaccines containing modified viral agents (smallpox, rabies) had long been in use, and yellow fever vaccine was on the verge of a highly successful career. Influenza vaccine was also about to take off with the American and British demonstrations of the immunogenicity in humans of cultivated live virus. Bacterial prophylactics for typhoid, the two paratyphoid fevers, whooping cough, plague, and tuberculous and hemorrhagic fever (leptospirosis), as well as vaccines for some predominately animal diseases (anthrax, fowl cholera), were well established. This was in addition to the therapeutic use of preformed antibodies to treat disease already in progress by passive immunization, i.e. diphtheria and tetanus antitoxin and antipneumoccal serum for lobar pneumonia. It was fully appreciated in the case of bacteria that certain surface antigens were required to induce protective immunity. As mentioned before, the pneumococcal type-specific capsular polysaccharides had been isolated, and successful efforts to use these for active vaccination were reported in 1935.

Immunization to rickettsial diseases, because the causative agents were not cultivable outside living tissues, was approached with attempts to exploit the microbes in their insect vectors. The fertile egg yolk sac was not discovered as a good pabulum for these organisms until 1938, and the culture of tissues, although long useful for other studies, had not been developed as a medium for growth of viruses or rickettsias. But ingenuity was not to be denied; for typhus fever, an imaginative vaccine had been devised by Weigl in Poland in the first years of the 1930s that entailed infecting lice per rectum, and a week later harvesting rickettsias by a kind of enema. The germs were then inactivated with alcohol or formalin; this vaccine was administered to almost 200,000 people. A few years later, Herald Cox discovered that the typhus bacteria would grow in yolk sacs of fertile eggs.

When one considers the array of vaccines that have arrived on the scene since those days of fifty years ago, some for major plagues of mankind, the status of prophylasis in those times seems modest. But from the point of view of principles, a great deal had already been established. Aside from the fact of the efficacy of immunization per se, toxins had been detoxified, the cellular requisites for effective bacterial vaccines had been clarified, in one case a chemical derivative of a bacterium had been readied for trial as a vaccine, and viruses modified from their original virulence were in use. I think no other chapter in man's ascent from the ooze has contributed comparable benefit to his physical welfare—and to that of other species—without exacting the customary price of progress: danger to the individual and disruption of his surroundings. This triumph is epitomized by the declaration that appeared in the Bulletin of the World Health Organization in May, 1980: It proclaimed that, consequent to an effort that began twenty years ago, the world is now free of smallpox. This conquest of a plague that had decimated mankind for millenia is without parallel in the history of human affairs; it deserves blasts of trumpets and an international official holiday as an annual observance.

In the mid-1930s, poliomyelitis remained one of the fearsome infections for which there was no prevention; laborious and expensive efforts were tried to limit its disastrous consequences. Because President Franklin Roosevelt had suffered this disease, he instigated the formation of the National Foundation for Poliomyelitis, which became a source of support for treatment and research. The Department of Bacteriology and Experimental Pathology at Stanford University was one center of laboratory investigation of this disease; this effort was headed by Professor Edwin W. Schultz, the Department's chairman. In 1935, he invited me to migrate from Baltimore for a year of postdoctoral fellowship under the aegis of the Foundation. In those days, such a transcontinental junket was quite an adventure. Although rail transportation was as fast and comfortable as it was ever to be, the trip was not lightly undertaken, in part because of the just subsiding depression. So the prospect of the journey, of a year in California, and of the opportunity for research on an interesting disease was an exhilirating one.

The department in 1935 was housed in a building that had survived the earthquake of 1906. It had been part of a large quadrangular museum of which intervening portions had collapsed, and hugh chunks of sandstone and stuccoed brick remained in the field between the department and the residual museum for many years after my arrival. This building had been assigned to the department in 1911 shortly after Hans Zinsser arrived from Columbia to become the first head of bacteriology as a Division of the medical school; for several preceding years, a course in bacteriology for medical students had been taught by Professor Robert Swain of the Chemistry Department. The quarters were intended to be temporary, a status terminated in June, 1981, when the department moved to handsome new quarters in the medical school complex.

Despite its rough treatment by natural forces and its primary design for a purpose very different from biomedical research, the old structure had lent itself well to modification, and by the time of my arrival there were a number of laboratories as well as teaching facilities accommodating a class of 65 medical students and about 30 graduates and undergraduates. On the lower of its two floors, some of the space housed animals; this facility was supplemented by an outside structure to take care of several hundred rhesus monkeys (a few of which occasionally escaped into neighboring Palo Alto) in addition to the more usual experimental animals.

The research laboratories contained the full range of current technical instruments, and also facilities unusual at the time. The outstanding example of this was a collection of air-driven ultracentrifuges. These were then being developed in this country by Professor Beams, a physicist at the University of Virginia. A local mechanician, Louis Grebmeir, then and for a number of succeeding years, manufactured ultracentrifuges with rotors of varying capacities and of different alloy compositions for the department in his tireless search for the ultimate machine. The rotors, whether large or small, were all air driven, and some attained speeds of up to 90,000 rpm. On one occasion a larger one tore loose from its mooring, which consisted of a thin wire shaft suspended from a vane that whirled on a bed of compressed air. The rotor sheared the lid bolts of the surrounding thick steel casing and tore a section out of the wall across the room. The assistant who customarily shepherded runs of ultracentrifugation was fortunately not in position at the moment, directly in the line of fire.

A room was maintained at constant temperature and humidity for the manufacture of cellulose membranes of predetermined average pore diameters. These were used for obtaining approximate dimensions of viruses.

At a somewhat later time Professor Schultz was among the sponsors of a University invitation to Professor L. Marton of Belgium, a pioneer in the development of the electron microscope. Professor Marton remained, I believe, for two years and assembled one microscope in the department. I recall well the photographs of *E. coli* bacteriophage made with this instrument, and the discussion about the figure-8 object in its head. I am not certain whether the question of chromosomal identity was seriously raised at the time; my recollection may be distorted by hindsight.

The department was a small one. In addition to Edwin W. Schultz, William Manwaring was a Research Professor with major interest in immunology. At the time of my arrival, Michael Doudoroff was a graduate student, as was Ryland Madison, who worked with Dr. Manwaring in the then new area of fibrinolysin, the streptokinase that causes fibrin to dissolve. Charles Clifton, who in 1945 became the first editor of this Annual Review, had preceded me in the department by a few years and researched and taught in the area of bacterial physiology. Paul Beard held a joint appointment in the Department of Engineering; he was concerned with sanitary bacteriology. Byron Olsen was a new instructor, charged primarily with teaching medical bacteriology.

This small group encompassed a good variety of research interests, and the department was a lively place for a young fellow. Dr. Schultz's interest in viruses extended to the bacteriophages, then relative newcomers to the field. In addition to trying to learn something about their nature, he set up a laboratory to determine whether or not these bacterial predators might be useful therapeutically. He advertised to physicians that the laboratory would try to adapt a bacteriophage—of which there was a large collection—to any bacterium isolated from a patient. This was done at cost to the laboratory, then about $2.00. The physician was expected to report briefly on the use of the phage and the subsequent fate of the patient. After several years the project was abandoned for lack of adequate reporting and because of the advent of sulfonamides.

The main effort in the department was in poliomyelitis research, including studies of the virus itself, of methods for detecting it or an immunologic response to it that would obviate the need for neutralization tests in monkeys, of establishing the infection in smaller animals, and of seeking an effective vaccine or some other way to ward off infection.

I participated in these activities along with Louis Gebhardt, who was then a graduate student, eventually to become head of the Department of Bacteriology at the University of Utah. Another young colleague was medical student Harold Pearson, later an instructor in the Department and subsequently Professor in various departments of the medical school of the University of Southern California.

Efforts to establish poliomyelitis in smaller animals failed, but about this time other investigators found that certain strains of the virus could be transmitted to small mammals—the cotton rat, and eventually hamsters and mice.

We had some success in characterizing the virus. Elford cellulose membranes had already "sized" it in the correct range of about 25 mM, and we were eventually able to sediment it in the air-driven ultracentrifuge. I have a vivid memory of Harold Pearson standing by the centrifuge with a pitchpipe, which, in the earliest days, was used for estimating speed. But reference to our publication indicates that during the course of this study, in 1936, we had become more sophisticated; Mr. Grebmeier had devised a stroboscopic light that picked up a marker on the rotor. In any case, we found that about 30,000 rpm for two hours sedimented pellets containing the virus.

We had been trying to arrange a serologic or other immunologic test for the presence of virus or of reactivity to it—to no avail. But, several years later, Hubert Loring of the Biochemistry Department sedimented larger amounts of virus, and together we demonstrated complement fixation with the sera of immunized rats that had resisted challenge infection. This protection of rats was premonitory of Salk's conquest of the disease some years later. It has been puzzling to me why our similar efforts in monkeys, which we had attempted many times to immunize with variously inactivated viral preparations, were never sufficiently convincing to presage Salk's later success. We worked with a single type of virus; antigenic distinctions among types were beginning to be appreciated, but we used the same viral preparations for vaccination and challenge. Perhaps the viral mass was not sufficient, or the strain employed was not sufficiently immunogenic.

It was thought by some investigators at that time, from work with monkeys, that the route for poliomyelitis infection was via the olfactory bulbs. There was a good deal of painstaking histological evidence to track the progress of virus to the central nervous system by that path in monkeys, which could readily be infected by lavaging their nasal passages with virus suspension. As fate would have it, this is not true for humans, who acquire their infections by the gastrointestinal route. But the rhesus monkey was considered so faithful a surrogate for man in poliomyelitis studies that this divergence that later became very apparent was not fully appreciated at the time.

Dr. Schultz devised a procedure intended to block entry of the virus into the olfactory nerves by spraying the nasal passages with tannic acid or zinc sulfate. Monkeys treated in this way subsequently resisted infection by this route, but the few hardy human volunteers who received the treatment succeeded only in losing some of their ability to smell.

TUBERCULOSIS

In the early 1930s, tuberculosis was still a serious problem, although a number of the more health-oriented countries of the world showed a progressive decline in incidence. A vaccine had been introduced in France in the early 1920s, which consisted of a living attenuated strain of bacilli of bovine origin, isolated for use as vaccine after long cultivation, and named for its originators, Bacille Calmette-Guérin, or BCG.

BCG has had a peculiarly checkered career in the annals of prophylactic immunology. There has never been question about its efficacy in experimental animals, but its usefulness for humans, although demonstrated many times in some controlled trials, has been found wanting in others. To confound the issue of use further, a major disaster occurred in Lübeck owing to a confusion of cultures, so that a group of infants was dosed with virulent bacteria by mouth. The fear of such a consequence was never completely overcome, though the negligence that led to it had been amply documented. In the US, an additional objection on the part of some pediatricians was that wide-spread use of the vaccine would obviate the diagnostic value of the tuberculin test, which depends on allergic reactivity to the bacillus. Hence, fifty years ago the US needed a broadly acceptable prophylatic agent, and a therapy more useful than the rest and mountain air then available to those with adequate financial resources.

My interest in tuberculosis evolved from a course in pathology at Johns Hopkins University with Arnold Rich, a man endowed with an exceptional ability to stimulate students. He was then writing *The Pathogenesis of Tuberculosis*—a thousand-page volume of information and concepts couched in an unusually attractive style.

In the early 1930s, much was known about this disease both clinically and pathologically, and there was a considerable store of information and conjecture about its immunologic features. Some of those conjectures are still with us. The fact that infection with the mycobacterium led to delayed hypersensitivity, as evidenced by reactivity to the bacilli, or to tuberculin, a concentrate of the medium in which they had grown, had been found by Robert Koch in the 1880s. There were those who equated this cellular hypersensitivity with immunity to the bacterium, i.e. with cellular- rather than antibody-mediated resistance. Rich could not believe nature would be so profligate as to evolve two different entities, cells and antibody molecules, both with the ability to recognize specific antigens. This conviction led him to try to abolish the cellular reactivity of delayed hypersensitivity in vaccinated guinea pigs by dosing them with increasing amounts of tuberculin, a process known to immunologists as desensitization. Such animals without sensitivity retained the ability to resist challenge infection, ergo, he believed his thesis was substantiated. But in fact, his share in the ultimate truth of this matter proved to be only partial: There is a cellular immunity to the tubercle bacillus as well as to other infectious agents and many tumors; the subsets of cells (lymphocytes) that participate in this are probably different from those responsible for delayed hypersensitivity. But the recognition units for antigen on these cells partake of antibody-like character, although their final definition has not yet been made.

I found Rich's lectures on this and related topics to be very stimulating and, subsequently, I thought a good deal about possible approaches to some of the immunologic facets of the disease. I surmised that if one could isolate from the bacillus its protection-inducing antigen(s), this might unravel the confusion regarding immunity and cellular reactivity while providing a vaccine of unimpugnable safety.

The California Tuberculosis Association at about that time, in the late 1930s, was in a state of nascent activation. A number of clinicians, among them Harold Trimble, Reginald Smart, Buford Wardrip, and Corwin Hinshaw, had a scholarly enthusiasm about the nature of tuberculosis along with their clincial interest in it. They had begun to assert their views in eastern meetings of the National Tuberculosis Association and succeeded in obtaining larger slices of funds to support research. They were good enough to provide my efforts with some of this.

I began a long continuing effort to disassemble *Mycobacterium tuberculosis* into components I hoped might be correlated with the body's responses to it. A number of my predecessors had devised procedures for fractioning the bacilli: notable among them were Anderson at Yale, Florence Sabin at the Rockefeller Institute, and Florence Seibert at the Phipps Institute of the University of Pennsylvania. They had prepared lipid, protein, and polysaccharide derivatives and had learned a good deal about responses to them, which led, in Seibert's work, to preparation of the widely used PPD (purified protein derivative) of the bacillus to supplant tuberculin for diagnostic skin testing.

I cultivated about a pound of virulent *M. tuberculosis* cells in a synthetic medium and began to extract these with organic solvents, intending to test each extractive and the bacillary residue for its capacity to induce protective immunity, delayed hypersensitivity, and antibodies to bacterial proteins.

The tubercle bacillus and its close relatives have the interesting ability to induce cellular delayed hypersensitivity not only to their own protein antigens, but also to any other extraneous antigen mixed with them for injection, particularly if the mixture is enclosed in an oily vehicle. Such mixtures also markedly promote antibody production to the extraneous antigens. These adjuvant abilities, first disclosed by French investigators in the 1930s, were brought to focus by Jules Freund in the early 1940s. Thenceforth, the bacillary-oil mixture has been known as Freund's adjuvant.

After a good deal of labor, we found that bacteria deprived of a group of chloroform-soluble lipids lost the ability to induce cellular delayed hypersensitivity. The extractive, furthermore, could replace the bacillus in this activity in association with a variety of antigens, with tuberculoproteins as well as with such extraneous substances as picryl chloride and egg albumin. Extraction of the lipids also deprived the bacteria of their immunizing ability, and of their antibody-stimulative capacity as constituents of Freund's adjuvant.

At this juncture in the research, in 1949, a Guggenheim Fellowship opened the way for a year of experience in Europe. I look back on that adventure as my closest brush with heroism, abetted by my wife's resolution and endurance. At the time, our five daughters were between ages ten and two. In 1950, airplanes flew, but hardly as family conveyances, at least not in our circles. It was boat and train all the way, replete with eighteen pieces of hand luggage and a number of auxiliary trunks and cartons that came along in their own time.

My scientific intinerary called for several months in Basel with Professor Josef Tomcsik at the Hygienic Institute of Basel University, and in Sweden with Dr. Gösta Widstrom, a colleague in tuberculosis research. These plans broadened in the event to include a short stay in Paris and several weeks in a fishing village between Nice and Cannes, where I settled my wife and daughters for a several-month period while I went off to Basel, to which I eventually fetched them. There I spent some months extracting lipids from various bacteria, writing a book on immunity, and enjoying the warm hospitality of our hosts, Josef and Olga Tomcsik,

The Tomcsiks had come from Budapest to succeed Robert Doerr, the well-known virologist and immunologist, who had headed the department.

Professor Doerr still came to his office daily; he was engaged in revising his monographs on the viruses, and—since I had shipped in cartons of current reprints on immunologic subjects as grist for my own textbook, and gave him full access to these—after a time we established a warm relationship. This overcame his general coolness toward America and its inhabitants. He had visited the US several years earlier, at a rather advanced age and without the benefit of conversational English or a companion. Nonetheless, he had crossed the country, and he told me in his forthright way that he found little to admire about it, with the exception of the Pacific Ocean, which he found to be first rate.

Professor Doerr spanned the immunologic era from Paul Ehrlich to the 1950s. He had worked with many of the contributors to the origins of immunology, and conversations with him were illuminating and entertaining. The word "dummkopf" was never far from his descriptive armamentarium.

We went then to Copenhagen for several months at the State Serum Institute, for an equal time to Stockholm, then to Britain, and finally to Paris and Le Havre for the return home.

During our stay in Basel, I received a note from Edgar Lederer of the Centre National de la Recherche Scientifique in Paris, suggesting a collaboration in working out the chemical characterization of the mycobacterial wax. Dr. Lederer was a pioneer in chromotographic analysis, and he hoped we might be able to distinguish which components in our lipoidal mixture were responsible for the biological effects. This collaboration continued for a number of years, during which Lederer's group chromatographed extractives, and we tested them in guinea pigs for their biologic capacities. For a time, these isolations made it appear that the salient substance concerned in the induction of cellular hypersensitivity was composed of an unsaturated fatty acid peculiar to mycobacteria, and called mycolic acid, esterified with a saccharide. But subsequent chemical studies by others showed that mycolic acid extracted from the bacillus is linked to muramic acid, which in turn is joined to four peptides. Eventually the important ingredient was identified as muramyl dipeptide—a very simple and ubiquitous component of bacterial cell walls. One wonders why a wider array of bacterial cells do not show the striking adjuvant and delayed hypersensitivity-inducing properties of mycobacteria if the simple muramyl-dipeptide is at the root of these. The possibility that mycolic acid may, after all, participate in the biologic events stimulated by these bacteria still intrigues some investigators.

The ignorance of the mid-1930s regarding the protection-inducing immunogen of the tubercle bacillus is not entirely relieved today. Perhaps the most promising recent lead into the solution of this has been proposed by my former student, Alfred Crowle, now at the Webb-Waring Institute of the University of Colorado Medical School. With great persistence Dr. Crowle has pursued evidence that he first uncovered as a student over twenty five years ago, that a trypsin extract of the bacillus contains a proteinacous protection-inducing substance. Recently, after many successful tests for protection in animals, this antigen has been administered to human subjects in whom it has shown immunogenicity. The World Health Organization is contemplating broader-scale tests at this writing.

With analogous persistence, another student of that era, Ivan Kochan, has elaborated evidence of mycobacterial stasis in animal sera to the point of defining a siderophore called mycobactin, which abstracts iron from the environment for the bacterium, and which may vary in effectiveness or quantity with the virulence of the organism.

Long experience with the striking immunologic events occasioned by the mycobacterium turned my attention to more general questions about cellular immunity.

Fifty years ago, cellular immunity and delayed hypersensitivity were well-developed concepts in respect to infectious diseases—such as tuberculosis, lymphogranuloma venereum, smallpox, and syphilis—for all of which diagnostic skin test materials were available. A lively investigational interest was also well developed in regard to the cellular hypersensitivity induced by plants and chemicals, such as poison oak, poison ivy, and various chemicals used in industry. But notions about the nature of these reactivities was obscure; they were definable mainly by the fact that reactivity could be demonstrated in the absence of antibodies, as for example in the demonstration back in 1910 by Bail that hypersensitivity to tuberculin in animals could be transferred via cells from the peritoneal cavity or the spleen, but not by antibody-containing serum.

The great problem besetting those times was, of course, that no one knew which cells were responsible for immune reactivities, either for antibodies or for cellular immunity. It was not until the 1950s that lymphocytes emerged as the central actors in the immunologic drama, and the unwinding of the different functional populations of these umbiquitous cells has been going on ever since, so that we know now that a subset of thymus-derived lymphocytes (T cells) are involved in delayed hypersensitivity, but that these may be different from another subset concerned with protective immunity, whereas another major population of cells derived from the bone marrow (B cells) are destined to be synthesizers of antibodies. But some of the important questions that plagued the immunologic generation of the 1930s are still unclarified, e.g. what precisely is the nature of the receptor that recognizes antigen on the delayed hypersensitive cell, and exactly how does the macrophage actived by Freund's adjuvant encourage cellular reactivity?

My interest in these questions led to a series of studies over some years, aimed mainly at the central one: What kind of stimulus determines whether or not an immunologic response will be predominately cellular or humoral and, once uncovered, can the stimulus be patterned to accommodate various situations in which one or the other kind of response would be desirable? An example would be the case of tumors, against which protective immunity is frequently of cellular type.

Experiments with students Margot Pearson, Michael Brunda, and Judith Britz showed that macrophages exposed to Freund's adjuvant and then mixed with an antigen in vitro for injection into animals gave rise to a predominately cellular immunity to the contained antigen. I hypothesized that this might be a result of degradation of antigen by the lysosomal enzymes of activated macrophages. This idea was to me a very attractive hypothesis because it fell in so neatly with reported examples of simple chemical antigenic entities that stimulate only cellular responses. However, Judith Britz showed rather convincingly that the effect depends upon a soluble factor from activated macrophages that in some way stimulates the appropriate T lymphocytes into responsiveness, a substance of the group now referred to as interleukins.

A good share of my professional time at Stanford University was spent in teaching—mostly of medical and graduate students, and episodically of undergraduate majors in microbiology. I was myself periodically a classmate of my own students in some of my earlier years. I came to the Department as a Fellow in 1935, but I was soon enthralled by the California ambience, the opportunities for research, and the University's public health nurse, Yvonne Fay, who later became a Raffel, and much to my joy has remained so for over forty years. Fortunately, Professor Schultz fell in with my views, and in 1937 I was made an Instructor. He further acceded to my developing notion that I be allowed periodic leaves for taking a medical degree. Some of the preclinical courses I had already completed at Johns Hopkins University as part of my doctoral requirements, so that between Stanford University and Duke University, which offered summer clinical clerkships, I was able to graduate from Stanford with the medical class of 1942.

Aside from the educational benefits of this training, I had the interesting experience of being cohorts with two classes at Stanford and with three at Duke University, and I formed some valued friendships while savoring my double role. Consequently, I felt an especially close rapport with medical students for many years, until the mid-1960s when some slippage occurred, occasioned by the prevalent student unrest, our own revision of teaching programs attendant upon a shift of teaching hospital from San Francisco to the University campus at Palo Alto, and probably my own changing outlook. In any case, for a quarter century or more I felt a warm association with medical students and thereon hangs the tale that follows. These events began about twenty years ago and unraveled over a period of more than two years.

One day in a lecture to the medical class about the anthrax bacillus and its relatives, I happened to mention that I had recently received a letter from Professor Ascoli, then a retired microbiologist living in Italy, in which he complained that in my recently published book on immunity, in a chapter devoted to anthrax, I had failed to take account of his report that extracts of the tissues of buried cattle could be used for the retrospective serologic diagnosis of anthrax. This reaction was ascribed to bacterial polysaccharides and was considered to be helpful for the diagnosis of herd infections.

I promptly forgot having mentioned this mild anecdote, nor did I associate it with what follows until a long time later.

One day a letter in scripted hand arrived from London. Its content was this:

My Dear Dr. Raffel,

I read with great interest your recent book *Immunity* and was greatly pleased to see a person of such youth make a fine contribution to the study of micro-organisms and serums.

I would like to protest, however, the very scanty reference to my research and especially the Jenner-Adams test (Proc. Royal Acad., 1799) for type III cowpox. I trust that in future editions, you will give credit where credit is due.

I remain

Yr. humble servant
Edw. Jenner

Some weeks later, a letter in German arrived from that country, posted from Kloster Lechfeld. Again, after pleasant introductory remarks, the writer went on: "Leider muss Ich aber zugeben, das Ich etwas enttaüischt bin, da Sie nichts über mein 'Serum gegen die Seminaria morbi' noch von meinem 'Gegen-Teufel Toxoid' geschrieben haben." This bore the signature of Fracastorius. Apparently he used German to express displeasure.

During the course of the following two years, at intervals of one to several months, there came a further series of letters. The next was mailed from Paris by Louis Pasteur:

Très cher et honoré confrère, . . .J'ai trouvé vos explications d'une très grande clarté et j'admets qu'une grande partie de la recherche enterprise par vous ne m'était pas connue. Je trouve pourtant que vous avez commis un péché d'oubli et si vous voulez bien me pardonner ma suggestion, je pense que vous devriez accorder plus de place à l'étude de la rage . . .

The series now took a different tack: A letter from Ferdinand Magellan from Spain, after the customary felicitous introduction in his language, took me to task for giving insufficient consideration to the intriguing disease syphilis. He countered my impression that the disease had been introduced to Europe by Columbus on his first return from America. According to this informant, syphilis was brought from the Philippines—he himself has had the disease ("no joking matter") for 400 years.

There followed a communication from J. P. Higginbottom of the Hertforshire and Bedfordshire Archeological Society accompanied by a box of Assyrian clay tablets "recently brought to light, in which an early dynasty immunologist, . . .although he favours your book as a whole, takes exception to your ideas on the acquired hemolytic anemia."

Then came a letter from Hong Kong, written in Chinese by Confucius, pointing out that in 498, when the letter was written, K. K. Chen had already found an herb that stimulated the pituitary gland and in turn the adrenals. Why had I not acknowledged this in speaking of cortisone and ACTH?

Next came an admonishing epistle from Mary Baker Eddy who did not think much of the book or of medicine in general. A subsequent note from Oa Mook in a completely foreign tongue (Lunarese, fortunately interlined with some translation) took umbrage at the fact that I had failed to refer to any lunar microbiologists and trusted that this omission would be remedied in future editions.

The final two letters became seriously scientific. The first from British Columbia—written by"E. Power Blake-Nutting, Director of the Spring Island Test Station"—stated that he had been struck by my discussion of cellular antigens and immunologic kidney disease. He in turn had some unpublished observations of studies undertaken with the Sooty Tern (*Aecleptis nigra*) and the Tufted Puffin (*Naris Wellanderi*). He injected macerated puffin kidney into the tern and derived an antiserum that on injection into the puffin, produced kidney damage. The injected puffin, in turn, revealed, in ultracentrifuged extracts of its kidneys, the presence of "a puffin antitern antipuffin kidney antibody." This was injected into a tern, and after two weeks, its globulin "plus previously prepared puffin antitern antipuffin kidney antibody" produced no damage in the puffin, "suggesting that the puffin antitern antipuffin kidney antibody had been neutralized by a tern antipuffin antitern antipuffin kidney antibody. We believe that this is the first demonstration antianti-antibody. . . . I hope that this may be of some help in clearing up the subject of cellular antigens."

Finally, Albert Smudge, PhD, wrote from Hawaii, from the Institute of Marine Biology, concerning homologues of the human atopic allergies seen in fish. These observations were based on studies of the homohomonukunukuapaoua, which, though it "serves as food for scores of larger fish, is strongly avoided by the Kualueluilui. When the two are placed in the same tank, about 80% of the Kualueluilui will show signs of distress: convulsive gill movements, increased body slime secretion, and in 24 hours a patchy necrosis of the skin . . . The Schultz-Dale reaction is negative . . . There seems to be a homohomonukunukuapaoua anti-Kualueluilui beta globulin antibody . . . " which "is the antibody of piscine atopic allergy."

This was the last of the epistolary series, and it happened to coincide with the graduation of that class of medicine that had been sophomores at the time of initiation of the series, when our course in microbiology and immunology was given. The author of these well-informed letters, and his operational methods in the use of several languages and always appropriate sites of posting, remained enigmas. At the seniors' farewell picnic of that year, the author was pointed out to me by a classmate, but the designated individual blandly refused either to acknowledge or disavow this claim.

Thus ended what was for me a delightful sequence, extending for more than two years, of first-rate wit and humor, and a warm feeling than at least one student had been moved by interest in the subject to have engaged in this protracted tour de force.

The conclusion of my vocation in formal teaching came far from home, in Iran, in 1977, the year before the Shah's approaching end had become obvious to all. But, in the first three months of that year, my wife and I lived in Shiraz, about 450 miles south of Teheran, and we had no hint of brewing troubles, nor for that matter apparently did the CIA.

The University of Shiraz was regarded as a symbol of the country's participation in modern education and scientific thought. It was a relatively large institution, second to the University of Teheran in size and prestige, and its special character lay in the fact that the teaching in all its branches was conducted in English. Since many of the faculty, at least in the School of Medicine with which I was associated, had spent considerable time in training in the US, and a few had done so in Britain, English presented no problem to most of them. The difficulty lay in the audience. The medical students, of whom there were only about 65 per class, were not equipped to cope with spoken English, virtually to a man—or woman, of whom there was a good representation, about a third of the total. The Iranian lecturers improved the situation by speaking slowly, putting on the board each point as it was made. But, most importantly, they were able to throw in a word or phrase of Farsi (i.e. Persian) at critical junctures.

My first lecture was to be an overview of immunology and, as it happened, this talk had to be delivered in the anatomy dissecting room. Perhaps fifty cadavers occupied the farther reaches of this spacious chamber. After about ten minuties of what I considered a spritely monolog, it occurred to me that students and cadavers were absorbing it with about equal avidity. Following this introductory effort of mine, one student confided to a faculty associate that I must have come from Texas (a state that I have visited only briefly on a few occasions) because he had heard cowboys on television, and I sounded just like them. This presumably related to form, not content.

During the remainder of my stay I spent a good part of my time in reducing my remaining seventeen lectures to a sprinkling of more easily conveyed concepts and bits of information. I should say that the students were courteous and attentive through all this, and they were probably entirely capable of grasping what I wished to say; we simply did not share a common medium of communication.

Thus concluded my years of professing and researching. I tasted the satisfactions of telling others about what interested me, and occasionally I had the thrill of fitting together notions in the laboratory. All this was done against the backdrop of a happy family life, a felicitous ambience in which to live it, and opportunities for extensive travels and stays in a number of other countries. As the run of destinies go, a fortunate one.

My association with the Annual Review of Microbiology was part of the background that contributed to the enjoyment of this destiny. In the days of my youth at Stanford, in 1945, I was invited to join its editorial staff as an associate editor to Charles Clifton, a faculty colleague who eventually wrote the first of these remembrances-of-things-past. The other first associate editor was Albert Barker and, among the three of us, we shared the reading of manuscripts for many volumes to come. The first editorial board, which met annually for a day with the editors to generate lists of topics and potential authors, was a particularly stimulating one—including as it did William Taliaferro, C. B. van Niel, M. D. Eaton, J. M. Sherman, E. C. Stakman, and W. E. Herrell—and succeeding boards have perpetuated a happy balance of accomplished men and women.

The early days with their associations were heady ones for me, and, over the years, this facet of my activities, and the friendship with Murray Luck and his wife Edoe, have been for my wife and me among the happiest of our relationships. Although my formal retirement from the *Annual Review of Microbiology* came with volume 33 in 1979, I am delighted that my affiliations with the Editorial Board continues, and apparently will until, as current editor Mort Starr puts it, I have lost my marbles or become otherwise disqualified.

Ann. Rev. Microbiol. 1982. 36:27–46

THE ECOLOGY AND ROLE OF PROTOZOA IN AEROBIC SEWAGE TREATMENT PROCESSES

Colin R. Curds

British Museum (Natural History), Cromwell Road, London, SW7 5BD, England

CONTENTS

INTRODUCTION

Two major aerobic biological processes are used commonly throughout the world to treat settled sewage: percolating filters and the activated-sludge process. Both processes rely upon the growth of microorganisms to remove unacceptable substances dissolved or suspended in sewage and in some cases to convert them into more acceptable compounds. In percolating filters, the sewage is allowed to trickle over inert surfaces (coke, clinker, plastics media, etc) upon which the microbes grow to form an attached microbial film. The activated-sludge process, however, is a truly aquatic process where the sewage and organisms are aerated together in tanks for several hours. The organisms form flocculent growths or activated sludge, which may then be easily separated from the effluent in sedimentation tanks to be recycled back to the aeration tank. Excess solids produced in these

0066-4227/82/1001-0027$02.00

aerobic processes may be treated in anaerobic digesters that use methanogenic bacteria. A full account of sewage-treatment processes is given elsewhere (39).

Although it has been known for many years that these two aerobic biological sewage-treatment processes contain animal and plant life from many phyla, the roles played by the various individual groups of organisms in the purification processes in many cases are not fully understood. The complete fauna and flora of percolating filters consist of a wide variety of organisms, including not only microorganisms, such as bacteria, fungi, algae, and protozoa, but also large populations of macroinvertebrate animals, such as insects, arachnids, worms, crustacea, rotifers, and others. Life in activated-sludge plants, however, is less varied, and bacteria in the form of sludge floc are generally the dominant organisms. Fungi are comparatively rare, but they and the actinomycetes occasionally become dominant. Algae rarely become established, presumably because of the lack of light. Rotifers, nematode worms, and more rarely oligochaete worms and chironomid larvae may be found.

PROTOZOAN FAUNA OF PERCOLATING FILTERS AND ACTIVATED-SLUDGE PLANTS

Protozoa are plentiful in both percolating filters and in activated-sludge plants, and it is common to find populations of these organisms in the order of 50,000 cells per ml in the mixed liquor of activated-sludge plants. Calculations based on such numbers indicate that protozoa may constitute approximately 5% (dry weight) of the mixed liquor-suspended solids in the aeration tank (63). Many authors have listed those protozoa found in aerobic waste-treatment processes, but these data are often given as secondary information; however some major lists have been published (7, 15, 17, 21, 27, 32, 38, 58, 61, 64, 65).

Taxonomy in general and identification in particular are difficult problems to solve; however, four keys written specifically about the protozoa found in treatment processes and polluted waters are now available (11, 15, 21, 61). Other more recent keys to amoebae (66) and ciliates (29, 37) should also help in the identification of these organisms, although neither is aimed specifically at sewage-treatment processes.

Five classes of protozoa are represented in aerobic sewage-treatment processes and one review (27) stated that 218 and 228 species of protozoa have been identified in percolating filters and activated-sludge plants, respectively. By far the largest number of species identified are ciliates, although flagellates and rhizopods are also well represented when considered together as a single group. However, when one is considering the importance of a group of organisms in the energy flow of a habitat, the numbers of species represented is of little significance; the biomass of individual populations is required for an assessment to be made. Unfortunately, little such quantitative data are available. The class of protozoa that contributes the greatest number of individuals to the protozoa population of a filter has been discussed by many authors, and it is generally agreed that although ciliates normally are numerically dominant, this may not necessarily be the case. Barker (6) listed the numbers of protozoa found in percolating filters per milliliter of liquor to be in the following range: rhizopods, 100–4600 ml^{-1}; flagellates, 200–1300 ml^{-1}; and ciliates, 500–10,000. He also states that ciliates are generally the most abundant. It is unfortunate that no data, however approximate, exist on biomass estimations; obviously it is not meaningful to compare numbers of organisms when their individual sizes are significantly different. Published information concerning protozoa in activated-sludge plants on the whole indicates that ciliates are the dominant protozoa in that process. However, some recent publications (5, 79, 83) suggest that various amoebae, both testate (e.g. *Cochliopodium* spp.) and small naked forms, are just as frequent and are often dominant organisms in terms of both numbers (79) and estimated biomass (83). The observations concerning amoebae so far have been limited to a few plants in the London area and one at Leicester, whereas other authors (32) have examined samples from percolating filters and activated-sludge plants situated all over England, Scotland, and Wales. That work (32) concluded that ciliates were normally the dominant class of protozoa in activated sludge, although on occasion amoebae and flagellates were seen in moderate numbers. Furthermore, another survey (12) of an activated-sludge plant in Surrey noted that over a complete year ciliate species were dominant, with the one exception of *Euglypha* sp., a testate amoeba that dominated for a short period (1 week) of time.

The relative abundance of a particular organism in a habitat often indicates its importance in the ecological structure of that habitat. To assess those species of ciliates that are commonly the most important in the two processes, both frequency and abundance data from the survey (32) were taken into account. From such an assessment the list of species in Table 1 was suggested as those of greatest importance. This list shows that only four species are common to both processes. However, the well-established view that the subclass Peritrichia is by far the most important protozoa subclass in these aerobic processes is confirmed. The reason is that all peritrich species found in these two processes are attached forms. It should be remembered that both percolating filters and the activated-sludge process rely on the presence of surfaces upon which the microorganism can grow. Thus an organism with the ability to attach itself to, or remain closely associated

Table 1 Suggested list of the most important protozoa in sewage-treatment processes in approximate order of importance (32)[a]

Protozoa	Habit	Food	Protozoa	Habit	Food
Opercularia microdiscum	A	B	*Aspidisca cicada*	C	B
Carchesium polypinum	A	B	*Vorticella convallaria*	A	B
Vorticella convallaria	A	B	*Vorticella microstoma*	A	B
Chilodonella uncinata	F/C	FB	*Trachelophyllum pusillum*	F/C	Ca
Opercularia coarctata	A	B	*Opercularia coarctata*	A	B
Opercularia phryganeae	A	B	*Vorticella alba*	A	B
Vorticella octava	A	B	*Carchesium polypinum*	A	B
Aspidisca cicada	C	B	*Euplotes moebiusi*	C	B
Cinetochilum margaritaceum	C	B	*Vorticella fromenteli*	A	B

[a] A, Attached; C, crawling; F/C, free-swimming and crawling; B, bacteriavore; Ca, carnivore; FB, filamentous algae and bacteria.

with, surfaces has a distinct advantage over organisms that swim freely in the liquid phase and are subject to washout in the effluent. In other words, both processes select for sedentary organisms, so perhaps it is not surprising that most of those in Table 1 are either attached or crawling forms.

Another important feature in Table 1 is the food of the ciliates present. The literature shows that most of the ciliates listed feed upon dispersed populations of bacteria; however, frequent reference to "browsing" hypotrichs such as *Aspidisca* and *Euplotes* implies that these ciliates can scrape bacteria from surfaces as a browsing limpet might when feeding on algal growths on rocks. In ciliates, none of the organelles present could enable browsing to take place. All bacteria-feeding ciliates rely upon ciliary currents to waft suspended bacteria into the oral region; however, since hypotrichs are in close proximity to surface growths, no doubt particles only lightly adhering to the substratum could become dislodged by the feeding currents. Only two ciliates listed in Table 1 do not feed upon dispersed bacterial populations: *Chilodonella,* which feeds upon the filamentous algae and bacteria that are more prevalent in percolating filters; and *Trachelophyllum,* which feeds upon peritrich ciliates such as *Opercularia.*

PROTOZOA AS INDICATORS OF PLANT PERFORMANCE

Many authors have investigated the presence of microorganisms in rivers with special reference to their degree of organic pollution, and one (60) described a large number of protozoan species as characteristic of a particular degree of pollution by sewage or saprobity. Many such schemes have been introduced since Kolkwitz & Marsson (56, 57) first suggested that rivers or sections of rivers could be classified according to their degree of organic pollution. A full account of the modern methods of zone description and measurement of saprobity is given elsewhere (80).

Many of the ciliated protozoa found in polluted rivers also occur in sewage-treatment processes. These organisms might be used to indicate the condition of an activated-sludge or percolating filter effluent. Thus, if a large proportion of the species present in an activated-sludge plant are oligosaprobic (i.e. species normally associated with relatively unpolluted rivers or sections of rivers), then one might expect the sludge to be in a "healthy" condition, whereas if most of the species found are normally associated with highly polluted, or polysaprobic, river conditions, it could mean that the sludge was not in a good condition. This method of assessment, although often applicable in extreme cases, is not satisfactory when a treatment plant contains mainly mesosaprobic species, as is often the case. Application of such a scheme to percolating filters is more difficult because of stratification through the filter depth. Gradual purification of the sewage takes place during its passage through the filter so that polysaprobic conditions occur at the top, with mesosaprobic zones of diminishing intensities occurring towards the base (89). It should be noted that the position of a particular zone in a filter is purely relative, since a change in the flow or strength of sewage can alter the precise depths of the zones. For example, an increase in sewage strength usually results in a downward shift in the saprobic grades, i.e. the polysaprobic and α-mesosaprobic zones would extend down further and could result in the extinction of the β-mesosaprobic zone. The change in zone positions would then be followed by the pertinent organisms (59). It follows, therefore, that the only possible method of using protozoa as indicator organisms in percolating filters is to sample the lowest zone, which in many full-scale works is not possible because of their mode of construction. The spatial zoning of organisms, characteristic of a filter, is not found in activated-sludge plants because of continual mixing and recirculation of the sludge (47). However, in activated sludge a marked temporal zonation or succession of protozoan types occurs during the development of a mature sludge. Several early publications (1, 9, 14, 20, 48, 55, 62) suggested there might be a distinct succession of protozoa during the establishment of an activated sludge. Flagellated protozoa are typically the first dominant group of protozoa and they are then replaced by free-swimming ciliates. The latter are later replaced by crawling hypotrichs such as *Aspidisca* spp. Finally, attached peritrich ciliates become established. Sometimes an inverse relationship between peritrichs and hypotrichs has been noted in full-scale (12, 42) and pilot-scale plants (20). Initially, nutrition was considered a major factor in determining successions (20, 62), but more recently Curds (24, 25) has been able to simulate, on a digital computer, successions similar to those described simply from theoretical consideration of growth kinetics, nutrition, and more particularly the settling properties of the organisms.

During the development and succession of the various microbial populations of an activated sludge, the quality of the effluent improves and it is this link between effluent quality and species of protozoa present that has been responsible for the suggestion by early authors (1–3) to use protozoa as indicators of activated sludge effluent quality. Several other authors (4, 41, 42, 67, 69, 77, 79, 84, 86) have associated various groups of protozoa with effluent quality, but most used a rather ad hoc qualitative approach based on intermittent observations of protozoa and effluent quality. The survey of Curds & Cockburn (32) was not aimed specifically at investigating indicator species, but their results suggested that there could be a correlation between the species structure of an activated sludge and the quality of the effluent being discharged. To test this hypothesis, the effluents from the plants investigated were divided (33) into four categories according to their biological oxygen demand (BOD): very high quality, BOD range 0–10 mg/liter; high quality, 11–20 mg/liter; inferior quality, 21–30 mg/liter; and low quality, above 30 mg/liter). Initially the frequencies of each species occurring in plants that deliver effluents within each of the four categories were calculated on a percentage basis. Many species were associated with all categories of effluent, but there was a tendency for a given species to occur more frequently in plants that deliver effluents within a particular category (Table 2): This indicated that the protozoan species found within the mixed liquor were in some way associated with the quality of the effluent. Table 2 shows that *Carchesium polypinum,* for example, was found principally in plants that produced good quality effluents, whereas flagellated protozoa were restricted to plants that produced inferior effluents. An arbitrary total of 10 points was awarded to each species and these points were distributed among the four effluent categories so that the greatest number of points was given to the effluent category with which a species was most frequently associated. For example, a ciliate with frequencies of 60, 80, 40, and 20% in plants that deliver effluents in the four categories would be awarded the 10 points in the ratio of 6:8:4:2, i.e. 3:4:2:1 in the respective categories. The points were called association ratings and more examples are given in Table 2.

To predict the effluent quality of a particular plant, the species of protozoa in the mixed liquor were identified and listed against the four effluent-quality categories. With the comprehensive version of Table 2 given by Curds & Cockburn (33), the appropriate number of points was awarded to each of the effluent categories for each species. The total number of points for each effluent category was then calculated and the category that gained the highest total number of points was the predicted effluent-quality category. The method (33) was shown to be 85% correct on the original data and 83% correct when tested on a further 34 sites not included in the original survey.

The apparent relationship between the species structure of an activated sludge and the quality of the effluent is of some practical value, but it should be emphasized that this method does not replace the standard 5-day BOD test. The method was not aimed specifically at predicting effluent BOD, but should be regarded as a way of assessing the efficiency of the activated sludge and the broad category of effluent likely to be achieved, giving at best only an indication of the average effluent BOD. It would be unreasonable to expect a correlation between the day-to-day variations in effluent BOD and the species structure of the protozoan populations, since at any instant they reflect these changes in physical, chemical, and biological environmental conditions over the preceding few days. However, the method might demonstrate a sudden drastic change in some operational factor, for example, the presence of a shock load of toxic discharge. This method is open to criticism since it does not take into account the numbers of members of an individual species. For example, it gives equal weight to the presence of few or many organisms of a particular species. It could be more quantitative by counting numbers of individual species, but this would be time consuming. Furthermore, the biomass of protozoa only reflects the biomass of food entering the system at steady state assuming the absence of toxins, and a far better approach would be to take the relative proportion of each species into account. However, even if the method were universally applicable, it has the major disadvantage that the numerous species of ciliate first have to be identified. It was stated earlier that although some keys are available to aid nonspecialists in the identification of ciliates in activated sludge, there is still a taxonomic problem, particularly with some of the more common peritrichous ciliates. Recently, a simpler approach that does not rely on precise identifications has been introduced (42): The microfauna are arranged into different trophic groups and are correlated with plant perfor-

Table 2 Percentage frequency of occurrence of some protozoa and their association ratings with plants producing effluents within four ranges of BOD (33)

BOD range (mg l^{-1} per liter)	Frequency of occurrence (%)[a]			
	0–10	11–20	21–30	730
Vorticella convallaria	63 (3)	73 (4)	37 (2)	22 (1)
Vorticella fromenteli	38 (5)	33 (4)	12 (1)	0 (0)
Carchesium polypinum	19 (3)	47 (5)	12 (2)	0 (0)
Aspidisca cicada	75 (3)	80 (3)	50 (2)	56 (2)
Euplotes patella	38 (4)	25 (3)	24 (3)	0 (0)
Flagellated protozoa	0 (0)	0 (0)	37 (4)	45 (6)

[a] Parentheses indicate association ratings awarded.

mances. Although this is easier for the nonspecialist, the user must still identify which organisms feed upon bacteria, flagellates, or ciliates. Furthermore, this quantitative approach (42) is not necessary (see above), nor is it particularly meaningful to compare flagellates and ciliates numerically when they have such different body sizes.

ROLE OF PROTOZOA IN THE ACTIVATED-SLUDGE PROCESS

Although a large amount of literature pertains to the role and importance of protozoa in the activated-sludge process, little work has been reported on their role in percolating filters. The major reason is that activated-sludge plants are easier to sample adequately and it is simpler to carry out small-scale experiments on these plants in the laboratory than on percolating filters. However, as was shown above, great similarity exists among the organisms found in these aerobic habitats, so it is reasonable to suppose they play a similar qualitative and quantitative part in the purification processes of both types of plant.

Protozoa were originally considered harmful to the activated-sludge process (43), but most authors have proposed that protozoa are of some benefit to the process for a variety of reasons and to a variable degree. Ardern & Lockett (2) were probably the first to carry out experimental work to investigate the possible role of protozoa in activated-sludge plants; they "selectively" killed the protozoa in one sludge with toluene vapor and estimated the effect of such a treatment on the efficiency of the sludge in purifying sewage. They showed that there was no significant difference in the 4-hr permanganate value of the two effluents and concluded "that from the standpoint of sewage treatment in general, protozoa play no important part in the process of purification by activated sludge although it is possible they facilitate the production of a more highly clarified effluent" (2). However, this work is immediately open to criticism, whatever the result, since toluene vapor is certainly not a "selective" toxin specific to protozoa and any changes noted could be attributable to unintended changes in other microorganisms present.

During the 1940s, Pillai et al (69–75) were prolific in their experiments on the role of ciliated protozoa in the activated-sludge process. Together they produced many papers that purported to show that ciliates were the agents by which sewage is purified and that bacteria were of secondary importance. However, none of the work was adequately controlled and all their results could be explained on a basis of contamination by bacterial populations, which effected the purification. Two of the papers (71, 74) demonstrated that in the presence of large numbers of peritrichous ciliates an increase in nitrate and nitrate concentration in the effluent was noted. Later (75), culture experiments showed that of 81 strains of bacteria separated from activated sludge, 37 produced nitrites; all increased their nitrite production in the presence of *Vorticella* sp., but no nitrates were produced. In the presence of *Epistylis* sp., however, the nitrite production increased to a higher level and nitrate production was noted. The authors concluded that the ciliates themselves were directly active in nitrification. Pillai et al (75) did not appear to be aware that only two bacterial genera, *Nitrosomas* and *Nitrobacter,* have been specifically implicated to date in nitrification; furthermore, the isolation methods they used were unlikely to select nitrifying bacterial strains, from which we must conclude that perhaps their methods of nitrite and nitrate analysis were at fault. Until this year no other organisms had been implicated in nitrification, but B. Finlay (personal communication) has presented strong circumstantial evidence that the ciliate *Loxodes* may be involved with nitrification in lakes; unfortunately, *Loxodes* has not yet been cultured in the laboratory, so appropriate direct experimental work has not yet been carried out. If it should prove that *Loxodes* is directly involved in nitrification, perhaps we would have to think again about the work of Pillai and his colleagues on nitrification and examine other ciliates for their potential abilities.

In all of the publications on the role of protozoa in the purification of sewage, one point is agreed upon by all authors, that is, when large numbers of ciliates are present there is a clear, sparkling effluent. Johnson (53) is one of the earliest to record that *Paramecium* in the sludge usually accompanied effluents of unusual clarity and suggests that protozoa help in the removal of fine particles of suspended matter and surplus bacteria. McKinney & Gram (62) found that although pure cultures of bacteria formed typical flocs in nutrient solutions, some bacteria always remained in suspension, producing turbidity and contributing to the biochemical oxygen demand of the effluent. The work of Curds et al (35) set out to demonstrate unequivocally whether or not the presence of ciliates had any effect upon effluent quality issuing from an activated-sludge plant. They designed and built six replicate activated-sludge units in which protozoa-free sludges could be grown so that the effect of subsequent controlled inoculations of ciliates on effluent quality could be determined. Each of the units, constructed of perspex and sealed by an airtight lid to prevent airborne contamination, first was sterilized chemically and then was closed with heat-treated (and hence protozoa-free) sewage. The protozoa-free sewage was then inoculated with mixed bacterial cultures derived from a full-scale activated-sludge plant. The sludges that grew in the six replicate units were kept free from protozoa by automatic dosing with cool heat-treated sewage. These methods made it possible for the first time to study the effects of the subsequent addition of protozoa to the plants and to assess the role of protozoa and to quantify the magnitude of their effect upon effluent quality. Initially, all six plants were operated under protozoa-free conditions and the sludge that grew in the units had the same macro- and microscopic appearances as do sludges in full-scale systems, except for the absence of protozoa. When 2000 mg of mixed liquor-suspended solids per liter had grown in each of the units the effluent quality was assessed with standard methods. Under protozoa-free conditions all six plants produced highly turbid effluents of inferior quality and the turbidity was found to be significantly related to the presence of very large numbers of viable bacteria suspended in the effluents (hundreds of millions per milliliter). Without protozoa, the BODs of the effluents were high, as were their contents of organic carbon, suspended solids, and other parameters (Table 3). Cultures of ciliated protozoa were then added to three of the plants while the other three continued to operate as control plants without protozoa. After a few days, during which time the protozoan population became properly established, plants that contained protozoa showed dramatic improvement in effluent quality. Clarity was greatly improved, which was associated with a significant decrease in the concentrations of viable bacteria in the effluents. Furthermore, the effluent BOD and concentrations of suspended solids decreased significantly (Table 3). The three units still operating without protozoa continued to deliver turbid, low-quality effluents.

The significant drop in numbers of bacteria and nonsettleable suspended solids in effluents after introduction of ciliates could be a result of two factors. Bacteria are the food organisms upon which ciliates prey and in addition certain ciliates have the ability to flocculate suspended matter and bacteria. The clarifying ability of ciliates by one process or another could explain much of the falls in total BOD, COD, 4-hr permanganate value, and organic nitrogen. However, it cannot easily explain the effects on filtered samples. Both filtered BOD and soluble organic carbon measurements decreased significantly after the introduction of ciliates which indicates that the removal of soluble organic complexes increased when they were present. It could be a result of two factors. It could be attributable directly to the activities of ciliates, since it is well known that certain ciliate species may be cultured upon organic solutions in the absence of bacteria. However, no published data give the amounts of soluble organic material utilized when the ciliate is feeding normally upon bacteria. The other alternative is that the predatory activities of the ciliates may stimulate uptake of the substrate by the bacteria. Such a phenomenon has been noted before in other systems; for example, Johannes (52) stated that protozoan predation would stimulate bacterial growth rates and hence substrate uptake rates. He (52) summarized the effects of predation: "Grazing ciliates prevent bacteria from reaching self-limiting numbers; the bacterial populations are thus kept in a prolonged state of physiological youth, and their rate of assimilation of organic materials is greatly increased." Such a possibility is conceivable if the substrate being measured is not the limiting factor; under these circumstances the reduction of the bacterial population by predation could be beneficial and could explain some of the observations. This idea has been expanded by Hunt et al (49), who developed a simulation model for the effect of predation on bacteria in continuous culture. They concluded that by lowering the bacterial biomass, predation increases the level of limiting nutrient, thereby increasing the growth rate of the bacteria and increasing the uptake of nonlimiting nutrient. This topic is reviewed by Stout (81) in his paper on the role of protozoa in nutrient recycling and energy flow.

The results of Curds et al (35) are in broad agreement with those of others (2, 62) and with some of the results of Pillai & Subramanyan (72, 73), and it would appear that the major role of the protozoa in the activated-sludge process is to aid in the clarification of the effluents. Furthermore, similar circumstantial evidence from full-scale plants shows that when protozoa are absent, the effluent obtained in turbid and of inferior quality (32, 33). As mentioned above, protozoa might cause the improvement in effluent quality in at least two ways: flocculation or predation. A considerable amount of evidence in the literature shows that protozoa in pure culture are able to flocculate suspended particulate matter and bacteria; this can aid both clarification of the effluent and formation of the sludge (2, 7, 8, 10, 18, 19, 45–47, 50, 51, 72, 82, 87, 88). In certain species flocculation is thought to be brought about by the secretion of a mucous-like substance from the peristome region (82, 88). Two authors (45, 88) found that the flagellate

Table 3 Effect of ciliated protozoa on the effluent quality of bench-scale, activated-sludge plants (35)[a]

Effluent analysis	Ciliates	
	Absent	Present
BOD	53–70	7–24
BOD after filtration	30–35	3–9
COD	198–250	134–142
Permanganate value	83–106	62–70
Organic carbon after filtration	31–50	14–25
Organic nitrogen	14–21	7–10
Suspended solids	86–118	26–34
Optical density at 620 μm	0.95–1.42	0.23–0.34
Viable bacteria count (millions per ml)	160–160	1–9

[a] Results shown in milligrams per liter unless otherwise noted.

Oicomones termo and the cilate *Balantiophorus minutus* were able to flocculate bacteria, whereas others (18, 19, 82) found that colloidal particles of India ink and DDT were flocculated by ciliates such as *Paramecium* and *Carchesium.* In the case of *Paramecium caudatum,* Curds (18, 19) found that at least two mechanisms were responsible: First, the ciliate was able to secrete a soluble polysaccharide (a polymer of glucose and arabinose) into the medium, which changed the surface charge of suspended colloidal particles present; and second, particles ingested were glued together during cyclosis by a mucin. In addition, it is well known that many bacteria are able to flocculate or grow in flocculant forms without the aid of protozoa.

The reduction in the numbers of bacteria from sewage to effluent after aerobic treatment has been noted by many authors (35, 36, 48, 78, 85, 87). It is also known from the literature that the dominant types of protozoa found in these processes feed mainly on bacteria, and so it has been suggested that the predatory activities of the protozoa might be responsible for the bacterial removal observed. Indeed, Guede (44) even suggested that grazing by protozoa could be a selection factor for activated-sludge bacteria. Curds & Vandyke (40) showed that the ciliates in activated sludge can feed upon a number of bacterial strains likely to be found in sewage-treatment processes, but Barritt (10) stated that since bacteria are only present as flocculated masses, they were unavailable as a food supply. Of course this latter work (10) is partly correct: Flocculant growths of bacteria are not available as food for protozoa since the latter organisms do not possess organelles capable of ingesting flocs. However, the work of Curds et al (35) has shown that nonflocculated bacteria do occur in large numbers in activated sludge, but only when protozoa are not present. The quantitative feeding studies of Curds & Cockburn (31, 34) and by later authors (see 30) who used batch and continuous-culture methods have led me to believe that if the protozoa in activated sludge feed at rates similar to those of the ciliate *Tetrahymena* in the pure culture, then the protozoan populations normally found in activated sludge could, by predation alone, easily remove sufficient quantities of bacteria to account for the reductions observed earlier (35).

Protozoa are known to feed upon pathogenic bacteria, including those that cause diseases such as diphtheria, cholera, typhus, and streptococcal infections, as well as fecal bacteria such as *Escherichia coli* (36). In the case of *E. coli,* Curds & Fey (36) found that 50% of those entering in the sewage were removed by unidentified processes when protozoa were absent, but after ciliates had been added a mean of 95% of the *E. coli* was removed from sewage to effluent. Since 50% of these bacteria were removed even in the absence of protozoa, then obviously other factors are also involved. In the work carried out neither phage nor *Bdellovibrio* could be isolated on lawns of the *E. coli* strain used, and perhaps simple death is the logical explanation since the coliforms soon reached steady-state populations and were not being accumulated in the sludge floc.

MICROBIAL POPULATION DYNAMICS OF ACTIVATED SLUDGE

Many of the observations made on protozoa in activated sludge can now be explained on the basis of growth kinetics, the settling properties of the organisms, and the operational characteristics of the plant. A number of workers have devised mathematical models for the activated-sludge process (chemostat with feedback), but all of them have considered sludge to consist of a single flocculating microorganism and few have considered the dynamics of different organisms and the effects they might have upon plant performance.

The use of computers as tools in the testing of general ecological principles, although relatively new, is gaining momentum and the easy access of microcomputers with large storage capacity is likely to encourage more of this approach. Bungay (13) applied analogue computer-simulation techniques to studies on the dynamics of continuous mixed-culture systems, and he included an example of a predator and its prey growing together in a chemostat. He found that the populations of organisms oscillated in a regular complementary manner. More recently, Canale (16) applied singular point analysis and linear approximation methods, and Curds (23) used a digital computer to simulate the dynamics of a similar culture system. Both found three possible solutions to the equations dependent upon the kinetic constants of the two organisms, although generally a stable limit-cycle oscillation would be expected. While developing continuous-culture techniques to study the consumption of bacteria by ciliated protozoa, Curds (22) found that when bacteria and ciliates were grown together in a single reactor a steady state was not achieved, although later Curds & Cockburn (34) showed that steady-state populations were possible when two separate chemostat reactors were used in series. Since that time many have modeled and carried out experimental work to show how organisms grown together in a reactor might react. These have been reviewed elsewhere (28, 30).

The mathematical modeling of mixed populations of microorganisms has now been extended so that their fates in an activated-sludge plant may be considered theoretically and their effects upon effluent quality may be judged (24). In this work (24) the plant modeled was a completely mixed reactor fitted with sludge feedback and continual wastage of surplus sludge at a constant fixed rate. Curds (24) considered the plant would contain and/or receive a number of microorganisms, which he defined by the following criteria.

1. Dispersed sewage bacteria are those borne in sewage in considerable quantities [30 mg (dry weight) per liter in an average British domestic sewage]. On entry into the reactor they do not flocculate; they remain in suspension in the sedimentation tank and are evenly dispersed throughout the effluent and recycle flows.
2. Sludge bacteria form the bulk of the sludge mass as a whole. They always flocculate and are considered to be in sufficiently small concentrations in the sewage as to be ignored. Although the division into two groups of bacteria may appear to be arbitrary, evidence now suggests that the bacteria comprising the sludge mass as a whole are not the same as those entering the system from largely fecal origin (68).
3. Soluble substrate is the limiting substrate considered in the model. (For the purposes of the model, the sewage was assumed to contain all elements required for the growth of the two bacteria in excess with the exception of one that was limited.)
4. Ciliated protozoa were assumed to feed entirely on sewage bacteria that remain in suspension since it was argued that the sludge bacteria would be present only as flocculated masses and hence would be unavailable as a food source. Ciliates were initially assumed to be of two types, according to their habit, free-swimming and attached. Later, Curds (25) included crawling forms in his model of successions. Free-swimming forms, because of their habit, never settle in the sedimentation tank but remain evenly dispersed throughout the liquid phase of effluent and recycled sludge. However, attached protozoa do settle on sedimentation because they attach themselves to the flocculated masses of sludge bacteria.

Of these populations, only dispersed sewage bacteria, free-swimming ciliates, and soluble substrate are able to leave the plant in the effluent; sludge bacteria and attached cilates never leave in the effluent but some are continually removed at a fixed rate after sedimentation in the sludge wastage flow. A further assumption made was that all organisms obeyed Michaelis-Menten growth kinetics, and that the specific growth sites of the bacteria were related to the concentration of soluble substrate whereas those of the ciliates were related to the concentration of available dispersed bacteria present in the reactor.

Initially, two separate simulations were carried out to ascertain whether or not the activated-sludge populations were dynamically stable. Both simulations included dispersed sewage and sludge bacteria, but in one the ciliate population consisted only of free-swimming forms and in the other the only ciliates present were of the attached type. The two simulations were carried out as a direct comparison of the effect of ciliate habit on the dynamics and concentrations of the various microbial populations. In each case the initial starting values for the population sizes were considered to be low and equal. In both simulations, as time proceeded the concentrations of sludge bacteria increased, whereas the concentrations of dispersed sewage bacteria and substrate concentrations decreased. The attached ciliate population increased with time, but the free-swimming populations decreased with time. All populations, however, asymptotically approached steady-state conditions without signs of oscillation, so it was concluded that the model was dynamically stable. Under these circumstances it is valid to solve manually the mass balance equations for simple steady-state solutions at various rates. This was done over a wide range of sludge-specific wastage rates and sewage dilution rates, keeping the wastage rate arbitrarily at one tenth the dilution rate under three situations: (*a*) where the protozoa are attached forms; (*b*) when the protozoa are free-swimming ciliates; and (*c*) when ciliated protozoa are not present. The results are illustrated in Fig. 1.

It can be shown by simple algebra that under steady-state conditions the specific growth rate of a settling organism (sludge bacteria and attached protozoa) is equal to the specific wastage rate of the sludge. This explains why in all three situations the concentration of soluble substrate is precisely the same at any given wastage rate. The microbial populations, however, were different for each situation. From the mathematics it seems that substrate concentration is independent of ciliates; it is clear, however, that the turbidity of the effluent, because of dispersed sewage bacterial populations, is completely dependent upon protozoa. Large numbers of dispersed bacteria were present when protozoa were absent (no predation), but numbers were fewer when free-swimming ciliates were present and were least when the ciliate population was composed of attached forms. The latter two cases are easily explained: The growth rate of attached forms is low and at steady state equals the specific wastage rate of the sludge, whereas the specific growth rate of a free-swimming cilate is always much higher and equals the sewage dilution rate. It follows from Michaelis-Menten kinetics that higher growth rates require greater concentrations of substrate (dispersed bacteria) to be present. In addition, free-swimming ciliates are easily washed out from the system even at relatively low sewage dilution rates whereas attached forms survive. From these theoretical observations an overall high-quality effluent would be expected when attached ciliates are dominant, a slightly worse effluent quality when free-swimming ciliates are dominant, and a low-quality effluent when no ciliates are present at all. This review has already shown that these ideas are in fact frequently expressed in the literature, and now mathematical modeling is supplying quantitative explanations for these observations.

Although the model presented by Curds (24) in many ways agreed qualitatively with what is found in practice, one major discrepency was that the model predicted steady-state conditions. However, it is well known that the protozoan populations in full-scale activated-sludge plants change from day to day (12) and so Curds (25, 26) went on to explore some possible factors that could contribute to the variations observed under practical conditions. The original model assumed, for example, that the strength, flow, and bacterial content of the sewage were constant whereas in practice at least two of these parameters, flow and strength, are known to vary in a reasonably well-defined diurnal rhythm. When these three parameters were varied independently (25) it was found that variation in sewage strength caused variation in the bacterial populations and soluble substrate concentration, variations in sewage bacteria affected the ciliates, and flow variations caused all populations to vary. When all three sewage parameters were varied simultaneously diurnally, all populations were affected and a summation effect was noted.

The diurnal variation of sewage parameters is a reasonably obvious factor that could cause variations in the microbial populations; however, the presence of carnivorous ciliates could also caused oscillatory behavior, which was demonstrated by computer simulation by Curds (25). He showed that the introduction of a free-swimming carnivorous ciliate such as *Hemiophrys,* which feed upon attached peritrichs, could induce violent oscillatory behavior in all of the microbial populations. Curds (26) later went on to investigate the operation of plants by comparing the conventional practice of keeping the solids concentration in the mixed liquor constant with a plant when the solids were allowed to vary but when their rate of wastage was kept constant. By using computer simulations he showed that from a theoretical point of view it was slightly better to use a constant wastage rate if any sewage parameter was likely to vary, as is usually the case, although he did point out that the slight improvement in practice was not likely to be measurable.

There is no doubt that mathematical models in ecological research can be of great value, but the validity of any model depends entirely upon the assumptions made. Two models have now been published that purport to describe the population dynamics of protozoa in the activated-sludge process. In one (16), the plant modeled was in fact a chemostat without feedback and all bacteria were assumed to remain in suspension and act on food for the protozoa. In the other (76), the protozoa were assumed to feed upon all bacterial populations present in the reactor. Limit-cycle oscillations were predicted in both models and on this basis Pirt & Bazin (76) suggested that it would be better to use a two-stage system. The first stage would be maintained protozoa free, whereas the second stage would be used to remove excess bacteria. Jones (54) discussed the erroneous assumptions made by Pirt & Bazin (76), even so a two-stage system might be of value but not for the reasons originally outlined (76). If the first-stage consisted of a chemostat without feedback, then a 2-hr sewage retention time would wash out protozoa but would still reduce the soluble BOD by about 70%. The object of this first stage would be to convert as much as possible of the

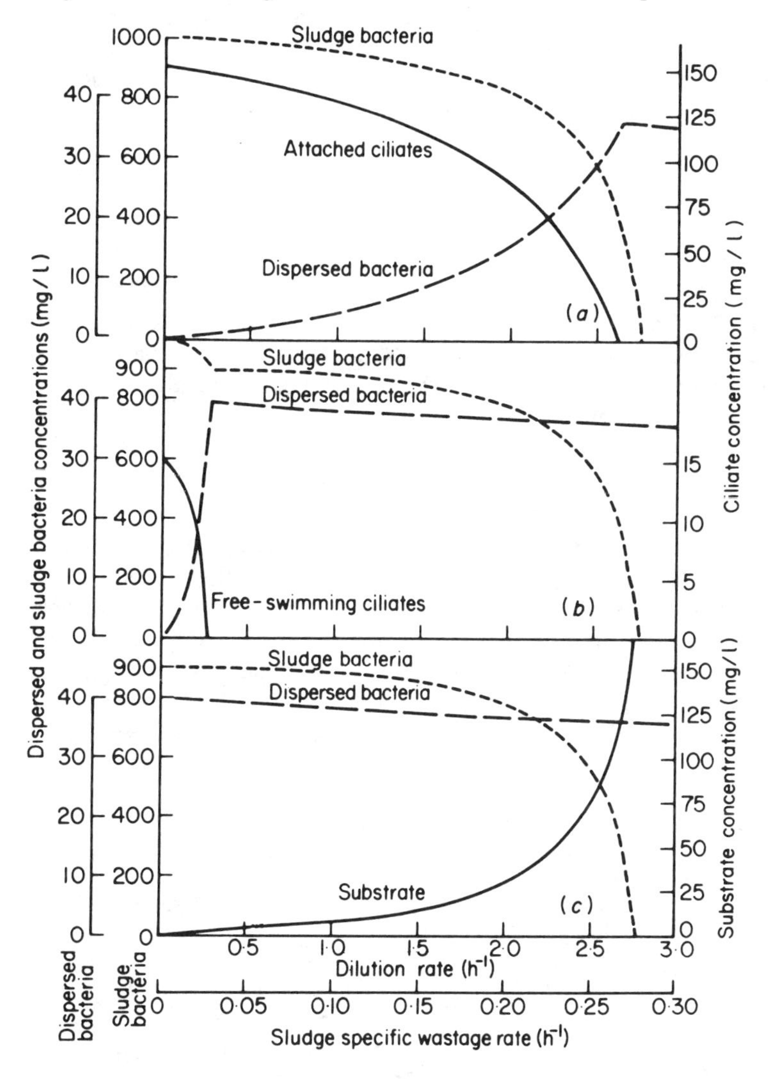

Figure 1 Steady-state populations of activated-sludge plants operating at different specific wastage rates when (*a*) attached ciliates are present, (*b*) free-swimming ciliates are present, and (*c*) no ciliates were present.

soluble organic substrates into nonflocculated bacteria. If the contents of the first stage then overflow into a conventional activated-sludge tank with feedback, this would allow for nitrification and the protozoa would clarify the effluent. The advantage of such a system would be to reduce the production of excess sludge solids (by about 20%), which would have cost advantages, since the disposal of solids accounts for approximately a third of the total costs of sewage treatment. Although this idea looks attractive on paper, the research and development work has not yet been carried out and cost of conversion or installation of major equipment would have to be taken into account.

CONCLUSIONS

It seems likely that the major role of the ciliated protozoa in aerobic waste-treatment processes is the removal of dispersed growths of bacteria by predation. This is supported by evidence from full-scale, small-scale, pure culture, and mathematical models. It is unlikely that protozoan-induced flocculation is of any real importance, although this would be difficult to prove conclusively. Flagellated protozoa and amoebae also feed upon bacteria and no doubt will play a similar role when present in sufficient numbers; furthermore, amoebae may have the ability to ingest flocculated bacteria, which could reduce sludge production. Although it is well known that protozoa may also utilize soluble organic complexes, how important this might be when bacteria are present as the primary food source is unknown and is worthy of further investigations. Similarly, the effect of predation upon bacterial activities in chemostat culture systems is ripe for investigation since the results would be applicable to all ecological situations where bacteria-feeding organisms are present. The use of protozoa as indicators of effluent quality looks promising and is of practical value. However, simpler methods should be sought and evaluated.

Literature Cited

1. Agersborg, H. P. K., Hatfield, W. D. 1929. *Sewage Works J.* 1:411–24
2. Ardern, E., Lockett, W. T. 1928. *Manchester Rivers Dept. Ann. Rep.* 1:41–46
3. Ardern, E., Lockett, W. T. 1936. *Water Pollut. Control,* 35:212–15
4. Baines, S., Hawkes, H. A., Hewitt, C. H., Jenkins, S. H. 1953. *Sewage Ind. Wastes* 25:1023–33
5. Bark, T. 1971. *Ann. Stn. Biol. Besse* 6–7:241–60
6. Barker, A. N. 1942. *Ann. Appl. Biol.* 29:23–33
7. Barker, A. N. 1943. *Nat. Hull.* 65–69
8. Barker, A. N. 1946. *Ann. Appl. Biol.* 33:314–25
9. Barker, A. N. 1949. *Water Pollut. Control.* 48:7–27
10. Barritt, N. W. 1940. *Ann. Appl. Biol.* 27:151–56
11. Bick, H. 1972. *Ciliated Protozoa,* Geneva: World Health Organization. 198 pp.
12. Brown, T. J. 1965. *Water Pollut. Control* 64:375–78
13. Bungay, H. R. 1968. *Chem. Eng. Prog. Symp.* 64:19–22
14. Buswell, A. M., Long, H. L. 1923. *J. Am. Water Works. Assoc.* 10:309–21

15. Calaway, W. T., Lackey, J. B. 1962. *Waste Treatment Protozoa-flagellata Fla. Eng. Ser.* No. 3, 140 pp.
16. Canale, R. P. 1969. *Biotechnol. Bioeng.* 11:887–907
17. Clay, E. 1964. *I.C.I. Paints Div. Res. Dept. Memo.* P.V.M. 48/A/782
18. Curds, C. R. 1963. *Studies on the ecology of ciliated protozoa in activated sludge and their role in the process of sewage purification.* PhD thesis. Univ. London. 237 pp.
19. Curds, C. R. 1963. *J. Gen. Microbiol.* 33:357–63
20. Curds, C. R. 1966. *Oikos* 15:282–89
21. Curds, C. R. 1969. *An illustrated key to the British freshwater ciliated protozoa commonly found in activated sludge.* Water Pollut. Tec. Pap. No. 12, London, H.M.S.O. 90 pp.
22. Curds, C. R. 1970. In *Proc. Symp. on Methods of Study of Soil Ecology,* ed. J. Philipson, pp. 127–29. Paris: UNESCO
23. Curds, C. R. 1971. *Water Res.* 5:793–812
24. Curds, C. R. 1971. *Water. Res.* 5: 1049–66
25. Curds, C. R. 1973. *Water. Res.* 7: 1269–84
26. Curds, C. R. 1973. *Water. Res.* 7: 1439–52
27. Curds, C. R. 1975. See Ref. 38, pp. 203–68
28. Curds, C. R. 1981. In *The Evolving Biosphere,* ed P. L. Forey, P. H. Greenwood, pp. 131–41. London: Br. Museum Nat. Hist. 311 pp.
29. Curds, C. R. 1982. *British and other freshwater ciliated protozoa. 1. Ciliophora: Kinetofragminophora. Synopsis of British Fauna (NS),* No. 22. 384 pp.
30. Curds, C. R., Bazin, M. J. 1977. *Adv. Aquatic Microbiol.* 1:115–76
31. Curds, C. R., Cockburn, A. 1968. *J. Gen. Microbiol.* 54:343–58
32. Curds, C. R., Cockburn, A. 1970. *Water Res.* 4:225–36
33. Curds, C. R., Cockburn, A. 1970. *Water Res.* 4:237–49
34. Curds, C. R., Cockburn, A. 1971. *J. Gen. Microbiol.* 66:95–108
35. Curds, C. R., Cockburn, A., Vandyke, J. M. 1968. *Water Pollut. Control* 67:312–29
36. Curds, C. R., Fey, G. J. 1969. *Water Res.* 3:853–67
37. Curds, C. R., Gates, M., Roberts, D. McL. 1983. *British and other freshwater ciliated protozoa. 2. Ciliophora: Oligohymenophora and Polyhymenophora. Synopsis of British Fauna (NS),* No. 23. In press
38. Curds, C. R., Hawkes, H. A., eds. 1975. *Ecological Aspects of Used-water Treatment I. The Organisms and their Ecology.* London: Academic. 414 pp.
39. Curds, C. R., Hawkes, H. A., eds. 1982. *Ecological Aspects of Used-water Treatment II. The Processes and their Ecology,* In press. London: Academic
40. Curds, C. R., Vandyke, J. M. 1966. *J. Appl. Ecol.* 3:127–37
41. Dixon, A. 1937. *Rep. Proc. 2nd Int. Congr. Microbiol.* 1936:250
42. Drakides, C. 1980. *Water Res.* 14:1199–207
43. Fairbrother, T. H., Renshaw, A. 1922. *J. Soc. Chem. Ind. London* 41:134–44
44. Guede, H. 1979. *Microb. Ecol.* 5(3): 255–37
45. Hardin, G. 1943. *Nature* 151:642
46. Hawkes, H. A. 1960. In *Waste Treatment,* ed. P.C.G. Isacc, pp. 52–98. Oxford: Pergamon
47. Hawkes, H. A. 1963. *The Ecology of Waste Water Treatment.* Oxford: Pergamon. 203 pp
48. Horosawa, I. 1950. *J. Jpn. Sewage Works. Assoc.* 148:62–67
49. Hunt, H. W., Cole, C. V., Klein, D. A., Coleman, D. C. 1977. *Microb. Ecol.* 3:259–78
50. Jager, G. 1938. *Z. Hyg. Infecktkrankh.* 120:620–25
51. Jenkins, S. H. 1942. *Nature* 150:607
52. Johannes, R. E. 1965. *Limnol. Oceanogr.* 10:434–42
53. Johnson, J. W. H. 1914. *J. Econ. Biol.* 9:105–24
54. Jones, G. L. 1973. *Nature* 243:546–47
55. Kolkwitz, R. 1926. *Kleine Mitt. Mitgl. Ver. WassVersorg.* 3:70–74
56. Kolkwitz, R., Marsson, M. 1908. *Ber. Dtsch. Bot. Ges.* 26a:509–19
57. Kolkwitz, R., Marsson, M. 1909. *Int. Rev. Ges. Hydrobiol. Hydrogr.* 2:126–52
58. Lackey, J. B. 1925. *Bull. N. J. Agric. Exp. Stn.* 417:1–39
59. Liebmann, H. 1949. *Vom Wass.* 17: 62–82
60. Liebmann, H. 1951. *Handbuch der Frischwasser- und Abwasser-biologie.* Jena: Fischer. 539 pp.
61. Martin, D. 1968. *Microfauna of Biological Filters.* Univ. Newcastle upon Tyne, Dept. Civil Eng. Bull. No. 39, Newcastle upon Tyne: Oriel. 81 pp.
62. McKinney, R. E., Gram, A. 1956. *Sewage Ind. Wastes* 28:1219–37
63. Ministry of Technology. 1968. *Notes on Water Pollution No. 43.* London: HMSO
64. Morishita, I. 1968. *Jpn. J. Hydrobiol.* 4:12–20

65. Morishita, I. 1970. *Jpn. J. Protozool.* 3:1–13
66. Page, F. C. 1976. *An Illustrated Key to Freshwater and Soil Amoebae.* Freshwater Biol. Assoc. Sci. Publ. No. 34. 155 pp.
67. Phelps, A. 1946. *J. Exp. Zool.* 102: 277–92
68. Pike, E. B. See Ref. 38, pp. 1–63
69. Pillai, S. C. 1941. *Curr. Sci. Bangalore* 10:84–85
70. Pillai, S. C. 1942. *Curr. Sci. Bangalore* 11:437
71. Pillai, S. C., Rajagopalan, R., Gurbaxani, M. I., Subrahmanyan, V. 1948. *Curr. Sci. Bangalore* 17:245–46
72. Pillai, S. C., Subrahmanyan, V. 1942. *Nature* 150:525
73. Pillai, S. C., Subrahmanyan, V. 1944. *Nature* 154:179–80
74. Pillai, S. C., Wadhwani, T. K., Gurbaxani, M. I., Subrahmanyan, V. 1947. *Curr. Sci. Bangalore* 16:340–41
75. Pillai, S. C., Wadhwani, T. K., Gurbaxani, M. I., Subrahmanyan, V. 1948. *Curr. Sci. Bangalore* 17:93–95, 122–23
76. Pirt, S. J., Bazin, M. J. 1972. *Nature* 239:290
77. Reynoldson, T. B. 1942. *Nature* 149: 608–9
78. Richards, H., Sawyer, G. C. 1922. *J. Soc. Chem. Ind. London.* 41:62–72
79. Schofield, T. 1971. *Water Pollut. Control* 70:32–47
80. Sladecek, V. 1969. *Verh. Int. Verein Limnol.* 17:546–59
81. Stout, J. D. 1980. *Adv. Microb. Ecol.* 4:1–50
82. Sugden, B., Lloyd, L. R. 1950. *Water Pollut. Control* 49:16–23
83. Sydenham, D. H. J. 1968. *The ecology of protozoa and other organisms in activated sludge.* PhD thesis. Univ. London. 262 pp
84. Szabo, Z. 1960. *Öst. Wasserw.* 12: 224–26
85. van der Drift, C., van Seggelen, E., Stumm, C., Hol, W., Tuinte, J. 1977. *Appl. Environ. Microbiol.* 34(3):315–19
86. Vedry, B. 1976. *L'Analyse Ecologique des Bouées Activées.* Paris: Bruker. 100 pp.
87. Viehl, K. 1937. *Z. Hyg. Infekt. Krankh.* 119:383–412
88. Watson, J. M. 1945. *Nature* 155:171–72
89. Yasuko, I. 1960. *Jpn. J. Ecol.* 10: 207–14

Ann. Rev. Microbiol. 1982. 36:47–73

THE EVOLUTION OF RNA VIRUSES

D. C. Reanney

Department of Microbiology, La Trobe University, Bundoora, Victoria 3083, Australia

---

CONTENTS

INTRODUCTION

Viruses with RNA genomes are found in prokaryotes, fungi, animals, and plants. They are an extraordinarily diverse group that includes isometric helical and enveloped members. Although RNA animal viruses have been classified into families and genera (71), the systematics of plant and fungal viruses is less advanced (71). At this stage little is known of the origins and evolution of these viruses. The complete nucleotide sequences of a few RNA viral genomes such as the ribophage, MS2 (32), the picornavirus, polio (55),

0066-4227/82/1001-0047$02.00

and the retrovirus Moloney murine leukemia virus (95) have been determined while the nucleotide sequences of some viral coat protein genes or the amino acid sequences of other coat proteins are known (24, 40). At the time of writing, however, not enough detailed sequence information exists to allow the construction of meaningful phylogenies from computer-based comparisons, except in the case of a few groups such as some tobamoviruses (37). Techniques such as molecular hybridization offer powerful tools to students of viral evolution, but their application to the comparative study of viral RNAs has been limited to date to a relatively small number of systematic studies [see, for example, the investigation into the molecular relatedness of the genomes of selected potyviruses made by Abu-Samah & Randles (1)].

Because of this patchy data base this review does not concentrate on the taxonomic aspects of RNA virology. What is unique about RNA viruses is that they are the only self-reproducing units in the biosphere to use a genetic material other than DNA. In evolutionary terms this means that the survival problems faced by RNA genomes have no parallels elsewhere in biology. It seems logical to link these survival problems to certain distinctive features of RNA viruses. These features include genomes of negative polarity and segmental genomes in which the genetic information is dispersed among separate species of RNA [such divided genomes are found among DNA viruses only in the geminivirus group (41)]. This review analyzes the selective pressures that may have directed the evolution of viral RNAs towards the particular genomic and translational strategies found today. Because of this emphasis on general issues of RNA virus evolution, no attempt is made to review individual viral taxa whose evolution has already been extensively studied; thus, the influenza viruses (Orthomyxoviridae)[1] and the RNA tumor viruses (Retroviridae) are mentioned only insofar as they relate to the overall theme of this paper.

RNA VERSUS DNA

The vicinal 2'-hydroxyl group in RNA endows ribopolynucleotides with bonding capacities and labilizing tendencies absent from deoxyribopolynucleotides (82, 102). These properties show why selection preferred DNA as the stable repository of genetic information and why roles that require defined molecular phenotypes were relegated to RNA. If, as is widely believed (30, 89), RNA preceded DNA as the genetic substance, the argument that DNA is better suited to an information-transmitting function is

[1]Family and genus names recommended by the International Committee for the Taxonomy of Viruses are used throughout this paper (71).

strengthened since it requires selection to favor a latecomer, DNA, over an established, self-replicating, RNA-based system.

The point is that once the genotypic and phenotypic functions had been committed to separate polynucleotides, any RNA molecules that retained —or acquired—a genetic character were forced to exploit replicative mechanisms fundamentally different from those used by DNA.

The Noisiness of RNA Replicases

DNA replication has associated error-suppressing and proofreading mechanisms (45, 58). In *Escherichia coli,* for example, the DNA polymerizing enzyme, DNA polymerase I, edits newly made DNA by excising mismatched bases and recopying the information (58). The efficiency of this editing is such that DNA replication in wild-type coliform bacteria has an error rate of about one false nucleotide incorporated for every 10^6–10^7 bases polymerized (31). Postreplicative mismatch correction may reduce this figure to about 10^{-10} (D. MacPhee, personal communication).

There are no reports of error-correcting mechanisms associated with any enzymes that make RNA as product or use RNA as template (30). Thus RNA synthesis is an intrinsically noisy process. The estimated error rate for RNA replicases lies between 10^{-3} and 10^{-4} (30, 31); DNA polymerases unable to carry out error-correcting functions have about the same replicative fidelity (31).

The noisiness of RNA replicases means that the genomes of RNA viruses should not be thought of as unique structures but as populations of different sequences clustered round a weighted average. Eigen & Schuster call this probabilistic distribution a quasi-species (31). The quasi-species concept is consistent with the studies of Domingo et al (27) who showed that about 15% of the clones arising from a multiply passaged population of the ribophage Qβ had fingerprint patterns that deviated from those of the RNA from the population as a whole. Domingo et al (27) proposed that the Qβ population is in a state of dynamic equilibrium, with viable mutants arising at a high rate on the one hand and being strongly selected against on the other. Further work has shown that the replication of other RNA viruses, e.g. foot-and-mouth disease virus and influenza virus, can be described in similar, statistical terms (26).

Segmentation and the Noisiness of RNA Synthesis

The noisiness of RNA replicases has undoubtedly had drastic affects on the evolution of RNA viruses. Table 1 shows that no RNA virus has achieved the genome size of such DNA viruses as the Herpesviridae, the Iridoviridae, the Adenoviridae, or the tailed bacteriophages. The largest continuous RNA genome described to date belongs to a coronavirus with a reported genome size of about 8×10^6 daltons (67). RNA molecules exceeding this length may not be able to sustain their identity in the face of the scrambling effect of mutation (clearly, for a fixed error level, the longer the RNA, the greater the possible loss of information). Indeed, when the total molecular weights of all the viral RNA molecules in Table 1 are averaged out, it is evident that the modal value of RNA genomes lies between $3\text{–}4 \times 10^6$ daltons. This figure may represent an optimal compromise between pressures to encode the minimum number of virus-specific functions needed for efficient reproduction and pressures to keep genome sizes within the stability limits set by the error proneness of RNA replicases and other factors.

Some RNA viruses may have been able to achieve or exceed this inferred optimal size by dividing their genomes into modules. This segmentation of genomes is a characteristic of certain RNA but not of DNA viruses (52, 65). Two strategies have been evolved: (*a*) Two or more RNA segments may be encapsidated in a single particle (e.g. the Orthomyxoviridae and the Arenaviridae); or (*b*) each RNA species may be encapsidated separately to give a multicomponent system (e.g. the bromoviruses). In the latter case infection is a multi-hit phenomenon (52).

At first sight it might seem that segmentation merely disperses the inaccuracy dilemma across several RNA species so that the error rate of the whole virus is merely the sum of the error rate of its component RNAs. However, such an additive effect may be compensated for in part by the way the segmental RNAs of many multicomponent viruses are assembled inside infected cells. These segmental RNAs are reproduced as units inside the infected cell and they are withdrawn from the intracellular pool of newly replicated RNAs at random (22, 65). This mixing of copied RNAs ensures that populations of sequences are shuffled into a spectrum of combinations, and this in turn may ensure that optimally adapted composite viruses arise at adequate rates at each round of replication. Genome segmentation also provides means of reassorting RNAs from different sources (as in the influenza group) and hence provides a substitute for the physical recombination that occurs in duplex DNA viruses (see below).

UV inactivation kinetics provide a model of the way segmentation may counteract the negative effects of inaccurate copying. It is possible to calibrate the amount of UV irradiation needed to introduce an average of one hit per molecule in a large population of RNAs of length x. If the RNA is divided into two modules of length $\frac{x}{2}$, the same UV dose will not inactivate all the viruses, since the physical division of the genetic information effectively uncouples the effects of the hits from each other. [Precisely because UV damage is introduced at random, a hit in module $A(\frac{x}{2})$ need not inactivate module $B(\frac{x}{2})$ and vice versa.] The mathematics are similar when mutations introduced by an error-prone polymerase are not a result of outside factors.

An interesting relationship between the segmentation of RNA genomes and the noisiness of RNA replicases is provided by the reoviruses. Reoviruses have double-stranded genomes divided into 10–12 segments with a summated molecular weight of $12–20 \times 10^6$, all of which are accommodated in single, isometric particles (71). Experiments have clearly established that reovirus maturation does not involve the random withdrawal of preformed duplex modules into viral heads (96): rather, each particle accumulates the correct combination of modules (96). The proper cónjunction of these various modules requires some specific recognition of defined nucleotide sequences in each separate species of RNA (96), and there may well be a hierarchy of cooperative maturation steps such that the correct formation of complex B depends on the prior formation of complex A (65). If this is so, then segmentation provides a multi-step checking device, which allows selection to reject RNA segments damaged in critical areas (or affected by mutations in these areas). In the case of the reoviruses, then, segmentation and selective interactions may combine to form a simple kind of proofreading of inherently unstable genetic information.

The existence of such a self-editing system may be one reason why reoviral genomes are so large in toto. It may not be coincidental that most double-stranded RNA viruses have segmental genomes with high total molecular weights, irrespective of whether they occur in animals (reo), plants (reo), or prokaryotes ($\phi 6$) (Table 1).

These considerations are not necessarily affected by the discovery of an (apparently) divided genome in bean golden mosaic virus (BGMV), a member of the DNA-containing geminivirus group (40). If the geminiviruses encode a DNA polymerase that lacks error-correcting functions, then BGMV would be operationally analogous to an RNA virus and its genome would constitute a quasi-species. If, as seems more likely, the DNA of BGMV is copied by host DNA polymerases, its small size may reflect a unique strategy of survival. More information is required to resolve this issue.

Unwinding Double-Stranded RNA Molecules

DNA replication occurs in a semi-conservative manner in which each complementary strand serves as template to assemble a full-length partner copy (58). Unwinding a molecule whose complementary strands are twisted round each other once every 20 base pairs presents formidable topological problems (58). These appear to be overcome by the concerted action of several proteins: (*a*) unwinding proteins that shift the equilibrium from a helix towards single strands (58); (*b*) DNA-dependent ATPases that couple unwinding with the hydrolysis of ATP (58); and (*c*) nuclease activities (58). It is noteworthy that an energy input of $1.5 \rightarrow 5$ kcal per mol base pair is needed to melt duplex DNA (58), and in some replicative systems two

Table 1 The relative sizes of the genomes of DNA and RNA viruses

Virus[a]	Segmental genome	Mol wt × 10^6	Host
Double-stranded DNA			
Pox	No	85–240	
Herpes	No	80–150	Animal
Irido	No	100–250	
Baculo	No	58–100	
Tailed phages	No	12–490	Prokaryotes
Adeno	No	20–25	Animals
Mycoplasma	No	7.6 (1.5)	Prokaryotes
Tecti	No	7.4–24	Prokaryotes
Cortico	No	6	
Caulimo	No	4.8–5.0	Plants
Single-stranded DNA			
Parvo	No	1.5–2.2	Animals
Micro	No	1.7	Prokaryotes
Ino	No	1.9–2.7	
Gemini	Yes (?)	0.7–0.9	Plants
Double-stranded RNA			
Reo	Yes	10–12 seg; 0.2–3.0/seg; total MW = 12–20	Animals
Fungal (e.g. penicillium, chrysogenum)	Yes	3 seg	Fungal
Cysto	Yes	2.3 & 3.1 & 5.0; total MW = 10.4	Prokaryotes
Single-stranded RNA			
Corona	No	5.5–6.1 (8)	
Paramyxo	No	5–7	Animals
Toga	No	4	
Orthomyxo	Yes	0.2–1.0/seg, 8 seg	
Rhabdo	No	3.5–4.6	
Buna	Yes	3, 2, 0.5	
Arena	Yes	2.1–3.2/seg	Plant/Animal;
Retro	Dimer	3/monomer	Animals
Picorna	No	2.5	
Calici	No	2.6–2.8	
Levi	No	1.2	Prokaryotes

Table 1 *(Continued)*

Virus[a]	Segmental genome	Mol wt × 10^6	Host
Single-stranded RNA (continued)			
Tymo	No	2	Plants
Luteo	No	2	
Tombus	No	1.5	
Clostero	No	2.3–4.3	
Carla	No	2.3	
Poty	No	3.0–3.5	
Potex	No	2.1	
Tobamo	No	2	
Nepo	Yes	2.4, 1.4–2.2	
Como	Yes	2.0, 1.4	
Tobra	Yes	2.4, 0.6–1.4	Plants
Cucumo	Yes	1.3, 1.1, 0.8	
Bromo	Yes	1.1, 1.0, 0.7	
Ilar	Yes	1.1, 0.9, 0.7	
Hordei	Yes	1.0–1.5	
Unclassified plant viruses (single-stranded)			
Tomato spotted wilt	Yes	2.6, 1.9, 1.7 & 1.3	
Southern bean mosaic	No	1.4	
Tobacco necrosis	No	1.3	
Maize chlorotic dwarf	No	3.2	
Pea enation mosaic	Yes	1.7, 1.3	
Alfalfa mosaic	Yes	1.1, 0.8, 0.7	

[a]Names according to the ICTV (see 71).

ATPs are consumed for every base pair melted. The energy cost of DNA synthesis is very high. No equivalent proteins take part in the replication of RNA molecules nor is ATP consumed (23, 68). In view of the size limit imposed on RNA genomes by copying inaccuracy it would be surprising if RNA viruses were able to encode RNA-unwinding proteins or ATPases. Also, since the synthesis of RNA on cellular RNA templates has never been demonstrated convincingly[2], it appears that RNA viruses cannot recruit cellular enzymes to copy their own sequences.

The point of these considerations can be focused by examining one of the longest double-stranded RNAs documented—one of the segments of the cystovirus $\phi 6$ (71). This contains about 7600 base pairs. Assuming a value

[2]However, several authors have reported the presence of an RNA-dependent RNA polymerizing activity in "uninfected" plants (see 106). The presence of such an activity could help explain the preponderance of RNA viruses in plants (86, 87).

of 22 base pairs/helical turn (the value for the A form of RNA) (80), this $\phi 6$ RNA contains about 345 twists. It is clearly impossible to unwind a double-stranded RNA of this length by any known molecular mechanism under physiological conditions. It follows that the so-called replicative strategies of RNA viral genomes (8, 68) should be viewed, at least in part, as compulsory responses to a constraint imposed on viruses by the physical impossibility of replicating their genomes in a semi-conservative manner.

Consider the viral groups one by one. First, take viruses such as the reovirus group whose genomes consist of duplex RNA. In reoviruses a single positive strand is displaced during copying from each duplex segment (96). These + strands serve as templates for single negative strands and the formation of the complementary strands regenerates the duplex (96). Parental RNA strands do not appear in progeny. The reoviruses then overcome the unwinding problem by avoiding it and directing replication through the separate synthesis of single stands. This mechanism has much more in common with transcription than with replication. Indeed, the synthesis of viroids (25) is actually mediated by the host DNA–dependent RNA polymerase (83).

The asymmetric character of RNA viral replication is even more evident in the replication of the single-stranded viruses. Take the negative-strand viruses such as the rhabdovirus, vesicular stomatitis virus, whose genome takes the form of a single strand of negative-sense RNA with a *Sw*20 value of 40 (68). A 40*S* positive-strand RNA has been detected in infected cells (68). Although the details are obscure, it seems virtually certain that the templating strands retain their single-stranded character, at least in part, throughout copying (68). Interestingly, the 40*S* RNAs occur as nucleocapsids whereas the mRNAs transcribed from the negative strands do not (68). It may be that the proteins in the replicative complex prevent the formation of extensively H-bonded helical regions, which would be difficult to separate.

The positive-strand RNA viruses show considerable diversity in their modes of reproduction. The genomes of leviviruses such as MS2 serve as mRNA to direct the synthesis of a replicase. This replicase then makes a negative copy. But, significantly, the replicating intermediate does not consist of a duplex structure: Rather, one negative strand directs the synthesis of multiple nascent positive strands, which appear to be held to the template by the replicase (103). If the replicase is removed by a protein denaturant, the resulting RNA contains double-stranded regions linked to a disorganized mass of single-stranded material.

Similar considerations apply to other positive-strand viruses. Tobacco mosaic virus (TMV) generates a double-stranded replicative form in infected plants (56). The replicative form exhibits a sharp thermal transition with a *Tm* of 97°C in standard saline citrate buffer (53). A duplex RNA of this length [equivalent to the predicted length of a double-stranded equivalent of the entire genome (53)] could not be melted under physiological conditions; hence it comes as no surprise to learn that the in vivo replicating complex (replication intermediate) contains single-stranded tails.

How are the newly made negative strands prevented from bonding with their templates in vivo? Inspection of the secondary structure for MS2 RNA (32), for example, shows that the high degree of predicted intramolecular folding confines continuous double-helical regions to about 20–40 bp (i.e. about 2–3 helical turns). This not only facilitates unwinding of the RNA by replicase, it also suggests that newly made negative-strand RNA, if prevented from bonding to the template immediately after synthesis, will spontaneously fold into compact structures that pose a kinetic barrier to the thermodynamically favored reannealing of the separate complementary strands.

Synthesis of the negative strand starts at the 3' end of the viral genome; hence, it may be especially important to prevent extensive duplex formation in this region. The highly ordered, tRNA-like secondary structures found at the 3' ends of so many positive-strand RNA viruses (43) may not only serve as signals for the replicase (43), they may also help deny newly made negative-strand RNA access to its complementary sequences. Other mechanisms also seem likely to prevent long regions of duplex RNA from forming in vivo. Nascent positive strands are often bound by ribosomes before their synthesis is completed (23) and, in the Qβ system, any one strand is normally being traversed by several replicases (23).

In light of these considerations, many peculiarities of RNA viral genomes stand out as more or less inevitable answers to a common problem imposed on the viruses by purely physico-chemical requirements. Single strandedness is seen as a means of preventing the information in an RNA viral genome from being trapped in a helical configuration that would be inert to its own replicase. Also, the absence from duplex RNA viruses of the circular genomes so characteristic of prokaryotic plasmids and viruses may be seen as a consequence of the restricted sizes of RNA viral genomes that do not allow RNA viruses to encode the proteins needed to melt topologically constrained structures. These topological constraints do not apply so forcibly to single-stranded RNA genomes. Accordingly, the genomes of the bunyaviruses can assume a circular configuration in which the 3' and 5' ends are united by H-bonding (71) and covalently circular RNAs have been found in three plant viruses (85, 104) and in viroids (25).

RNA Topologies

RNA strands in water automatically collapse into a minimum free-energy state that maximizes stacking interactions between the bases and optimizes the internal pairing capacity of the molecule (82). Short RNAs, therefore, have reproducible three dimensional phenotypes like proteins (102). Have these phenotypic potentialities been exploited during the evolution of RNA viruses? Although there is not enough information to answer this question for most RNA viruses, in the case of the ribophages there can be no doubt that topological aspects play a key role in the controlled read-out of viral genes. The ribosome binding site of the *A* protein gene, for example, is normally buried in the folded structure of the RNA and becomes transiently accessible only during synthesis of the + strands (23). In this way translation of the *A* protein gene is severely limited. The importance of topology to the phage's survival is shown by the evolutionary retention of sequences that do not encode proteins, but which contribute to the maintenance of a preferred RNA geometry by nucleotide:nucleotide interactions (32). Indeed, the need to preserve specific aspects of three dimensional configuration in target areas of the genome may be another reason why individual viral RNA molecules seldom exceed 10^4 nucleotides in size (Table 1).

COMPACTING THE GENETIC INFORMATION IN RNA VIRAL GENOMES

If the infidelity of RNA replicases and other factors limits the optimal size of RNA genomes to about 3–4 X 10^6 daltons, it follows that RNA viruses are under strong pressure to devise ways of increasing the amount of information they can encode without exceeding this limit. Several features of RNA viruses can be rationalized by assuming that these features result from pressures to compact the genetic information.

Overlapping Genes: Different Reading Frames

One of the most striking ways to increase information content economically is to use a single nucleotide sequence to encode more than one protein. The discovery of overlapping genes in the DNA phage ϕX174 (Microviridae) has been followed by the discovery of similar multiple-choice systems in ribophages (Leviviridae). Model et al (74) described a conditional lethal mutant of f2 that grows in nonsuppressing strains without lysing them. Since three known cistrons are required for the intracellular accumulation of infectious phages, the existence of this mutant implied the presence of a fourth complementation group. This fourth gene, which encodes a lysis protein, has been identified as a 75-amino acid-long polypeptide translated in the +1 phase relative to both the coat protein and replicase cistrons (6).

From an evolutionary viewpoint it is interesting that this lysis protein in f 2 appears to have some homology with the inferred lysis protein of the DNA phage ϕX176 (13). Likewise, segment 8 of the influenza virus genome contains two unique polypeptides, NS_1 and NS_2 (61). Lamb & Lai (62) have shown that these two proteins overlap by 70 amino acids translated from different reading frames. A similar situation occurs in segment 7, which encodes at least two proteins, M_1 and M_2 (60, 63). A nucleotide body region about 271 nucleotides long in M_2 RNA can be translated in the +1 reading frame relative to M_1 mRNA and the sequence data indicate that M_1 and M_2 overlap by 14 amino acids (64). The retrovirus Moloney murine leukemia virus contains a short (60 bp) overlap between the *pol* protein and *env* polyprotein coding regions (95): If the ATG triplet at position 5777 is indeed used to initiate translation of the *env* polyprotein, these two genes must be in different reading frames (95). A multiple choice system also occurs in at least one plant virus. Data presented by Ghosh et al (35) suggest that the gene for the coat and P4 proteins of Southern bean mosaic virus overlap, and since these proteins lack common tryptic peptides it appears they are encoded in different reading frames (35).

The presence of genes read in different frames in viruses of plants, animals, and bacteria suggests that this strategy has been separately evolved in the three groups. To date there are no reports of gene overlaps due to phase shifts in the DNA of cellular organisms.

Overlapping Genes: In-Phase Overlaps

In the well-studied ϕX174 system one gene *G* generates four different peptides by using internal initiation signals and by modulating the efficiency with which suppressor tRNA's neutralize STOP codons (81). Similarly, the ribophage Qβ exploits the nucleotide sequence of the coat protein gene to generate the A1 protein by sometimes reading the UGA codon at the end of the coat protein gene (13). As a result, the A1 protein shares a common N terminal sequence with the coat protein but differs in its carboxy terminus (13).

Some plant RNA viruses may exploit a similar mechanism. Tobacco rattle virus is a multicomponent virus whose two particles each contain a discrete RNA species. Translation of one of these RNAs yields two proteins whose combined molecular weights exceed the coding capacity of the RNA (65, 79). Pelham has presented evidence that the larger of the two proteins arises from leaky readthrough of an opal (UGA) codon (79). Whether or not the readthrough protein has any functional significance in vivo remains to be established. Overlapping proteins have also been detected in in vitro studies of the translation products of the RNAs of turnip yellow mosaic virus (72, 75). Other overlapping genes may exist in TMV and cowpea mosaic virus (22).

In-phase gene overlaps are not unique to viruses. Smith & Parkinson (98) have described an overlapping gene at the *cheA* locus in *E. coli;* in this case two peptides appear to be translated from the same coding sequence in the same reading frame but from different initiation sites (98). Geller & Rich (34) have purified a tryptophanyl tRNA from rabbit reticulocytes that suppresses the UGA codon of β-hemoglobin mRNA. They suggest that natural readthrough proteins arising from the suppression of UGA or UAG but not UAA serve essential cellular functions (34). It seems likely that in-phase overlaps offer a general strategy for generating new proteins by conserving the amino acid sequence in one part of the protein and varying the other(s) and thus combining tested sequences with trial sequences. This kind of permutations-on-a-theme evolution may be especially important in the development of proteins that interact with each other. It has been suggested, for example, that the in-phase translation of the overlapping sequences of the *C* and *Nu3* genes in phage λ results in proteins whose shared structural domains are important for normal prohead formation (94).

RNA:RNA Splicing

The sequence overlaps in the phage and plant viral RNAs described above were achieved by the use of alternative start and stop signals in the relevant mRNA's. The sequence overlaps in the retroviruses and the orthomyxoviruses are unique because these two taxa, unlike all other known RNA viruses, have an obligatory nuclear phase (11).

The primary transcripts of most nuclear structural genes in eukaryotes undergo one or more rounds of selective RNA processing before they are transported to the cytoplasm for translation (4, 19, 93). During these processing steps intervening sequences, introns, are excised from precursor RNAs and the remaining coding modules, exons, are religated (19). The nuclear involvement of the retroviruses and the orthomyxoviruses means that RNAs transcribed from the genomes of these groups are potential targets for this processing machinery. The role of RNA splicing in the maturation of viral mRNA's has been intensively studied in the case of the influenza viruses. RNA transcribed from segment 8 of the composite influenza viral genome codes for two overlapping proteins, NS_1 and NS_2, that are translated from distinct mRNA species (62). As shown in Figure 1, these mRNA's share 56 nucleotides at their 5' ends, after which the sequences diverge because of the absence from NS_2 RNA of a 473-nucleotide-long module present in NS_1 (62). Of the two overlapping proteins in segment 7 (64), M_2 mRNA contains an inferred intron 689 nucleotides long whereas a further potential mRNA ($mRNA_3$) is interrupted by a 729-base-long sequence (64). The nucleotide sequences at the boundaries of the NS_2/NS_1 and $M_2/mRNA_3$ introns are similar to those found at the splice points of other eukaryote nuclear structural genes (64, 93). The 3' nucleotides of the intervening sequences in the M_1 RNA cannot be so readily accommodated in this standard canonical sequence because the positions of the inferred splice point and the expected pyrimidine tract do not occupy the same relative positions as in other split genes (64, 93).

Figure 1 illustrates the dramatic way in which the selective processing of identical RNAs transcribed from a common nucleotide sequence can generate proteins that contain fundamentally different amino acid sequences (90).

The only other group of RNA viruses in which some members have a nuclear association is the rhabdovirus group. A few plant rhabdoviruses mature within the nucleus. Plant nuclear genes are probably split and the consensus sequences that surround introns in the bean storage protein can be accommodated largely within the canonical sequence determined for animal genes (101). It seems quite possible then that plant rhabdoviral RNAs may be processed by the splicing machinery of the plant nucleus. This idea gains credence when one remembers that rhabdoviruses, like orthomyxoviruses, contain negative-strand genomes from which positive-strand mRNA's are transcribed.

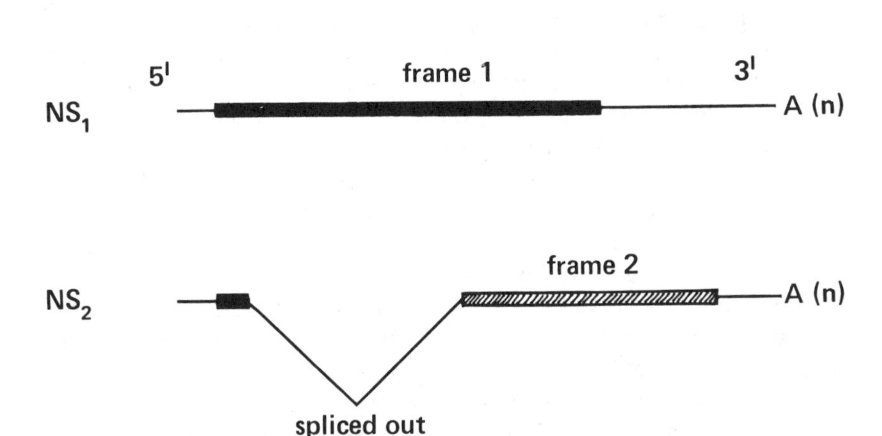

Figure 1 A proposed arrangment of sequences in the mRNA's for NS_1 and NS_2 influenza proteins. Thin lines at each end of the molecules represent noncoding regions; thick lines represent coding segments; the cross-hatched segment in NS_2 mRNA indicates this sequence is translated in a different reading frame from the corresponding sequence of NS_1. The v-shaped thin line represents the intron believed to be excised by RNA splicing. (From 62.)

Other Pressures to Compact Information?

The genetic compression described above is readily interpreted as a result of selection to maximize the reproductive fitness of viral genes by increasing the numbers of viral progeny per infected cell. However, various considerations make this inference unsatisfactory as a general explanation: For one thing, as noted, only the strategy of reading genes in different frames seems unique to viruses.

The problem can be focused by looking at another self-replicating intracellular unit, the mitochondrion. The entire base sequence of human mitochondrial DNA has recently been determined (5). This mitochondrial genome represents a remarkable example of genetic economy with hardly a single nucleotide being wasted. Indeed, even the final nucleotide of most UAA codons has been trimmed off since the required A(A) is regenerated by polyadenylation (15). It is difficult to see how such fine tuning could be rationalized in terms of selection for smaller DNAs. Not only is the loss of a single nucleotide insignificant in terms of metabolic load, but mitochondria in other organisms such as yeast contain large amounts of apparently functionless extra DNA (33).

A different (supplementary?) reason for the genetic compactness of human mitochondrial DNA and viral RNA lies in their possible evolutionary origins. Mitochondria are believed to be endosymbiotic prokaryotes (69) whereas many eukaryote RNA viruses have features reminiscent of prokaryotes (e.g. polycistronic mRNA's). Whatever their origins, what is incontestable is that both sets of extrachromosomal polynucleotide are organized and expressed differently to nuclear genes (39, 68). It follows that they must regulate the expression of their genomes through mechanisms that differ from those that control nuclear genes.

One example suffices to make the point. Transcripts of most nuclear genes contain interruptions (introns) that are excised to form mRNA (19): These intron sequences appear to represent functionless "junk" sequences (28). By contrast, the split *cob-box* gene of yeast mitochondria contains a sequence (*box3*) that appears to act as an exon in the primary RNA transcript but as an intron in later processing steps (17, 51). Because of this interconvertibility of introns and exons, the *cob-box* locus constitutes an overlapping gene. The organization and expression of the *cob-box* gene in mitochondria is similar to that of the NS complex in influenza virus (Figure 1). The primary transcript of the *cob-box* gene appears to encode a splicing enzyme that regulates the synthesis of its overlapping gene product (apocytochrome *b*) in a self-adjusting way (51). Similarly, by *coupling* the production of NS_2 to the production of NS_1 by gene overlap selection ensures that the NS_2 protein can only appear after the mRNA for the NS_1 protein has accumulated in the nucleus. In both cases the use of shared sequences seems pegged more to genetic regulation than to genetic economy.[3]

VARIATIONS IN GENOME SIZE

Other things being equal, a virus *A* that can produce more viable progeny than a second virus *B* has a selective advantage over *A* in a competitive situation. Since smaller genomes usually generate more copies than do larger genomes, there appears to have been pressure towards small genomes during the evolution of RNA viruses.

In vitro studies provide models of the natural processes. Spiegelman devised a test-tube system for replicating purified Qβ RNA in the presence of the nucleotide triphosphate precursors and the purified Qβ replicase (99). With this system it was possible to obtain infectious viral RNA after many serial transfers. When pressure for abbreviated RNAs was applied by shortening the intervals between transfers, a spectrum of mini-variants emerged: For example, after 74 serial transfers a variant that had lost 83% of the wild-type viral RNA was detected (99). This variant replicated 15 times faster than did its parent.

Eigen et al have shown that when the precursors are incubated with the Qβ replicase alone, after a long lag period, de novo RNAs form (30). By subjecting these de novo RNAs to a variety of different selective pressures, e.g. exposure to RNase (30), it was possible to generate novel RNAs with different genotypes. The adaptive capacity of these RNAs is quite remarkable.

Similar events may well have contributed to the evolution of two kinds of RNA often described in RNA virology. The first is defective interfering (DI) RNA. When cells are infected with RNA viruses at high multiplicities of infection and serially passaged for a number of generations, small RNAs normally emerge and these often have the capacity to inhibit the multiplication of the parental viral RNA (23, 47). The second class is represented by the RNA of satellite viruses; these small RNAs exploit the replicases of associated, independently viable viruses for their own reproduction (65). The examples of both DI RNAs and satellite RNAs show that short RNAs are readily selected from large populations of molecules under natural

[3]Recent work on ribophage MS2 strongly supports the above hypothesis on regulatory functions for overlapping genes (see 52a). Kastelein et al have shown that the lysis cistron, which overlaps the coat and replicase cistrons in the +1 reading frame, is translated by that fraction of ribosomes that arrive out of phase from the upstream coat protein gene. They suggest that the tight coupling of the coat and lysis genes by use of shared sequences guarantees that an infected cell cannot by lysed unless enough coat proteins have been made to produce a good crop of progeny phages.

conditions when an appropriate replicase is supplied from another source.

An extreme example of RNA simplification may be represented by viroids. These are single-stranded RNAs with a very high degree of intramolecular self-complementarity. The prototype viroid, potato spindle tuber viroid, recruits the DNA-dependent RNA polymerase II of the host cell's nucleus for its own replication (83). Evidently the rod-like rigidity of the viroid is sufficiently DNA-like to trick the cellular enzyme into using a quasi-duplex RNA as template rather than a duplex DNA. At present the origin(s) of viroids is obscure, but recent studies with velvet tobacco mottle virus (VTMoV) and solanum nodiflorum mottle virus may provide clues (85). These serologically related viruses contain several RNA species: a single-stranded RNA of mol wt about 1.5×10^6 (RNA-1); a highly base-paired circular RNA of mol wt 1.2×10^5 (RNA-2); and a linear form of RNA-2 (RNA-3) (38, 85). RNA-2 is structurally analogous to a viroid. Gould et al (38) have shown that both RNA-1 and RNA-2 of VTMoV are essential for the initiation of infection and that their interdependence is highly specific: The function(s) of the viroid-like RNA for RNA-1 replication could not be supplied by a range of conventional viroids, and reciprocally, the viroid-like RNA failed to replicate when co-innoculated with a range of viruses with molecular weights similar to VTMoV RNA-1 (38). Gould et al speculate that viruses like VTMoV may represent a stage in viroid evolution from an RNA virus (38).

The VTMoV example can also be used as an example of how bigger RNA viruses may have originated. Gould et al (38) speculate that VTMoV may have originated from an *association* between single-stranded RNA viruses and viroids. This is a special example of a general principle spelled out elsewhere (77), namely, that segmental RNA viruses may arise whenever there occurs in a common host cell two RNA molecules each of which *separately* contains genes *jointly* needed for intracellular replication and extracellular survival (77). It is possible to envisage how an insect vector, for example, could introduce a large population of RNA viruses into plant or animal tissue seeded with a resident population of satellite RNAs. The DI RNAs generated from the introduced viruses may occasionally lead to a chance combination of DI and satellite RNAs—nonviable by themselves —that could form a viable reassortant virus on encapsidation.

Alternatively, subgenomic RNAs from different parental viruses might combine with genomic RNAs to form a novel multicomponent virus.

Parent 1	Parent 2
genomic $RNA_{a\&b}$ (replicase b)	$RNA_{x\&y}$ (replicase y)
subgenomic RNA_a (coat)	RNA_x (coat)

Possible reassortants

Genomic $RNA_{a\&b}$	$RNA_{x\&y}$
RNA_x	RNA_a

Under appropriate selection the redundant old coat protein genes might be trimmed from the genomic RNA. This would be especially likely if the new subgenomic coat protein RNA produced capsomere proteins more efficiently than the original gene did or if the presence of the new subgenomic RNA somehow interfered with the production of the previous coat protein subgenomic RNA. It is important to remember here that some subgenomic RNAs such as those of tobacco necrosis virus may replicate independently of the genome (18), whereas some subgenomic RNAs may be separately encapsidated to form, for example, the short rods found in cells infected with helical viruses like TMV (12, 44).

The evolution of new viruses through this kind of accidental reassortment would be especially favored by changes in the host range of a virus as consequence of new (insect) vectors or by alterations in the biogeography of existing vectors since both these factors could bring different viruses into common hosts. Mixed infection experiments with divided genome viruses lend some support to this thesis: The RNA-3 of cowpea chlorotic mottle virus, for example, can substitute for the RNA-3 of brome mosaic virus, although for some reason the reverse does not occur (10). The Q strain of cucumber mosaic virus and the V strain of tomato aspermy virus (TAV) form a remarkably vigorous hybrid that contains RNA-3 of CMV and RNA-1 and RNA-2 of TAV (42). Although in general segmental RNAs are normally not readily interchangeable among serologically unrelated viruses (65), or even among distantly related viruses of the same group (65), an enormous number of permutational combinations must have been tried across the evolutionary timescale. It seems reasonable to conclude that many existing segmental viruses arose from mixings of the types described above.

Although the conjunction of segmental RNAs may have reversed the natural trend toward smaller genomes, it is less easy to see how longer continuous RNAs could have evolved from smaller ones during evolution. Despite occasional claims (9) that recombination occurs generally among RNA viruses, physical recombination among most RNA viral molecules must be considered unlikely unless or until a plausible mechanism is experimentally demonstrated. One must note, however, that there is no lack of candidate RNA ligating functions. The joining of RNA sequences takes place routinely in the nuclei of plant and animal cells as part of the splicing process (19), and a recent report suggests that the nuclear enzyme, RNA polymerase II, has a ligating capacity (84). An RNA ligase able to close linear RNAs into circles has been demonstrated in wheat germ (57). RNA ligating enzymes may also exist in cytoplasmic organelles such as mitochondria (73) and chloroplasts (3), both of which sometimes contain splicing systems. Whether or not any of these ligating functions can mediate the covalent joining of independently evolved RNAs is another matter.

Whereas physical recombination between preformed RNA molecules remains a case to be proven, the transfer of information between RNA molecules could conceivably be achieved during RNA synthesis by a process like copy choice. This may be what happens in the case of recombination between picornaviral genomes.

TRANSLATIONAL STRATEGIES OF RNA VIRUS GENOMES

Eukaryotic cells process the transcripts of all known genes as monocistronic mRNA's that possess a 5' m^7Gppp cap and, in many but not all cases, a 3' poly A tail (59). Yet a eukaryote virus must encode a minimum of two proteins (a replicase and a coat protein) to achieve replicative autonomy and an extracellular phase.

It is interesting to view the genomic strategies of RNA viruses in the light of this dilemma. In several groups of viruses (the Reoviridae and the Orthomyxoviridae) the division into unit-length mRNA equivalents has taken place at the level of the genome itself. Transcription of the negative strand of the double-stranded segments of the reoviruses or of negative-sense segments of the orthomyxoviruses automatically produces monocistronic mRNA's that are capped and translated in the usual way. In the case of the negative-strand viruses with continuous genomes (e.g. the Paramyxoviridae and the Rhabdoviridae), initiation and termination signals appear to be encoded in the negative strand itself.

The translation problem is especially severe for positive-strand RNA viruses with continuous genomes. Eukaryotic ribosomes recognize the 5' regions of mRNA's (59) and cannot normally accommodate internal AUG signals, as do prokaryotes. There is no problem with internal stop signals, which are recognized by both pro- and eukaryotes. (If this were not so then chain-terminating mutants could not be isolated from eukaryotic cells.) It is therefore the initiation of translation of cistrons downstream from the 5' gene that poses the problem for a polycistronic mRNA in a higher cell.

One solution is represented by the coronaviruses. Avian infectious bronchitis virus directs the synthesis of six species of polyadenylated RNA in infected cells (100). Stern & Kennedy (100) and others have established that these RNAs comprise a nested set with common 3' sequences and variable-length 5' sequences. Here the translation problem is overcome by a non-overlapping system in which only the 5' end unique to each RNA is translated (100). The mechanism by which these subgenomic RNAs are generated is not known at the time of writing, but UV target size measurements on another coronavirus, mouse hepatitis virus, suggest that each RNA is independently initiated on a negative-strand template and is not derived by processing a precursor species (50).

The strategy of using subgenomic RNAs is employed by the alphaviruses and by a variety of plant viruses including TMV (48), tobacco rattle virus (14), Southern bean mosaic virus (35), barley stripe mosaic virus (49), and turnip rosette virus (76).

A quite different strategy is employed by viruses like the Picornaviridae. Here the whole polycistronic mRNA is treated by the cell as one unit and is translated into a polyprotein. Virus-specific peptides are derived from this precursor by post-translational processing (2). Viruses that employ this strategy normally have a protein covalently linked to their 5' ends (7, 105). These strategies are not mutually exclusive. Turnip rosette virus, for example, seems to combine the formation of subgenomic RNAs with protein processing (76).

These various translational strategies obviously hold clues to the evolution of RNA viruses. Consider two (extreme) possibilities. The first is that many RNA viruses of eukaryotes evolved from cellular eukaryotic polynucleotides (77). The simplest way for this to happen would be for a group of monocistronic RNAs to become separately encapsidated to form a di- or tripartite system. If this in fact occurred, segmental monocistronic RNAs could be considered ancestral. An inspection of the known 5' and 3' termini of viral RNAs shows that although RNAs from some segmental viruses (e.g. barley stripe mosaic virus) have a 5' m^7Gppp cap and a 3' poly(A) tail, many (most?) have one feature or the other, but not both. Why short viral RNAs are unable to simultaneously mimic cellular mRNAs in these two key aspects is unclear.

An alternative theory envisages the common ancestor of many contemporary eukaryote RNA viruses as already having a polycistronic genome (70). It has been pointed out (22) that bacteria and their phages, both of which have polycistronic mRNA's, appeared much earlier in evolution than animals or plants, hence—at least in terms of time—the polycistronic format for viruses must be considered primal. For a polycistronic genome to multiply in the alien context of a plant or animal cell certain conditions must be fulfilled. First, the viral replicase gene must be translated. This may not pose an impossible problem as putative RNA polymerizing activities seem to be encoded towards the 5' end of most eukaryote viral RNAs even today (106).

A much more severe barrier is posed by the need to translate the coat protein cistrons that lie toward the 3' ends of contemporary eukaryote viral RNAs (106).

These problems can be overcome by recalling that the natural tendency in viral evolution is towards smaller molecules (see above). The DI RNAs that accumulate so quickly in multiply passaged viruses could serve as models or prototypes for the evolution of subgenomic RNAs. Indeed, formally, DI molecules are subgenomic RNAs. Early steps in the evolution of a eukaryote RNA virus from a phage-like ancestor might involve (*a*) the infection of a eukaryote cell by a polycistronic RNA whose replicase cistron could be translated by host ribosomes, (*b*) an accumulation of defective short RNAs, (*c*) acquisition by a shortened RNA of the capacity to initiate separate translation of capsomere proteins, and (*d*) joint or separate packaging of these two RNAs (original infecting RNA and abbreviated derivative). In essence the evolution of the modern-style virus would occur as selection removed the randomness from the process in which the coat protein subgenomic RNA was generated in subsequent infections; this could happen by mutational changes in the topology of the original RNA, which increasingly caused incomplete replication to generate fragments of reproducible size or which preferentially exposed internal sites in a complete RNA to processing by a resident cleaving enzyme.

Davies (22) recognizes nine different translational strategies of positive-strand RNA (plant) viruses. He suggests that the ancestral virus had a phage-like polycistronic RNA and that evolution has proceeded through the formation of divided genomes and subgenomic RNAs towards the situation exemplified in alfalfa mosaic virus and brome mosaic virus where viral information is carried on separate short RNAs that closely resemble host messengers.

The purpose of these speculations is chiefly to suggest that the evolution of positive-strand RNA viruses from a prokaryote-like ancestor is in accord with our present knowledge. However, it is difficult to account for the evolution of some other classes of RNA virus in this way. A genetic element organized along prokaryotic lines would face severe problems in colonizing a eukaryotic nucleus (90). Thus, the segmented, negative-strand orthomyxoviruses that multiply, in part, in the nucleus and whose RNAs are processed by the nuclear splicing system (see above) seem unlikely to be descendents of polycistronic RNAs. Orthomyxoviruses may have arisen from nuclear RNAs (90). It is noteworthy that whereas most positive-strand RNA viruses have isometric capsids, all negative-strand viruses have helical nucleocapsids enclosed in lipoprotein envelopes (71). RNA viruses seem to encompass several independent lines of evolution (9).

THE EVOLUTION OF HOST RANGE AND THE ORIGIN OF NEW VIRAL SPECIES

The only RNA viruses more or less ubiquitous among bacteria are the leviviruses (71). These tiny, positive-strand RNA viruses adsorb to the pili (or other surface components) specified by various resident plasmids. The only other type of RNA phage is the cystovirus taxon represented by a single-type member, $\phi 6$. $\phi 6$ is unique among phages in containing three segments of double-stranded RNA enclosed in a lipoprotein membrane (71). From many points of view $\phi 6$ seems to have much more in common with the segmental, double-stranded RNA viruses of animals, plants, and fungi than with other phages. Its unique character and highly restricted distribution (it has only been reported in *Pseudomonas phaseolicola*) raise the possibility that the distant ancestor of $\phi 6$ was a eukaryote virus that somehow acquired the ability to propagate in bacteria.

This idea seems less dramatic when one remembers that duplex RNA viruses such as the orbivirus bluetonge can multiply in both vertebrates and invertebrates (71), as can members of several other RNA virus taxa (71). The orbivirus example focuses attention on something seldom given sufficient weight in treatments of viral evolution, i.e. ecological opportunity. The chief reason why orbiviruses can propagate in hosts separated by over 300 million years of separate evolution is probably not because viruses of insects find it easier to multiply in vertebrates than in other hosts or vice versa, but rather because the feeding habits of insects vectors ensure a continuing, high-frequency input of insect viruses into vertebrate tissue. Thus the spread of certain viruses will tend to follow ecological routes mapped out by available transmission mechanisms and may have little relationship to the evolutionary relatedness of eventual hosts. This may explain why rhabdoviruses of insects can multiply in vertebrates or plants (71), but not both. No insects feed on both plant and animal tissue: The specialized mouth parts needed to penetrate phloem, etc, are not easily adaptable to feeding on animal materials.

In the face of the evolutionary distances between their arthropod and vertebrate hosts it is surprising that various arboviruses retain stable identities. What happens when such wide-host-range viruses are denied access to one of their hosts (which is in some ways equivalent to geographic isolation)? Studies with wound tumor virus (WTV) by Reddy & Black (91) showed that when WTV was maintained in sweet clover plants for long periods of time (up to 24 years) without passage through an insect vector, mutants arose that could no longer colonize the arthropod (91). Some of these mutants lacked whole genome segments; in particular some mutants lacked segments 2 or 5 entirely, but retained the ability to replicate effectively in plants (92). If a virus like WTV arose in an insect, prolonged separation from its original host could result in the evolution of a new plant virus. The converse is also true. This example cautions against taking the existence of broadly similar viruses in different kingdoms as evidence of independent origins or convergent evolution.

Reddy & Black's work reveals an amazing degree of RNA polymorphism among WTV populations (91). It is evidently not difficult to obtain clones that differ in their gel patterns from wild-type isolates. Although the WTV example shows how a new virus could arise by loss of genome segments and subsegments and/or mutation in the retained RNAs, a new virus could also arise by gaining foreign RNAs from a co-infecting virus (see above).

VIRAL RNAs AS SELFISH GENES?

Campbell (16) has pointed out that all extrachromosomal genetic elements (or accessory polynucleotides) are characterized by a capacity for overreplication relative to the chromosome. Accessory DNAs have been dubbed "selfish DNA" by Doolittle & Sapienza (28) and Orgel & Crick (78). The selfish DNA concept removes the need to invoke selective arguments to explain the prevalence of nomadic DNA modules such as insertion sequences or transposons in bacteria. Since these modules have a capacity for autonomous replication they will replicate and spread automatically until some counterbalancing (selective) force brings about an equilibrium.

Despite its anthropomorphism, the selfish gene concept is useful because it focuses attention on the self-replicating element itself and not on the phenotypic effects it has on infected cells. The traditional view of plasmids, for example, was that they conferred some selective advantage (e.g. antibiotic resistance) on cells under certain conditions, and hence it was natural to regard the host ranges of conjugative plasmids as an adaptive device that increased the survival chances of the bacterial cells (87). The selfish gene concept showed that a broad host range is very much to the advantage of the plasmid and that any benefits to the host are accidental side effects of the plasmid's own survival strategy (21).

Genes in an RNA virus, unlike genes in a plasmid, must at some stage move laterally from one cell to another to survive. It follows that the survival chances of selfish RNA genes will be greatly expanded if the virus can propagate in several species of host, especially if these hosts occur in the same habitat. Accordingly, many RNA viruses have quite extraordinary host ranges: Alfalfa mosaic virus, for example, propagates in 222 different plant species (20) whereas Tobacco ringspot virus infects 246 species distributed among 54 families (97). Even if 200 plant species became extinct, alfalfa mosaic virus genes could still survive in the remaining 22 susceptible hosts. From this point of view the survival strategy of the selfish RNA genes must be considered as more effective than that of DNA chromosomal genes. The survival of RNA viral genes is also aided by their defining ability to alternate between a cellular and an extracellular state. In the extracellular, particle state genes may survive stresses that would kill cells, e.g. high concentrations of toxic chemicals, periods of intense cold, or long periods without nutrients (54, 90).

The spread of some selfish DNA genes has had a hitch-hiker effect on some selfish RNA genes. Many ribophages can only propagate in cells that contain plasmids of specific incompatibility (Inc) groups. Leviviruses, like PRR1, propagate in cells that contain plasmids of the IncP group (88). IncP plasmids have an extraordinary host range, which encompasses such diverse groups as the enterobacteria, *Pseudomonas, Agrobacterium, Azotobacter,* etc (88). By spreading their survival options across so many taxonomic baskets, IncP plasmids have inadvertently widened the survival potential of the RNA genes that parasitize them.

As Campbell (16) and others have pointed out, over-replication of any parasitic polynucleotide may be harmful to the host and hence, in the long run, to the parasite itself. Therefore one would expect selection to damp down the harmful effects of infection with RNA viruses until a *modus vivendi* with the host was reached. Subclinical persistent infections can in fact be achieved by many RNA viruses. One mechanism that restrains the over-replication of an RNA virus is the accumulation of DI RNAs (46). Other moderating mechanisms exist (23). In some cases infection with RNA viruses may actually have a selective advantage for the host. For example, Gibbs (36) has shown that plants infected with Kennedya yellow mosaic virus are less likely to be eaten by cropping animals than are virus-free plants of the same species.

CONCLUSIONS

The synthesis of RNA molecules on RNA templates is achieved by a small number of proteins, often a single replicase enzyme. This is in sharp contrast to DNA replication, where a variety of different proteins is employed (58). Consequently RNA synthesis lacks the error-correcting functions associated with DNA synthesis and RNA replication has been channeled through mechanisms that avoid the formation of long double-stranded regions, since RNA-unwinding mechanisms do not exist. These aspects of RNA replication have profoundly affected the evolution of RNA viruses.

The absence of an RNA unwinding mechanism has favored the selection of single-strand genomes of either positive or negative polarity. The noisiness of RNA replicases and other factors has prevented RNA genomes from exceeding an ill-defined size threshold and has facilitated the development of multicomponent systems in which the genetic information is divided among several small RNA strands. This strategy is especially noticeable among RNA viruses of plants (87). Segmental genomes have the added bonus that reassortment between them is possible, thus expanding the pool of variation in interbreeding populations.

The evolution of RNA viruses, especially those with divided genomes, has been accelerated and promoted by the evolution of vectors like insects. By providing ecological links between phylogenetically distant hosts, insects and other vectors have greatly widened the survival base of the selfish genes of RNA viruses. These expanded host ranges of RNA viruses also ensured that different (segmental) viruses would increasingly be mixed in common hosts and so made possible the permutational combinations among ancestral RNA viruses that have undoubtedly contributed to the present-day diversity of RNA viral genomes.

Viruses may be the most numerous genetic objects in the biosphere (86). Although DNA viruses predominate in prokaryotes, RNA viruses are abundant in animals and are preponderant in plants (87). RNA viruses then are one of evolution's major success stories. The example of MS2 ribophage shows why. With a genome of only 3569 nucleotides (32), MS2 encodes four proteins, three in phase and one out of phase (13). The expression of these genes is sensitively controlled by two interacting features: (*a*) the topology of the RNA, which modulates the frequency with which individual genes are read; and (*b*) the virus-specified proteins themselves, which affect the translation of their own coding genes. Early in infection translation of the replicase gene is boosted by a host factor (*i*), which promotes ribosome binding to this gene; however, the replicase itself binds *i* (23) so that as concentrations of replicase increase, synthesis of replicase shuts down. Translation of the replicase gene is also repressed by the phage coat protein, which complexes with the ribosome-binding site of the replicase cistron (29). In this way early protein synthesis is directed towards production of the protein needed to replicate viral RNA, whereas late protein synthesis is channeled into production of the protein needed to encapsidate viral RNA. The net result of this remarkable combination of genetic economy and feedback control is that a single MS2-infected cell, under ideal conditions, can produce about 20,000 copies of the infecting input phage (66). This high-amplification factor together with the error rate of the MS2 replicase ensures that the population of progeny RNAs contains enough faithful replicas to maintain the identity of the phage while at the same time offering a clustered spread of variant sequences (i.e. survival options) to the test of selection. In this way the very infidelity of RNA replication is turned to the advantage of the virus. RNA viruses may be small, but they're not dumb.

ACKNOWLEDGMENTS

I am grateful to R. I. B. Francki, R. E. F. Matthews, and J. W. Randles for critical reading of the draft of this manuscript.

Literature Cited

1. Abu-Samah, N., Randles, J. W. 1981. *Virology* 110:436–44
2. Agol, V. I. 1980. *Prog. Med. Virol.* 26:119–57
3. Allet, B., Rochaix, J. D. 1979. *Cell* 18:55–60
4. Aloni, Y. 1981. *Progr. Nucleic Acid Res. Mol. Biol.* 25:1–31
5. Anderson, S., Bankier, A. T., Barrell, B. G., de Bruijn, M. H. L., Coulson, A. R., Drouin, J., Eperon, I. C., Nierlich, D. P., Roe, B. A., Sanger, F., Schreier, P. H., Smith, A. J. H., Staden, R., Young, I. G. 1981. *Nature* 290:457–65
6. Atkins, J. F., Steitz, J. A., Anderson, C. W., Model, P. 1979. *Cell* 18:247–56
7. Babich, A., Wimmer, E., Toyoda, H., Nomoto, A. 1980. *Intervirology* 13: 192–99
8. Bachrach, H. L. 1978. *Adv. Virus Res.* 22:163–86
9. Baltimore, D. 1980. *Ann. NY Acad. Sci.* 354:491–97
10. Bancroft, J. B. 1972. *J. Gen. Virol.* 14:223–28
11. Barrett, T., Wolstenholme, A. J., Mahy, B. W. J. 1979. *Virology* 98:211–25
12. Beachy, R. N., Zaitlin, M. 1977. *Virology* 81:160–69
13. Beremand, M. N., Blumenthal, T. 1979. *Cell* 18:257–66
14. Bisaro, D. M., Siegel, A. 1982. *Virology*. In press
15. Borst, P., Grivell, L. A. 1981. *Nature* 290:443–44
16. Campbell, A. 1981. *Ann. Rev. Microbiol.* 35:55–83
17. Church, G. M., Gilbert, W. 1980. *Mobilisation and Reassembly of Genetic Information,* ed. D. R. Joseph, J. Schultz, W. A. Scott, R. Werner, pp. 379–95. New York: Academic
18. Condit, C., Fraenkel-Conrat, H. 1979. *Virology* 97:122–30
19. Crick, F. 1979. *Science* 204:264–71
20. Crill, P., Hagendorn, D. J., Hanson, E. W. 1970. *WD Agric. Exp. Sta. Res. Bull. No. 280*
21. Davey, R. B., Reanney, D. C. 1980. *Evol. Biol.* 13:113–47
22. Davies, J. W. 1979. In *Nucleic Acids in Plants,* ed. T. C. Hall, J. W. Davies, pp. 111–49. Boca Raton: CRC Press
23. Davis, B. D., Dulbecco, R., Eisen, H. N., Ginsberg, H. S. 1980. *Microbiology,* 3rd ed. New York: Harper & Row. 1355 pp.
24. Dayhoff, M. O. 1972. *Atlas of Protein Sequence and Structure.* Silver Spring, MD: Natl. Biochem. Res. Found. 246 pp.
25. Diener, T. O. 1979. *Science* 205:859–66
26. Domingo, E., Majera, R., Sobrino, F., Perez-Brena, P., Lopez-Galindez, C., Davila, M., Ortin, J. 1981. *5th Int. Congr. Virol.* Strasbourg (Abstr.)
27. Domingo, E., Sabo, D., Taniguchi, T., Weissman, C. 1978. *Cell* 13:735–44
28. Doolittle, W. F., Sapienza, C. 1980. *Nature* 284:601–3
29. Eggen, K., Nathan, D. J. 1969. *J. Mol. Biol.* 39:293–305
30. Eigen, M., Gardiner, W., Schuster, P., Winkler-Oswatitsch, R. 1981. *Sci. Am.* 244:88–118
31. Eigen, M., Schuster, P. 1977. *Naturwissenschaften* 64:541–65
32. Fiers, W., Contreras, R., Duerinck, F., Haegeman, G., Iserentant, D., Merregaert, J., Min Jou, W., Molemans, F., Raeymaekers, A., Van den Berghe, A., Volckaert, G., Ysebaert, M. 1976. *Nature* 260:500–7
33. Fox, T. D. 1981. *Nature* 292:109–10
34. Geller, A. I., Rich, A. 1980. *Nature* 283:41–46

35. Ghosh, A., Rutgers, T., Mang, K.-Q., Kaesberg, P. 1981. *J. Virol.* 39:87–92
36. Gibbs, A. 1980. *Intervirology* 13:42–47
37. Gibbs, A. 1980. *Intervirology* 14:101–8
38. Gould, A. R., Francki, R. I. B., Randles, J. W. 1981. *Virology* 110: 420–26
39. Grivell, L. A., Arnberg, A. C., Boer, P. H., Borst, P., Bos, J. L., van Bruggen, E. F. J., Groot, G. S. P., Hecht, N. B., Hensgens, L. A. M., van Ommen, G.J.B., Tabak, H.F. 1979. In *Extrachromosomal DNA*, ed. D. Cummings, P. Borst, I. David, S. Weissman, C. F. Fox, 15:305–24, NewYork: Academic
40. Guilley, H., Briand, J. P. 1978. *Cell* 15:113–22
41. Haber, S., Ikegami, M., Bajet, N. B., Goodman, R. M. 1981. *Nature* 289: 324–26
42. Habili, N., Francki, R. I. B. 1974. *Virology* 61:443–49
43. Hall, T. C. 1979. *Int. Rev. Cytol.* 60:1–26
44. Higgens, T. J. V., Goodwin, P. B., Whitfield, P. R. 1976. *Virology* 71:486–97
45. Hopfield, J. J. 1974. *Proc. Natl. Acad. Sci. USA* 71:4135–39
46. Huang, A. S., Baltimore, D. 1970. *Nature* 226:325–27
47. Huang, A. S., Baltimore, D. 1977. In *Comprehensive Virology,* ed. H. Fraenkel-Conrat, R. R. Wagner, pp. 73–116, New York: Plenum
48. Hunter, T. R., Hunt, T., Knowland, J., Zimmerman, D. 1976. *Nature* 260: 759–64
49. Jackson, A. O., Gustafson, G. D., McFarland, J. E., Milner, J. J. 1981. *5th Int. Congr. Virol. Strasbourg* (Abstr.)
50. Jacobs, L., Spaan, W. J. M., Horzinek, M. C., van der Zeijst, B. A. M. 1981. *J. Virol.* 39:401–6
51. Jacq, C., Lazowska, J., Slonimski, P. P. 1980. *CR Acad. Sci. Ser. C* 290:89–92
52. Jaspars, E. M. J. 1974. *Adv. Virol. Res.* 19:37–149
52a. Kastelein, R. A., Remaut, E., Fiers, W., van Duin, J. 1982. *Nature* 295: 35–41
53. Keilland-Brandt, M. C., Nilsson-Tillgren, T. 1973. *Mol. Gen. Genet.* 121: 219–28
54. Kelly, W. J., Reanney, D. C. 1978. *1st Int. Symp. Microbial Ecol.,* ed. M. Loutit, J. Miles, pp. 113–20. Berlin: Springer
55. Kitamura, N., Semler, B. L., Rothberg, P. G., Larsen, G. R., Adler, C. J., Dorner, A. J., Emini, E. A., Hanecak, R., Lee, J. J., van der Werf, S., Anderson, C. W., Wimmer, E. 1981. *Nature* 291:547–53
56. Koenig, R. 1971. *J. Gen. Virol.* 10: 111–14
57. Konarska, M., Filipowicz, W., Domdey, H., Gross, H. J. 1981. *Nature* 293:112–16
58. Kornberg, A. 1980. *DNA Replication.* San Francisco: Freeman. 724 pp.
59. Kozak, M. 1978. *Cell* 15:1109–23
60. Lamb, R. A., Choppin, P. W. 1981. *Virology* 112:729–37
61. Lamb, R. A., Choppin, P. W., Chanock, R. M., Lai, C. J. 1980. *Proc. Natl. Acad. Sci. USA* 77:1857–61
62. Lamb, R. A., Lai, C. J. 1980. *Cell* 21:475–85
63. Lamb, R. A., Lai, C. J. 1981. *Virology* 112:746–51
64. Lamb, R. A., Lai, C. J., Choppin, P. W. 1981. *Proc. Natl. Acad. Sci. USA* 78:4170–74
65. Lane, L. C. 1979. In *Nucleic Acids in Plants,* ed. T. C. Hall, J. W. Davies, 2:65–110. Boca Raton: CRC Press
66. Loeb, T., Zinder, N. D. A. 1961. *Proc. Natl. Acad. Sci. USA* 47:282–89
67. Lomniczi, B., Kennedy, I. 1977. *J. Virol.* 24:99–107
68. Luria, S. E., Darnell, J. E. Jr., Baltimore, D., Campbell, A. 1978. *General Virology,* New York: Wiley. 578 pp. 3rd ed.
69. Margulis, L. 1970. *Origin of Eukaryotic Cells.* New Haven, Conn: Yale Univ. Press
70. Matthews, R. E. F. 1978. *4th Int. Congr. Virol. The Hague* (Abstr.)
71. Matthews, R. E. F. 1979. *Intervirology* 12:129–296
72. Mellema, J. R., Benicourt, C., Haenni, A. L., Noort, A., Pleij, C. W. A., Bosch, L. 1979. *Virology* 96:38–46
73. Merten, S., Synenki, R. M., Locker, J., Christianson, T., Rabinowitz, M. 1980. *Proc. Natl. Acad. Sci. USA* 77:1417–21
74. Model, P., Webster, R. E., Zinder, N. D. 1979. *Cell* 18:235–46
75. Morch, M.-D., Benicourt, C. 1980. *J. Virol.* 34:85–94
76. Morris-Krsinich, B. A. M., Hull, R. 1981. *Virology* 114:98–112
77. Nahmias, A. J., Reanney, D. C. 1977. *Ann. Rev. Ecol. Syst.* 8:29–49
78. Orgel, L. E., Crick, F. 1980. *Nature* 284:604–7
79. Pelham, H. R. B. 1979. *Virology* 97: 256–65
80. Perutz, M. F. 1961. *Proteins and Nucleic Acids.* Amsterdam: Elsevier 211 pp.

81. Pollock, T. J., Tessman, I., Tessman, E. S. 1978. *Nature* 274:34–37
82. Quigley, G. J., Wang, A. H. J., Seeman, N. C., Suddath, F. L., Rich, A., Sussman, J. L., Kim, S. H. 1975. *Proc. Natl. Acad. Sci. USA* 72:4866–70
83. Rackwitz, H. R., Rohde, W., Sänger, H. L. 1981. *Nature* 291:297–301
84. Rackwitz, H. R., Rohde, W., Mülhbach, H. P., Sanger, H. L. Submitted for publication
85. Randles, J. W., Davies, C., Hatta, T., Gould, A. R., Francki, R. I. B. 1981. *Virology* 108:111–22
86. Reanney, D. C. 1974. *Int. Rev. Cytol.* 37:21–52
87. Reanney, D. C. 1976. *Bacteriol. Rev.* 40:552–90
88. Reanney, D. C. 1977. *Genetic Interaction and Gene Transfer, Brookhaven Symp. Biol.* 29:248–71
89. Reanney, D. C. 1979. *Nature* 277:598–600
90. Reanney, D. C. 1981. In *The Human Herpesviruses,* ed. A. J. Nahmias, W. R. Dowdle, R. F. Schinazi, pp. 519–36. New York: Elsevier
91. Reddy, D. V. R., Black, L. M. 1974. *Virology* 61:458–73
92. Reddy, D. V. R., Black, L. M. 1977. *Virology* 80:336–46
93. Sharp, P. A. 1981. *Cell* 23:643–46
94. Shaw, J. E., Murialdo, H. 1980. *Nature* 283:30–35
95. Shinnick, T. M., Lerner, R. A., Gregor Sutcliffe, J. 1981. *Nature* 293:543–48
96. Silverstein, S. C., Christmas, J. K., Acs, G. 1976. *Ann. Rev. Biochem.* 45:375–408
97. Smith, K. M. 1972. *A Textbook of Plant Virus Diseases.* London:Longman 3rd ed.
98. Smith, R. A., Parkinson, J. S. 1980. *Proc. Natl. Acad. Sci. USA* 77:5370–74
99. Spiegelman, S. 1970. In *The Neurosciences,* ed. F. O. Schmitt, pp. 927–45. New York: Rockefeller Univ. Press
100. Stern, D. F., Kennedy, S. I. T. 1980. *J. Virol.* 36:440–49
101. Sun, S. M., Slightom, J. L., Hall, T. C. 1981. *Nature* 289:37–41
102. Sussman, J. L., Kim, S.-H. 1976. *Science* 192:853–58
103. Thach, S. S., Thach, R. E. 1973. *J. Mol. Biol.* 81:367–80
104. Tien-Po, Davies, C., Hatta, T., Francki, R. I. B. 1981. *FEBS Lett.* 132:353–56
105. Wimmer, E. 1979. In *The Molecular Biology of Picornaviruses,* ed. Perez-Bercoff, pp. 175–90. New York: Plenum
106. Zaitlin, M. 1979. In *Nucleic Acids in Plants,* ed. T. C. Hall, J. W. Davies, 2:31–64. Boca Raton: CRC Press

Ann. Rev. Microbiol. 1982. 36:75–99

LOW-MOLECULAR-WEIGHT ENZYME INHIBITORS OF MICROBIAL ORIGIN

Hamao Umezawa

Institute of Microbial Chemistry, Shinagawa-ku, Tokyo, Japan

CONTENTS

0066-4227/82/1001-0075$02.00

INTRODUCTION

A main objective of research on antibiotics is to find new active agents. In this research, it is most important to explore new research areas where new microbial products with potential usefulness can be found. By 1965, extensive advancement took place in the chemistry of natural products and made possible the elucidation of the structure of low-molecular-weight compounds in a short period of time. Moreover, the research in enzymology of biological functions and disease processes began to show rapid progress. It seems logical, therefore, that research on antibiotics be extended to enzyme inhibitors that have various pharmacological activities. Thus, the study of low-molecular-weight enzyme inhibitors was initiated by Umezawa (95). Since the discovery of a protease inhibitor was reported in 1969, nearly 50 inhibitors of various enzymes have been found in microbial culture filtrates. Their structures were elucidated, most of them were chemically synthesized, and new types of structures for inhibition were disclosed. These are reviewed by Umezawa (95–97, 99) and Aoyagi (2, 12).

It is often said that the reason microorganisms in nature produce antibiotics is to suppress the growth of the competitors. However, most enzyme inhibitors produced by microorganisms have no significant antimicrobial activity. Moreover, as shown by a study on leupeptin, low-molecular-weight enzyme inhibitors released in culture filtrates are secondary metabolites with no obvious function in the growth of microbial cells. Antibiotics are also secondary metabolites. The genetics of secondary metabolites are being studied in an effort to determine why an almost unlimited number of secondary metabolites of various structures are produced by microorganisms. Antibiotics can be divided into various groups. Each group contains a common characteristic structural part. Antibiotics of the same group are produced by strains belonging to different species, genera, or families, which indicates a wide distribution of a gene set responsible for the biosynthesis of that characteristic structural moiety. Such common molecular structures display no cytotoxicity. It has been proposed (97) that a gene set directing the biosynthesis of the common structural group was generated whose product is modified to the final products. The final products are released extracellularly. In different strains, the product is modified differently, which results in different final products. It is also possible that such a gene whose product is responsible for the modification was transferred to cells of other strains. Plasmids may have been involved in this transfer. It is possible that numerous genes involved in the biosynthesis of secondary metabolites have been generated in the natural environment and many of them transferred into cells of other strains.

Low-molecular-weight inhibitors released from microbial cells are different from macromolecular endogenous inhibitors in that the latter have functional roles in the cells, whereas the former do not.

The success of determining the presence of the target product in culture filtrates is dependent upon the testing method. In general, in the screening for enzyme inhibitors, the product of the enzymatic reaction is quantitatively measured, with quantity being determined in both culture broths and extracts (95). It may be interesting to note that I could find inhibitors but not stimulators of enzymes. The data screening for various enzyme inhibitors suggest that to find a compound that stimulates an enzyme activity, it is necessary to search for an inhibitor of the intrinsic inhibitor of this enzyme.

Enzyme inhibitors, especially protease inhibitors, have been used for identification of enzymes and analysis of the roles of enzymes in biological functions or disease processes. They have also been used as ligands for affinity chromatography. For instance, renin was first purified by affinity chromatography with pepstatin as the functional group (65). Inhibitors of enzymes involved in noradrenaline biosynthesis have a hypotensive effect (95). An inhibitor of mevalonate synthesis shows a therapeutic effect in hypercholesterolemia (94a). Inhibitors of cell surface enzymes enhance immune responses (98). Parallel to progress in the biochemistry of biological functions and disease processes, research on low-molecular-weight enzyme inhibitors produced by microorganisms will be expanded.

This chapter reviews the structures and activities of low-molecular-weight inhibitors produced by various microorganisms.

PROTEASE INHIBITORS

Protease inhibitors in animal and plant cells or in blood plasma are macromolecular peptides and are involved in the control of a variety of cell functions, inflammation, blood coagulation, etc. The inhibitors obtained by extraction from microbial cells are also macromolecular peptides, but many protease inhibitors released extracellularly are low-molecular-weight peptides of unusual structures.

Inhibitors of Serine and Thiol Proteases

Leupeptin, which inhibited plasmin, trypsin, papain, and cathepsin B, was discovered by testing for plasmin-inhibiting activity in streptomyces culture filtrates (6, 10, 46). Two leupeptins were isolated, and their structures were determined to be acetyl or propionyl-L-leucyl-L-leucyl-L-argininal (41). Trypsin hydrolyzes an arginyl or lysyl bond in peptides, and leupeptin contains an argininal group. This aldehyde structure for inhibition of serine • thiol proteases was first found in leupeptin.

R-CO-NH-CH-CO-NH-CH-CO-NH-CH-CHO

Leupeptin R = CH_3-, CH_3CH_2-

HOOC-CH-NH-CO-NH-CH-CO-NH-CH-CO-NH-CH-CHO

Antipain

HOOC-CH-NH-CO-NH-CH-CO-NH-CH-CO-NH-CH-CHO

Chymostatin

HOOC-CH-NH-CO-NH-CH-CO-NH-CH-CO-NH-CH-CHO

Elastatinal

The study of the cell free synthesis of leupeptin (34, 86) has led to the isolation of a multifunctional enzyme that catalyzes the synthesis of leupeptin acid (acetyl-L-leucyl-L-leucyl-L-arginine) in a reaction mixture containing acetyl-L-leucine, leucine, L-arginine, and ATP. The sequence of the synthesis of leupeptin acid by this enzyme is as follows: acetylleucine → acetylleucylleucine → acetylleucylleucylarginine. Leupeptin acid is reduced by a very labile enzyme to leupeptin (34). An enzyme that transfers the acetyl group from acetyl-CoA to leucine is found in leupeptin-producing strains (87).

Leupeptin is produced by strains of soil streptomyces that belong to more than 11 species (95), which indicates that the gene involved in the biosynthesis of leupeptin acid is widely distributed among various species of *Streptomyces.* The ability to produce leupeptin is transferred by conjugation from a leupeptin-producing methionine-requiring mutant to a leupeptin-nonproducing arginine-requiring mutant in high frequency (108), which suggests plasmid involvement in leupeptin production.

Leupeptin-nonproducing mutants grow as well as the leupeptin-producing parent strain, and no difference has been observed between proteolysis in cells of leupeptin-producing mutants and that of nonproducing mutants. Protein autodegradation in cell homogenates of leupeptin-producing strains is inhibited much more strongly by chymostatin (a chymotrypsin inhibitor) or ethylenediaminetetraacetic acid (EDTA) than by leupeptin (88). Studies on the distribution of leupeptin acid and leupeptin in cells and media indicate that leupeptin acid, which lacks antiprotease activity, is produced within the cell; and leupeptin, produced from leupeptin acid, is rapidly released extracellularly (88).

Antipain produced by *Streptomyces michigaensis, Actinomyces violasceus, Streptomyces mauvecolor,* and *Streptomyces yokosukaensis,* chymostatin, produced by *Streptomyces hygroscopicus* and *Streptomyces lavendulae,* and elastatinal, produced by *Streptomyces griseoruber,* were discovered by screening for inhibitors of papain, chymotrypsin, or pancreatic elastase, respectively. Antipain (82. 113) inhibits papain, trypsin, cathepsin B; chymostatin (91, 102) inhibits chymotrypsin; and elastatinal (74, 104) inhibits pancreatic elastase. All these inhibitors have a C-terminal aldehyde structure for inhibition of serine or thiol proteases, as does leupeptin. Kinetic data of the inhibitors are shown in Table 1.

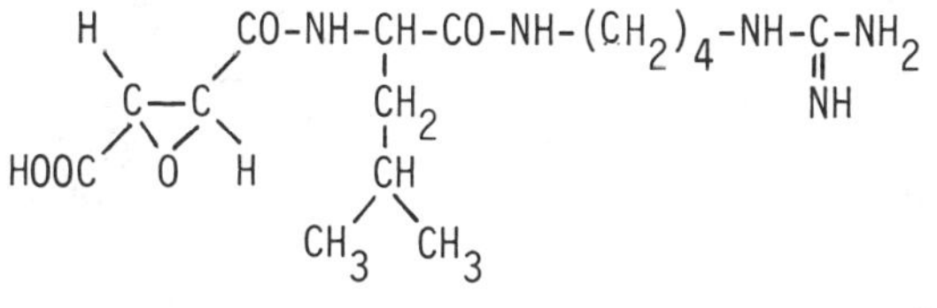

H, HOOC, O, H — C—C — $CO-NH-CH-CO-NH-(CH_2)_4-NH-C(=NH)-NH_2$, CH_2, CH, CH_3 CH_3

E-64

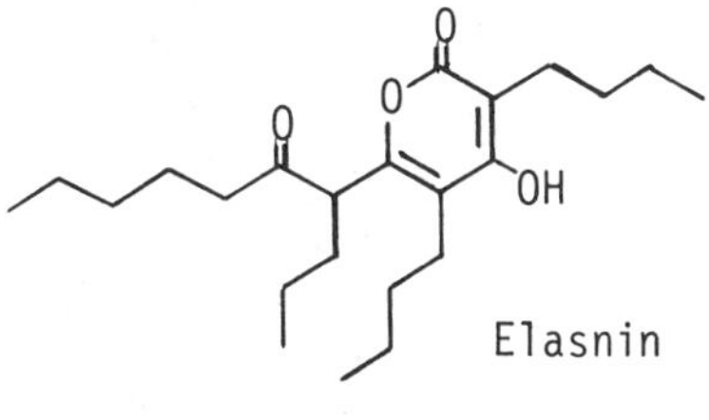

Elasnin

The screening for papain inhibitors in culture filtrates of fungi has resulted in the finding of a type of inhibitor designated as E-64. As the result of the reaction of the epoxide group with the sulfhydryl group, this inhibitor binds covalently to thiol proteases (30).

An inhibitor of human granulocyte elastase has been isolated from *Streptomyces noboritoensis* and named elasnin. It strongly inhibits human granulocyte elastase, but it is only mildly inhibitory for pancreatic elastase (75, 76).

All of these inhibitors have low toxicity. Leupeptin, antipain, and chymostatin inhibit carageenin-induced edema. Leupeptin ointment (1%) applied immediately after a burn suppresses pain and blister formation. Potential usefulness of leupeptin in the treatment of muscle dystrophy has been suggested by its effect on dystrophy mice (81). It has been reported that leupeptin inhibits carcinogenesis by dibenzanthracene and croton oil in rat skin and metastasis in an experimental animal model (35). Leupeptin and other protease inhibitors have also been reported to inhibit the growth of *Plasmodium beghei* (58, 59).

Inhibitors of Carboxyl Proteases

```
                            CH3                                  CH3
                            |                                    |
     CH3        CH3         CH-CH3                               CH-CH3
     |          |           |                                    |
     CH-CH3     CH-CH3      CH2 OH              CH3              CH2 OH
     |          |           |   |               |                |   |
RCO-NH-CH-CO-NH-CH-CO-NH-CH—CH-CH2-CO-NH-CH-CO-NH-CH—CH-CH2-COOH
       (S)        (S)         (S) (S)            (S)             (S) (S)
Pepstatin
                            CH3                                  CH3
                            |                                    |
     CH3        CH3         CH-CH3                               CH-CH3
     |          |           |                                    |
     CH-CH3     CH-CH3      CH2 OH              CH3              CH2 O
     |          |           |   |               |                |   ||
RCO-NH-CH-CO-NH-CH-CO-NH-CH—CH-CH2-CO-NH-CH-CO-NH-CH—C-CH3
Pepstanone
                            CH3                                  CH3
                            |                                    |
     CH3        CH3         CH-CH3                               CH-CH3
     |          |           |                                    |
     CH-CH3     CH-CH3      CH2 OH              CH2OH            CH2 OH
     |          |           |   |               |                |   |
RCO-NH-CH-CO-NH-CH-CO-NH-CH—CH-CH2-CO-NH-CH-CO-NH-CH—CH-CH2-COOH
Hydroxypepstatin

           R = (CH3)2>CH-(CH2)n-, CH3-(CH2)n-     (n=0, 1, 2 .....20)
```

Specific inhibitors of pepsin have long remained unknown, although it was imagined that such inhibitors might be useful in the treatment of gastric ulcer. By testing for antipepsin activity, pepstatin was discovered in culture filtrates of various species of antinomycetes (64, 103). A pepstatin-producing strain classified as *Streptomyces testaceus* produced various pepstatins that differed from one another in the fatty acid moiety (C_2-C_{20}) (13). A pepstatin containing an isovaleryl group has been most widely used for biological and biochemical studies (K_i values are shown in Table 1). Moreover, as minor components, pepstanone (62), containing (S)-3-amino-5-methylhexane-2-one instead of the C-terminal (3S, 4S)-4-amino-3-hydroxy-6-methylheptanoic acid (AHMHA), and hydroxypepstatin (107), containing L-serine instead of L-alanine, were isolated. Pepstatin containing an acetyl group and pepstatins containing propionyl or isobutyryl groups were isolated from *Streptomyces naniwaensis* (67) and from *Streptomyces* No. 2907 (40).

Table 1 K_i values of inhibitors of proteases, alkaline phosphatase, and esterase

Inhibitor	Enzyme	Substrate[a]	K_i ($\times 10^{-7}$ M)	Type of inhibition
Leupeptin	Trypsin	BAEE	1.3	Competitive
Elastatinal	Elastase	$Ac(Ala)_3NA$	2.4	Competitive
Pepstatin	Pepsin	Phe-Gly-His-Phe (NO_2)-Phe-Ala-PheMe	0.001	Competitive
Phosphoramidon	Thermolysin	Z-Gly-$LeuNH_2$	0.28	Competitive
Amastatin	Aminopeptidase A	Glu-NA	1.5	Competitive
Bestatin	Aminopeptidase B	Arg–NA	0.6	Competitive
Forphenicine	Alkaline phosphatase	PNPP	1.64	Uncompetitive
Esterastin	Esterase	PNPA	0.002	Competitive
Ebelactone A	Esterase	PNPA	0.92	Competitive
Ebelactone B	Esterase	PNPA	0.005	Competitive

[a] BAEE, N^{α}-benzoyl-L-arginine methyl ester hydrochloride; $Ac(Ala)_3NA$, acetyl-alanyl-alanyl-alanine *p*-nitroanilide; Z-Gly-$LeuNH_2$, carbobenzoxy-glycyl-leucineamide; Glu-NA, L-glutamic acid β-naphthylamide; Arg-NA, L-arginine β-naphthylamide; PNPP, *p*-nitrophenyl phosphate; PNPA, *p*-nitrophenyl acetate.

Pepstatins, pepstanones, and hydroxypepstatins have almost identical activity against pepsin and cathepsin D. Pepstatin is more effective against renin than are pepstanone or hydroxypepstatin (5, 7). The activity of pepstatin against renin increases with increasing numbers of carbon atoms in the fatty acid moiety (13). As shown by the activity of pepstanone, the terminal carboxyl group of pepstatin is not involved with binding to pepsin. Esters and amides of pepstatin, pepstatinal, and pepstatinol have an activity against pepsin similar to pepstatins. The hydroxyl group in the inner AHMHA moiety is necessary for the activity.

Pepstatin analogs have also been synthesized. AHMHA and its N-acyl derivative have no activity. N-Acetylvalyl-AHMHA is active, and the addition of another valine between the acetyl and valyl groups does not increase the activity. Addition of L-alanine to the C-terminal group increases activity about 100 times. This suggests that the acyl-valyl-AHMHA-L-alanine is the smallest molecular structure that can exhibit inhibition against pepsin and cathepsin D to the same degree as does pepstatin (7). Acetyl-L-valyl-L-valyl-[(3S, 4R)-4-amino-3-hydroxy-6-methyl] heptanoic acid prepared by chemical synthesis has no activity (42). This suggests that the 4S-configuration of AHMHA is required for activity.

Pepstatin may be biosynthesized in a multifunctional enzyme system as leupeptin. ^{14}C-labeled AHMHA added to the fermentation medium is not incorporated into pepstatin, but [^{14}C]leucine and [^{14}C]acetate are incorporated into this amino acid moiety of pepstatin (63). The ability of the *Streptomyces* strain to produce pepstatin is eliminated by treatment with acriflavine and plasmid involvement in the production of pepstatin has been suggested.

A bacterial enzyme that hydrolyzes the isovaleryl bond in pepstatin containing the isovaleryl group has been found, and from the residual peptide, benzoyl-L-valyl-AHMHA-L-alanyl-AHMHA and L-lactyl-L-valyl-AHMHA-L-alanyl-AHMHA have been synthesized. These analogs are more water soluble than is pepstatin and have almost identical activity against pepsin and cathepsin D as does pepstatin. However, these water-soluble analogs have much weaker activity against renin than does pepstatin (60, 94). The addition of aspartic acid or arginine to the C-terminus of pepstatin increases its water solubility. Such water-soluble analogs have almost the same activity against renin as does pepstatin and also have a hypotensive action (P. Corvol, personal communication). Renin was first purified by affinity chromatography with a pepstatin column (23, 65).

Pepstatin inhibits carageenin-induced edema. It also suppresses the generation of Shay rat ulcer. Therapeutic effects on stomach ulcers in man have also been observed. Pepstatin has been reported to be effective against experimental muscle dystrophy and enhances the effect of leupeptin (18, 61). Pepstatin also inhibits leukokinin formation and ascites accumulation in ascites carcinoma of mice (29). Pepstatin inhibits the growth of *Plasmodium beghei* (58, 59). It has also been reported to inhibit focus formation by murine sarcoma virus (118).

Inhibitors of Metallo-Proteases

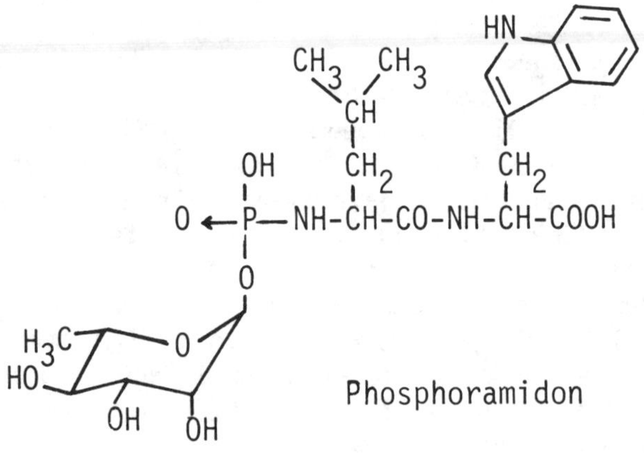

Phosphoramidon

Phosphoramidon produced by *Streptomyces tanashiensis* inhibits metallo-proteases (83, 114). Its L-rhamnose moiety is not involved in this inhibitory action. *N*-phosphate of L-leucyl-L-tryptophan shows a stronger action than phosphoramidon (45).

Phosphoramidon and *N*-phosphate of L-leucyl-L-tryptophan inhibit thermolysin, metalloendopeptidases of *Bacillus subtilis* and *Streptomyces griseus*. Kinetic data are shown in Table 1. Affinity chromatography with phosphoramidon is useful in purification of these enzymes (44). Phosphoramidon inhibits a metalloprotease that has been called an elastase of *Pseudomonas aeruginosa* (62a).

INHIBITORS OF HYDROLYTIC ENZYMES LOCATED ON CELL SURFACES

In 1972, it was found that the administration of a very small dose of diketocoriolin B, an antitumor antibiotic, increased the number of mouse spleen cells producing antibody to sheep red blood cells (38). Additionally, diketocoriolin B inhibited Na^+-K^+-ATPase, a membrane-bound enzyme (55). This suggested that the binding of diketocoriolin B to ATPase in membranes promoted the blastogenesis of lymphocytes producing antibodies to sheep red blood cells. Later diketocoriolin B was confirmed to act directly on B lymphocytes (39).

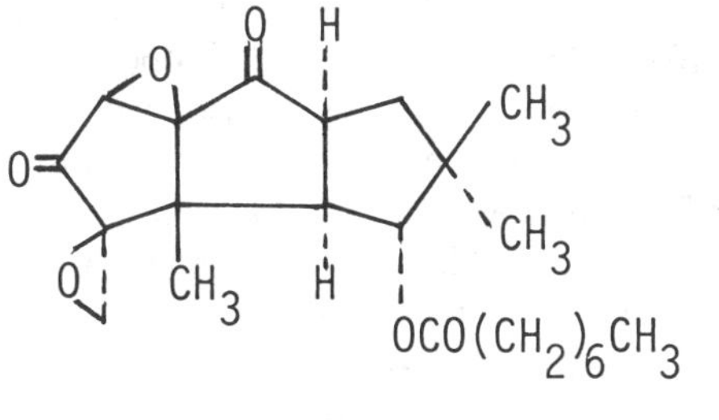

Diketocoriolin B

In view of these findings, I assumed the screening of compounds that bind to cell surface membranes would result in the discovery of immunomodifiers; a screening for inhibitors of cell surface enzymes was initiated.

Inhibitors of Aminopeptidases, Alkaline Phosphatase, and Esterase

All aminopeptidases were found not only within cells but also on surfaces (8, 9). These enzymes are not released extracellularly. Alkaline phosphatase and esterase were also found on cell surfaces. The screening of streptomyces culture filtrates for inhibitors of these enzymes led to the discovery of bestatin, produced by *Streptomyces olivoreticuli* (85, 98, 105), amastatin, produced by *Streptomyces* sp. ME98-M3 (11, 93, 98), forphenicine, produced by *Actinomyces fulvoviridis* var. *acarbodicus* (15, 98, 116), esterastin,

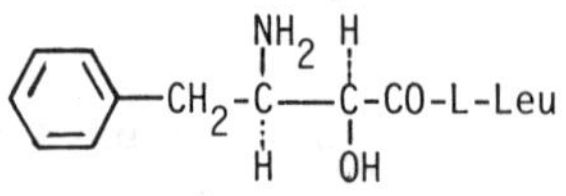

Bestatin

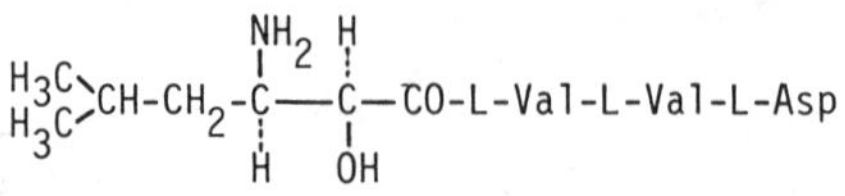

Amastatin

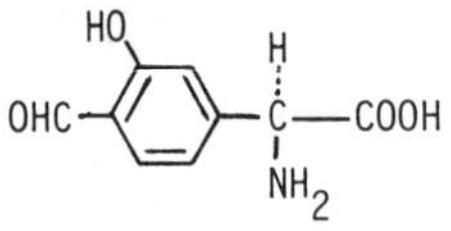

Forphenicine

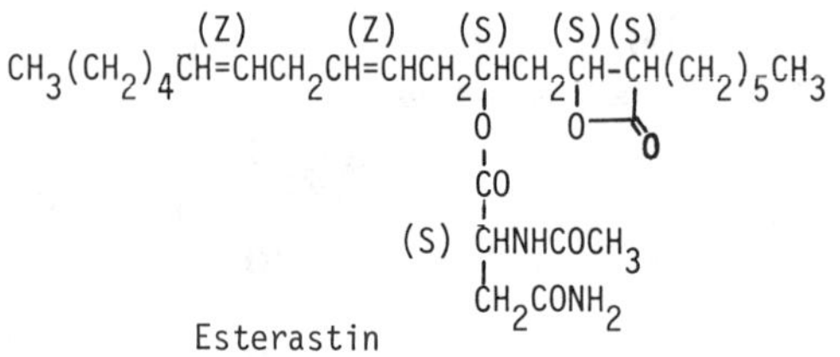

Esterastin

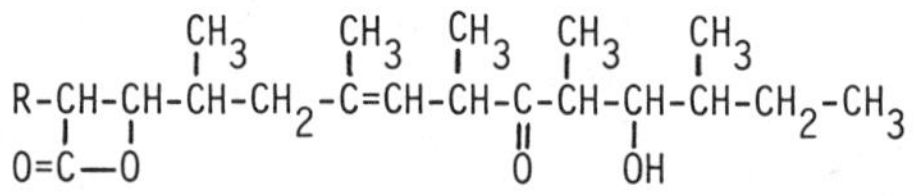

Ebelactone

Ebelactone A: R=CH_3
Ebelactone B: R=CH_3CH_2

produced by *Streptomyces lavendulae* (47, 98, 100), and ebelactone, produced by *Streptomyces aburaviensis* (98, 106). The K_i values of these inhibitors are shown in Table 1. All of these inhibitors except for esterastin enhanced immune responses in mice. Bestatin inhibited aminopeptidase B and leucine aminopeptidase. At low doses (1–100 μg/mouse) it enhanced delayed-type hypersensitivity (DTH) to sheep red blood cells, and at higher doses (1 mg/mouse) it increased the number of antibody-forming cells in spleen. Amastatin, an inhibitor of aminopeptidase A, increased the number of antibody-forming cells. Forphenicine, which inhibited chicken intestine alkaline phosphatase, enhanced DTH (1–100 μg/mouse) and increased the number of antibody-forming cells (10-1000 μg/mouse). Ebelactone enhanced DTH. In contrast, esterastine suppressed DTH and reduced the number of antibody-forming cells. The K_i value of esterastin was about 10^{-10} M, whereas those of the others were about 10^{-6}-10^{-8} M (98).

Isomers with the same configuration (S) as bestatin at the carbon atom adjacent to the carbonyl group of the 3-amino-2-hydroxy-4-phenylbutanoyl moiety showed similar activity as bestatin in inhibiting aminopeptidase B and enhancing DTH, but isomers with the R configuration at this carbon atom had much weaker activities than did bestatin. The type of inhibition of aminopeptidases by bestatin was competitive with respect to the substrates (106a).

Forphenicine inhibits chicken intestine alkaline phosphatase, but its action against other alkaline phosphatases is very weak. The type of forphenicine inhibition of chicken intestine alkaline phosphatase is very interesting;

it shows uncompetitive inhibition with the substrate. Its derivative, forphenicinol, in which the aldehyde group of forphenicine is reduced to alcohol, was prepared. It does not inhibit alkaline phosphatase, but it does bind to animal cells including lymphocytes. Forphenicinol, when given at 0.1–100 μg/mouse by intraperitoneal injection or 0.1–1000 μg/mouse by oral administration, enhanced DTH (106a).

In oral administration, both bestatin and forphenicinol enhanced DTH and showed antitumor effects against mice tumors sensitive to immune-enhancing agents. Bestatin has been studied clincially in detail. Bestatin, at doses of 30 or 60 mg daily, increases T-cell percentage and natural killer cell activity, which were reduced in cancer patients, and its possible usefulness in cancer treatment has been suggested (98).

GLYCOSIDASE INHIBITORS

Inhibitors of Sialidase

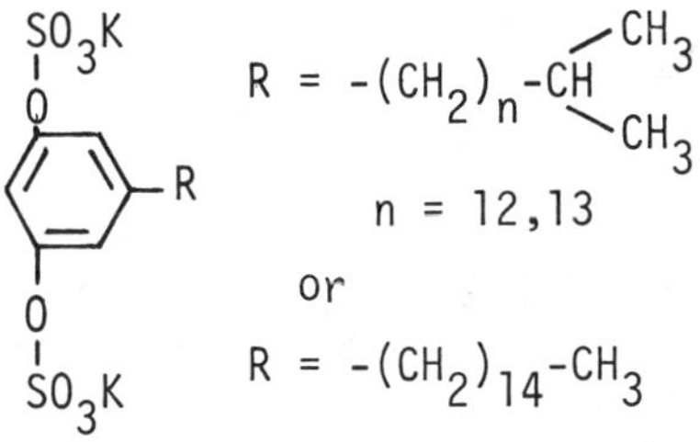

$R = -(CH_2)_n-CH(CH_3)_2$

$n = 12,13$

or

$R = -(CH_2)_{14}-CH_3$

Panosialin

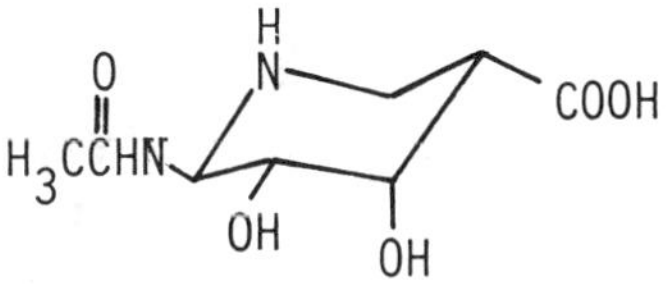

Siastatin B

Sialidase is an exo-type α-ketosidase (α-glycosidase). It liberates *N*-acetylneuraminic acid, *N*,O-diacetylneuraminic acid, *N*-glycolylneuraminic acid, etc, from glycoconjugates. In screening for an inhibitor of this enzyme prepared from influenza virus, panosialin (14) was found in cultured broths of *Streptomyces pseudoverticillus* var. *panosialinus.* It was obtained as a mixture of 5-alkylbenzene-1,3-disulfates. The three major components were 5-isopentadecylbenzene-1,3-disulfate, 5-*n*-pentadecylbenzene-1,3-disulfate, and 5-isohexadecylbenzene-1,3-disulfate (54). Panosialin is an anionic surface active agent. Panosialin exhibits a strong inactivating action on influenza viruses. It destroys the virion structure (71).

Siastatin, produced by *Streptomyces verticillus* var. *quintum,* was found by screening for an inhibitor of the sialidases from *Clostridium perfringens* (101). Siastatin A was more potent against sialidases from *C. perfringens* and chorioallantoic membrane than against sialidases from other sources including streptomyces. Siastatin B had a broader spectrum than did siastatin A. Neither siastatin A nor B inhibited sialidases prepared from two kinds of viruses and *Vibrio cholerae.* K_i values of siastatin B against sialidase of *C. perfringens* and streptomyces were 1.7×10^{-5} M and 4.3 X

10^{-5} M, respectively. The action of siastatin B against both enzymes is competitive with respect to sialolactose.

Inhibitors of β-Galactosidase

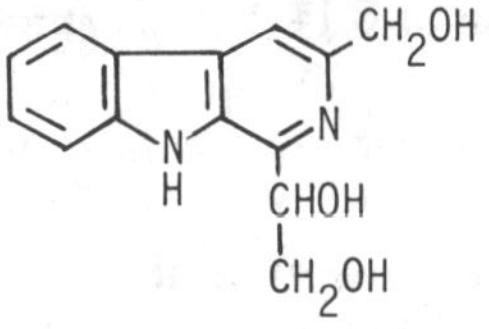

Pyridindolol

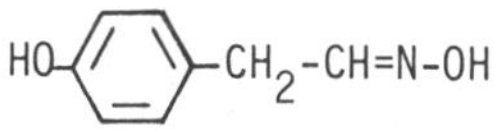

p-Hydroxyphenylacetaldoxome; HPAAO

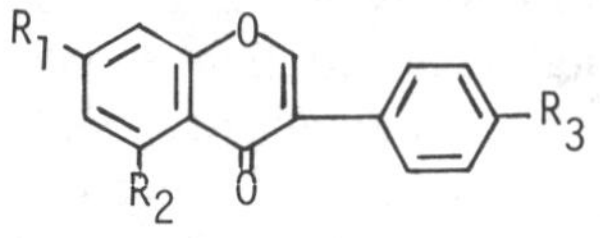

	R_1	R_2	R_3
II:	O-rhm	H	O-rhm
III-1:	OH	OH	OH
III-2:	O-rhm	H	OH
III-3:	O-rhm	OH	O-rhm
IV-1:	OH	H	OH
IV-2:	O-rhm	OH	OH

rhm = rhamnosyl

Isoflavonoid

In screening for an inhibitor of β-galactosidase prepared from bovine liver, pyridindolol was found in culture filtrates of *Streptomyces alboverticillus* MD401–C5 (4, 53). The activity of pyridindolol against β-galactosidase in acidic and neutral conditions indicates it is a specific inhibitor of β-galactosidase under acidic conditions. The study of pyridindolol and its analogs suggests that both hydroxy groups in the side chain and the β-carboline skeleton are essential for the activity in inhibiting β-galactosidase under acidic conditions. Pyridindolol inhibits β-galactosidase in a noncompetitive manner, with the K_i value being 2 X 10^{-6} M. Pyridindolol also inhibits β -glucosidase, but not sialidase, α-D-mannosidase, or *N*-acetyl-β-D-glucosaminidase (52). Pyridindolol inhibits the activation of peripheral lymphocytes by phytohemagglutinin.

Six isoflavone compounds active under both acidic and neutral conditions (3) were also found by screening for inhibitors of β-galactosidase. They are produced by *Streptomyces xanthophaeus.* Inhibitors II, III-2, III-3, and IV-2 inhibit β-galactosidase and β-fucosidase prepared from various bovine and porcine organs and rat kidney. Inhibitors II and III-3 inhibit β-glucuronidase prepared from bovine liver. The types of inhibition are competitive. K_i value for III-2 is 7.1 X 10^{-6} M.

p-Hydroxyphenylacetaldoxime (HPAAO; *anti* and *syn* forms) were also found by screening for inhibitors of β-galactosidase (32). It was produced by *Streptomyces nigellus* MD824–CG2. These inhibitors and their analogs have been chemically synthesized. HPAAO inhibits bovine liver β-galactosidase, β-glucosidase, and β-fucosidase, all of which have pH optima between 6.0 and 8.0. The *syn* form of HPAAO is more active than the *anti* form against bovine liver neutral β-galactosidase. The oxime moiety of HPAAO is essential for inhibition of β-galactosidase. The activities of aromatic and aliphatic oxime derivatives have also been studied (31).

Inhibitors of Amylases

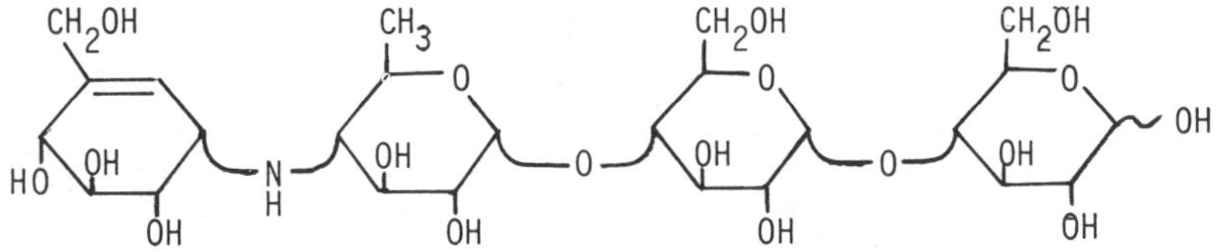

BAYg5421

BAYg5421 (78) is an inhibitor of α-amylase and is produced by *Actinoplanes.* BAYg5421 is very effective in inhibiting starch digestion in vivo. Clinical studies have shown its possible usefulness in preventing or treating obesity. Other α-amylase inhibitors have also been found in *Streptomyces flavochromogenes* (43) and *Streptomyces diastaticus* subsp. *amylostaticus* No. 2476 (66).

INHIBITORS OF ENZYMES INVOLVED IN BIOSYNTHESIS AND METABOLISM OF NORADRENALINE

Inhibitors of Tyrosine Hydroxylase, Dopa Decarboxylase, Dopamine β-Hydroxylase, Catechol O-Methyltransferase, Monoamine Oxydase

In testing the effect of culture filtrates of mushrooms, fungi, streptomyces, and bacteria, inhibitors of tyrosine hydroxylase, dopa decarboxylase, and dopamine β-hydroxylase have been found. A tyrosine hydroxylase inhibitor found in a cultured mushroom (*Oudemansiella radicata*) was named oudenone (110). It has low toxicity and has a hypotensive effect against hypertention of spontaneously hypertensive rats. Oudenone and its analogs have been chemically synthesized (72).

Five isoflavones that inhibit dopa decarboxylase have been found in culture filtrates of aspergillus and streptomyces (112). They are psitectorigenin, genistein, a new isoflavone (New Ifv I, 3', 4', 5, 7-tetrahydroxy-8-methoxyisoflavone), orobol, and 8-hydroxygenistein. This was the first time these isoflavones had been isolated from microorganisms. Orobol was also isolated from culture filtrates of streptomyces and mushrooms. Two other new isoflavones (New Ifv II, 3', 5, 7-trihydroxy-4', 6-dimethoxyisoflavone,

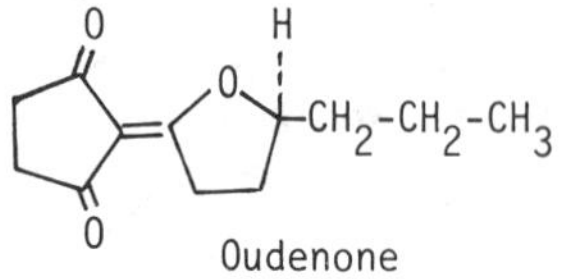

Oudenone

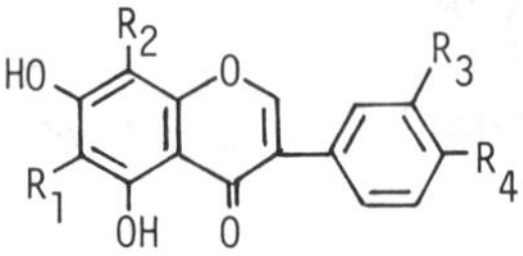

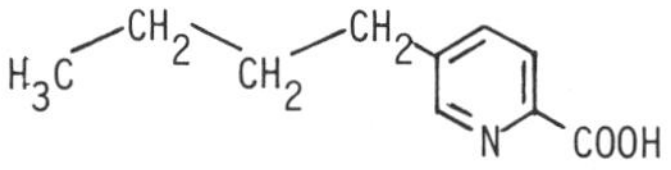

Fusaric acid

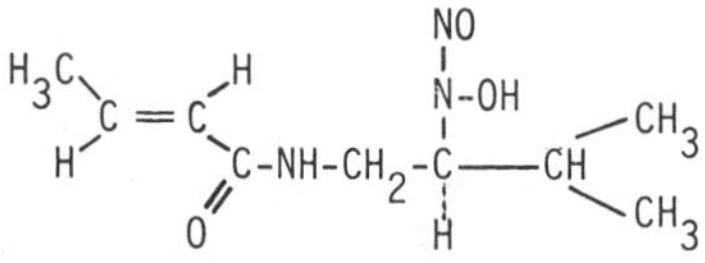

Dopastin

psi-Tectorigenin:	$R_1, R_3 = H$, $R_2 = OCH_3$, $R_4 = OH$
Genistein:	$R_1, R_2, R_3 = H$, $R_4 = OH$
New Ifv I:	$R_1 = H$, $R_2 = OCH_3$, $R_3, R_4 = OH$
Orobol:	$R_1, R_2 = H$, $R_3, R_4 = OH$
8-Hydroxygenistein:	$R_1, R_3 = H$, $R_2, R_4 = OH$
New Ifv II:	$R_1, R_4 = OCH_3$, $R_2 = H$, $R_3 = OH$
New Ifv III:	$R_1 = H$, $R_2, R_4 = OCH_3$, $R_3 = OH$

and New Ifv III, 3', 5, 7-trihydroxy-4', 8-dimethoxyisoflavone), which were found in streptomyces by screening for inhibitors of catechol O-methyltransferase, also inhibited dopa decarboxylase (20). The action of these isoflavones is not limited to dopa decarboxylase. They also inhibit histidine decarboxylase. These isoflavones inhibit neither tyrosine hydroxylase nor dopamine β-hydroxylase. Among these isoflavones, hypotensive action was tested for orobol, New Ifv II, and New Ifv III. All had a hypotensive effect in spontaneously hypertensive rats; the degree of activity was similar to α-methyldopa.

An active agent (33, 84) that inhibited dopamine β-hydroxylase, found in culture filtrates of a fungus, was identified as fusaric acid (also called fusarinic acid, 5-butylpicolinic acid), which had been found by Yabuta et al in 1934 (115a) as a plant toxin produced by *Fusarium* sp. Dopastin (36, 37), which inhibits dopamine β-hydroxylase, was found in a bacterium classified as a *Pseudomonas.*

Among 5-alkyl analogs of fusaric acid, there was a parallel relationship between hypotensive effect and dopamine β-hydroxylase-inhibiting activity (84). Both effects were dependent on the number of carbon atoms in the 5-alkyl group, and butyl and pentylpicolinic acid were most potent both in dopamine β-hydroxylase-inhibiting action and in hypotensive action.

Fusaric acid and dopastin both had a marked hypotensive effect in spontaneously hypertensive rats. The calcium salt of fusaric acid has been studied clinically and has been confirmed to be a useful hypotensive agent. Administered fusaric acid is metabolized and the amount excreted in urine is very slight (95). Therefore, fusaric acid analogs more resistant to metabolism in vivo have been synthesized (111). Fusaric acid has been suggested to be useful in the treatment of alcoholism.

Feeding of 1.0% sodium chloride instead of water to SH-rats 18–23 weeks old causes cerebral bleeding in 80% (202/252) by 35 days (68). Daily oral administration of fusaric acid (25–50 mg/kg/day), pyratrione (an oudenone analog, 12.5–50 mg/kg/day), or α-methyldopa (12.5–50 mg/kg/day) completely prevented this bleeding.

Isoflavones, which inhibit dopa decarboxylase, as already described, also inhibit catechol O-methyltransferase and histidine decarboxylase. However, another new isoflavone that shows a specific inhibition against catechol O-methyltransferase has been found in *Streptomyces* (20). This compound, 3', 8-dihydroxy-4', 6, 7-trimethoxyisoflavone, shows the strongest inhibition against catechol O-methyltransferase (IC_{50} 5.8 × 10^{-7} M). This specific inhibitor of catechol O-methyltransferase has no hypotensive effect.

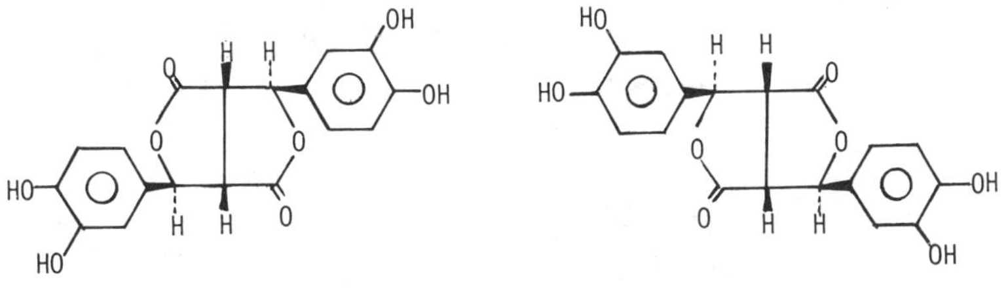

(+)-Dehydrodicaffeic Acid Dilactone (-)-Dehydrodicaffeic Acid Dilactone

Dehydrodicaffeic acid dilactone was also found by screening for inhibitors of catechol O-methyltransferase. It is produced by a cultured mushroom classified as *Inonotus* sp. Both (+)- and (–)-dehydrodicaffeic acid dilactones were produced and they had the same activity (49). The structures were elucidated both chemically and by X-ray crystal analysis of (+)- and (–)-tetra-O-methyldehydrodicaffeic acid dilactones (69). They are biosynthesized from two molecules of caffeic acid. The enzyme involved in this biosynthesis has been purified (51). (+)- and (–)- Dehydrodicaffeic acid dilactones inhibit also dopa decarboxylase, dopamine β-hydroxylase, and cAMP phosphodiesterase and show hypotensive effect against hypertension of spontaneously hypertensive rats (49, 50). Their derivatives have been synthesized (50).

Besides isoflavone compounds and dehydrodicaffeic acid dilactone, naphthoquinone compounds have also been found by screening for catechol

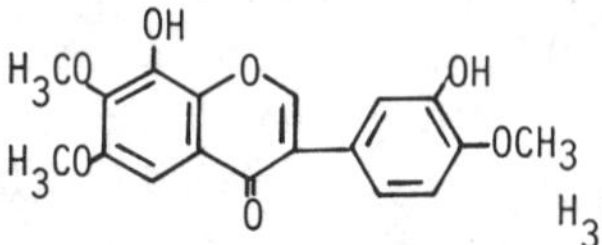

3',8-Dehydroxy-4',6,7-trimethoxy-isoflavone

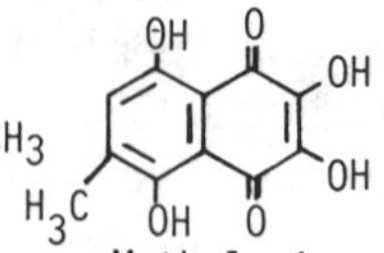

Methylspinazarin

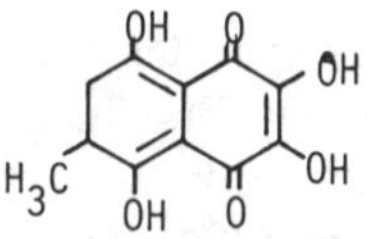

6,7-Dihydromethyl-spinazarin

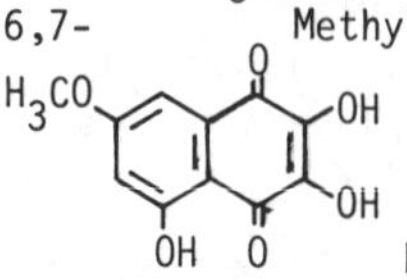

7-O-Methylspinochrome B

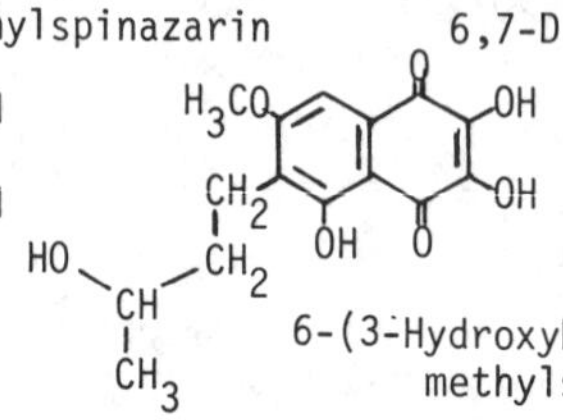

6-(3-Hydroxybutyl)-7-O-methylspinochrome B

O-methyltransferase inhibitors. Methylspinazarin and 6, 7-dihydromethylspinazarin were isolated from cultured broths of actinomycetes (21) and 7-O-methylspinochrome B and 6-(3-hydroxybutyl)-7-O-methylspinochrome B (22) were isolated from fungi. These compounds also showed a significant inhibition against tyrosine hydroxylase and dopamine β-hydroxylase and exhibited a hypotensive effect.

Inhibitors of monoamine oxidases found by the screening were identified to be phenethylamine, pimprinine, *trans*-cinnamic acid amide, and harman (90). The former two were found in streptomyces culture filtrates, and the latter two were found in a cultured mushroom classified as *Coriolus* sp.

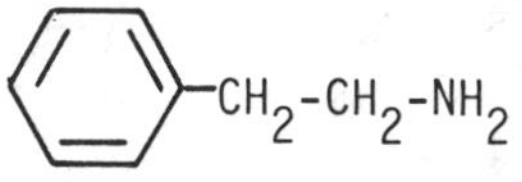

Phenethylamine

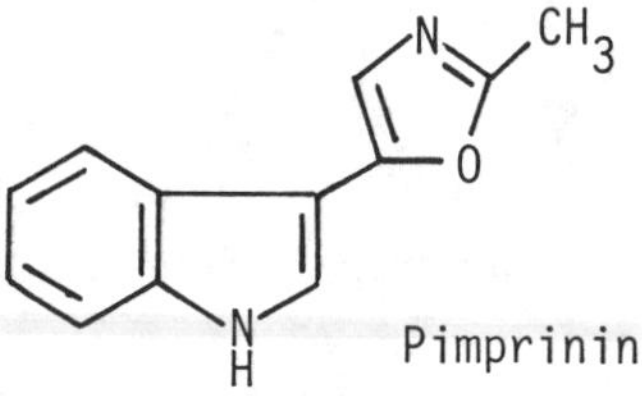

Pimprinine

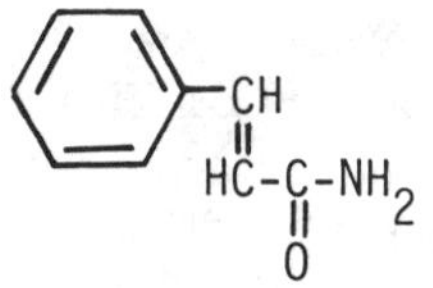

trans-Cinnamic acid amide

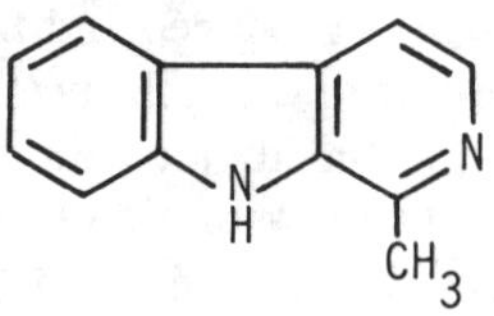

Harman

In the screening of tyrosine hydroxylase inhibitors, two pigment antibiotics, aquayamycin (79) and chrothiomycin (16), were discovered. Another antibiotic, frenolicin, was also found to show a similar effect (92). These antibiotics inhibit both tyrosine hydroxylase and dopamine β-hydroxylase.

INHIBITORS OF HISTIDINE DECARBOXYLASE

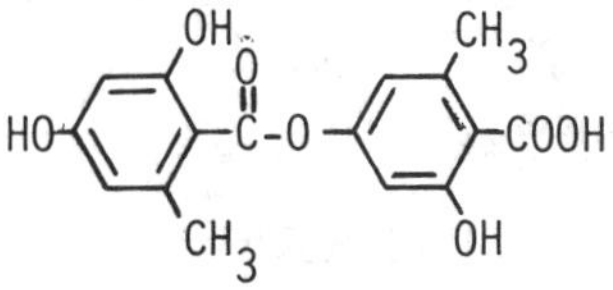

Lecanoric acid

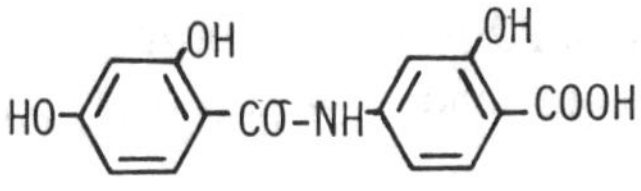

N-(2,4-Dihydroxybenzoyl)-4-aminosalicylic acid

Isoflavone compounds described as inhibitors of dopa decarboxylase also inhibit histidine decarboxylase. A specific inhibitor of histidine decarboxylase found in culture filtrates of *Pyricularia* Fl-*178* (109) was identified as lecanoric acid, which had been known to be a metabolite of lichen. Although no structural relationship exists with histidine, the type of inhibition by lecanoric acid is competitive with histidine ($K_i = 6.9 \times 10^{-7}$ M) and noncompetitive with pyridoxal phosphate ($K_i = 1.1 \times 10^{-5}$ M). The ester bond in lecanoric acid is easily hydrolyzed and is not effective in vivo. But an analog with a peptide bond rather than ester bond is more resistant to metabolism and shows greater activity. For example, *N*-(2, 4-dihydroxybenzoyl)-4-aminosalicylic acid shows competitive inhibition with histidine ($K_i = 3.3 \times 10^{-8}$ M).

INHIBITOR OF NONSPECIFIC N-METHYLTRANSFERASE

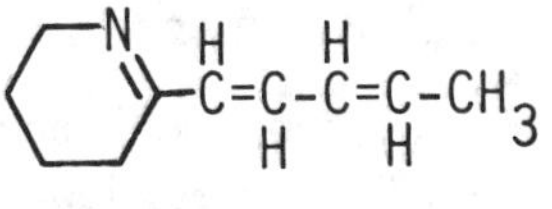

An inhibitor of non-specific N-methyltransferase

Although I endeavored to find an inhibitor of *N*-methyltransferase for synthesis of adrenaline from noradrenaline, this study was not successful. However, in testing the inhibition of *N*-methylation of tryptamine by *N*-methyltransferase prepared from rabbit lung, an inhibitor of nonspecific

N-methyltransferase was found in culture filtrates of *Actinomyces* MD736–C5 (48). Its structure was elucidated to be 1-[2-(3,4,5,6- tetrahydropyridyl)]-1, 3-pentadiene. All inhibitors of enzymes involved in animal physiology described above have low toxicity, but this inhibitor displays a strong toxicity, with an LD_{50} of 6 mg/kg. The inhibition was competitive both with tryptamine (K_i = 2.4 X 10^{-5} M) and *S*-adenosylmethionine (K_i =3.0 X 10^{-5} M). Although it competes with *S*-adenosylmethionine, it does not inhibit catechol-O-methyltransferease.

INHIBITORS OF cAMP PHOSPHODIESTERASE

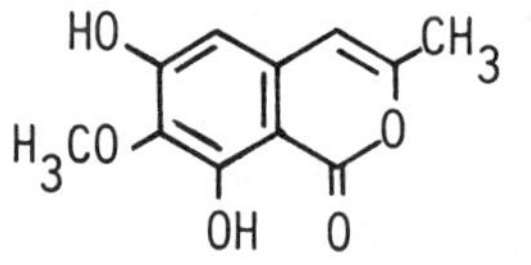

Reticulol

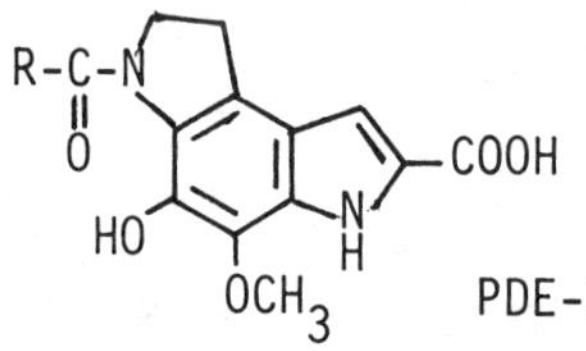

PDE-I : R = NH_2
PDE-II: R = CH_3

Inhibitors of cAMP phosphodiesterase prepared from rabbit brain have been found in microbial culture filtrates. One of them found in culture filtrate of *Streptomyces* MD611–C6 was identified to be reticulol (6, 8-dihydroxy-7-methoxy-3-methylisocumarin) (27), which had been isolated from actinomycetes as a metabolite. Two others were identified as known isoflavones (orobol and genistein). And two inhibitors designated as PDE-I and PDE-II were found in culture filtrates of *Streptomyces* MD611-C6 (26) and had interesting new structures. PDE-I and PDE-II had stronger activity than reticulol.

INHIBITORS OF 3-HYDROXY-3-METHYLGLUTARYL COENZYME A REDUCTASE

The screening for inhibitors of mevalonic acid synthesis in a rat liver enzyme system has led to the findings of inhibitors with potential usefulness. An inhibitor found in culture filtrates of *Pythiums ultimum* was identified as a known fungus antibiotic, citrinin. The other inhibitors (ML-236A, ML-236B, and ML-236C) were found in culture filtrates of *Penicillium citrinum* (25). ML-236B was also found by screening for antifungal metabolites and was named compactin (17). ML-236A, ML-236B, and ML-236C inhibit 3-hydroxy-3-methylglutaryl CoA reductase. ML-236B is the most inhibitory, with ML-236A being least effective (24). The acid forms of ML-236A and ML-236B prepared by hydrolysis show a stronger activity than the respective lactone forms. Two forms of ML-236B are more potent than those of A. K_i values were 0.22 μM for ML-236A sodium salt and 0.01 μM for ML-236B sodium salt. K_m values for two substrates were D,L-3-hydroxy-3-methylglutarate, 33 μM; NaNADH, 40 μM. A clinical study of ML-236B (compactin) has shown its usefulness in the treatment of primary hypercholesterolemia (117).

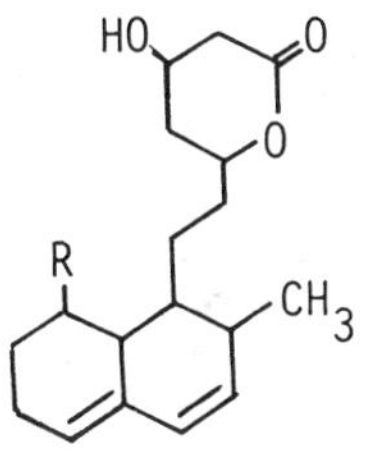

ML-235A (R = OH),
ML-236B (R = $OCOCH(CH_3)CH_2CH_3$)
ML-236C (R = H)

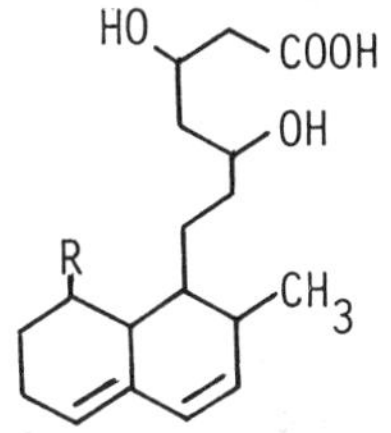

Acid forms of
ML-236A,
ML-236B,
ML-236C

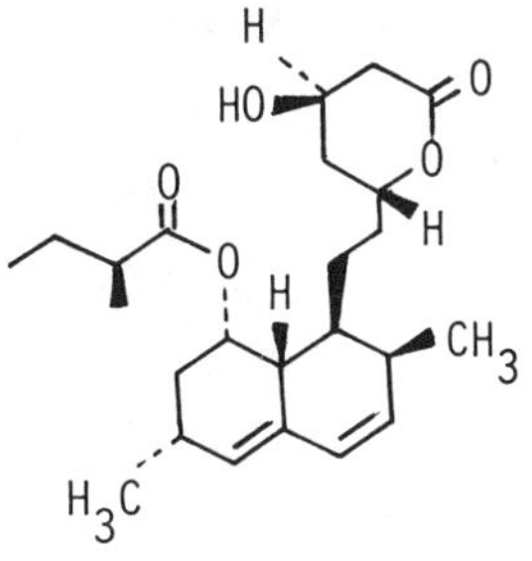

Mevinolin

6-Methyl-ML-236B (6-methylcompactin) was found in *Aspergillus terreus* and was named mevinolin. Its acid form is called mevinolinic acid and it shows the strongest inhibition against 3-hydroxy-3-methylglutaryl CoA reductase (1).

INHIBITORS OF GLYOXALASE

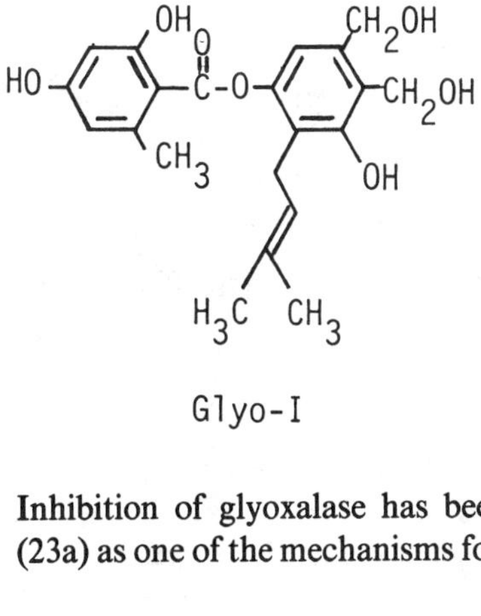

Glyo-I

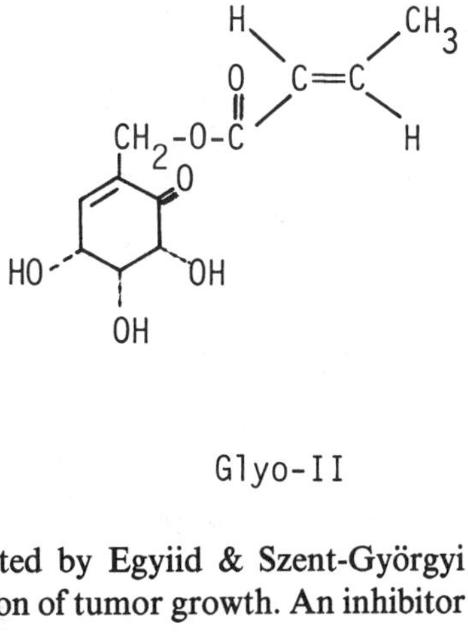

Glyo-II

Inhibition of glyoxalase has been suggested by Egyiid & Szent-Györgyi (23a) as one of the mechanisms for inhibition of tumor growth. An inhibitor

(Glyo-I) was found in a cultured mushroom (56). The type of inhibition was competitive with the hemimercaptal adduct ($K_i = 4.6 \times 10^{-6}$ M) (57). It inhibited the growth of Yoshida rat sarcoma cells, with 50% inhibition at 50–100 μg/ml. The ester bond in this compound is easily hydrolyzed, and this compound does not show any in vivo effect.

Another inhibitor (Glyo-II) was found in actinomycetes, and its structure was shown as (4R,5R,6R)-2-crotonyloxymethyl-4,5,6- trihydroxycyclohex-2-enone. The inhibition of glyoxalase by this compound occurs as follows (19, 89):

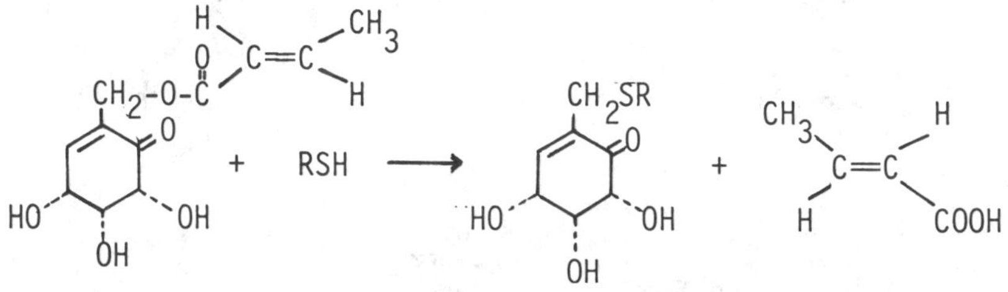

H, CH3, O, C=C, CH2-O-C, H, O, HO, OH, OH + RSH ⟶ CH2SR, O, HO, OH, OH + CH3, H, C=C, H, COOH

Glyo II showed no antibacterial activity at 100 μg/ml, but it did inhibit Yoshida rat sarcoma cells in tissue culture (IC_{50} at 18 μg/ml). Growth of both ascites and solid forms of Ehrlich carcinoma in mice was suppressed by daily intraperitoneal injection of 10 mg/kg for 10 days. It also prolonged the survival period in mouse leukemia 1210.

INHIBITORS OF ADENOSINE DEAMINASE

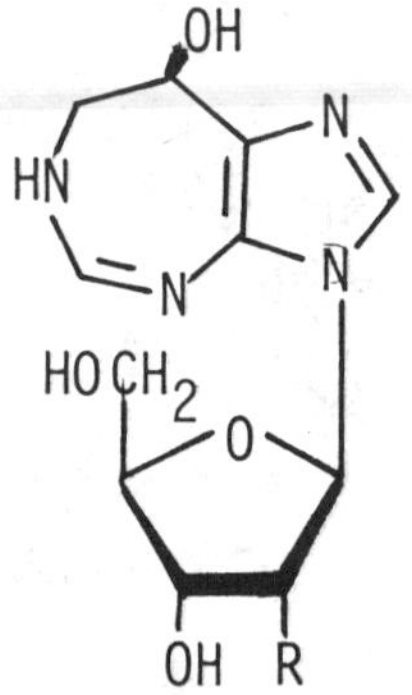

OH, HN, N, N, N, HOCH2, O, OH R

Coformycin (R = OH)
Deoxycoformycin (R = H)

During a study of formycin, it was observed that antibacterial and antitumor activities of crude formycin were much stronger than the purified formycin. This was due to an impurity that inhibited the inactivation of formycin by adenosine deaminase (77). This impurity was purified and named coformycin. It is an unusual nucleoside, containing a seven-membered ring (70). Coformycin exhibits strong inhibition of adenosine deaminase, which can also deaminate formycin ($K_i = 6.5 \times 10^{-8}$ M). Later, deoxycoformycin was found in *Streptomyces antibioticus* (115) by screening for adenosine deaminase inhibitors.

CONCLUDING REMARKS

Low-molecular-weight compounds found by screening for enzyme inhibitors have been described. The following compounds have also been found by the screening of microbial culture filtrates: A tryptophan hydroxylase inhibitor produced by actinomycetes was identified as 2,5- dihydro-L-phyenylalanine ($K_i = 4.4 \times 10^{-5}$ M) (73); 5-formyluracil, which inhibits xanthine oxidase was produced by actinomycetes (95); and poly-L-malic acid (molecular weight, 5000), an inhibitor of acid proteases, was found in *Penicillium cyclopium* (80).

From the viewpoint of the mechanism of action, antibiotics are inhibitors of those enzymes involved in the growth process. It may be helpful to list the books where mechanisms of antibiotic action are reviewed. The ones by Gottlieb & Shaw (28) and Corcoran & Hahn (22a) can be recommended.

All enzyme inhibitors described above, with the exception of inhibitors of nonspecific *N*-methyltransferase, coformycin, and deoxy-coformycin, have low toxicity, and most of them have no significant antimicrobial activity.

In most cases, the screening for enzyme inhibitors in microbial culture filtrates has resulted in finding the sought-after compound. The biosynthesis of this wide variety of compounds by microorganisms is actively being studied from a genetic viewpoint. It can be said that in the next 10 years, progress in the biochemistry of physiological and disease processes will be greatly enhanced by expansion of research into low-molecular-weight inhibitors produced by microorganisms.

Literature Cited

1. Alberts, A. W., Chen, J., Kuron, G., Hunt, V., Huff, J., Hoffman, C., Rothrock, J., Lopez, M., Joshua, H., Harris, E., Patchett, A., Monaghan, R., Currie, S., Stapley, E., Alberts-Schonberg, G., Hensens, O., Hirshfield, J., Hoogsteen, K., Liesch, J., Springer, J. 1980. *Proc. Natl. Acad. Sci. USA* 77:3957–61
2. Aoyagi, T. 1978. In *Bioactive Peptides Produced by Microorganisms,* ed. H. Umezawa, T. Shiba, T. Takita, pp. 129–51. Tokyo: Kodansha Sci. 275 pp.
3. Aoyagi, T., Hazato, T., Kumagai, M., Hamada, M., Takeuchi, T., Umezawa, H. 1975. *J. Antibiot.* 28:1006–8
4. Aoyagi, T., Kumagai, M., Hazato, T., Hamada, M., Takeuchi, T., Umezawa, H. 1975. *J. Antibiot.* 28:555–57
5. Aoyagi, T., Kunimoto, S., Morishima, H., Takeuchi, T., Umezawa, H. 1971. *J. Antibiot.* 24:687–94
6. Aoyagi, T., Miyata, S., Nanbo, M., Kojima, F., Ishizuka, M., Takeuchi, T., Umezawa, H. 1969. *J. Antibiot.* 22: 558–68

7. Aoyagi, T., Morishima, H., Nishizawa, R., Kunimoto, S., Takeuchi, T., Umezawa, H. 1972. *J. Antibiot.* 25: 689–94
8. Aoyagi, T., Nagai, M., Iwabuchi, M., Liaw, W. S., Andoh, T., Umezawa, H. 1978. *Cancer Res.* 38:3505–8
9. Aoyagi, T., Suda, H., Nagai, M., Ogawa, K., Suzuki, J., Takeuchi, T., Umezawa, H. 1976. *Biochim. Biophys. Acta* 452:131–43
10. Aoyagi, T., Takeuchi, T., Matsuzaki, A., Kawamura, K., Kondo, S., Hamada, M., Umezawa, H. 1969. *J. Antibiot.* 22:283–86
11. Aoyagi, T., Tobe, H., Kojima, F., Hamada, M., Takeuchi, T., Umezawa, H. 1978. *J. Antibiot.* 31:636–38
12. Aoyagi, T., Umezawa, H. 1975. In *Proteases and Biological Control,* ed. E. Reich, D. Rifken, E. Shaw, pp. 429–54. Cold Spring Harbor: Cold Spring Harbor Lab. 1021 pp.
13. Aoyagi, T., Yagisawa, Y., Kumagai, M., Hamada, M., Morishima, H., Takeuchi, T., Umezawa, H. 1973. *J. Antibiot.* 26:539–41
14. Aoyagi, T., Yagisawa, M., Kumagai, M., Hamada, M., Okami, Y., Takeuchi, T., Umezawa, H. 1971. *J. Antibiot.* 24:860–69
15. Aoyagi, T., Yamamoto, T., Kojiri, K., Kojima, F., Hamada, M., Takeuchi, T., Umezawa, H. 1978. *J. Antibiot.* 31: 244–46
16. Ayukawa, S., Hamada, M., Kojiri, K., Takeuchi, T., Hara, T., Nagatsu, T., Umezawa, H. 1969. *J. Antibiot.* 22:303–8
17. Brown, A. G., Smale, T. C., King, T. J., Hasenkamp, R., Thompson, R. H. 1976. *J. Chem. Soc. Perkin I* pp. 1165–70
18. Chelmicka-Schorr, E. E., Arnason, B. G. W., Astrom, K. E., Darzynkiewicz, Z. 1978. *J. Neuropathol. Exp. Neurol.* 37:263–68
19. Chimura, H., Nakamura, H., Takita, T., Takeuchi, T., Umezawa, H., Kato, K., Saito, S., Tomisawa, T., Iitaka, Y. 1975. *J. Antibiot.* 28:743–48
20. Chimura, H., Sawa, T., Kumada, Y., Naganawa, H., Matsuzaki, M., Takita, T., Hamada, M., Takeuchi, T., Umezawa, H. 1975. *J. Antibiot.* 28: 619–26
21. Chimura, H., Sawa, T., Kumada, Y., Nakamura, F., Matsuzaki, M., Takita, T., Takeuchi, T., Umezawa, H. 1973. *J. Antibiot.* 26:618–20
22. Chimura, H., Sawa, T., Takita, T., Matsuzaki, M., Takeuchi, T., Nagatsu, T., Umezawa, H. 1973. *J. Antibiot.* 26: 112–14
22a. Corcoran, J. W., Hahn, F. E., eds. 1975. *Antibiotics III, Mechanisms of Action of Antimicrobial and Antitumor Agents.* New York: Springer
23. Corvol, P., Davaux, C., Menard, J. 1973. *FEBS Lett.* 34:189–92
23a. Együd, L. G., Szent-Györgi, A. 1966. *Natl. Acad. Sci. USA* 55:388–93
24. Endo, A., Kuroda, M., Tanzawa, K. 1976. *FEBS Lett.* 72:323–26
25. Endo, A., Kuroda, M., Tsujita, Y. 1976. *J. Antibiot.* 29:1346–48
26. Enomoto, Y., Furutani, Y., Naganawa, H., Hamada, M., Takeuchi, T., Umezawa, H. 1978. *Agric. Biol. Chem.* 42:1331–36
27. Furutani, Y., Shimada, M., Hamada, M., Takeuchi, T., Umezawa, H. 1977. *Agric. Biol. Chem.* 41:989–93
28. Gottlieb, D., Shaw, P. D., eds. 1967. Antibiotics I, *Mechanisms of Action.* New York: Springer
29. Greenbaum, L. M., Grebow, P., Johnston, M., Prekash, A., Semente, G. 1975. *Cancer. Res.* 35:706–10
30. Hanada, K., Tamai, M., Tamagishi, M., Ohmura, S., Sawada, J., Tanaka, I. 1978. *Agric. Biol. Chem.* 42:523–28
31. Hazato, T., Aoyagi, T., Umezawa, H. 1979. *J. Antibiot.* 32:212–16
32. Hazato, T., Kumagai, M., Naganawa, H., Aoyagi, T., Umezawa, H. 1979. *J. Antibiot.* 32:91–93
33. Hidaka, H., Nagatsu, T., Takeya, K., Takeuchi, T., Suda, H., Kojiri, K., Matsuzaki, M., Umezawa, H. 1969. *J. Antibiot.* 22:228–30
34. Hori, M., Hemmi, H., Suzukake, K., Hayashi, H., Uehara, Y., Takeuchi, T., Umezawa, H. 1978. *J. Antibiot.* 31: 95–98
35. Hozumi, M., Ogawa, M., Sugimura, T., Takeuchi, T., Umezawa, H. 1972. *Cancer Res.* 32:1725–28
36. Iinuma, H., Takeuchi, T., Kondo, S., Matsuzaki, M., Umezawa, H., Ohno, M. 1972. *J. Antibiot.* 25:497–500
37. Iinuma, H., Yagisawa, N., Shibahara, S., Suhara, Y., Kondo, S., Maeda, K., Takeuchi, T., Ohno, M., Umezawa, H. 1974. *Agric. Biol. Chem.* 38:2099–105
38. Ishizuka, M., Iinuma, H., Takeuchi, T., Umezawa, H. 1972. *J. Antibiot.* 25: 320–21
39. Ishizuka, M., Takeuchi, T., Umezawa, H. 1981. *J. Antibiot.* 34:95–102
40. Kakinuma, A., Kanamaru, T. 1976. *J. Takeda Res. Lab.* 35:123–27

41. Kawamura, K., Kondo, S., Maeda, K., Umezawa, H. 1969. *Chem. Pharm. Bull.* 17:1902–1909
42. Kinoshita, M., Aburaki, S., Hagiwara, A., Imai, J. 1973. *J. Antibiot.* 26:249–51
43. Koba, Y., Najima, M., Ueda, S. 1976. *Agric. Biol. Chem.* 40:1167–73
44. Komiyama, T., Aoyagi, T., Takeuchi, T., Umezawa, H. 1975. *Biochem. Biophys. Res. Commun.* 65:352–57
45. Komiyama, T., Suda, H., Aoyagi, T., Takeuchi, T., Umezawa, H., Fujimoto, K., Umezawa, S. 1975. *Arch. Biochem. Biophys.* 171:727–31
46. Kondo, S., Kawamura, K., Iwanaga, J., Hamada, M., Aoyagi, T., Maeda, K., Takeuchi, T., Umezawa, H. 1969. *Chem. Pharm. Bull.* 17:1896–901
47. Kondo, S., Uotani, K., Miyamoto, M., Hazato, T., Naganawa, H., Aoyagi, T., Umezawa, H. 1978. *J. Antibiot.* 31:797–800
48. Kumada, Y., Naganawa, H., Hamada, M., Takeuchi, T., Umezawa, H. 1974. *J. Antibiot.* 27:726–28
49. Kumada, Y., Naganawa, H., Iinuma, H., Matsuzaki, M., Takeuchi, T., Umezawa, H. 1976. *J. Antibiot.* 29: 882–89
50. Kumada, Y., Naganawa, H., Takeuchi, T., Umezawa, H., Yamashita, K., Watanabe, K. 1978. *J. Antibiot.* 31: 105–11
51. Kumada, Y., Takeuchi, T., Umezawa, H. 1977. *Agric. Biol. Chem.* 41:869–76
52. Kumagai, M., Aoyagi, T., Umezawa, H. 1976. *J. Antibiot.* 29:696–703
53. Kumagai, M., Naganawa, H., Aoyagi, T., Umezawa, H. 1975. *J. Antibiot.* 28:876–80
54. Kumagai, M., Suhara, Y., Aoyagi, T., Umezawa, H. 1971. *J. Antibiot.* 24:870–75
55. Kunimoto, T., Hori, M., Umezawa, H. 1973. *Biochim. Biophys. Acta* 298: 513–25
56. Kurasawa, S., Naganawa, H., Takeuchi, T., Umezawa, H. 1975. *Agric. Biol. Chem.* 39:2009–14
57. Kurasawa, S., Takeuchi, T., Umezawa, H. 1976. *Agric. Biol. Chem.* 40:559–66
58. Levy, M. R., Chow, S. C. 1974. *Biochim. Biophys. Acta* 334:423–30
59. Levy, M. R., Chow, S. C. 1975. *Experientia* 31:52–54
60. Matsushita, Y., Tone, H., Hori, S., Yagi, Y., Takamatsu, A., Morishima, H., Aoyagi, T., Takeuchi, T., Umezawa, H. 1975. *J. Antibiot.* 28: 1016–18
61. McGowan, E. B., Shafiq, S. A., Stracher, A. 1976. *Exp. Neurol.* 50: 649–57
62. Miyano, T., Tomiyasu, M., Iizuka, H., Tomisaka, S., Takita, T., Aoyagi, T., Umezawa, H. 1972. *J. Antibiot.* 25:489–91
62a. Morihara, K., Tsuzuki, H. 1975. *Agric. Biol. Chem.* 39:1123–28
63. Morishima, H., Sawa, T., Takita, T., Aoyagi, T., Takeuchi, T., Umezawa, H. 1974. *J. Antibiot.* 27:267–73
64. Morishima, H., Takita, T., Aoyagi, T., Takeuchi, T., Umezawa, H. 1970. *J. Antibiot.* 23:263–65
65. Murakami, K., Inagami, T., Michelakis, A. M., Cohen, S. 1973. *Biochem. Biophys. Res. Commun.* 54:482–87
66. Murao, S., Ohyama, K., Ogura, S. 1977. *Agric. Biol. Chem.* 41:919–24
67. Murao, S., Satoi, S. 1970. *Agric. Biol. Chem.* 34:1265–67
68. Nagatsu, T., Kato, T., Numata, Y., Ikuta, K., Kuzuya, H., Umezawa, H., Matsuzaki, M., Takeuchi, T. 1975. *Experientia* 31:767–68
69. Nakamura, H., Iitaka, Y., Kumada, Y., Takeuchi, T., Umezawa, H. 1977. *Acta Cryst.* 33(B):1260–63
70. Nakamura, H., Koyama, G., Iitaka, Y., Ohno, M., Yagisawa, N., Kondo, S., Maeda, K., Umezawa, H. 1974. *J. Am. Chem. Soc.* 96:4327–28
71. Nerome, K., Kumagai, M., Aoyagi, T. 1972. *Arch. Ges. Virusforsch.* 39: 353–59
72. Ohno, M., Okamoto, M., Kawabe, N., Umezawa, H., Takeuchi, T., Iinuma, H., Takahashi, S. 1971. *J. Am. Chem. Soc.* 93:1285–86
73. Okabayashi, K., Morishima, H., Hamada, M., Takeuchi, T., Umezawa, H. 1977. *J. Antibiot.* 30:675–77
74. Okura, A., Morishima, H., Takita, T., Aoyagi, T., Takeuchi, T., Umezawa, H. 1975. *J. Antibiot.* 28:337–39
75. Ōmura, S., Nakagawa, A., Ohno, H. 1979. *J. Am. Chem. Soc.* 101:4386–88
76. Ōmura, S., Ohno, H., Saheki, T., Yoshida, M., Nakagawa, A. 1978. *Biochem. Biophys. Res. Commun.* 83: 704–9
77. Sawa, T., Fukagawa, Y., Honma, I., Takeuchi, T., Umezawa, H. 1967. *J. Antibiot. Ser. A* 20:227–31
78. Schmidt, D. D., Frommer, W., Junge, B., Müller, L., Wingender, W., Truscheit, E. 1977. *Naturwissenschaften* 64:535–36
79. Sezaki, M., Hara, T., Ayukawa, S., Takeuchi, T., Okami, Y., Hamada, M.,

Nagatsu, T., Umezawa, H. 1968. *J. Antibiot.* 21:91–97
80. Shimada, K., Matsushima, K., Fukumoto, J., Yamamoto, T. 1969. *Biochem. Biophys. Res. Commun.* 35: 619–24
81. Stracher, A., McGowan, E. B., Shafiq, S. A. 1978. *Science* 200:50–51
82. Suda, H., Aoyagi, T., Hamada, M., Takeuchi, T., Umezawa, H. 1972. *J. Antibiot.* 25:263–66
83. Suda, H., Aoyagi, T., Takeuchi, T., Umezawa, H. 1973. *J. Antibiot.* 26: 621–23
84. Suda, H., Takeuchi, T., Nagatsu, T., Matsuzaki, M., Matsumoto, I., Umezawa, H. 1969. *Chem. Pharm. Bull.* 17:2377–80
85. Suda, H., Takita, T., Aoyagi, T., Umezawa, H. 1976. *J. Antibiot.* 29: 100–1
86. Suzukake, K., Fujiyama, T., Hayashi, H., Hori, M., Umezawa, H. 1979. *J. Antibiot.* 32:523–30
87. Suzukake, K., Hayashi, H., Hori, M., Umezawa, H. 1980. *J. Antibiot.* 33: 857–62
88. Suzukake, K., Takada, M., Hori, M., Umezawa, H. 1980. *J. Antibiot.* 33: 1172–76
89. Takeuchi, T., Chimura, H., Hamada, M., Umezawa, H., Yoshioka, O., Oguchi, N., Takahashi, Y., Matsuda, A. 1975. *J. Antibiot.* 28:737–42
90. Takeuchi, T., Ogawa, K., Iinuma, H., Suda, H., Ukita, K., Nagatsu, T., Kato, M., Umezawa, H., Tanabe, O. 1973. *J. Antibiot.* 26:162–67
91. Tatsuta, K., Mikami, N., Fujimoto, K., Umezawa, S., Umezawa, H., Aoyagi, T. 1973. *J. Antibiot.* 26:625–46
92. Taylor, R. J. Jr., Stubbs, C. S. Jr., Ellenbogen, L. 1970. *Biochem. Pharmacol.* 19:1737–41
93. Tobe, H., Morishima, H., Naganawa, H., Takita, T., Aoyagi, T., Umezawa, H., 1979. *Agric. Biol. Chem.* 43:591–96
94. Tone, H., Matsushita, Y., Yagi, Y., Takamatsu, A. 1975. *J. Antibiot.* 28: 1012–45
94a. Tsujita, Y., Kuroda, M., Tanzawa, K., Kitano, N., Endo, A. 1979. *Atherosclerosis* 32:307–13
95. Umezawa, H. 1972. *Enzyme Inhibitors of Microbial Origin,* pp. 1–114. Tokyo: Univ. Tokyo. 114 pp.
96. Umezawa, H. 1976. *Methods Enzymol.* 45:678–95
97. Umezawa, H. 1977. *Jap. J. Antibiot.* 30:S138–63 (Suppl.)
98. Umezawa, H. 1981. *Small Molecular Immunomodifiers of Microbial Origin,* pp. 1–254 Oxford: Pergamon. 254 pp.
99. Umezawa, H., Aoyagi, T. 1977. In *Proteinases in Mammalian Cells and Tissues,* ed. A. J. Barett, pp. 637–62. Amsterdam: North-Holland. 735 pp.
100. Umezawa, H., Aoyagi, T., Hazato, T., Uotani, K., Kojima, F., Hamada, M., Takeuchi, T. 1978. *J. Antibiot.* 31:639–41
101. Umezawa, H., Aoyagi, T., Komiyama, T., Morishima, H., Hamada, M., Takeuchi, T. 1974. *J. Antibiot.* 27: 963–69
102. Umezawa, H., Aoyagi, T., Morishima, H., Kunimoto, S., Matsuzaki, M., Hamada, M., Takeuchi, T. 1970. *J. Antibiot.* 23:425–28
103. Umezawa, H., Aoyagi, T., Morishima, H., Matsuzaki, M., Hamada, M., Takeuchi, T. 1970. *J. Antibiot.* 23: 259–62
104. Umezawa, H., Aoyagi, T., Okura, A., Morishima, H., Takeuchi, T., Okami, Y. 1973. *J. Antibiot.* 26:787–89
105. Umezawa, H., Aoyagi, T., Suda, H., Hamada, M., Takeuchi, T. 1976. *J. Antibiot.* 29:97–99
106. Umezawa, H., Aoyagi, T., Uotani, K., Hamada, M., Takeuchi, T., Takahashi, S. 1980. *J. Antibiot.* 33:1594–96
106a. Umezawa, H., Ishizuka, M., Aoyagi, T., Takeuchi, T. 1976. *J. Antibiot.* 29:857–59
107. Umezawa, H., Miyano, T., Murakami, T., Takita, T., Aoyagi, T., Takeuchi, T., Naganawa, H., Morishima, H. 1973. *J. Antibiot.* 27:615–17
108. Umezawa, H., Okami, Y., Hotta, K. 1978. *J. Antibiot.* 31:99–102
109. Umezawa, H., Shibamoto, N., Naganawa, H., Ayukawa, S., Kono, K., Sakamoto, T. 1974. *J. Antibiot.* 27: 587–96
110. Umezawa, H., Takeuchi, T., Iinuma, H., Suzuki, K., Ito, M., Matsuzaki, M., Nagatsu, T., Tanabe, O. 1970. *J. Antibiot.* 23:514–18
111. Umezawa, H., Takeuchi, T., Miyano, K., Koshigoe, T., Hamano, H. 1973. *J. Antibiot.* 26:189
112. Umezawa, H., Tobe, H., Shibamoto, N., Nakamura, F., Nakamura, K., Matsuzaki, M., Takeuchi, T. 1975. *J. Antibiot.* 28:947–52
113. Umezawa, S., Tatsuta, K. Fujimoto, K., Tsuchiya, T., Umezawa, H., Naganawa, H. 1972. *J. Antibiot.* 25:267–70
114. Umezawa, S., Tatsuta, K., Izawa, O.,

Tsuchiya, T., Umezawa, H. 1972. *Tetrahedron Lett.* 97–100
115. Woo, W. K., Dion, H. W., Lange, M., Dahl, L. F., Durham, L. J. 1974. *J. Heterocycl. Chem.* 11:641–43
115a. Yabuta, T., Kambe, K., Hayashi, T. 1934. *J. Agric. Chem. Soc. Jpn.* 10:1059–68
116. Yamamoto, T., Kojiri, K., Morishima, H., Naganawa, H., Aoyagi, T., Umezawa, H. 1978. *J. Antibiot.* 31: 483–84
117. Yamamoto, A., Sudo, H., Endo, A. 1980. *Atherosclerosis* 35:259–66
118. Yuasa, Y., Shimojo, H., Aoyagi, T., Umezawa, H. 1975. *J. Natl. Cancer Inst.* 54:1255–56

Ann. Rev. Microbiol. 1982. 36:101–23

PRIMARY AMEBIC MENINGOENCEPHALITIS AND THE BIOLOGY OF *NAEGLERIA FOWLERI*

David T. John

Department of Microbiology, Medical College of Virginia, Virginia Commonwealth University, Richmond, Virginia 23298

CONTENTS

INTRODUCTION

Primary amebic meningoencephalitis (PAM) is a rapidly fatal human disease caused by the ameboflagellate *Naegleria fowleri.* The disease first was detected in man 17 years ago by Fowler & Carter in Australia (56). A year later in 1966 three fatal infections were described from Florida by Butt (10).

0066-4227/82/1001-0101$02.00

The symptomatology of these cases was remarkably similar to that observed in Australia. Although it was not apparent then, the seven cases in Australia and Florida provided almost a complete array of the important clinical and pathological features of the disease.

Notable, also, was the indication that infection was acquired by intranasal instillation during swimming. Butt (10) recognized the discovery of a new disease in Australia and Florida by contributing a new name, primary amebic meningoencephalitis.

Naegleria fowleri was named in honor of Dr. Malcom Fowler who first recognized the disease it caused (13). This has been the only species of *Naegleria* to be isolated from victims of PAM. Synonyms for *N. fowleri* are *N. aerobia* (114) and *N. invades* (24). Nonpathogenic species of *Naegleria* include *N. gruberi* (see 58), possibly the most common ameba in fresh water (101), *N. thorntoni* (113), and *N. lovaniensis* (118). *N. jadini,* isolated from swimming pool water in Belgium, is only slightly pathogenic for mice (140).

Naegleria is an ameboflagellate; it belongs to a family of amebae (Vahlkampfiidae) whose members can transform from amebae into flagellates (101). The *Naegleria* flagellate is a transient, nonfeeding, nondividing form. The life cycle of *Naegleria* also includes cyst formation and excystment by amebae (Figure 1). *Naegleria* flagellates do not encyst; only the amebae divide and are able to encyst.

HUMAN INFECTION

Amebic infections of the central nervous system may be caused by the parasitic ameba *Entamoeba histolytica* or by the opportunistic free-living amebae *Naegleria fowleri* or *Acanthamoeba* spp. *E. histolytica* may produce a brain abscess after extraintestinal invasion and subsequent hematogenous

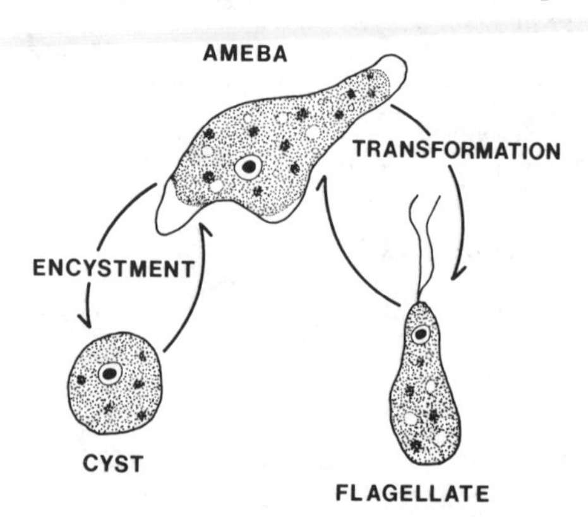

Figure 1 Life cycle of *Naegleria fowleri.*

spread. *Acanthamoeba* spp. produce an illness known as granulomatous amebic encephalitis (91). Disease usually occurs in chronically ill or debilitated individuals, some of whom may be undergoing immunosuppressive therapy. Invasion of the central nervous system appears to be hematogenous, arising from a primary lesion of the skin, lungs, or kidneys. In this review, discussion is limited to *N. fowleri* and the disease it produces.

Clinical Features

PAM typically occurs in healthy children or young adults with a recent history of swimming in freshwater lakes or pools. The disease is rapidly fatal, usually producing death within 72 hr after the onset of symptoms. Infection follows inhalation of water containing amebae or flagellates. It also has been suggested that inhaling cysts, during dusts storms, for example, could lead to infection (84, 108).

Amebae penetrate the nasal mucosa and the cribiform plate and travel along the olfactory nerves to the brain. Amebae first invade the olfactory bulbs and then spread to the more posterior regions of the brain. Within the brain they provoke inflammation and cause extensive damage to the tissue (13, 93).

The clinical course is dramatic. Symptoms begin with severe frontal headache, fever (39–40°C), and anorexia. This is followed by nausea, vomiting, and signs of meningeal irritation. Involvement of the olfactory lobes may cause disturbances in the sense of smell or taste and may be noted early in the course of the disease. Visual disturbances may occur. The patient may experience confusion, irritability, and restlessness and may become irrational before lapsing into coma. Generalized seizures also may be present. In order of frequency of occurrence, the more important symptoms include headache, anorexia, nausea, vomiting, fever, and stiff neck (49, 93).

Pathology

The gross pathologic findings in PAM are remarkably constant. The cerebral hemispheres usually are edematous and swollen. Meninges are diffusely hyperemic with a slight purulent exudate. The cortex contains many focal superficial hemorrhages. The olfactory bulbs exhibit marked involvement with hemorrhage, necrosis, and purulent exudate (14, 49, 93).

Microscope examination reveals many amebae in the subarachnoid and perivascular spaces. Presumably, the perivascular spaces provide a path of migration for the amebae, and the blood vessels supply the oxygen needed by these aerobic organisms. In fewer numbers, amebae are found clustered within the brain tissue and in the purulent exudate of the meninges and brain substance. Within the exudate some amebae may be seen engulfed by macrophages. Many amebae are observed to contain phagocytosed cellular debris and erythrocytes. The purulent exudate contains numerous polymorphonuclear and mononuclear leukocytes (14, 93).

The cortical gray matter is a preferred site for amebae development; consequently, severe involvement occurs in the cerebral hemispheres, cerebellum, brain stem, and upper portions of the spinal cord. Encephalitis ranges from slight amebic invasion and inflammation to massive invasion with purulent, hemorrhagic necrosis. Typically, the olfactory bulbs exhibit extensive amebic invasion, hemorrhage, and an inflammatory exudate; the involvement here is greater than in other areas of the brain. Infection of the central nervous system with *N. fowleri* may be described best as an acute, hemorrhagic, necrotizing meningoencephalitis (14, 93).

Focal demyelination in the white matter of the brain and spinal cord has been described (27, 51). Curiously, demyelination occurred in the absence of amebae or cellular infiltration. Chang (27) suggests that demyelination is caused by a phospholipolytic enzyme or enzyme-like substance produced by actively growing amebae present in the adjacent gray matter.

Myocarditis has been described in some patients dying of PAM (14, 89). It has been suggested (89) that the observed myocarditis may be caused by a circulating myotoxin either produced by amebae in the brain or released by rapidly degenerating invading amebae. However, evidence has not yet been produced to substantiate either hypothesis.

Diagnosis

The diagnosis of PAM is made by microscope identification of living or stained amebae in patient cerebrospinal fluid. Motile amebae are readily seen in simple wet-mount preparations of spinal fluid. Amebae can be distinguished from other cells by their limax (L. sluglike) shape and progressive movement. It is not necessary to warm the slide since amebae remain fully active at room temperature. Refrigeration of the spinal fluid is not recommended as this may kill the amebae (13, 93).

Spinal fluid smears may be stained with Wright or Giemsa stains (108). The bacterial Gram stain is of little positive value since heat fixing causes the amebae to stain poorly and to appear as degenerate cells. Giemsa- or Wright-stained amebae have considerable amounts of sky-blue cytoplasm and relatively small, delicate, pink nuclei. Mononuclear leukocytes, on the other hand, have large purplish nuclei with only small amounts of sky-blue cytoplasm. Acridine orange has also been used to distinguish *N. fowleri* amebae from leukocytes (97). Using acridine orange and ultraviolet microscopy, amebae stain brick-red with pale green nuclei in contrast to leukocytes, which are bright green.

Amebae also may be cultured by placing some of the spinal fluid on non-nutrient agar (1.5%) spread with a lawn of *Enterobacter aerogenes* or

Escherichia coli and incubated at 37°C. The amebae will grow on the moist agar surface and will utilize the bacteria as a source of food.

Clinically, PAM very closely resembles fulminating bacterial meningitis, and the laboratory findings are also similar. The cerebrospinal fluid is purulent or sanguinopurulent with leukocyte counts, predominantly neutrophils, ranging from a few hundred to over 20,000 cells/mm^3. Spinal fluid glucose levels are low, and generally, protein content is increased. Typically, Gram-stained smears and cultures of spinal fluid are negative for bacteria (14, 93).

Treatment

At present, no satisfactory treatment for PAM exists. The antibiotics used to treat bacterial meningitis are ineffective in naeglerial infection, as are the antiamebic drugs. Amphotericin B, a drug of considerable toxicity, is an antinaeglerial agent for which there is some evidence of clinical effectiveness. Both known survivors of PAM were treated with amphotericin B, given intravenously and intrathecally (2, 15); the only patient in the United States known to have survived naeglerial infection (15) also was given parenteral miconazole and oral rifampin. The in vitro testing of a highly virulent human isolate of *N. fowleri* demonstrated that amebae were extremely susceptible to amphotericin B [minimal inhibitory concentration (MIC), 0.15 μg/ml], somewhat susceptible to miconazole (MIC, 25 μg/ml), and resistant to rifampin (MIC,≤100 μ/ml) (119). Mice were protected by treatment with amphotericin (7.5 mg/kg/day) but not by treatment with lower doses of amphotericin B alone or in combination with miconazole (100 mg/kg) or rifampin (220 mg/kg). It appears, then, that amphotericin B currently is the most effective treatment for PAM. Amphotericin B is administered intravenously at a dose of 1 mg per kg of body weight daily, with intrathecal inoculations of 0.1 mg on alternate days (14).

Amphotericin B is a polyene compound that acts upon the plasma membrane, disrupting its selective permeability, and causing leakage of cellular components (82). When exposed to amphotericin B, amebae round up and do not form pseudopodia. Membrane-related changes, evident by electron microscopy, include enhanced nuclear plasticity, increased amount of smooth and rough endoplasmic reticulum, decreased food vacuole formation, and production of blebs of the plasma membrane (110).

Lipopolysaccharide has been shown to afford mice some protection for several days after challenge with *N. fowleri* (1). Slightly better protection was provided by treatment of mice with Δ9-tetrahydrocannabinol (105).

Tetracycline (127) and rifampin (126) have been shown to act synergistically with amphotericin B to protect mice against *N. fowleri* infection. In the tetracycline study, chemotherapy was started 72 hr after the mice had been infected intranasally. Survival was 38% for amphotericin B-treated mice and 88% for mice treated with the amphotericin B-tetracycline combination (127). Combination therapy is a reasonable approach to treatment for patients with PAM.

EPIDEMIOLOGY AND ECOLOGY

Geographic Distribution

Although it is a relatively rare disease, PAM has been reported worldwide. Most of the reports have been from developed rather than from developing nations, perhaps because of a greater awareness of the disease in these countries and not because of greater incidence.

Cases have been reported from Belgium, Czechoslovakia, Great Britain, Northern Ireland, India, Australia, New Zealand, New Guinea, Uganda, Nigeria, Venezuela, Panama, Puerto Rico, and the United States. As of October 1981, 108 cases of *Naegleria* infection have been described worldwide; 49 of these have been from the United States. The states reporting cases have been New York, Virginia, North Carolina, South Carolina, Georgia, Florida, Mississippi, Arkansas, Texas, Arizona, Nevada, and California.

The majority of patients with naeglerial infection have had a history of recent swimming in fresh water during hot summer weather. In Richmond, Virginia, infection in 14 of the 16 cases probably was acquired in two man-made lakes located within a few miles of each other (12, 51, 108).

Over a 3-year period, 16 young people died in Czechoslovakia after swimming in the same heated, chlorinated, indoor swimming pool (23). Similar fatal cases have been reported following swimming in swimming pools in Belgium (68), England (11), and New Zealand (34); in hot springs in California (15) and New Zealand (31); in lakes in Florida (10, 16), Texas (16), Arkansas (143), and South Carolina (39); and in streams in Belgium (129), New Zealand (32), and Mississippi (87).

Infection has not always been acquired through swimming, though. The South Australia and northern Nigeria cases have occurred in rather arid regions where swimming is not often indulged. The proposed means of infection for these areas has been face washing and bath-related activity (2, 85) and inhalation of dust-borne cysts (84).

Environmental Isolations

Naegleria fowleri has been isolated from a variety of environmental sources. During the summer of 1971, Nelson (98) isolated *N. fowleri* from a water sample taken from a pond where a victim of PAM swam in 1967. This constituted the first isolation of *N. fowleri* from an environmental source. Today, the pond no longer is used for swimming but has since been converted into an ornamental pool in a suburban housing development. However, another man-made lake in the immediate neighborhood still attracts swimmers daily throughout the summer months despite the fact that five children have died of PAM after swimming there; the last case occurred 14 years ago in 1968 (12, 51, 108).

Additional environmental isolations of *N. fowleri* have been made from tap water in South Australia (69); heated indoor swimming pools in Czechoslovakia (79) and England (11); lakes in Florida (139), Nigeria (86), and Poland (80); warm effluents from power plants in Florida (121), Texas (121), and France (47) and from factories in Belgium (41); sewage and sludge samples in India (115), Korea (117), and Nigeria (86); soil samples in South Australia (3), New Zealand (34), and Nigeria (86); and river water in Spain (95).

N. fowleri clearly is cosmopolitan in its environmental distribution. Curiously, though, most of the isolations have been made from habitats manipulated or altered in some way by man or, and this appears to be an important factor, from habitats subjected to warming, whether man-made or natural.

On two occasions, *N. fowleri* has been isolated from the nasal passages of young children. Amebae pathogenic for mice were isolated by nasal swab from a 7-year-old boy in Richmond, Virginia (28, 112), and from two young children, under 10 years of age, in Zaria, Nigeria (84). Although the Nigerian children had symptoms of upper respiratory tract infection, none of the children had signs or symptoms of meningitis, and the Richmond child had no evidence of infection of any sort.

Animal Infection

There are only a few reports, apart from those describing human infections, in which *Naegleria* has been isolated from animals in nature. They include the recovery of *Naegleria* from the gills of fish (122), from snails (81, 103), and from various reptiles (57). *Acanthamoeba,* a related freshwater ameba, has been detected in visceral lesions of a dog (4), in pulmonary lesions of a buffalo (53), in a bull (96), in renal granulomas in a gold fish (131), and cultured from clinical specimens and tissues of a variety of diseased domestic animals (77).

There is no evidence for the existence of an animal reservoir or carrier host for *N. fowleri.* Nonetheless, until adequate surveys have been carried out, the possibility cannot be excluded. Animals that need to be examined include aquatic insects (and insect larvae), crustaceans and mollusks, fishes, amphibians and aquatic reptiles, waterfowl and shore birds, and aquatic or diving mammals such as beavers, otters, and muskrats. All the above animals could encounter *N. fowleri* naturally and, particularly the migratory species, could serve as transport hosts. Perhaps this accounts for the observation made by DeJonckheere & Van De Voorde (45) that *N. fowleri* could be isolated from waters where old factories discharged thermal waste, but not from the discharges of newer factories. The animal(s) involved may require a period of adjustment before investigating a newly built facility.

Researchers have used several laboratory animal species as models for their investigations of experimental PAM. *Naegleria fowleri* produces fatal meningoencephalitis in mice (13, 18, 29, 92, and others), guinea pigs (18, 30, 104), rabbits (29), monkeys (29, 141), and sheep (144). The recent significant finding that sheep are susceptible to *N. fowleri* following intranasal instillation of amebae suggests the possible occurrance of PAM among domestic livestock.

Chick embryos also are susceptible to *N. fowleri* infection (66). Amebae inoculated on the chorioallantoic membrane disseminate to the brain, liver, spleen, and lungs. Inoculation with as few as 10 amebae will cause death of the chick embryo within 5 days.

The mouse is uniquely appropriate as an experimental model for studying PAM. The basic features of the disease in man and mouse are the same with respect to portal of entry, incubation period, and invasion of the nasal mucosa, cribiform plate, and olfactory bulbs with subsequent spread to more distant areas of the brain (92)

Fatal naeglerial meningoencephalitis in mice is route and dose dependent. Most of the animal studies have employed either intranasal or intracerebral routes of inoculation, which require fewer amebae to establish infection than do the other routes. Clinical symptoms and death were produced in mice inocluated intravenously and intraperitoneally with large doses of amebae (1, 13). A dose of 5×10^6 amebae per mouse was administered subcutaneously without deaths occurring (1). However, since the outcome of an infection with *N. fowleri* is dose dependent, it is reasonable to propose that a larger dose (or a more virulent strain) could produce the disease in mice via subcutaneous inoculation.

It is also possible to produce infection in mice after intranasal instillation with the flagellate stage of *N. fowleri* (116). However, since flagellates are rather unstable, they undoubtedly reverted to the ameba stage in the nasal passages, and the amebae subsequently invaded the nasal mucosa. Flagellates were not seen in brain smears of mice dying from meningoencephalitis; only amebae were noted (116).

Cysts of *N. fowleri* never have been observed in human brain tissue or in tissues from experimentally produced animal infections. This is in direct contrast to infection with *Acanthamoeba* in which cysts are routinely detected.

Control and Prevention

Because of the swimming-related nature of naeglerial infection, many swimming areas have been subjected to intense investigation. And although it is true that *N. fowleri* has been isolated from some swimming areas (described above), there are several reports in which environmental sampling of swimming pools has failed to produce *N. fowleri* (19, 42, 48, 70, 88). Obviously, some factors favor the development of *N. fowleri* in swimming areas. These include warm temperature, presence of an adequate food source (organic matter or bacteria), insufficient residual free chlorine, minimal competition from other protozoans, and probably optimal pH and oxygen levels.

With the present limited understanding of the ecology of *N. fowleri,* practical measures for prevention and control of naeglerial infection include education of the public, awareness within the medical community, and adequate chlorination (10 ppm) of public swimming facilities (44). Chang (26) has demonstrated that *N. fowleri* amebae are sensitive to drying, high temperature (>51°C), low temperature (<10°C), and especially freezing. Cysts are sensitive to desiccation (nonviable in <5 min), survive poorly at 0°C, but tolerate high temperatures (51–65°C). Cysts of *N. fowleri* have been stored for up to 8 months with excysting amebae retaining virulence (132). Effectiveness of chlorination is dependent upon residual free chlorine, temperature, and pH (26).

Legionella pneumophila, the agent of Legionnaires' disease, has been shown to be pathogenic for *N. gruberi* and *N. jadini* (107). It has been suggested that *Naegleria* (and *Acanthamoeba*) possibly may be natural hosts for *L. pneumophila,* and that human infection may be acquired not by inhaling free legionellae but by inhaling amebae full of legionellae (50–1000 or more bacteria per ameba) (107). Our own studies (D. T. John, N. C. Mobley, unpublished data) indicate that *N. fowleri* is more susceptible to infection by *L. pneumophila* than is *N. gruberi.* Perhaps *Legionella* (and/or other bacteria or viruses) serves as a natural biological control of *N. fowleri* and accounts for the relative scarcity of *N. fowleri* in the environment.

Considering the millions of persons who swim each summer in streams, lakes, and swimming pools, the probability of becoming infected is rather remote. For example, despite the extensive distribution of *N. fowleri* in Florida's freshwater lakes (121, 139), estimates are that the risk of acquiring naeglerial infection is about one case per 2.6 million exposures (138).

PHYSIOLOGY/METABOLISM

Nutrition and Growth

Perhaps the simplest way to grow *N. fowleri* is on the surface of non-nutrient agar (1.5%) spread with living or dead *Enterobacter aerogenes* or *Escherichia coli.* Under these conditions, the amebae feed upon the bacteria, and as growth enters stationary phase and the food supply is used up, they begin to encyst. Cysts, if kept from drying out, will remain viable for months, possibly years.

Unfortunately, the presence of bacteria too often hampers a variety of quantitative studies. Therefore, several liquid media have been developed for the axenic cultivation of *N. fowleri* (5, 17, 26, 99). In general, these media contain a phosphate buffer and either liver extract, yeast extract, peptone, or casein derivative bases supplemented with serum.

Liquid axenic media developed by Balamuth (5), Červa (17), Chang (26), and Nelson & Jones (99) have been used for the cultivation of *N. fowleri* (20, 40, 52, 133). Growth of *N. fowleri* under agitated and unagitated culture conditions was compared using these four axenic media (63). The less enriched media of Červa (17) and Nelson & Jones (99) supported greater cell yields under both agitated and unagitated culture conditions.

The liver infusion-proteose peptone-yeast extract-glucose-calf serum medium of Balamuth (5) originally was developed for the axenic cultivation of *N. gruberi;* hence, it is not surprising that it does not support good growth of *N. fowleri.* The two species are distinct organisms with different nutritional requirements. In contrast to *N. gruberi, N. fowleri* grows best in less enriched media (63, 110). Červa medium contains only casitone and horse serum (17) and Nelson medium has liver digest, glucose, and calf serum (133).

By using agitated cultures and Nelson medium, it is possible to obtain large quantities (3×10^9 amebae/liter) of *N. fowleri* (133). At 37°C, the mean generation time is 5.5 hr for exponentially growing cells. There is only slight utilization of glucose, and amino acids appear to serve as carbon and energy sources.

Physical factors shown to affect the growth of *N. fowleri* in liquid axenic cultures include pH, temperature, viscosity, and dissolved inorganic salts. The pH optimum for growth initiation in agitated cultures is 5.5 (133), and 6.5 in unagitated cultures (21, 133). The pH of Nelson culture medium increases about 2 U during 96 hr of growth (133). However, if the medium is adjusted to maintain a constant pH, there is no change in growth of *N. fowleri,* indicating that the pH increase of the culture medium does not limit ameba growth (83).

Optimal temperature for the growth of *N. fowleri* is dependent upon the composition and concentration of the culture medium, viability of the amebae in the inoculum, and size of the inoculum (20). It has been suggested that a threshold inoculum is necessary for growth initiation of *N. fowleri* in broth cultures (40). The suggestion was made to account for inconsistent ameba growth in cultures inoculated with less than 10^4 amebae/ml. However, the results of a study in which flasks are inoculated with 10^2–10^6 ameba/ml suggest that a threshold inoculum level does not exist (83). Rather, there was a maximum population density or carrying capacity for the medium. Cultures inoculated with 10^2 grew about 10 generations before entering stationary growth phase, whereas cultures inoculated with 10^6 amebae/ml grew only one generation.

A 0.5% concentration of methylcellulose in liquid medium does not inhibit the growth of *N. fowleri* (21). By comparison, *N. gruberi* growth is inhibited by concentrations of methylcellulose greater than 0.2%.

Serum appears to be an important component of the liquid media used for axenic cultivation of *N. fowleri.* Various kinds and concentrations of serum have been used. Several forms of calf serum and sera from other vertebrate species have been evaluated for their ability to support growth of *N. fowleri* in Nelson medium (64, 73). Of the 17 sera tested, calf serum supported the greatest cell yields (1.48×10^6 amebae/ml), whereas fetal calf serum produced the lowest cell yields (2.09×10^5 amebae/ml). Between these two and ranked in order of decreasing cell yield are pig, dialyzed calf, monkey, newborn calf, lamb, turtle, dog, chicken, mouse, rabbit, frog, horse, gamma globulin-free calf, fish, and human sera.

Hemin (1 μg/ml) has been shown to replace serum as a growth requirement for *N. gruberi* in axenic liquid cultures (6). However, the same is not true for cultures of *N. fowleri,* although hemin, in addition to the serum, will enhance growth (75). A semi-defined medium without serum and containing hemin has been described (37). A curious observation about the Richmond, Virginia, cases is that 14 of the 16 infections were acquired in two man-made lakes, which occur in the vicinity of the first iron smelter to be built in the United States. The water in some of the small streams in the area often appears reddish because of its iron content. Perhaps the presence of iron or other heavy metals in nature provides an environment favorable to the growth of *N. fowleri.*

Cell Differentiation

Naegleria fowleri is an ameboflagellate and, therefore, it is able to transform from ameba to flagellate and revert to ameba, and to encyst and under favorable conditions to excyst (Figure 1). Yet, apart from testing to verify the identity of *N. fowleri* isolates, there are no definitive physiological studies of differentiation in *N. fowleri.* Cell differentiation in *N. gruberi* has been reviewed by Fulton (58).

Trophozoites of *N. fowleri* are long and slender (8–15 μm) and have progressive flowing movement. The resting nucleus is spherical with a sharply defined nuclear membrane and large centrally located nucleolus. Nuclear division is promitotic, in which the nucleolus elongates and divides into two polar masses and the nuclear membrane remains intact (101).

Cysts are spherical, often clumped closely together, and 7–10 μm in diameter (13). Ultrastructure examination reveals an average of less than

2 mucoid-plugged pores per cyst and a relatively thin cyst wall, a feature that makes *N. fowleri* cysts susceptible to desiccation (109).

When trophozoites are suspended in distilled water or non-nutrient buffer (58) they transform into temporary flagellated forms. The flagellate has an elongate, pear-shaped body, usually two flagella of equal length, a nucleus in the narrower anterior region, and no cytostome (13, 101). The ultrastructure of *N. fowleri* flagellates is that of a typical eucaryotic protist (102). There is a distinct nuclear membrane and prominent nucleolus, numerous vacuoles and cytoplasmic inclusions, pleomorphic mitochondria, and some rough endoplasmic reticulum. Amebae of *N. fowleri* (LEE strain) became flagellates 150 to 180 min after transfer to non-nutrient buffer. Basal bodies, a rootlet, and flagella are formed quickly after an initial lag of 90 min.

Carter (13) observed that transformation occurred after 20 hr in distilled water, and that the flagellates persisted until 48 hr. Transformation was unreliable below 21°C and was better, with 50% flagellates, at 37°C.

We have noted (D. T. John, N. C. Mobley, unpublished data) that several of our axenically cultured *N. fowleri* isolates no longer differentiate into flagellates when amebae are placed in non-nutrient buffer (Table 1). Appar-

Table 1 Differentiation of *Naegleria fowleri* amebae to flagellates[a]

Strain of ameba	Isolation		Ref.	Flagellates (%)	Maximum transformation[b]
	Location	Date			
Lovell	Florida	1974	40	65	5
KUL	Belgium	1973	129	53	4
LEE(M-11)[c]	Virginia	1968	51	52	4
LEE(ATCC-30894)	Virginia	1968	51	31	4
TY	Virginia	1969	51	22	5
6088	California	1978	15	20	5
WM	Virginia	1969	51	17	4
HB-5	Texas	1977	None	14	5
CJ	Virginia	1967	51	12	5
0 359	Belgium	1970	68	2	4
GJ	Florida	1972	139	<1	4
HB-4	North Carolina	1977	None	0	
HB-4(M-8)[c]	North Carolina	1977	None	0	
NF66	Australia	1966	13	0	
NF69	Australia	1969	13	0	

[a]Amebae were grown at 37°C in Nelson medium with 2% calf serum. At 72 hr, medium was removed, amebae were rinsed three times with Page saline, covered with cold (4°C) Page saline, and vortexed. Amebae (~10^6 cells/ml) were placed in 125-ml Erlenmeyer flasks and shaken at 37°C in a gyrotory shaker (New Brunswick) at 100 rpm. Flasks were examined hourly for 8 hr by placing samples in D'Antoni's iodine and counting the percentage of flagellates. Cultures not producing flagellates were held and examined periodically to 36 hr.

[b]Time in hours after the initiation of differentiation (time zero).

[c]M-11 and M-8 denote amebae that had been serially mouse-passaged 11 and 8 times, respectively.

ently, some axenic populations of *N. gruberi* amebae fail to transform (100) and must be maintained with bacteria to do so. Fulton (58), however, states that axenically grown *N. gruberi* differentiate to flagellates but do so more slowly than bacteria-grown cells. Whatever the reason, it would be instructive to examine transformation in our non-differentiating *N. fowleri* isolates after growth with bacteria.

Macromolecular Composition

Changes in cell composition of *N. fowleri* are related to culture age (135). For agitated axenic cultures, average cell dry mass remained constant during log growth at 150 pg/ameba, but decreased 30% during stationary growth at 96 hr. During log growth 80–85% of the cell dry mass was protein (120 pg/ameba). Cell dry mass and protein of *N. fowleri* are about 70% of values reported for *N. gruberi* (134).

During log and stationary growth phases, carbohydrate content averaged 15 pg/ameba, and RNA was about 18 pg/ameba (135). RNA content for *N. gruberi* was about 8 pg/ameba (134). Perhaps the over twofold greater RNA value for the smaller pathogenic *N. fowleri* reflects different biosynthetic capabilities by maintenance of a larger ribosome complement.

Total DNA content was 0.2 pg/ameba during log growth; it doubled during transition from log phase to stationary phase and then gradually decreased to nearly initial levels. The peak in DNA content corresponded to an increase in the average number of nuclei per ameba; nuclear number then decreased as cells entered stationary growth phase (135).

Cell Membrane and Agglutination

Concanavalin A (Con A) agglutinates *N. gruberi* but does not agglutinate *N. fowleri* (76, 120). Agglutination is time and temperature dependent and Con A concentration and ameba concentration dependent over certain ranges. At least 10^6 amebae/ml are needed for maximum agglutination, and Con A concentrations higher than 100 μg/ml do not appreciably increase agglutination. These results indicate that only *N. gruberi* has appreciable quantities of N-acetyl glucosamine and mannose-like residues manifested on the cell surface. *Naegleria lovaniensis,* which is nonpathogenic for mice, is also agglutinated by Con A (118).

Respiratory Metabolism

As an opportunistic parasite, *N. fowleri* lives in an oxygen-rich environment (the brain) and so one would expect it to have an aerobic metabolism. The synonym *N. aerobia* (114) recognized the aerobic nature of the organism, in contrast to the anaerobic nature of strictly parasitic amebae. Unlike *Entamoeba histolytica,* an anaerobic parasitic ameba that lacks mitochondria (59), *N. fowleri* lives in aerobic aqueous environments and has many mitochondria (90, 111).

Whole cell respiration rates were measured polarographically throughout the growth cycle of *N. fowleri* (136). Under agitated culture conditions, amebae consumed 30 ng of O/min/mg of cell protein during log growth. Under similar conditions *N. gruberi* amebae consumed 80 ng of O/min/mg of cell protein (137). The lower oxygen consumption, and most likely oxygen requirement, by *N. fowleri* probably explains the presence of the pathogen in heated waters where dissolved oxygen concentrations are substantially reduced.

Respiratory rate gradually declined during stationary growth phase. The reduction in respiratory rate may involve respiratory control since further increases in respiratory rate did not occur in spite of fresh oxygen supplies (136).

The respiratory process of isolated *N. fowleri* mitochondria is similar to classical mammalian cell mitochondria. Oxidation was "coupled" to phosphorylation (ATP formation) as shown by the two- to threefold increase in respiration upon addition of phosphate acceptor or uncoupling agent. Difference spectra of oxidized and dithionite-reduced mitochondria showed distinct absorption bands of flavins, *c*-type, *b*-type, and *a*-type cytochromes (136).

VIRULENCE AND IMMUNITY

Mechanisms of Pathogenesis

The determinants of virulence, invasiveness, and pathogenicity of *Naegleria fowleri* are largely unknown. Electron microscope studies of experimentally induced PAM in mice have demonstrated phagocytic activity by amebae (90, 130).

N. fowleri causes a destructive cytopathic effect in cultured mammalian cells, and it is generally supposed that this cytopathic activity is associated with trophozoite pathogenicity. Opinion, though, is divided on the mechanism of cell damage. Chang (25) reported that supernatant medium from *N. fowleri*-infected cultures of mammalian cells induced cell degeneration when fresh cultures were inoculated with it, which suggests that amebae secrete cytolytic or cytotoxic enzyme-like substances. He concluded (27) that the pathogenicity of cytopathic effect of *N. fowleri* can be attributed to a phospholipolytic enzyme released by the amebae during active growth.

Wong et al (142) observed differences in cytolytic enzyme synthesis between highly virulent and low-virulent strains of *N. fowleri;* for example, highly virulent strains produced a magnitude more catalase than low-virulent strains. Cursons et al (33) found that *N. fowleri* produced a greater amount of phospholipase A than did nonpathogenic *N. gruberi.* Hysmith et al (67) have measured sphingomyelinase levels in *Naegleria* culture media and determined that sphingomyelinase activity was approximately 100-fold greater in culture medium of virulent *N. fowleri* than it was in culture media of low-virulent *N. fowleri* or *N. gruberi.* Increased levels of sphingomyelinase activity in culture medium may be directly associated with the demyelination in brain and spinal cord that has been described in PAM (27, 51).

Brown (7, 8) suggests that the cytopathogenicity of *N. fowleri* in secondary mouse-embryo cells does not involve ameba-associated cytotoxic activity, but depends on normal phagocytic function. Amebae that were immobilized and agglutinated by specific antiserum exhibited no cytopathic activity, although they remained alive and were in constant contact with the host cells. Cytochalasin B, shown to inhibit phagocytosis in amebae, inhibited the cytopathogenicity of *N. fowleri* when it was added to cell culture medium. Brown concludes that *N. fowleri* amebae attack and destroy cultured mouse-embryo cells by a phagocytosis-like mechanism alone, without the aid of ameba-associated cytotoxic or cytolytic agents.

We have examined the susceptibility of various mammalian cell lines to the cytopathic activity of a highly virulent strain (HB-4) of *N. fowleri.* With a multiplicity of infection of 1, complete destruction of the monolayers occurred from 24 hr (BHK-21, Vero, WI-38 cells) to 72 hr (L929, Ktk$^-$, Neuro-2a cells) after inoculation of amebae for cultures incubated at 37°C (Table 2). Cytopathic effect occurs to a lesser extent and later for cultures inoculated with a low-virulent strain of *N. fowleri.*

Brown (9) also has described the cytopathogenicity of nonpathogenic *N. gruberi* in several mammalian cell lines, cultured at 30 rather than 37°C. Again, cytopathic effect was caused by phagocytosis, and he suggests that temperature sensitivity has been a significant factor in the reported differences in cytopathogenicity between *N. fowleri* and *N. gruberi* amebae.

Naegleria-induced cytopathic effect has also been attributed to the transmission of infectious cytopathogenic material from ameba to susceptible avian and mammalian cultured cells (52). The infectious material, present in both *N. fowleri* and *N. gruberi,* appears to be a protein with an estimated molecular weight of 50,000. It is capable of sustaining itself in cell culture and in serial passage through multiple dilutions.

Several factors have been shown to affect the virulence of *N. fowleri* for mice; they include incubation temperature, growth phase, and strain of ameba (61). Amebae cultured at 30 and 37°C were more virulent than amebae growth at 23 and 44°C. Mortality was greater for mice inoculated with amebae harvested at late logarithmic and early stationary growth

Table 2 *Naegleria fowleri* (HB–4 strain)-induced cytopathic effect (CPE) in cultured mammalian cells[a]

Cell line[b]	Animal species and tissue	Time (hr) for maximum CPE[c]		
		24	48	72
BHK-21	Baby golden hamster kidney	IV		
Vero	African green monkey kidney	IV		
WI-38	Human lung	IV		
CF-3	Human foreskin		IV	
CHO-K1	Chinese hamster ovary		IV	
HeLa S3	Human epithelioid carcinoma, cervix		IV	
HEp-2	Human epidermoid carcinoma, larynx		IV	
Mv 1 Lu	Mink lung		IV	
NB41A3	Mouse neuroblastoma		IV	
L929	Mouse connective tissue			IV
Ltk⁻	Mouse connective tissue			IV
Neuro-2a	Mouse neuroblastoma			IV

[a]HB–4 is a highly virulent strain of *N. fowleri* isolated by R. B. Finley in July 1977 from patient spinal fluid, before death, at Winston-Salem, NC.

[b]Cells were grown and maintained in MEM with fetal calf serum. All cell lines were obtained from the American Type Culture Collection (Rockville, MD) except CF-3 (Noble Fdn., Ardmore, OK) and Ltk⁻ (S. Kit, Baylor, Houston, TX), a thymidine kinase deficient L929 derived cell line.

[c]Multiplicity of infection was 1; four 25-cm^2 tissue culture flasks per cell line; 37°C incubation temperature; CPE of IV represents complete breakdown of monolayer so that culture flask contains only amebae and cellular debris.

phases than it was for amebae harvested at early logarithmic and late stationary growth phases.

The incubation of *N. fowleri* amebae with mouse anti-*N. fowleri* serum has been shown to have a rather dramatic effect upon ameba virulence and subsequent mouse mortality after intravenous inoculation (62). The mortality rate was 15% for mice inoculated with immune serum-treated amebae compared with 95% mortality rate for mice inoculated with amebae incubated in normal (nonimmune) mouse serum. Viability testing, using trypan blue exclusion, immediately after incubation of amebae with serum, showed that the immune mouse serum caused greater damage to the amebae, with ~5% nonviable cells as compared to less than 1% nonviable amebae after incubation with normal mouse serum.

Virulence varies greatly among the different human isolates of *N. fowleri.* Table 3 gives the mortality for mice inoculated intranasally with 5×10^3 amebae/mouse of 13 different strains (isolates) of *N. fowleri.* Cumulative percent mortality at 28 days ranged from 0–100% and mean time to death

Table 3 Mortality for mice inoculated intranasally with 5 × 10^3 amebae/mouse of 13 strains of *Naegleria fowleri*[a]

Strain of ameba	Initial isolation	Date of isolation	Ref.	Cumulative percent dead[b]	Mean time to death (hr)
CJ	Virginia	1967	51	0[c]	—
KUL	Belgium	1973	129	0[c]	—
TY	Virginia	1969	51	10	14.0
0 359	Belgium	1970	68	15	13.3
LEE(ATCC-30894)	Virginia	1968	51	20	17.8
NF66	Australia	1966	13	75	16.8
Lovell	Florida	1974	40	85	15.5
HB-4	North Carolina	1977	None	85	7.2
WM	Virginia	1969	51	90	11.1
NF69	Australia	1969	13	90	10.6
GJ	Florida	1972	139	90	9.4
6088	California	1978	15	90	9.6
HB-5	Texas	1977	None	95	8.1
LEE(M-10)[d]	Virginia	1968	51	100	14.0

[a]There were 20 male DUB/ICR mice per group (13–18 g).
[b]Recorded to 28 days after inoculation.
[c]Mortality occurred with increased dosage.
[d]The M–10 denotes LEE strain amebae that had been serially mouse-passaged 10 times.

for the 13 strains was from 7.2 to 17.8 days. Kadlec (78) describes the great variation that occurred in virulence for 33 strains of *N. fowleri* isolated from the water of an indoor swimming pool. *Naegleria fowleri* amebae have been shown to differ in their susceptibility to trimethoprim, and the difference appears to be related to ameba virulence (22). The growth of virulent strains was unaffected by trimethoprim (400 μg/ml of medium), whereas avirulent isolates were completely inhibited by the drug at concentrations of 4 μg/ml. Trimethoprim susceptibility may be a useful aid in the selection of virulent environmental isolates of *N. fowleri.*

Ameba virulence appears to be related to the length of time a strain has been maintained in axenic culture (61). Axenic cultivation gradually decreases the virulence of *N. fowleri.* However, virulence can be enhanced and perhaps restored to original levels by passage of amebae in mammalian cell culture and by serial mouse passage (43, 142).

Resistance and Susceptibility

The factors responsible for innate resistance or susceptibility to naeglerial infection are undefined. Relatively few human infections have occurred even though large numbers of individuals have been exposed to similar environmental conditions. Most cases of PAM have occurred in previously healthy children or young adults.

When examining naeglerial infection in mice, it is readily apparent that the age of the test animals dramatically affects the outcome of *N. fowleri* exposure. Young mice are uniformly susceptible, but as they become older they also become increasingly more resistant to infection. Female mice have been shown to be more resistant to naeglerial infection than male mice of the same age and strain (60). Interestingly, there have been more male than female victims of PAM. This generally has been attributed to the vigorous swimming and diving and adventuresomeness of males and, thus, the probability of greater exposure to *N. fowleri* rather than to the possibility of greater susceptibility. Perhaps males, indeed, are more susceptible to infection, and specific hormones (or hormone levels) are involved. To date this area of research has not been investigated.

Different mouse strains vary greatly in their susceptibility to infection with *N. fowleri* (43, 60). In both these studies, C_{57}Bl mice were the most resistant of the strains tested. One especially susceptible mouse strain (A/-HeCr) is deficient in the C'5 component of complement (60). Obviously, it is premature to suggest that complement is involved in host resistance. Nonetheless, *N. fowleri* has been shown to be an efficient activator of serum complement by either the classical or alternative pathways (65, 106).

Serum obtained before death from a victim of PAM was assayed for immunoglobulin (Ig) levels (38). Although IgG and IgM levels were within normal limits, the IgA value was low. However, a second report from England (11) revealed normal levels of IgA for serum taken from a patient on day 3 of a fatal infection. In both patients, serum was measured, not secretory IgA levels. Because serum IgA levels may not be an accurate reflection of secretory IgA concentrations, both patients may have had deficient levels of secretory IgA. Serum IgA levels are greatly increased in mice given three intranasal inoculations of living *N. fowleri* (62, 72).

Immunization and Protection

Mice have been immunized with living, formalinized, and freeze-thawed *N. fowleri* via subcutaneous, intraperitoneal, intravenous, and intranasal routes of inoculation (74, 128). Protection of mice against lethal intranasal or intravenous challenge ranged from 6 to 55%. Intravenous immunization with *N. gruberi* afforded 65% protection (74), and three intranasal immunizing doses with *N. gruberi* produced 88% protection against intranasal challenge with *N. fowleri* (71). Nonpathogenic *N. gruberi* appears to be a better immunogen than *N. fowleri.* Perhaps man's unwitting exposure to *N. gruberi,* abundant in freshwater environments, affords protection against what normally would be a lethal exposure to *N. fowleri.*

Modest protection has been achieved by intraperitoneal immunization of mice with *N. fowleri* culture supernatant (124). Controls, receiving equal volumes of fresh culture filtrate, did not exhibit protection. Presumably, protection was afforded by antigenic material derived from the amebae during cultivation. To date, all immunization attempts have produced only incomplete protection; solid immunity has not yet been achieved.

The mechanisms for protective immunity only recently have been given consideration. Protection against naeglerial infection can be transferred in mice by immune serum but not by sensitized spleen cells (125). Antibody to *N. fowleri* has been detected in surveys of normal human sera. Using a radioimmunoassay, the response against intracellular antigens was higher than the response against cell surface antigen (123). The indirect fluorescent-antibody test demonstrated titers ranging from 1:5 to 1:20 for 93 serum samples tested (35).

The role of cell-mediated immunity has been partly examined. A delayed hypersensitivity reaction has been described for guinea pigs sensitized by the soluble antigens of freeze-thawed *N. fowleri* (46) and by freeze-thawed amebae plus complete Freund adjuvant (142). Macrophage inhibition by the lymphokine macrophage inhibition factor has been described for *N. fowleri* (36).

Naegleria fowleri amebae have been shown by immunofluorescence to cap and remove or internalize surface-bound antibody (54). The ability of *N. fowleri* to remove antibody from the cell surface may enable the amebae to counter the host's immune defenses.

Neutrophils from *N. fowleri*-immunized mice are capable of killing amebae (55). One method of killing is for a group of neutrophils to surround an ameba and destroy it, presumably by contact and release of enzymes onto the ameba cell membrane. However, a novel phagocytic process has been described in which neutrophils pinch off portions of an ameba. Although unable to phagocytose an entire ameba, several neutrophils are able to rupture an ameba by pinching off and engulfing portions of it (55).

CONCLUDING REMARKS

Primary amebic meningoencephalitis, a fatal human disease caused by *Naegleria fowleri,* is known only as a somewhat rare or exotic disease, in short, a medical curiosity. However, circumstantial evidence suggests that human meddling with water resources may exaggerate the problem. If pollution and/or treatment of water increases the population of *N. fowleri* in public waters, then man and animals dependent upon impounded waters may be at serious and growing risk.

The determinative factors in virulence and host resistance to naeglerial infections are unclear in both human and experimentally induced PAM. Relatively few human infections have occurred, even though large numbers of individuals have been exposed to similar environmental conditions. And most of these infections have occurred in previously healthy children or young adults. It is not known whether the higher incidence of disease among young males reflects behavioral attributes (adventurous, vigorous swimming and diving) or physiological factors (hormone levels, immunological competence).

The virulence factors that contribute to the pathogenesis of *N. fowleri* are largely undefined. In vitro studies have implicated toxins and cytopathic enzymes, infectious cytopathogenic material, and phagocytosis, a natural function of all amebae. Perhaps it is not unreasonable to expect all of these factors to be involved in pathogenesis within the infected host.

Two recent cases of fatal meningoencephalitis resulting from free-living amebae have been described in which the organisms could not be identified positively as either *Naegleria* or *Acanthamoeba* (50, 94). The genus *Vahlkampfia* has been suggested as an alternative. The significance of these two cases is that given the right conditions, serious disease in man (and possibly animals) may be produced by free-living amebae other than those we have come to regard as pathogenic.

Acknowledgments

I am pleased to dedicate this review to Dr. Cecil B. Hamann of Asbury College who first introduced me to the subject of medical parasitology. My research has been supported by grant 771019 from A. H. Robins Co., Richmond, Virginia, and by Public Health Service grant AI 18788 from the National Institute of Allergy and Infectious Diseases.

Literature Cited

1. Adams, A. C., John, D. T., Bradley, S. G. 1976. *Infect. Immun.* 13:1387–91
2. Anderson, K., Jamieson, A. 1972. *Lancet* 1:902–3
3. Anderson, K., Jamieson, A. 1972. *Lancet* 2:379
4. Ayers, K. M., Billups, L. H., Garner, F. M. 1972. *Vet. Pathol.* 9:221–26
5. Balamuth, W. 1964. *J. Protozool.* 11: 19–20
6. Band, R. N., Balamuth, W. 1974. *Appl. Microbiol.* 28:64–65
7. Brown, T. 1979. *J. Med. Microbiol.* 12:355–62
8. Brown, T. 1979. *J. Med. Microbiol.* 12:363–71
9. Brown, T. 1980. *Cytopathogenicity of pathogenic and non-pathogenic Naegleria spp. in mammalian cell cultures.* Presented at Int. Conf. Biol. Pathogen. Small Free-Living Amoebae, 2nd, Univ. Fla., Gainesville, p. 88 (Abstr.)
10. Butt, C. G. 1966. *N. Engl. J. Med.* 274:1473–76
11. Cain, A. R. R., Mann, P. G., Warhurst, D. C. 1979. *Lancet* 1:441
12. Callicott, J. H. Jr. 1968. *Am. J. Clin. Pathol.* 49:84–91
13. Carter, R. F. 1970. *J. Pathol.* 100: 217–44
14. Carter, R. F. 1972. *Trans. Roy. Soc. Trop. Med. Hyg.* 66:193–213
15. Center for Disease Control. 1978. *Morbid. Mortal. Weekly Rep.* 27:343–44
16. Center for Disease Control. 1980. *Morbid. Mortal. Weekly Rep.* 29:405–7
17. Červa, L. 1969. *Science* 163:576
18. Červa, L. 1971. *Folia Parasitol. Prague* 18:171–76
19. Červa, L. 1971. *Hydrobiology* 38:141–61

20. Červa, L. 1977. *Folia Parasitol. Prague* 24:221–28
21. Červa, L. 1978. *Folia Parasitol. Prague* 25:1–8
22. Červa, L. 1980. *Science* 209:1541
23. Červa, L., Novák, K., Culbertson, C. G. 1968. *Am. J. Epidemiol.* 88:436–44
24. Chang, S. L. 1971. *Curr. Top. Comp. Pathobiol.* 1:201–54
25. Chang, S. L. 1974. *CRC Crit. Rev. Microbiol.* 3:135–59
26. Chang, S. L. 1978. *Appl. Environ. Microbiol.* 35:368–75
27. Chang, S. L. 1979. *Folia Parasitol. Prague* 26:195–200
28. Chang, S. L., Healy, G. R., McCabe, L., Shumaker, J. B., Schultz, M. G. 1975. *Hlth. Lab. Sci.* 12:1–7
29. Culbertson, C. G., Ensminger, P. W., Overton, W. M. 1968. *J. Protozool.* 15:353–63
30. Culbertson, C. G., Ensminger, P. W., Overton, W. M. 1972. *Am. J. Clin. Pathol.* 57:375–86
31. Cursons, R. T. M., Brown, T. J. 1975. *NZ Med. J.* 82:123–25
32. Cursons, R. T. M., Brown, T. J., Burns, B. J., Taylor, D. E. M. 1976. *NZ Med. J.* 84:479–81
33. Cursons, R. T. M., Brown, T. J., Keys, E. A. 1978. *J. Parasitol.* 64:744–45
34. Cursons, R. T. M., Brown, T. J., Keys, E. A., Gordon, E. H., Leng, R. H., Havill, J. H., Hyne, B. E. B. 1979. *NZ Med. J.* 90:330–31
35. Cursons, R. T. M., Brown, T. J., Keys, E. A., Moriarty, K. M., Till, D. 1980. *Infect. Immun.* 29:401–7
36. Cursons, R. T. M., Brown, T. J., Keys, E. A., Moriarty, K. M., Till, D. 1980. *Infect. Immun.* 29:408–10
37. Cursons, R. T. M., Donald, J. J., Brown, T. J., Keys, E. A. 1979. *J. Parasitol.* 65:189–91
38. Cursons, R. T. M., Keys, E. A., Brown, T. J., Learmonth, J., Campbell, C., Metcalf, P. 1979. *Lancet* 1:223–24
39. Darby, C. P., Conradi, S. E., Holbrook, T. W., Chatellier, C. 1979. *Am. J. Dis. Child.* 133:1025–27
40. DeJonckheere, J. 1977. *Appl. Environ. Microbiol.* 33:751–57
41. DeJonckheere, J. F. 1978. *Protistology* 14:475–81
42. DeJonckheere, J. F. 1979. *Ann. Microbiol. B. Inst. Pasteur* 130:205–12
43. DeJonckheere, J. F. 1979. *Pathol. Biol.* 27:453–58
44. DeJonckheere, J. F., Van De Voorde, H. 1976. *Appl. Environ. Microbiol.* 31: 294–97
45. DeJonckheere, J. F., Van De Voorde, H. 1977. *Am. J. Trop. Med. Hyg.* 26:10–15
46. Diffley, P., Skeels, M. R., Songandares-Bernal, F. 1976. *Z. Parasitenk.* 49:133–37
47. Dive, D. G., Leclerc, H., DeJonckheere, J., Delattre, J. M. 1981. *Ann. Microbiol. A Inst. Pasteur* 132:97–105
48. Dive, D., Leclerc, H., Picard, J. P., Telliez, E., Vangrevelinghe, R. 1978. *Ann. Microbiol. B Inst. Pasteur* 129:225–44
49. Duma, R. J. 1978. In *Handbook of Clinical Neurology, Infections of the Nervous System, Part III,* ed. H. L. Klawans, 35:25–65. Amsterdam: North-Holland
50. Duma, R. J., Helwig, W. B., Martinez, A. J. 1978. *Ann. Intern. Med.* 88: 468–73
51. Duma, R. J., Rosenblum, W. I., McGehee, R. F., Jones, M. M., Nelson, E. C. 1971. *Ann. Intern. Med.* 74: 923–31
52. Dunnebacke, T. H., Schuster, F. L. 1977. *Am. J. Trop. Med. Hyg.* 26:412–21
53. Dwivedi, J. N., Singh, C. M. 1965. *Ind. J. Microbiol.* 3:31–34
54. Ferrante, A., Thong, Y. H. 1979. *Int. J. Parasitol.* 9:599–601
55. Ferrante, A., Thong, Y. H. 1980. *Immunol. Lett.* 2:37–41
56. Fowler, M., Carter, R. F. 1965. *Br. Med. J.* 2:740–42
57. Frank, W., Bosch, I. 1972. *Z. Parasitenk.* 40:139–50
58. Fulton, C. 1977. *Ann. Rev. Microbiol.* 31:597–629
59. Griffin, J. L. 1971. *Arch. Pathol.* 91:271–80
60. Haggerty, R. M., John, D. T. 1978. *Infect. Immun.* 20:73–77
61. Haggerty, R. M., John, D. T. 1980. *Proc. Helminthol. Soc. Wash.* 47:129–34
62. Haggerty, R. M., John, D. T. 1982. *J. Protozool.* 29:117–22
63. Haight, J. B., John, D. T. 1980. *Folia Parasitol. Prague* 27:207–12
64. Haight, J. B., John, D. T. 1982. *Proc. Helminthol. Soc. Wash.* 49:127–34
65. Holbrook, T. W., Boackle, R. J., Parker, B. W., Vesely, J. 1980. *Infect. Immun.* 30:58–61
66. Holbrook, T. W., Parker, B. W. 1979. *Am. J. Trop. Med. Hyg.* 28:984–87
67. Hysmith, R. M., Franson, R. C., John, D. T. 1981. *Abstr. 56th Ann. Meet. Am. Soc. Parasitol.,* p. 27
68. Jadin, J. B., Hermanne, J., Robyn, G., Willaert, E., Van Maercke, Y., Stevens,

W. 1971. *Ann. Soc. Belge Med. Trop.* 51:255–66
69. Jamieson, A., Anderson, K. 1973. *Pathology* 5:55–58
70. Janitschke, K., Werner, H., Muller, G. 1980. *Zbl. Bakt. I. Abt. Orig. B* 170:108–22
71. John, D. T., Bush, L. E. 1980. *Protection of mice against Naegleria fowleri following intranasal immunization.* Presented at Int. Conf. Biol. Pathogen. Small Free-Living Amoebae, 2nd, Univ. Fla., Gainesville, p. 54 (Abstr.)
72. John, D. T., Haggerty, R. M. 1978. *J. Protozool.* 25:18A
73. John, D. T., Haight, J. B. 1978. *Abstr. 53rd Ann. Meet. Am. Soc. Parasitol.,* p. 94
74. John, D. T., Weik, R. R., Adams, A. C. 1977. *Infect. Immun.* 16:817–20
75. Josephson, S. L., John, D. T. 1976. *J. Protozool.* 23:23A
76. Josephson, S. L., Weik, R. R., John, D. T. 1977. *Am. J. Trop. Med. Hyg.* 26:856–58
77. Kadlec, V. 1978. *J. Protozool.* 25: 235–37
78. Kadlec, V., 1981. *Folia Parasitol. Prague* 28:97–103
79. Kadlec, V., Skvárová, J., Červa, L., Nebázniva, D. 1980. *Folia Parasitol. Prague* 27:11–17
80. Kasprzak, W., Mazur, T. 1974. *Proc. III Int. Congr. Parasitol.* 1:188
81. Kingston, N., Taylor, P. C. 1976. *Proc. Helminthol. Soc. Wash.* 43:227–29
82. Kobayashi, G. S., Medoff, G. 1977. *Ann. Rev. Microbiol.* 31:291–308
83. Koch-Weik, S. M., Weik, R. R., John, D. T. 1980. *Proc. Helminthol. Soc. Wash.* 47:270–72
84. Lawande, R. V., Abraham, S. N., John, I., Egler, L. J. 1979. *Am. J. Clin. Pathol.* 71:201–3
85. Lawande, R. V., Macfarlane, J. T., Weir, W. R. C., Awunor-Renner, C. 1980. *Am. J. Trop. Med. Hyg.* 29:21–25
86. Lawande, R. V., Ogunkanmi, A. E., Egler, L. J. 1979. *Ann. Trop. Med. Parasitol.* 73:51–56
87. Lockey, M. W. 1978. *Laryngoscope* 88:484–503
88. Lyons, T. B., Kapur, R. 1977. *Appl. Environ. Microbiol.* 33:551–55
89. Markowitz, S. M., Martinez, A. J., Duma, R. J., Shiel, F. O. M. 1974. *Am. J. Clin. Pathol.* 62:619–28
90. Martinez, A. J., Duma, R. J., Nelson, E. C., Moretta, F. L. 1973. *Lab. Invest.* 29:121–33
91. Martinez, A. J., García, C. A., Halks-Miller, M., Arce-Vela, R. 1980. *Acta Neuropathol. Berlin* 51:85–91
92. Martinez, A. J., Nelson, E. C., Duma, R. J. 1973. *Am. J. Pathol.* 73:545–48
93. Martinez, A. J., dos Santos, J. G., Nelson, E. C., Stamm, W. P., Willaert, E. 1977. In *Pathology Annual, Part 2,* ed. S. C. Sommers, P. P. Rosen, 12:225–50. New York: Appleton-Century-Crofts
94. Martinez, A. J., Sotelo-Avila, C., Alcalá, H., Willaert, E. 1980. *Acta Neuropathol. Berlin* 49:7–12
95. Mascaró, C., Fluviá, C., Osuna, A., Guevara, D. 1981. *J. Parasitol.* 67:599
96. McConnell, E. E., Garner, F. M., Kirk, J. H. 1968. *Vet. Pathol.* 5:1–6
97. Medley, S. 1980. *Med. J. Aus.* 2:635
98. Nelson, E. C. 1972. *Va. J. Sci.* 23:145
99. Nelson, E. C., Jones, M. M. 1970. *J. Parasitol.* 56:248
100. Outka, D. E. 1965. *J. Protozool.* 12:85–93
101. Page, F. C. 1976. *An Illustrated Key to Freshwater and Soil Amoebae.* Culture Centre of Algae and Protozoa, Cambridge. Freshwater Biological Assoc., Sci. Publ. No. 34, 155 pp.
102. Patterson, M., Woodworth, T. W., Marciano-Cabral, F., Bradley, S. G. 1981. *J. Bacteriol.* 147:217–26
103. Payne, W. L., Jackson, G. J. 1978. *J. Parasitol.* 64:44
104. Phillips, B. P. 1974. *Am. J. Trop. Med. Hyg.* 23:850–55
105. Pringle, H. L., Bradley, S. G., Harris, L. S. 1979. *Antimicrob. Agents Chemother.* 16:674–79
106. Rowan-Kelly, B., Ferrante, A., Thong, Y. H. 1980. *Trans. Roy. Soc. Trop. Med. Hyg.* 74:333–36
107. Rowbotham, T. J. 1980. *J. Clin. Pathol.* 33:1179–83
108. dos Santos, J. G. 1970. *Am. J. Clin. Pathol.* 54:737–42
109. Schuster, F. L. 1975. *J. Protozool.* 22: 352–59
110. Schuster, F. L. 1979. *Biochemistry and Physiology of Protozoa,* Vol. 1, pp. 215–85. New York: Academic. 2nd ed.
111. Schuster, F. L., Rechthand, E. 1975. *Antimicrob. Agents Chemother.* 8:591–605
112. Shumaker, J. B., Healy, G. R., English, D., Schulta, M., Page, F. C. 1971. *Lancet* 2:602–3
113. Singh, B. N. 1952. *Philos. Trans. Roy. Soc. London B* 236:405–61
114. Singh, B. N., Das, S. R. 1970. *Philos. Trans. Roy. Soc. London B* 259:435–76
115. Singh, B. N., Das, S. R. 1972. *Curr. Sci.* 41:277–81

116. Singh, B. N., Das, S. R. 1972. *Curr. Sci.* 41:625–28
117. Soh, C. T., Im, K. I., Chang, B. P., Hwang, H. K. 1977. *Abstr. V Int. Congr. Protozool.*, p. 412
118. Stevens, A. R., DeJonckheere, J., Willaert, E. 1980. *Int. J. Parasitol.* 10:51–64
119. Stevens, A. R., Shulman, S. T., Lansen, T. A., Cichon, M. J., Willaert, E. 1981. *J. Infect. Dis.* 143:193–99
120. Stevens, A. R., Stein, S. 1977. *J. Parasitol.* 63:151–52
121. Stevens, A. R., Tyndall, R. L., Coutant, C. C., Willaert, E. 1977. *Appl. Environ. Microbiol.* 34:701–5
122. Taylor, P. W. 1977. *J. Parasitol.* 63:232–37
123. Tew, J. G., Burmeister, J., Greene, E. J., Pflaumer, S. K., Goldstein, J. 1977. *J. Immunol. Methods* 14:231–41
124. Thong, Y. H., Ferrante, A., Rowan-Kelly, B., O'Keefe, D. E. 1979. *Trans. Roy. Soc. Trop. Med. Hyg.* 73:684–85
125. Thong, Y. H., Ferrante, A., Shepherd, C., Rowan-Kelly, B. 1978. *Trans. Roy. Soc. Trop. Med. Hyg.* 72:650–52
126. Thong, Y. H., Rowan-Kelly, B., Ferrante, A. 1979. *Scand. J. Infect. Dis.* 11:151–53
127. Thong, Y. H., Rowan-Kelly, B., Ferrante, A. 1979. *Trans. Roy. Soc. Trop. Med. Hyg.* 73:336–37
128. Thong, Y. H., Shepherd, C., Ferrante, A., Rowan-Kelly, B. 1978. *Am. J. Trop. Med. Hyg.* 27:238–40
129. Van Den Driessche, E., Vandepitte, J., Van Dijck, P. J., DeJonckheere, J., Van De Voorde, H. 1973. *Lancet* 2:971
130. Visvesvara, G. S., Callaway, C. S. 1974. *J. Protozool.* 21:239–50
131. Voelker, F. A., Anver, M. R., McKee, A. E., Casey, H. W., Brenniman, G. R. 1977. *Vet. Pathol.* 14:247–55
132. Warhurst, D. C., Carman, J. A., Mann, P. G. 1980. *Trans. Roy. Soc. Trop. Med. Hyg.* 74:832
133. Weik, R. R., John, D. T. 1977. *J. Parasitol.* 63:868–71
134. Weik, R. R., John, D. T. 1977. *J. Protozool.* 24:196–200
135. Weik, R. R., John, D. T. 1978. *J. Parasitol.* 64:746–47
136. Weik, R. R., John, D. T. 1979. *J. Parasitol.* 65:700–8
137. Weik, R. R., John, D. T. 1979. *J. Protozool.* 26:311–18
138. Wellings, F. M. 1977. *J. Fla. Med. Assoc.* 64:327–28
139. Wellings, F. M., Amuso, P. T., Chang, S. L., Lewis, A. L. 1977. *Appl. Environ. Microbiol.* 34:661–67
140. Willaert, E., LeRay, D. 1973. *Protistology* 9:417–26
141. Wong, M. M., Karr, S. L. Jr., Balamuth, W. B. 1975. *J. Parasitol.* 61:199–208
142. Wong, M. M., Karr, S. L. Jr., Chow, C. K. 1977. *J. Parasitol.* 63:872–78
143. Yamauchi, T., Jimenez, J. F., McKee, T. W., Euler, A. R., White, P. C. 1979. *J. Ark. Med. Soc.* (Sept.) 5 pp.
144. Young, M. D., Willaert, E., Neal, F. C., Simpson, C. F., Stevens, A. R. 1980. *Am. J. Trop. Med. Hyg.* 29:476–77.

Ann. Rev. Microbiol. 1982. 36:125–44

COLICINS AND OTHER BACTERIOCINS WITH ESTABLISHED MODES OF ACTION

Jordan Konisky

Department of Microbiology, University of Illinois, Urbana, Illinois 61801

CONTENTS

INTRODUCTION

The literature abounds with examples of antagonism between particular bacterial strains. In many instances the mediating agent is a bacteriocin. In the most general sense, these substances are defined by two criteria—they are proteins or complexes of proteins, and they are not active against the

0066-4227/82/1001-0125$02.00

producer bacterium. Such a definition is based entirely on function and avoids more restrictive considerations, such as whether (*a*) the toxin is encoded by plasmid or chromosomal genes, (*b*) the host range is narrow or broad, or (*c*) the bacteriocin is a simple protein (such as a colicin) or a complex structure composed of several distinct subunit species (such as a phage tail-like pyocin).

The bacteriocins have been the subject of several reviews (53, 92, 128, 141). These have considered many aspects of these antibiotic-like toxins, including structure, mode of action, genetics, evolution, and ecology. In contrast to these comprehensive treatments, this review focuses on a discussion of the mode of action of those few bacteriocins whose primary target has been identified. There is no treatment of the less well understood bacteriocins.

COLICIN E2 AND E3 AND CLOACIN DF13

Mode of Action

The toxins colicin E2 and E3 and cloacin DF13 are the most thoroughly understood bacteriocins (for reviews see 27, 66, 73). Their modes of action and immunity systems have been determined to the molecular level, and a great deal of information is available concerning their functional architecture. Because of their many similarities, they are discussed together. These proteins are enzymes—E3 and DF13 have RNase activity and E2 has endonuclease activity on DNA. They are toxins because their amino acid sequences endow them with some structural feature that allows them to traverse the bacterial envelope.

Early experiments demonstrating that treatment of *Escherichia coli* cells with colicin E3 led to specific inhibition of protein synthesis suggested that the cellular target was a component of the machinery of protein synthesis (108). This was verified when it was shown that 70*S* ribosomes isolated from E3-treated cells are defective in their capacity to support in vitro protein synthesis (95). A similar situation was found in the case of cloacin DF13-treated *Enterobacter cloacae,* with the exception that the cloacin caused, in addition, the leakage of cellular potassium (30). The basis for such K^+ efflux has not been subsequently investigated. In both cases ribosome damage is localized to the 16*S* RNA of the 30*S* ribosomal subunit, which sustains a single nucleolytic cleavage at the same position near the 3'-terminus of the molecule, thereby leading to the generation of a fragment of 49 nucleotides (10, 29, 137). Since an identical cleavage occurs when isolated ribosomes are treated with highly purified E3 or DF13 (8, 11, 29), the in vivo action of these bacteriocins most certainly involves penetration of all or part of these molecules through the bacterial surface layers, thereby allowing direct interaction of toxin with ribosomes.

Although early in vitro studies allowed the possibility that these bacteriocins acted indirectly (for example, by activating a latent RNase of the ribosome), there is now compelling evidence that colicin E3 (and, thus, presumably cloacin DF13) has endogenous RNase activity (113).

Treatment of *E. coli* with colicin E2 leads to the specific inhibition of DNA synthesis and induces DNA degradation (108). It seems likely that the additional effect of colicin E2 on active transport of cells lysogenic for phage lambda (6) is secondary to the primary action of this colicin. In vitro, highly purified E2 exhibits DNase, but not RNase, activity (133, 135). There is little doubt that the ability of this colicin to act as a DNA endonuclease is the basis for its action against cells. Here again, the colicin must somehow penetrate the cell envelope, at least to the extent that it is able to interact with the bacterial chromosome.

General Structure

Each of these bacteriocins is released from producing cells as a 1 : 1 complex of two polypeptides with a composite molecular weight of 60,000 for colicin E2 and E3, and 67,000 for the cloacin (28, 74, 75, 135). Upon dissociation, E2 and E3 yield polypeptides of 50,000 and 10,000 daltons, whereas DF13 is composed of proteins of 58,000 and 9,000 daltons. In each case the larger polypeptide, termed E2*, E3*, or DF13*, is the active moiety of the complex (28, 74, 135). Since E2*, E3*, and DF13* are also active against whole cells and display the same host specificity as the complex, each of these subunits must also contain the structural information required for receptor selection (see below) and penetration through the outer and inner bacterial membranes.

Each of the smaller subunits has no bacteriocin-like activity but instead inhibits the nuclease activity of the larger polypeptide (28, 78, 135). The native complexes are only weakly active in vitro and addition of purified small subunit to E2*, E3*, and DF13* completely prevents their action on isolated ribosomes. Such inhibition is absolutely specific. For example, the smaller polypeptide derived from native colicin E2 neutralizes the in vitro action of E2* but not E3*, and vice versa (75, 135). The same specific interaction with only the homologous catalytic subunit is displayed by the smaller subunits of the E3 or DF13 complex (120). Clearly, any model for explaining the in vivo mode of action sequence must provide for removal of the inhibitor subunit to activate the bacteriocin.

Immunity System

The highly specific interaction between homologous catalytic and inhibitor subunits is the basis for the immunity system of these bacteriocins. The structural genes for the inhibitory subunits have been shown to reside on the bacteriocin-determining plasmids (90, 139), and expression is such that even under uninduced conditions, in which only a small fraction of plasmid-carrying cells are producing bacteriocin, all the cells in the population contain some free immunity protein.

In considering a general model for immunity to these three bacteriocins, workers in the field have settled on the following scheme (27, 56, 73, 93). Upon synthesis, E2*, E3*, and DF13* interact with their respective immunity proteins to form inactive complexes. Thus, the producing organism is protected from the action of endogenous bacteriocin and maintains synthetic capacity for continuous toxin synthesis until eventual release of the complex into the medium. The mechanism whereby immunity protein is removed from the complex to generate active nuclease is unknown. Although there have been several suggestions that such activation might be mediated by interaction of complex with the cellular outer membrane receptor, in vitro attempts to verify this mechanism have not been successful (121). Nevertheless, it seems likely that removal of the immunity protein occurs during penetration through the cell surface layer and that only active molecules enter the cell interior. Immunity from exogenous bacteriocin would result from neutralization of such activated bacteriocin by free intracellular immunity protein produced under the direction of the relevant immunity-determining plasmid.

Functional and Structural Domains

The analysis of fragments derived from protease digestion of these bacteriocins has proven to be a very useful approach in delineating their functional and structural domains. The general pattern obtained is supported by studies on mutant colicin E3 and DF13 (1, 44, 104). The nuclease activity of these proteins resides in the C-terminal portion of E2*, E3*, and DF13*, comprising about 25% of the total polypeptide (31, 105, 114, 116, 162). Although this region of each molecule is very basic and quite sensitive to trypsin digestion, its interaction with each corresponding acidic immunity protein not only neutralizes the activity of each polypeptide, but renders the region resistant to protease attack (31, 140, 162). This may serve to protect these bacteriocins against inactivation by proteases in the cell envelope (9, 15, 18, 151) and, thus, explains several reports that the bacteriocin complex is more active in vivo than the corresponding E3* and DF13* subunits (28, 56, 74).

The N-terminal region (about 25% of the polypeptide) of each E2*, E3*, and DF13* is thought to be involved in translocation across the cell membranes (17, 116). Its rather hydrophobic character may reflect this function (31, 67, 115). The structural domain that interacts with each cognate outer membrane receptor occupies the central portion of each catalytic subunit (27, 116).

Digestion of colicins E2* and E3* with trypsin leads to the generation of a N-terminal fragment that represents approximately 70% of the intact polypeptide and a C-terminal fragment that comprises the rest of the molecule. Analysis of these fragments shows that the similarity of amino acid composition and antigenic specificity of native colicin E2 and E3 (54) is entirely a result of the almost identical amino acid composition of the N-terminal fragment (115). These results are expected since these two colicins bind to the identical outer membrane receptor protein (43, 102) and probably have similar modes of entry into cells (82). In contrast, the C-terminal tryptic fragments of E2* and E3* are significantly different in composition (115) reflecting both the difference in the catalytic function of the two colicins and the fact that this is the region that binds homologous, but not heterologous, immunity protein. Although E2- and E3-immunity proteins are of similar size, are very acidic in character, and bind to common functional domains of their respective homologous E2* and E3* polypeptides, they display no obvious structural relatedness (115).

Studies on the protein chemistry of DF13* have not revealed sequence similarities to colicins E2* and E3*. This is not surprising since the cloacin absorbs to a different cell surface receptor. In spite of the fact that the immunity proteins of colicin E3 and cloacin interact only with their respective homologous E3* and DF13* subunit, thesc inhibitors do have extensive regions of primary sequence homology (149). It is possible that a direct comparison of amino acid sequence in the catalytic domain of these bacteriocins will reveal similarities.

COLICINS A, E1, Ia, Ib, AND K

Mode of Action

The proteins colicin A, E1, Ia, Ib, and K form ion-permeable channels in the bacterial cytoplasmic membrane. It seems likely that the many reported structural and metabolic changes seen in treated cells are a secondary result of the colicin-induced collapse of the membrane proton motive force. This discussion focuses on the primary action of these colicins. Readers desiring a more comprehensive treatment of the various biochemical changes observed in affected cells should consult recent reviews (58, 92).

Early mode of action studies implicated the cytoplasmic membrane as a primary target of colicins A, E1, K, Ia, and Ib. These bacteriocins inhibit protein and nucleic acid biosynthesis and uncouple electron transport from active transport of thiomethyl-β-D-galactoside and potassium (37, 38, 45, 106). Treated cells leak potassium (22, 34, 45, 55, 106, 109, 161), and in the case of colicin E1 and K, it has been shown that affected cells become more permeable to magnesium and cobalt (101). This loss of cellular potassium and magnesium has been implicated as the primary cause of cell death (97).

In common with several antimicrobial agents that act by dissipating the transmembrane potential, colicin-treated cells display enhanced transport of glucose via the phosphoenolpyruvate-phosphotransferase system (45, 76). Colicin E1 or K-treated cells[1] show reduced activity of membrane-associated ATP-linked reverse transhydrogenase (131). Although these colicins cause a lowering of intracellular ATP levels, this decrease is not responsible per se for the effects of these molecules on membrane function (34). The inhibition of macromolecular synthesis by these toxins probably derives from several factors, including (*a*) low levels of ATP, (*b*) inability to accumulate substrates by active transport, and (*c*) inability to maintain sufficient levels of cofactors, such as cations, etc.

Although the pattern of changes seen in inhibited cells led several groups to conclude that these colicins act by collapsing the energized membrane state, it was not until Gould et al (46, 47) observed that colicin E1 or K treatment leads to a more rapid rate of proton extrusion and a higher amplitude of the H^+/O ratio that more direct evidence was obtained. These results suggested that upon colicin treatment the cell membrane becomes freely permeable to counterions, leading to a collapse in the membrane electrical potential. This was verified when it was shown that colicins Ia, Ib, and A_k inhibit uptake of triphenylmethyl phosphonium cation, whose accumulation is a measure of $\Delta\psi$ (145, 154, 160). Since the proton motive force is composed solely of the $\Delta\psi$ term at physiological pH ($\Delta pH = 0$ at pH 7.5), the colicin-induced depolarization of the cytoplasmic membrane leads to dissipation of the sole driving force for many active transport systems, as well as ATP production by oxidative phosphorylation.

The finding that addition of colicins Ia, Ib, E1, A_1, and K to planar bilayer membranes prepared from soybean phospholipids increases the electrical conductance across such artifical membranes by forming voltage-dependent channels marked a critical advance (136). Significantly, colicins E2 and E3 are without effect in this system. Thus, colicins that collapse $\Delta\psi$ in whole cells are also able to induce transmembrane ion flow in a protein-free artificial membrane. It was proposed that in vivo membrane depolarization results from insertion of colicin molecules into the cytoplasmic membrane where they form aqueous channels. Although this scheme awaits in vivo verification, several other in vitro studies are supportive. When added to liposomes of heterogeneous composition (soybean or *E. coli* phospholipids), colicins E1, Ia, and A_k (but not E2 or E3) induce the rapid transmembrane flux of several ions, as well as the slower leakage of small

[1]It has recently become known that several papers that purport to study colicin K action actually dealt with colicin A (100a). This resulted from a mixup in colicin-producing strains. Where it was relevant, I used the notation A_k to designate such instances.

molecules such as choline, sucrose, leucine, and glucose-6-phosphate (88, 146). This effect is not a result of general disruption of the membrane, since neither intravesicular inulin nor dextran is released. Although colicin E1 treatment increases the permeability of liposomes of homogeneous composition (dimyristoyl-phosphatidyl-choline), treated vesicles do not leak nonelectrolytes of the size of glycerol or larger (147). The fact that colicin-treated liposomes leak a variety of molecules suggests that these colicins do not function as mobile carriers in this system, but do form nonspecific channels of finite size. Direct support for such a channel function was obtained when it was determined that colicin E1 depolarizes dimyristoyl-phosphatidyl-choline vesicles both above and below the membrane-phase transition temperature (147). In contrast to the voltage dependence for channel formation in planar artificial membranes (136), the action of these colicins on liposomes is not dependent on a transmembrane electrical potential (88, 146). The basis for this discrepancy has not been clarified.

The colicin Ia-induced proton permeability seen in liposomes (146) is different from the situation seen in whole cells or membrane vesicles (outside pH adjusted to 5.5 to allow the generation of ΔpH), which upon colicin Ia treatment display a transient enhancement in ΔpH (144, 145). Indeed, it seems likely that the colicin does promote proton permeability in whole cells or vesicles, but that the flux of protons is simply very small compared to the rate of flux of other ions in the system, which are present in much higher concentration. It is also possible that the colicin-induced proton flux across the membrane is too small to collapse ΔpH because of the low proton concentration compared to the large internal buffering capacity (136). The observed slight enhancement of ΔpH is in keeping with the many observations that a collapse in $\Delta\psi$ is accompanied by an increase in ΔpH (127).

The formation of single channel in the cytoplasmic membrane having the same selectivity and conductance properties of those channels formed in the artificial planer membrane system would account for the observation that these colicins display single-hit killing. Such a channel would elicit sufficient ion flow to depolarize the membrane within a few minutes (136).

Structural Features

In aqueous solution colicins exhibit physical properties indicative of overall structural asymmetry (91, 92). However, colicins E1 (56,000 daltons), Ia (79,000), Ib (80,000), and K (45,000) are by far the most elongated, a feature that may be related to their mechanism of action. With prolate-shaped molecules, estimated axial ratios are 15, 11.8, 10.8, and 9.6 for E1, Ia, Ib, and K, respectively. Assuming oblates, the corresponding calculated ratios are 20, 14.9, 13.4, and 11.8. Although such estimates are mere approximations, the high proportion of polar amino acids found in these colicins dictates that they assume elongated forms in solution to maximize their interaction with the aqueous environment. Obviously, these colicins may take on a very different overall structure when integrated into the hydrophobic environment of the cell envelope membranes.

Protease digestion of colicin E1 yields a C-terminal fragment (18,000 daltons), which is enriched in nonpolar amino acids (24). This fragment depolarizes both whole cell-derived membrane vesicles and liposomes composed of dimyristoylphosphatidyl-choline (24; W. Cramer, personal communication). In contrast, a 40,000-dalton N-terminal fragment is thought to contain the domain associated with receptor recognition. There has been a suggestion that the channel-forming colicins share a common functional domain of similar primary structure (150). However, this work is difficult to evaluate, since some of the reported relevant colicin E1 amino acid sequence data conflicts with results obtained in three separate laboratories (personal communication from S. Luria, W. Cramer, and J. Lebowitz).

Immunity System

As in the case of colicins E2 and E3 and cloacin DF13, immunity to the channel-forming colicins does not involve an alteration in receptors (94, 102). However, in contrast to these enzymes, there is no evidence that colicins E1, K, Ia, Ib, or A are released from cells in complex with another polypeptide.

Although the immunity system for any one of these colicins has not been clarified, the molecular interaction involved must be highly specific. For example, strains harboring the Col Ia plasmid are immune to colicin Ia, yet sensitive to colicin Ib, and vice versa, even though these colicins share many physical and chemical properties, exhibit extensive homology in primary structure, and adsorb to a common receptor (91). Furthermore, no bacterial mutants described are insensitive to one but not the other (17, 26). Thus, it seems most reasonable that the immunity system involves some specific interaction between a plasmid-coded gene product and the colicin, rather than a plasmid-determined alteration in their mode of action sequence. It has recently been shown that immunity to colicin Ia is mediated by a plasmid-determined inner membrane protein of 14,500 daltons. Immunity operates at the level of the cytoplasmic membrane since membrane vesicles prepared from Ia-immune cells can be depolarized by colicins E1 and Ib, but not Ia (155). These results raise the possibility that immunity derives from neutralization of colicin by this immunity protein and that the association takes place at or within the cytoplasmic membrane. The formation of such stoichiometric complexes might well provide the explanation for the observation of immunity breakdown that occurs at high Ia or Ib concentration (100). Breakdown would occur when the cytoplasmic membrane is challenged with an amount of colicin in excess of the amount of immunity protein. Although several studies suggest the Col E1 plasmid gene that determines immunity to this colicin encodes a polypeptide of approximately 13,000–14,500 daltons (33, 68, 118, 123), its role in mediating immunity has not been explored.

COLICIN L

Colicin L is a bacteriocin produced by *Serratia marcescens* strain JF246 and is active against certain *E. coli* strains, but not those *Serratia* strains tested (42). This toxin (64,000 daltons) (40) inhibits synthesis of protein, DNA, and RNA and induces the efflux of accumulated leucine (39). General membrane damage does not occur, since treated cells can transport α-methyl-D-glucoside. Inhibited cells suffer a reduction in ATP levels. Although no outer membrane receptor protein has been identified for this colicin, mutants lacking a major outer membrane polypeptide, the *ompA* protein, are insensitive to the colicin. Colicin tolerance in these strains can be overcome by treatments that affect the outer membrane or peptidoglycan layers of the cell envelope (41). Furthermore, the colicin inhibits active transport in vesicles prepared from both *ompA*$^+$ and *ompA* cells. These results indicate that the *ompA* protein may play a role in mediating access of the colicin L molecule to the cytoplasmic membrane, which may be the primary cell target (42). Although the action of colicin L is reminiscent of the effects of colicins Ia, Ib, A, E1, and K, it remains to be established whether or not this reflects a similar action on the cytoplasmic membrane.

COLICIN M

Colicin M (27,000 daltons) (134) causes cell lysis, and under conditions of osmotic protection induces the formation of spheroplasts (14). Although the colicin exhibits no murein hydrolase activity in vitro, treatment of cells does lead to inhibition of murein synthesis and promotion of murein hydrolysis (cited in 134). This suggests that the cellular target may be the enzymes involved in peptidoglycan formation. Since this enzyme system is probably located at the outer surface of the inner membrane, action of this colicin may require only partial penetration of the cell envelope (134). There is experimental support in favor of this notion (134). Further delineation of the mode of action of this colicin will undoubtedly require the development of an in vitro system that responds to its action.

TO THE TARGET

Role of Receptors

Colicin action is initiated by adsorption of each toxin to a specific outer membrane receptor. The presence or absence of such receptors is a critical factor in defining the activity spectrum of a particular colicin against members of the *Enterobacteriacae* (48, 129). However, since the presence of receptors is not sufficient to ensure strain sensitivity (see below), other strain-specific properties undoubtedly contribute.

Many of the colicin receptors have been shown to be involved in outer membrane-mediated nutrient uptake (see 13, 82, 93 for reviews). Thus, the polypeptide that serves as the receptor for colicins E1, E2, and E3 functions in uptake of vitamin B12, whereas the colicin K receptor serves as a specific diffusion pathway for nucleosides (3, 32, 50, 98). Several colicin receptors are involved in iron uptake, serving as siderophore-binding proteins. Thus, the *E. coli tonA* protein is receptor for colicin M and ferrichrome (51, 152, 153), whereas enterochelin and colicins B and D (primary cell targets unknown) utilize a common polypeptide for adsorption (52, 60, 125, 126, 153). There is indirect, but suggestive, evidence that the colicin Ia, Ib receptor may also be involved in iron accumulation (93). These important physiological functions undoubtedly exert selective pressure for the maintenance of active receptors on the surface of sensitive organisms.

The molecular events between the initial adsorption to receptors and final interaction with a particular cellular target is not known for any colicin. However, there are indications that such translocation may be energy dependent (77, 107, 119, 124, 130). Fortunately, the lack of information in this area has stymied neither discussion nor speculations, and several reviews have dealt with this unsettled aspect of colicin action (13, 59, 82, 93).

In general, receptors are thought to provide a means whereby colicins are able to overcome the outer membrane barrier, which excludes access of exogenously added proteins to the periplasm and cytoplasmic membrane. This view derives from several studies in which it has been possible to demonstrate colicin-mediated de-energization of membrane vesicles prepared from resistant cells that lack receptors (7, 80, 142, 144, 154). Similarly, osmotic shock alleviates the need for receptors in cells challenged with colicins E3 or M (12, 148).

Possible Role of Proteases

Over the last several years, almost every discussion of colicin action has entertained a possible scheme involving toxin cleavage with subsequent translocation of an active fragment to the cell target. Receptor-mediated fragmentation seemed particularly intriguing. Although support for such an hypothesis was claimed in a study that purported to demonstrate receptor-dependent cleavage of colicins E1, E2, E3, and K with the generation of an active polypeptide fragment (151), this work has been retracted (D. Sherratt, personal communication). Similary, the determination that cleavage of colicin E4 by whole cells is receptor dependent (18) is complicated by the later realization that the colicin studied was actually colicin A. The original conclusion of receptor dependence has proven to be incorrect (C. Lazdunski, personal communication).

Although there is no little doubt that under certain conditions addition of colicins to whole cells or isolated outer membranes leads to fragmentation, it has been convincingly established that receptors have no obligatory role in the proteolysis of colicin Ia, Ib, A, or E1 (9, 15). Furthermore, since colicin A activity is actually enhanced under conditions that prevent cleavage, fragmentation of the kind observed is clearly not required for action of this colicin (15). There is much need for further work in this area.

Colicin Uptake

Although one might expect that dissection of the steps intervening between adsorption and interaction with target might be amenable to genetic analysis, this general approach has proven disappointing. Although many such tolerant mutants have been described, and even in some cases characterized enough to identify the altered gene product as an envelope component, in no case has it been possible to define in molecular terms how that component facilitates colicin action.

According to its activity spectrum against a variety of mutants, a particular colicin can be unambiguously assigned to one of two groups (25, 26). Although type B colicins (B, Ia, Ia, V, D, and M) are inactive on strains that have a lesion in the *tonB* gene, but active against strains mutant in *tolA* or *tolB* genes, the type A colicins (E1, E2, E3, K, A, L) show the opposite specificity. This distinction is independent of mode of action, but it is thought to reflect two different modes of colicin uptake. There has been discussion (82) of the possibility that some A-type colicins gain entry or access by utilizing those sites of adhesion between inner and outer membranes described by Bayer (4) in a process that depends on the *tolA* and *tolB* gene function. In the case of colicins E2 and E3, supporting evidence has been presented (2, 3, 82). Although the exact function of the *tonB* gene product has not been determined, available information has led to the consensus that it plays a role in mediating transfer of outer membrane-bound nutrients (such as vitamin B12 and siderophores) or type B colicins from their respective surface receptors to the cytoplasmic membrane (13, 82, 93). Such transfer might occur with or without direct interaction of the two membranes.

A model has been considered that proposes that channel-forming colicins are able to interact with the cytoplasmic membrane while remaining adsorbed to their respective outer membrane receptors. For the type B colicins Ia and Ib, such a transenvelope orientation has been proposed to occur at hypothetical sites of apposition between inner and outer membrane (93, 96). In was further hypothesized that the formation of such regions of apposition is *tonB* dependent. According to this model, the A-type colicins E1, A, and K would span the envelope at Bayer adhesion sites. Based on calculated dimensions, these colicins are all of sufficient size to span the cell envelope. The earlier described assignment of receptor and action domains of colicin E1 to respective N-terminal and C-terminal regions (24) would predict the simultaneous attachment of these regions to outer and inner membrane, respectively.

Support for this model derives from the observation that addition of trypsin to inhibited cells reverses the effect of colicins Ia and K on macromolecular synthesis and the inhibition of active transport by colicin E1 (23, 100, 110). Presumably, digestion of colicin exposed at the cell surface suffices to disrupt or destroy that structural domain of the molecule forming channels in the inner membrane. However, an alternative mechanism involving a requirement for continuous translocation of surface colicins (trypsin sensitive) from receptor to cytoplasmic membrane (trypsin insensitive) cannot be ruled out. Such a mechanism involving transient channel formation for any particular colicin molecule would manifest trypsin reversibility. Although the finding that immobilized colicin E1 but not E2 or E3 kills sensitive cells (99) supports the idea that this colicin can span the cell envelope, cleavage of the Sephadex-bound molecule to yield an active penetrating fragment was not ruled out.

In the case of colicin E3, it is possible to bypass the need for receptors by subjecting cells to osmotic shock (143). Thus, receptors are not absolutely essential for translocation of this molecule across the inner membrane. However, efficient translocation through adhesion sites in whole cells may require that the colicin be presented to the inner membrane in an optimal orientation. This alignment might be assured by a fixed spatial relationship between the outer membrane receptor and the cytoplasmic membrane. Although the hydrophobic domains of colicins E2 and E3 and cloacin DF13 most likely play a role in their translocation across the inner membrane, there is absolutely no hint of how this is brought about.

PESTICIN A1122

Pesticin A1122 is a 65,000-molecular-weight polypeptide produced by *Yersinia pestis* (63). The activity spectrum of the toxin is defined by the presence of a specific outer membrane receptor elaborated by sensitive organisms, which include serotype I strains of *Yersinia pseudotuberculosis*, some isolates of *Yersinia enterocolitica*, nonpesticinogenic *Y. pestis*, and *E. coli* ϕ, but not *E. coli* K12 (36).

Studies on the mode of pesticin action have been limited to its effects on *E. coli*. Since *ton B* derivatives of *E. coli* ø are insensitive to pesticin (36), its mode of uptake by cells probably has some feature in common with the B-type colicins. Although the pesticin receptor is distinct from those used by colicins I, B, D, or M, evidence suggests that it may operate as a component in some as yet undefined iron uptake system (16, 36, 64).

Pesticin A1122 is an enzyme whose toxic action results from its ability to degrade cellular murein (35). Addition of this toxin to either *E. coli* ϕ or sensitive Yersinieae induces the formation of osmotically stable spheroplast-like forms, which is paralleled by a normal increase in cell mass. There is no inhibition of DNA, RNA, or protein synthesis (35, 49). Its mode of action was established by the finding that in vitro the purified pesticin catalyzes the hydrolysis of the β-1,4 bond between N-acetyl glucosamine and N-acetylmuramic acid in the glycan backbone of the bacterial cell wall (35). Murein preparations from a wide variety of naturally insensitive gram-negative strains, as well as from the immune-producing organism and resistant mutants of sensitive strains, are degraded by the pesticin in vitro. These results demonstrate that the activity spectrum of this bacteriocin is not dictated by the cell target, but more likely reflects the ability of the toxin to gain access to the murein of the challenged organism.

STAPHYLOCOCCIN 1580

The bacteriocin staphylococcin 1580, which is produced by *Staphylococcus epidermidis* 1580, is active against many gram-positive, but not gram-negative, bacteria (79). Although early mode of action studies (78, 80) were carried out with pure material, which was shown to be a 150,000–400,000 dalton complex of subunits that contain protein, carbohydrate, and lipid (81), later studies utilize a less characterized but apparently pure preparation (157–159). The effect of staphylococcin 1580 on the energy metabolism of sensitive *Staphylococcus aureus* or *Bacillus subtilis* cells is similar to what is seen in *E. coli* cells treated with the channel-forming colicins. There is a rapid inhibition of macromolecular synthesis and active transport, depletion of cellular ATP, and efflux of preaccumulated rubidium ion and glutamic acid. In contrast, electron transport is not significantly inhibited and is even stimulated in pyrurate grown *S. aureus* cells.

Further study of staphylococcin action has made use of membrane vesicles prepared from both *S. aureus* and *B. subtilis* (80, 159). With fluorescent dyes to monitor the membrane potential in such membrane preparations, it has been shown that the toxin is able to collapse the $\Delta\psi$, whether generated from respiration or by a potassium diffusion potential. This finding strongly implicates the cytoplasmic membrane as the primary cell target and suggests that the many changes seen in staphylococcin 1580-treated cells result from its ability to depolarize the energy-transducing cytoplasmic membrane. This work is at a stage that would greatly benefit from the use of artificial membrane systems.

There is an energy requirement for the initiation of staphylococcin 1580 action that is thought to reflect some energy-dependent step in the killing mechanism (156). These steps have not been defined, and delineation will probably require a genetic approach. Two potentially useful mutant classes have already been reported (80). Although active transport in membrane vesicles prepared from resistant *S. aureus* mutants is inhibited by the staphylococcin, similar membrane preparations isolated from tolerant mutants of stable *S. aureus* L-forms are not affected by the toxin. Thus, resistant and tolerant mutants may define alterations in surface receptor and other components in the uptake process, respectively.

BUTYRICIN 7423

Clostridium butyricum 7423 produces the rather hydrophobic bacteriocin butyricin 7423 (mol wt 32,500), which is active against *Clostridium pasteurianum* (21). Treated cells are inhibited in DNA, RNA, and protein synthesis. In addition there is a lowering of ATP levels and an induced efflux of cellular K^+ (20, 21). Although the butyricin inhibits the F_1F_0-ATPase of vegetatively growing cells in vitro (19), the primary action of the toxin is thought not to be due to stoichiometric inhibition of this enzyme. The observed inhibition of $\Delta\psi$ is thought not to derive from the formation of ion-permeable channels in the cytoplasmic membrane, but to derive from some, as yet uncharacterized, interaction with this membrane (J. G. Morris personal communication).

PYOCIN R1

The particulate bacteriocins produced by various *Pseudomonas aeruginosa* strains are grouped in five classes according to differences in receptor specificity (84). Each producer strain is immune to the pyocin it produces. These toxins resemble contractile bacteriophage tails in structure, each being composed of a contractile sheath, a core, and fibers (69, 71). Their structures are very similar as viewed in the electron microscope, and they display antigenic relatedness. Of each of the pyocin's over 20 distinct subunit proteins only that polypeptide that comprises the main component of the fiber differs from pyocin to pyocin (117). Since these fibers function in the adsorption of each bacteriocin to its cognate lipopolysaccharide receptor on the surface of sensitive organisms (71, 81, 103), such differences may well form the basis for the host range of each R-type pyocin.

All the evidence suggests that the R-type pyocins originated from a common temperate bacteriophage, which by mutation became defective in assembly of normal particles. Subsequent divergence of the fiber structure would have generated the various R-type pyocins. Such an origin is supported by the finding that several *P. aeruginosa* strains produce phages that display immunologic cross-reaction with the R-type pyocins (61, 62, 70, 87). Furthermore, the isolated tail of one of these phages has pyocin-like bacteriocidal action (138). There is genetic evidence that these bacteriocins are coded by allelic chromosomal genes (83, 86).

With regard to mode of action against sensitive *P. aeruginosa* strains, pyocin R1 is the most thoroughly studied R-type pyocin. Treatment leads to an immediate cessation of DNA, RNA, and protein synthesis with no cell lysis or DNA degradation (89). Ribosomes isolated from R1-treated cells are physically degraded (89). It has been proposed that the arrest in macromolecular synthesis may derive from the observed inability of treated cells to take up or/and maintain amino acids, nucleosides, and ions (65). The finding that pyocin R1 causes an uncoupling of respiration to solute transport implies that the pyocin causes damage to the energy-transducing membrane, rather than acting to inhibit those reactions that lead to oxygen reduction (85). The enhanced binding of hydrophobic fluorescent probes to R1-treated cell, with accompanying changes in certain parameters of fluorescence (149), is similar to the studies on the channel-forming colicins and, here too, probably reflects structural changes in the cell envelope as a result of membrane de-energization. The sum of these observations strongly suggests that the cytoplasmic membrane is the primary target of this bacteriocin. Whether or not the killing mechanism involves a direct interaction of the pyocin, or some component of its composite structure with this membrane, is unknown. It seems likely that the other R-type pyocins have a similar action mechanism (117).

PYOCIN AP41

The bacteriocin pyocin AP41 has been isolated as a complex of two polypeptides (90,000 and 6000–7000 daltons). Killing activity against sensitive *P. aeruginosa* strains resides in the larger component (132). In vivo the pyocin inhibits DNA synthesis preferentially and induces production of resident pyocins or phages. These results suggest a colicin E2-like mode of action and that the small subunit of the pyocin complex might correspond to an immunity protein. This scheme is supported by a report (M.

Kageyama, personal communication) that trypsinolysis of native AP41 leads to the generation of a 16,000-dalton fragment that displays in vitro DNA endonuclease activity. This activity is inhibited by addition of the small subunit.

MEGACIN A-216

Megacin A-216 (51,000 daltons) is a phospholipase (57, 72, 122). Bacteriocin-treated *Bacillus megaterium* leak intracellular material but do not lyse, and treated protoplasts are converted to cell ghosts. In vitro, the purified megacin catalyzes the formation of lysolecithin from lecithin and, thus, can be classified as an A-2 type phospholipase.

Immunity of the meg$^+$-producing *B. megaterium* strain is mediated by a proteinaceous inhibitor found in the culture medium (111, 112). This substance not only prevents the action of the megacin on protoplasts, but it inhibits megacin-dependent conversion of lecithin to lysolecithin. Furthermore, megacinogenic mutants displaying increased immunity produce higher levels of inhibitor when compared to the parental meg$^+$ strain, whereas mutants showing increased sensitivity produce less. These results, together with the finding that the megacin inhibitor is not a general inhibitor of phospholipase A activity and is not produced by nonmegacinogenic strains, establish its role in the megacin immunity system.

CONCLUDING REMARKS

Although it has been possible to discuss the mode of action of several bacteriocins in terms of interaction with primary cell targets, a complete understanding of their killing mechanism at the molecular level requires a more thorough delineation of how each toxin reaches its target. This particular aspect of bacteriocin action has proven quite refractory to experimental attack. A major technical problem in following the fate of bound bacteriocin is that although killing is single hit, not all bacteriocin-receptor interactions are functionally effective. Although the basis for such heterogeneity is not known, it might well derive from differences in molecular environment. For example, only a minority of receptors may be able to interact with envelope components that can function to mediate toxin uptake. A further complication is that any manipulation that involves cell breakage, fractionation of envelopes, etc, may well cause a redistribution of bacteriocin molecules or fragments. The possibility of nonspecific modification of absorbed colicin by envelope proteases must also be taken into account. Clearly, the sorting out of physiologically functioning toxin from bacteriocins not taking part in the mode of action scheme is very difficult. Ideally, one would like some means to distinguish "killers" from "non-killers" and an approach that will allow the freezing of "killer molecules" at various steps in the uptake process.

The voltage dependence of the channels formed by certain of the colicins in the planar membrane system provides the membraneologist with a potentially very useful system for probing the nature of gated channel formation. Furthermore, the ability to assay a functional interaction (channel formation) between a membrane and an exogenously added protein soluble in aqueous solutions may allow one to discern certain features of a process that may be of general relevance to membrane biogenesis and protein secretion.

Although a start has been made in elucidating the functional domains of several colicins, progress has relied, for the most part, on the good fortune that protease cleavage of intact molecules has yielded useful fragments. What are now needed are more defined studies in which the investigator is able to dictate specific changes in protein structure. Such an approach is now feasible through DNA technology. Indeed, the cloning and sequence analysis of several colicin structural genes have been undertaken by several groups, and it should not be long before primary sequences are known. This kind of information should make possible the use of directed in vitro mutagenesis to generate interesting molecules for both in vitro and in vivo analysis. Undoubtedly this approach can be applied to all the bacteriocins.

Literature Cited

1. Andreoli, P. M., Overbeeke, N., Veltkamp, E., van Embden, J. D. A., Nijkamp, H. J. J. 1978. *Mol. Gen. Genet.* 160:1–11
2. Bassford, P. J., Kadner, R. J., Schnaitman, C. A. 1977. *J. Bacteriol.* 129:265–75
3. Bassford, P. J., Schnaitman, C. A., Kadner, R. J. 1977. *J. Bacteriol.* 130:750–58
4. Bayer, M. 1979. In *Bacterial Outer Membranes*, ed. M. Inouye, pp. 167–202. New York: Wiley
5. Beppu, T., Arima, K. 1976. *J. Bacteriol.* 93:80–85
6. Beppu, T., Yamamoto, H., Arima, K. 1975. *Antimicrob. Agents Chemother.* 8:617–26
7. Bhattacharyya, P., Wendt, L., Whitney, E., Silver, S. 1970. *Science* 168:998–1000
8. Boon, T. 1971. *Proc. Natl. Acad. Sci. USA* 68:2421–25
9. Bowles, L. K., Konisky, J. 1981. *J. Bacteriol.* 145:668–71.
10. Bowman, C. M., Dahlberg, J. E., Ikemura, T., Konisky, J., Nomura, M. 1971. *Proc. Natl. Acad. Sci. USA* 68:964–68
11. Bowman, C. M., Sidikaro, J., Nomura, M. 1971. *Nature London New Biol.* 234:133–37
12. Braun, J., Frenz, J., Hantke, K., Schaller, K. 1980. *J. Bacteriol.* 142:162–68
13. Braun, V., Hantke, K. 1977. In *Microbial Interactions*, ed. J. L. Reissig, pp. 101–37. London: Chapman and Hall
14. Braun, V., Schaller, K., Wabl, M. R. 1974. *Antimicrob. Agents Chemother.* 5:520–33
15. Brey, R. N. 1981. *J. Bacteriol.* 149:306–15
16. Brubaker, R. R., Surgalla, M. J. 1961. *J. Bacteriol.* 82:940–49
17. Cardelli, J., Konisky, J. 1974. *J. Bacteriol.* 119:379–85
18. Cavard, D., Lazdunski, C. 1979. *Eur. J. Biochem.* 96:525–33
19. Clarke, D. J., Fuller, F. M., Morris, J. G. 1979. *Eur. J. Biochem.* 98:597–612
20. Clarke, D. J., Morris, J. G. 1976. *J. Gen. Microbiol.* 95:67–77
21. Clarke, D. J., Robson, R. M., Morris, J. G. 1975. *Antimicrob. Agents Chemother.* 7:256–64

22. Dandeu, J. P., Billault, A., Barbu, E. 1969. *C. R. Acad. Sci.* 269:2044–47
23. Dankert, J., Hammond, S. M., Cramer, W. A. 1980. *J. Bacteriol.* 143:594–602
24. Dankert, J. R., Uratani, Y., Grabau, C., Cramer, W. A., Hermodson, M. 1981. *Fed. Proc.* 1844 (Abstr.)
25. Davies, J. K., Reeves, P. 1975. *J. Bacteriol.* 123:96–101
26. Davies, J. K., Reeves, P. 1975. *J. Bacteriol.* 123:102–17
27. DeGraaf, F. K. 1979. *Zentralbl. Bakteriol. Parasitenkd. Infektionskr. Hyg. Abt. 1 Orig. Reibe A* 244:121–34
28. DeGraaf, F. K., Klaasen-Boor, P. 1977. *Eur. J. Biochem.* 73:107–14
29. DeGraaf, F. K., Niekus, H. G. D., Klootwijk, J. 1973. *FEBS Lett.* 35: 161–65
30. DeGraaf, F. K., Planta, R. J., Stouthamer, A. H. 1971. *Biochim Biophys. Acta* 240:122–36
31. DeGraaf, F. K., Stukart, M. J., Boogerd, F. D., Metselaar, K. 1978. *Biochemistry* 17:1137–42
32. DiMasi, D. R., White, J. C., Schnaitman, C. A., Bradbeer, C. 1973. *J. Bacteriol.* 115:506–13
33. Ebina, Y., Kishi, F., Nakazawa, T. 1979. *Nucleic Acids Res.* 7:639–49
34. Feingold, D. S. 1970. *J. Membr. Biol.* 3:372–86
35. Ferber, D. M., Brubaker, R. R. 1978. *J. Bacteriol.* 136:495–501
36. Ferber, D. M., Fowler, J. M., Brubaker, R. R. 1981. *J. Bacteriol.* 146:506–11
37. Fields, K. L., Luria, S. E. 1969. *J. Bacteriol.* 97:57–63
38. Fields, K. L., Luria, S. E. 1969. *J. Bacteriol.* 97:64–77
39. Foulds, J. 1971. *J. Bacteriol.* 107: 833–39
40. Foulds, J. 1972. *J. Bacteriol.* 110: 1001–9
41. Foulds, J., Chai, T-J. 1978. *J. Bacteriol.* 133:158–64
42. Foulds, J., Shemin, D. 1969. *J. Bacteriol.* 99:655–60
43. Fredericq, P. 1958. *Symp. Soc. Exp. Biol.* 12:104–22
44. Gaastra, W., Oudega, B., DeGraaf, F. K. 1978. *Biochim. Biophys. Acta* 540: 301–12
45. Gilchrist, M. J. R., Konisky, J. 1975. *J. Biol. Chem.* 250:2457–62
46. Gould, J. M., Cramer, W. A. 1977. *J. Biol. Chem.* 252:5491–97
47. Gould, J. M., Cramer, W. A., van Thienen, G. 1976. *Biochem. Biophys. Res. Commun.* 72:1519–25
48. Graham, A. C., Stocker, B. A. D. 1977. *J. Bacteriol.* 130:1214–23
49. Hall, P. J., Brubaker, R. R. 1978. *J. Bacteriol.* 136:786–89
50. Hantke, K. 1976. *FEBS Lett.* 70:109–12
51. Hantke, K., Braun, V. 1975. *FEBS Lett.* 49:301–5
52. Hantke, K., Braun, V. 1975. *FEBS Lett.* 59:277–81
53. Hardy, K. G. 1975. *Bacteriol. Rev.* 39:464–515
54. Herschman, H. R., Helinski, D. R. 1967. *J. Biol. Chem.* 242:5360–68
55. Hirata, H. S., Fukui, S., Ishikawa, S. 1969. *J. Biochem.* 65:843–47
56. Hirose, A., Kumagai, J., Imahori, K. 1976. *J. Biochem.* 79:305–11
57. Holland, I. B. 1962. *J. Gen. Microbiol.* 29:603–14
58. Holland, I. B. 1975. *Adv. Microb. Physiol.* 12:56–139
59. Holland, I. B. 1976. In *Specficity and Action of Animal, Bacterial and Plant Toxins,* ed. P. Cuatrecasas, pp. 99–127. London: Chapman and Hall.
60. Hollifield, W. C., Neilands, J. B. 1978. *Biochemistry* 17:1922–28
61. Homma, J. Y., Shionoya, H. 1967. *Jpn. J. Exp. Med.* 37:395–421
62. Homma, J. Y., Watabe, H., Tanabe, Y. 1968. *Jpn. J. Exp. Med.* 38:213–24
63. Hu, P. C., Brubaker, R. R. 1974. *J. Biol. Chem.* 249:4749–53
64. Hu, P. C., Yang, G. C. H., Brubaker, R. R. 1972. *J. Bacteriol.* 112:212–19
65. Iijima, M. 1978. *J. Biochem.* 83:395–402
66. Imahori, K. 1979. *TIBS* 4:212–13
67. Imahori, K., Ohno, S., Ohno-Iwashita, Y., Suzuki, K. 1978. In *Versatility of Proteins,* ed. C. H. Li, pp. 167–82. New York: Academic
68. Inselburg, J., Applebaum, B. 1978. *J. Bacteriol.* 133:1444–51
69. Ishii, S., Nishi, Y., Egami, F. 1965. *J. Mol. Biol.* 13:428–31
70. Ito, S., Kageyama, M. 1970. *J. Gen. Appl. Microbiol.* 16:231–40
71. Ito, S., Kageyama, M., Egami, F. 1970. *J. Gen. Appl. Microbiol.* 16:205–14
72. Ivanovics, G. E., Afoldi, L., Nagy, E. 1959. *J. Gen. Microbiol.* 21:51–60
73. Jakes, K. 1982. In *The Molecular Actions of Toxins and Viruses,* ed. S. van Heyninger, P. Cohen. North Holland: Elsevier. In press
74. Jakes, K. S., Zinder, N. D. 1974. *Proc. Natl. Acad. Sci. USA* 71:3380–84
75. Jakes, K., Zinder, N. D., Boon, T. 1974. *J. Biol. Chem.* 249:438–44
76. Jetten, A. M. 1976. *Biochim. Biophys. Acta* 440:403–11

77. Jetten, A. M., Jetten, M. E. R. 1975. *Biochim. Biophys. Acta* 387:12–22
78. Jetten, A. M., Vogels, G. D. 1972. *Antimicrob. Agents Chemother.* 2:456–63
79. Jetten, A. M., Vogels, G. D. 1972. *J. Bacteriol.* 112:243–50
80. Jetten, A. M., Vogel, G. D. 1973. *Biochim. Biophys. Acta* 311:483–95
81. Jetten, A. M., Vogels, G. D., DeWindt, F. 1972. *J. Bacteriol.* 112:235–42
82. Kadner, R. J., Bassford, P. J., Pugsley, A. P. 1979. Zentralbl. Bakteriol. Parasitenkd. Infektionskr. Hyg. Abt. I. Orig. Reihe A 244:90–104
83. Kageyama, M. 1974. *J. Gen. Appl. Microbiol.* 20:269–75
84. Kageyama, M. 1975. In *Microbial Drug Resistance,* ed. S. Mitsuhashi, H. Hashimoto, pp. 291–305. Tokyo: Univ. Tokyo
85. Kageyama, M. 1978. *J. Biochem.* 84:1373–79
86. Kageyama, M., Inagaki, A. 1974. *J. Gen. Appl. Microbiol.* 20:257–67
87. Kageyama, M., Shinomiya, T., Aihara, Y., Kobayaski, M. 1979. *J. Virol.* 32:951–57
88. Kayalar, C., Luria, S. E. 1979. In *Membrane Bioenergetics,* ed. C. P. Lee, G. Schatz, L. Ernster, pp. 297–306. Ncw York: Addison-Wesley
89. Kaziro, Y., Tanaka, M. 1965. *J. Biochem.* 57:689–95
90. Konings, R. N. H., Andreoli, P. M., Veltkamp, E., Nijkamp, H. J. J. 1976. *J. Bacteriol.* 126:861–68
91. Konisky, J. 1973. In *Chemistry and Functions of Colicins,* ed. L. Hager, pp. 41–58. New York: Academic
92. Konisky, J. 1978. In *The Bacteria,* ed. L. N. Ornston, J. R. Sokatch, 6:319–59. New York: Academic
93. Konisky, J. 1979. In *Bacterial Outer Membranes,* ed. M. Inouye, pp. 319–59. New York: Wiley
94. Konisky, J., Cowell, B. S. 1972. *J. Biol. Chem.* 247:6524–29
95. Konisky, J., Nomura, M. 1967. *J. Mol. Biol.* 26:181–95
96. Konisky, J., Tokuda, H. 1979. *Zentralbl. Bakteriol. Parasitenkol. Infektionsk. Hyg. Abt. 1 Orig. Reihe A* 244:105–20
97. Kopecky, A. L., Copeland, D. P., Lusk, J. E. 1975. *Proc. Natl. Acad. Sci. USA* 72:4631–34
98. Krieger-Brauer, H. J., Braun, V. 1980. *Arch. Microbiol.* 124:233–42
99. Lau, C., Richards, F. M. 1976. *Biochemistry* 15:666–71
100. Levisohn, R., Konisky, J., Nomura, M. 1968. *J. Bacteriol.* 96:811–21
100a. Luria, S. E. 1982. *J. Bacteriol.* 149:386
101. Lusk, J. E., Nelson, D. L. 1972. *J. Bacteriol.* 112:148–60
102. Maeda, A., Nomura, M. 1966. *J. Bacteriol.* 91:685–94
103. Meadow, P. M., Wells, P. L. 1978. *J. Gen. Microbiol.* 108:339–43
104. Mock, M., Schwartz, M. 1980. *J. Bacteriol.* 142:384–90
105. Mooi, F. R., DeGraaf, F. K. 1976. *FEBS Lett.* 62:304–8
106. Nagel de Zwaig, R. 1969. *J. Bacteriol.* 99:913–14
107. Nieva-Gomez, D., Konisky, J., Gennis, R. B. 1976. *Biochemistry* 15:2747–53
108. Nomura, M. 1963. *Cold Spring Harbor Symp. Quant. Biol.* 28:315–24
109. Nomura, M., Maeda, A. 1965. *Zentralbl. Bakteriol. Parasitenkd. Infektionskr. Hyg. Abt. 1 Orig.* 196:216–39
110. Nomura, M., Nakamura, M. 1962. *Biochem. Biophys. Res. Commun.* 7:306–9
111. Ochi, T., Higashi, Y., Inoue, K., Amano, T. 1970. *Biken J.* 13:63–76
112. Ochi, T., Yano, K., Amano, T. 1971. *Biken J.* 14:423–24
113. Ohno, S., Imahori, K. 1978. *J. Biochem.* 84:1637–40
114 Ohno, S., Ohno-Iwashita, Y., Suzuki, K., Imahori, K. 1977. *J. Biochem.* 82:1045–53
115. Ohno-Iwashita, Y., Imahori, K. 1979. *FEBS Lett.* 100:249–52
116. Ohno-Iwashita, Y., Imahori, K. 1980. *Biochemistry* 19:652–59
117. Ohsumi, M., Shinomiya, T., Kageyama, M. 1980. *J. Biochem.* 87:1119–26
118. Oka, A., Nomura, N., Morita, M., Sugisaki, H., Sugimoto, K., Takanami, M. 1979. *Mol. Gen. Genet.* 172:151–59
119. Okamoto, K. 1975. *Biochim. Biophys. Acta* 389:370–79
120. Oudega, B., Klaasen-Boor, P., DeGraaf, F. K. 1975. *Biochim. Biophys. Acta* 392:184–95
121. Oudega, B., van der Molen, J., DeGraaf, F. K. 1979. *J. Bacteriol.* 140:964–70
122. Ozaki, M. Y., Higaski, Y., Saito, H., An, I., Amano, T. 1966. *Biken J.* 9:201–13
123. Patient, R. K. 1979. *Nucleic Acids Res.* 6:2647–65
124. Plate, C. A. 1973. *Antimicrob. Agents Chemother.* 4:16–24
125. Pugsley, A., Reeves, P. 1977. *Antimicrob. Agents Chemother.* 11:345–58
126. Pugsley, A. P., Reeves, P. 1977. *Biochem. Biophys. Res. Commun.* 74:903–11
127. Ramos, S., Schuldiner, S., Kaback, H.

R. 1976. *Proc. Natl. Acad. Sci. USA* 73:1892–96
128. Reeves, P. R. 1972. *The Bacteriocins.* New York: Springer. 142 pp.
129. Reeves, P. 1979. *Zentralbl. Bakteriol. Parasitenkd. Infektionskr. Hyg. Abt. 1 Orig. Reihe A* 244:7889
130. Reynolds, B. L., Reeves, P. R. 1969. *J. Bacteriol.* 100:301–9
131. Sabet, S. F. 1976. *J. Bacteriol.* 126:601–8
132. Sano, Y., Kageyama, M. 1981. *J. Bacteriol.* 146:733–39
133. Saxe, L. S. 1975. *Biochemistry* 14:2058–63
134. Schaller, K., Dreher, R., Braun, V. 1981. *J. Bacteriol.* 146:54–63
135. Schaller, K., Nomura, M. 1976. *Proc. Natl. Acad. Sci. USA* 73:3989–93
136. Schein, S. J., Kagan, B. L., Finkelstein, A. 1978. *Nature* 276:159–63
137. Senior, B. W., Holland, I. B. 1971. *Proc. Natl. Acad. Sci. USA* 68:959–63
138. Shinomiya, T., Shiga, S. 1979. *J. Virol.* 32:958–67
139. Sidikaro, J., Nomura, M. 1975. *J. Biol. Chem.* 250:1123–31
140. Suzuki, K., Imahori, K. 1978. *J. Biochem.* 84:1021–29
141. Tagg, J. R., Dajani, A. S., Wannamaker, L. W. 1976. *Bacteriol. Rev.* 40:722–56
142. Takagaki, Y., Kunugita, K., Mitsuhashi, M. 1973. *J. Bacteriol.* 113:42–50
143. Tilby, M., Hindennach, I., Henning, U. 1978. *J. Bacteriol.* 136:1189–91
144. Tokuda, H., Konisky, J. 1978. *J. Biol. Chem.* 253:7731–37
145. Tokuda, H., Konisky, J. 1978. *Proc. Natl. Acad. Sci. USA* 75:2579–83
146. Tokuda, H., Konisky, J. 1979. *Proc. Natl. Acad. Sci. USA* 76:6167–71
147. Uratani, Y., Cramer, W. A. 1981. *J. Biol. Chem.* 256:4017–23
148. Uratani, Y., Kageyama, M. 1977. *J. Biochem.* 81:333–41
149. Van den Elzen, P. J. M., Gaastra, W., Spelt, C. E., DeGraaf, F. K., Veltkamp, E., Nijkamp, H. J. J. 1980. *Nucleic Acid Res.* 8:4349–63
150. Watson, D. H. 1980. *Biochim. Biophys. Acta* 622:287–96
151. Watson, D. H., Sherratt, D. 1979. *Nature* 278:362–64
152. Wayne, R., Frick, K., Neilands, J. B. 1976. *J. Bacteriol.* 126:7–12
153. Wayne, R., Neilands, J. B. 1975. *J. Bacteriol.* 121:497–503
154. Weaver, C. A., Kagan, B. L., Finkelstein, A., Konisky, J. 1981. *Biochim Biophys. Acta* 645:137–42
155. Weaver, C. A., Redborg, A., Konisky, J. 1981. *J. Bacteriol.* 148:817–28
156. Weerkamp, A., Geerts, W., Vogels, G. D. 1978. *Biochim. Biophys. Acta* 539:372–85
157. Weerkamp, A., Heinen-von Borries, U. J., Vogels, G. D. 1978. *Antonie van Leeuwenhoek J. Microbiol. Ser.* 44:35–48
158. Weerkamp, A., Vogels, G. D. 1978. *Antimicrob. Agents Chemother.* 13:146–53
159. Weerkamp, A., Vogels, G. D. 1978. *Biochim. Biophys. Acta* 539:386–97
160. Weiss, M. J., Luria, S. E. 1978. *Proc. Natl. Acad. Sci. USA* 75:2483–87
161. Wendt, L. 1970. *J. Bacteriol.* 104:1236–41
162. Yamamoto, H., Nishida, K., Beppu, T., Arima, K. 1978. *J. Biochem.* 83:827–34

Ann. Rev. Microbiol. 1982. 36:145–72

IMMOBILIZED MICROBIAL CELLS

Saburo Fukui and Atsuo Tanaka

Laboratory of Industrial Biochemistry,
Department of Industrial Chemistry, Faculty of Engineering, Kyoto University,
Yoshida, Sakyo-ku,
Kyoto 606, Japan

CONTENTS

INTRODUCTION

Immobilization of biocatalysts—enzymes, organelles, microbial cells, and other living systems—is attracting worldwide attention. Immobilized biocatalysts catalyze biochemical reactions under more stabilized conditions than their free counterparts; moreover, they can be reused economically. At present, applications of immobilized biocatalysts include (*a*) production of useful compounds by stereospecific and/or regiospecific reactions under mild conditions, (*b*) production of energy by biological processes, (*c*) analyses of various compounds with high sensitivity and high specificity, and (*d*) utilization in new types of drugs, in making artificial organs, etc. These processes require immobilization not only of single enzyme systems that catalyze simple reactions, such as hydrolysis, oxidore-

0066-4227/82/1001-0145$02.00

duction, isomerization, addition, etc, but also of multienzyme systems that mediate more complicated reactions for the synthesis and conversion of numerous compounds. These systems often involve the regeneration of ATP and/or coenzymes that participate in oxidation-reduction reactions.

Microbial cells and cellular organelles contain metabolic systems that catalyze such complicated reactions. Therefore, immobilization of microbial cells and organelles in turn permits immobilization of multistep and cooperative enzyme systems. Furthermore, with immobilization of microbial cells, the procedure of extracting enzymes from the cells is no longer required. This avoids inactivation of enzymes during tedious and time consuming purification procedures and it increases stability of many membrane-associated enzymes unstable in a solubilized state. These examples illustrate the advantages of immobilized microbial cells over immobilized enzymes in various fields.

However, immobilized microbial cells have several disadvantages that must be overcome before commercial application is possible. (*a*) Microbial cells contain many enzymes that catalyze undesirable reactions. Mutation or proper treatment of cells may lower the levels of enzymes participating in side reactions and may increase the activities of desirable enzymes. Recent progress in genetic engineering may enable cells to have high levels of specific enzymes or activities of foreign enzymes. (*b*) Cell wall and cell membrane of intact cells often prevent the permeation of substrates, products, and other reaction components into and out of cells. In these cases, permeability barriers must be destroyed by proper treatment of cells before or after immobilization.

Immobilized cells were first used for industrial production of L-aspartic acid (9, 120). Thereafter, numerous enzymatic reactions were carried out with immobilized microbial cells; some processes, such as production of L-malic acid and high fructose syrup, have been industrialized. In these cases, enzymatic reactions are catalyzed not only by a single enzyme, but also by a multistep enzyme system conjugated with regeneration of ATP or other coenzymes. The cells are often utilized in a dead state with appropriate treatment before or after immobilization, although the desired enzymes are in active and stable states. These cells are classified here as immobilized treated cells.

A variety of reactions can also be achieved by immobilized living cells, whether they are resting or growing in gel matrices. Definition of immobilized resting cells is often difficult because the state of immobilized cells was not mentioned in the literature in many cases. Therefore, immobilized resting cells sometimes mean immobilized nontreated cells.

If immobilized cells are kept in a growing state within gel matrices by a continuous supply of suitable nutrients and if their biological functions are useful, as in the case of conventional fermentation procedures, this new technique of immobilized growing cells seems very advantageous. Immobilized growing cells serve as renewable or self-proliferating biocatalysts, which are located in a certain defined region of space and are protected against unfavorable circumstances outside of their small domain. Physicochemical interactions between gels and the cells often give favorable effect(s) on the stability of entrapped cells, as is mentioned later.

However, immobilized resting cells and immobilized growing cells seem to have some disadvantages: (*a*) Yield of products may be lowered by the consumption of substrates as carbon and energy sources to maintain the cells living or growing state(s); and (*b*) in the case of immobilized growing cells, products may be contaminated by cells leaked from gel matrices. Nevertheless, immobilized living cells are most promising because they retain enzyme activities for a very long time.

Immobilized cell systems have been reviewed by Chibata & Tosa (7, 8), Abbott (1), and Dunnill (15). Recent examples, together with previous important works on immobilized treated cells, immobilized resting cells, and immobilized growing cells are summarized in Tables 1, 2, and 3, respectively.

In this review, we describe several new problems of fundamental and practical importance in the area of immobilized microbial cells.

NEW TECHNIQUES FOR IMMOBILIZATION OF MICROBIAL CELLS

Entrapment methods are the most common for the preparation of immobilized microbial cells. In addition to natural polymers—such as alginate, cellulose, gelatin, collagen, and agar—polyacrylamide has been employed as a gel matrix by many researchers. Although the polyacrylamide gel method is well defined and is used most widely for the entrapment of microbial cells, reactive monomers and, in some cases, reactive cross-linking reagents are liable to inactivate enzymes, even those within the cells, during the entrapment procedure. Also, there will be fear of a poisonous action caused by the monomer remaining in the polymer gels. To avoid such unfavorable effects of monomer and to develop more convenient immobilization procedures, continuous efforts are being made to find new materials and techniques for gel formation. This section deals with several novel methods for immobilizing microbial cells.

Takata and his co-workers screened an ideal general method to immobilize microbial cells using *Streptomyces phaeochromogenes* with glucose isomerase activity as the test organism (110). Among various natural and synthetic polymers, κ-carrageenan, a kind of polysaccharide, was found to

Table 1 Application of immobilized treated cells

Microorganism	Immobilization method	Substrate	Product	Reference
Escherichia coli	Entrapment in polyacrylamide gel	Ammonium fumarate	L-Aspartate	9, 120
E. coli	Entrapment with κ-carrageenan	Ammonium fumarate	L-Aspartate	96
E. coli	Entrapment with κ-carrageenan and locust bean gum	Ammonium fumarate	L-Aspartate	80
Pseudomonas dacunhae	Entrapment with κ-carrageenan	Ammonium L-aspartate	L-Alanine	131
Pseudomonas putida	Entrapment in polyacrylamide gel	L-Arginine	L-Citrulline	See 1, 7
Streptococcus faecalis	Entrapment in polyacrylamide gel	L-Arginine	L-Citrulline	See 1
Aspergillus ochraceus	Mycelium pellets	*N*-Acetyl-DL-methionine	L-Methionine	31
Erwinia herbicola	Entrapment in collagen	Pyruvate + phenol + NH_3	L-Tyrosine	130
E. coli	Entrapment in polyacrylamide gel	L-Glutamate + L-cysteine + glycine	Glutathione	74
Saccharomyces cerevisiae	Entrapment in polyacrylamide gel	L-Glutamate + L-cysteine + glycine	Glutathione	72
E. coli + *S. cerevisiae*	Entrapment with κ-carrageenan	L-Glutamate + L-cysteine + glycine	Glutathione	73
Brevibacterium ammoniagenes	Entrapment in polyacrylamide gel	Fumarate	L-Malate	See 1, 7
Brevibacterium flavum	Entrapment with κ-carrageenan	Fumarate	L-Malate	111
Achromobacter liquidum	Entrapment in polyacrylamide gel	L-Histidine	Urocanic acid	See 1, 7
Micrococcus luteus	Immobilization on carbodiimide activated carboxymethylcellulose	L-Histidine	Urocanic acid	33
Actinoplanes missouriensis	Entrapment in cellulose fiber	Glucose	Fructose	61

Streptomyces phaeochromogenes	Entrapment by radiation polymerization of 2-hydroxyethyl methacrylate	Glucose	Fructose	56
Lactobacillus bulgaricus, E. coli, or *Kluyveromyces lactis*	Entrapment in polyacrylamide gel	Lactose	Glucose + galactose	85
Gluconobacter melanogenus	Entrapment in polyacrylamide gel	L-Sorbose	L-Sorbosone	See 1
G. melanogenus + *Pseudomonas* sp.	Entrapment in polyacrylamide gel	L-Sorbose	2-Keto-L-gulonic acid	See 1
Bakers' yeast	Entrapment with photo-cross-linkable resin prepolymer	Adenosine + glucose + P_i	ATP	3
S. cerevisiae	Microencapsulation with ethylcellulose	AMP + glucose + P_i	ATP	95
Hansenula jadinii	Entrapment with photo-cross-linkable resin prepolymers	CMP + choline + P_i + glucose	CDP-choline	45, 46
S. cerevisiae	Microencapsulation with ethylcellulose	CMP + choline + P_i + glucose	CDP-choline	95
Achromobacter aceris	Entrapment in polyacrylamide gel	NAD + ATP	NADP	122
B. ammoniagenes	Entrapment in polyacrylamide gel	NAD + polyphosphate	NADP	71
S. cerevisiae + *B. ammoniagenes*	Entrapment in polyacrylamide gel or Encapsulation with cellulose acetate-butylate	NAD + glucose + P_i	NADP	76
B. ammoniagenes	Entrapment in polyacrylamide gel	Pantothenate + cysteine + ATP	CoA	95, 101
B. ammoniagenes + *S. cerevisiae*	Entrapment in polyacrylamide gel	Pantothenate + cysteine + glucose + P_i	CoA	95

Table 1 *(Continued)*

Microorganism	Immobilization method	Substrate	Product	Reference
Enterobacter aerogenes	Entrapment with photo-cross-linkable resin prepolymer or urethane prepolymer	Uracil arabinoside + adenine	Adenine arabinoside	134
Achromobacter butyri	Entrapment in polyacrylamide gel	Glucose + metaphosphate	Glucose-6-phosphate	77
Kluyvera citrophila	Entrapment in polyacrylamide gel	6-Aminopenicillanic acid + D-phenylglycine methyl ester	Ampicillin	70
Arthrobacter simplex	Entrapment with photo-cross-linkable resin prepolymers or urethane prepolymers	Hydrocortisone	Prednisolone	107, 108
A. simplex	Entrapment in collagen	Hydrocortisone	Prednisolone	11
Nocardia rhodocrous	Entrapment with photo-cross-linkable resin prepolymers	4-Androstene-3,17-dione	1,4-Androstadiene-3,17-dione	132
N. rhodocrous	Entrapment with photo-cross-linkable resin prepolymers or urethane prepolymers	3β-Hydroxy-Δ^5-steroids	3-Keto-Δ^4-steroids	87, 89
N. rhodocrous	Entrapment with urethane prepolymers	Testosterone	Δ^1-Dehydrotestosterone or 4-androstene-3,17-dione	16
Rhodotorula minuta	Entrapment with photo-cross-linkable resin prepolymer or urethane prepolymers	*dl*-Menthyl succinate	*l*-Menthol	88

be a suitable matrix for entrapment of the cells. A solution of κ-carrageenan and a suspension of cells were mixed at 37°C, and the mixture was cooled to 10°C and was allowed to stand for 30 min. The gel formed was soaked in cold potassium chloride solution to increase gel strength. Although carrageenan-entrapped cells showed glucose isomerase activity comparable to that of polyacrylamide gel-entrapped cells, mechanical strength of carrageenan gel was lower than that of polyacrylamide gel. Hardening of carrageenan gel with hexamethylenediamine and glutaraldehyde significantly improved not only the gel strength, but also the operational stability of the enzyme without reduction of the enzyme activity during the procedure. Locust bean gum was also used with κ-carrageenan to enhance gel strength and to lower gelling temperature (80, 110). In addition to *S. phaeochromogenes, Escherichia coli* (aspartase) and *Brevibacterium ammoniagenes* (fumarase) were also entrapped by this method (121). Entrapped *E. coli* cells showed 693 days of half-life of aspartase in the continuous production of L-aspartate from ammonium fumarate (96). Immobilized growing cells such as *Serratia marcescens, Acetobacter suboxydans,* and *Saccharomyces cerevisiae,* were prepared by this method and were used for the production of L-isoleucine, L-sorbose, and ethanol, respectively (123).

The advantage of the carrageenan method is that the immobilization of microbial cells can be conveniently carried out under very mild conditions. In the case of entrapment of living cells, the survival ratio of microorganisms after immobilization is high, and numbers of living cells can be counted easily by the serially diluted drop-plate method, since the gel becomes a cell suspension by removing the gel-inducing agent.

Cellulose derivatives, such as acetate, triacetate, and nitrate, have been widely used for immobilization of enzymes and microbial cells. Recently, Linko et al (60, 62) have developed a new method to entrap microbial cells in α-cellulose beads. Microbial cells with glucose isomerase, invertase, β-galactosidase, or urease activity were immobilized with an activity yield of 22–85%, depending on the kind of enzyme. Proteinic gels formed from bovine albumin and glutaraldehyde were also employed to entrap *E. coli* cells that had β-galactosidase activity (91). Membranes and porous particles were prepared by this method.

Inorganic compounds, such as metal hydroxides, were interesting supports for cell immobilization because of their low cost, convenient preparation, the absence of any need for pretreatment, etc. Kennedy et al (43) developed a method to immobilize living *E. coli* and *S. cerevisiae* cells with titanium (IV) hydroxide and zirconium (IV) hydroxide. Microbial cells were immobilized by mixing metal hydroxides prepared from their tetrachlorides. Hydroxyl groups on the surface of the metal hydroxide seem to

Table 2 Application of immobilized resting cells and immobilized nontreated cells

Microorganism	Immobilization method	Substrate	Product	Reference
Streptococcus faecalis	Entrapment in polyacrylamide gel	Arginine	Ornitine + putrescine	See 1
Escherichia coli	Entrapment in polyacrylamide gel	Indole + pyruvate + NH_3	L-Tryptophan	65
Bacillus sp.	Entrapment in polyacrylamide gel	Hydantoins	N-Carbamyl-D-amino acids	129
Saccharomyces cerevisiae	Entrapment in polyacrylamide gel	L-Glutamate + L-cysteine + glycine	Glutathione	72, 75
Proteus mirabilis	Entrapment in formaldehyde-cross-linked gelatin	2-Keto acids	2R-Hydroxy acids	119
Clostridium sp.	Entrapment in formaldehyde-cross-linked gelatin or polyacrylamide gel	2-Enoates	Saturated acids	119
Saccharomyces pastorianus	Entrapment with agar	Sucrose	Glucose + fructose	See 1
Gluconobacter melanogenus + *Pseudomonas* sp.	Entrapment in polyacrylamide gel	L-Sorbose	2-Keto-gulonic acid	See 1
Acetobacter suboxydans	Entrapment in polyacrylamide gel or intercellular cross-linking with glutaraldehyde	Polyhydric alcohols	Ketoses	99
Escherichia freundii	Entrapment in polyacrylamide gel	Glucose + *p*-nitrophenyl-phosphate	Glucose-6-phosphate + glucose-1-phosphate	94
Mastigocladus laminosus	Entrapment in calcium alginate	ADP + P_i + light	ATP	97
Brevibacterium ammoniagenes	Entrapment by radiation copolymerization of acrylamide and metal salt of acrylic acid	NAD + ATP	NADP	25

E. coli	Entrapment in polyacrylamide gel	Penicillin G	6-Aminopenicillanic acid	See 7
Penicillium chrysogenum	Entrapment in polyacrylamide gel	Glucose + NH_3 + phenylacetate	Penicillin G	68
Pseudomonas testosteroni	Entrapment in polyacrylamide gel	Reichstein's compound S	Δ^1-Dehydro-Reichstein's compound S	133
Curvularia lunata	Entrapment in polyacrylamide gel	Reichstein's compound S	Hydrocortisone	See 1
Mycobacterium phlei	Entrapment in polyacrylamide gel	4-Cholesten-3-(O-carboxymethyl)-oxime	4-Androstene-17-on-3-(O-carboxymethyl)-oxime	4
S. cerevisiae	Entrapment in calcium alginate	Glucose	Ethanol	44
Acetobacter xylinum	Entrapment in polyacrylamide gel	Glycerol	Dihydroxyacetone	78
Clostridium butyricum	Entrapment in polyacrylamide gel	Glucose	H_2	See 1
C. butyricum	Entrapment with agar	Wastewater	H_2	109
Anabaena cylindrica	Adsorption on glass beads	H_2O + light	H_2	57
Anabaena sp.	Entrapment with agar	H_2O + light	H_2	41
M. laminosus	Entrapment in calcium alginate	H_2O + light	Photocurrent	81
Candida tropicalis	Entrapment in polystyrene or aluminum alginate	Phenol	(Degradation)	See 7
Achromobacter guttatus	Entrapment in polyacrylamide gel	ϵ-Aminocaproic acid cyclic dimer	(Hydrolysis)	See 7
Azotobacter vinelandii	Adsorption on anion exchange cellulose	N_2	(Nitrogen-fixation)	100
Micrococcus denitrificans	Encapsulation with liquid membrane	NO_3^-, NO_2^-	(Reduction)	See 7
Photobacterium phosphoreum	Entrapment in calcium alginate	—	Bioluminescence	64

Table 3 Application of immobilized growing cells

Microorganism	Immobilization method	Substrate	Product	Reference
Corynebacterium glutamicum	Entrapment in polyacrylamide gel	Glucose	L-Glutamate	103
Brevibacterium flavum	Complexation with collagen	Glucose	L-Glutamate	12
Serratia marcescens	Entrapment with κ-carrageenan	Glucose + D-threonine	L-Isoleucine	126
Acetobacter sp.	Immobilization on titanium (IV) hydroxide	Ethanol	Acetic acid	42
Lactobacillus delbrückii	Entrapment in calcium alginate	Glucose	Lactate	59
Enterobacter aerogenes	Entrapment with κ-carrageenan	Glucose	2,3-Butanediol	10
Bacillus sp.	Entrapment in polyacrylamide gel	Peptone	Bacitracin	69
Fusarium sp.	Various methods	Penicillin V	6-Aminopenicillanic acid	5
Penicillium ulticae	Entrapment with κ-carrageenan	Glucose	Patulin	13
Rhizopus nigricans	Entrapment with agar	Progesterone	11α-Hydroxy-progesterone	63
Arthrobacter simplex	Entrapment in polyacrylamide gel or calcium alginate	Hydrocortisone	Prednisolone	58, 83, 84
Arthrobacter globiformis	Entrapment in polyacrylamide gel	Hydrocortisone	Prednisolone	52
Curvularia lunata	Entrapment in polyacrylamide gel or calcium alginate	Reichstein's compound S	Hydrocortisone	82
C. lunata	Entrapment with photo-cross-linkable resin prepolymer	Reichstein's compound S	Hydrocortisone	105, 106
Bacillus subtilis	Entrapment in polyacrylamide gel	Meat extract + yeast extract	α-Amylase	50
Streptomyces fradiae	Entrapment in polyacrylamide gel	Meat extract + starch	Protease	51
Saccharomyces carlsbergensis	Entrapment with κ-carrageenan	Glucose	Ethanol	6, 8
Saccharomyces cerevisiae	Adsorption on gelatin-coated Raschig rings	Glucose	Ethanol	102
S. cerevisiae	Entrapment with κ-carrageenan	Glucose	Ethanol	123–125
S. cerevisiae	Entrapment in aluminum alginate	Glucose	Ethanol	20
S. cerevisiae	Immobilization on Carrier A	Molasses	Ethanol	23
S. cerevisiae	Entrapment in calcium alginate	Barley wort	Beer	59
Methanogenic bacteria	Entrapment with agar	Wastewater	Methane	35
Methanosarcina barkeri	Entrapment in calcium alginate	Methanol	Methane	98
Pseudomonas putida	Entrapment in polyacrylamide gel	Benzene	(Degradation)	104
Candida tropicalis	Entrapment in polyacrylamide gel	Phenol	(Degradation)	47

be replaced by suitable ligands from cells, resulting in the formation of partial covalent bonds. Ligands could be the hydroxyl residue of serine or the ϵ-amino residue of lysine. *Acetobacter* sp. was immobilized on titanium (IV) hydroxide to continuously produce acetic acid from ethanol (42).

For the development of immobilized microbial cell techniques, gel matrices that have a variety of physico-chemical properties are required, depending on the nature of reactions. However, it is difficult to satisfy such demand by using natural polymer gels because of the difficulty in preparing their derivatives. Synthetic polymer gels that have different characteristics may be prepared by simple procedures from certain monomers or from prepolymers that have different properties, although the conventional method of polyacrylamide gel sometimes inactivates enzymes in cells.

Kumakura et al immobilized glucose isomerase-containing *S. phaeochromogenes* with a hydrophilic monomer (2-hydroxyethyl methacrylate) (53, 55) or hydrophobic monomers (butanediol monomethacrylate,1,3-butyleneglycol dimethacrylate, neopentylglycol dimethacrylate, and trimethyrolpropane triacrylate) (54) by γ-irradiation at –78°C.

More simple and convenient methods are the prepolymer methods developed by the present authors by using photo-cross-linkable resin prepolymers (18, 19) and urethane prepolymers (21). These methods have the following merits. (*a*) The formation of polymer matrices can be performed under extremely mild conditions with simple procedures. Photo-cross-linkable resin prepolymers are polymerized in the presence of an appropriate photosensitizer by illumination with near-UV light for several minutes. Urethane prepolymers can be polymerized only by mixing with an aqueous solution of enzymes or an aqueous suspension of organelles or microbial cells. Thus, the reactions could be achieved without heating, shifting pH to extreme values, or the use of chemicals that might modify the structures of biocatalysts. (*b*) Prepolymers containing multiple photosensitive functional groups or isocyanate groups at a desired distance can be prepared previously, in the absence of biocatalysts. This makes it possible to select the size of network of gel matrices. Furthermore, a suitable ionic, hydrophobic, or hydrophilic property can be introduced at this stage.

These prepolymers have been widely applied to the immobilization of not only enzymes (17, 32, 113, 114, 116), but also cellular organelles such as yeast peroxisomes (microbodies) (115, 117, 118), yeast mitochondria (112) and bacterial chromatophores (22), and microbial cells (45, 46, 87, 89, 107, 132). Water-insoluble, hydrophobic prepolymers were also utilized successfully to entrap enzymes and microbial cells (90). Advantages of hydrophobic gels in the transformations of lipophilic compounds by immobilized cells under hydrophobic conditions are discussed later. Structures of several photo-cross-linkable resin prepolymers and structure and properties of urethane prepolymers are shown in Figures 1 and 2, respectively. Polyurethane was also utilized to entrap *E. coli* having penicillin acylase (48).

Only a few works have been reported concerning covalent binding of microbial cells to a suitable support. One example is the immobilization of yeast cells to hydroxyalkyl methacrylate gels reported by Jirků et al (34) and Gulaya et al (24). They immobilized *Zygosaccharomyces lactis,* which had β-galactosidase activity, and *Saccharomyces paradoxus,* which showed secodione reduction activity, on these gels, respectively.

ATP REGENERATION

Recently, extensive studies on bioreactor, combining ATP-consuming and ATP-generating systems with immobilized biocatalysts, have been carried out to produce complex compounds. The study on synthesis of gramicidin S conjugated with ATP-regenerating system using acetylphosphate is well known. However, from the viewpoint of practical usage, enzyme activities and their operational stabilities in such systems are sometimes unequal, and it is very difficult to maintain whole enzyme systems in a stable state for long periods of time. Moreover, supplies of acetylphosphate at a low price cannot be expected because its mass production has not been attempted.

Immobilized microbial cells, whether they are living or dead, can be utilized as a tool to regenerate ATP. For example, immobilized yeast cells (heat treated or acetone dried) phosphorylate adenosine, AMP, or ADP to yield ATP using glycolytic energy (3, 46, 95). Acetone-dried bakers' yeast was immobilized with photo-cross-linkable resin prepolymer and was equipped in a reactor designed for loading sheet-type immobilized yeast cells. In the continuous system supplied with glucose, phosphate, mag-

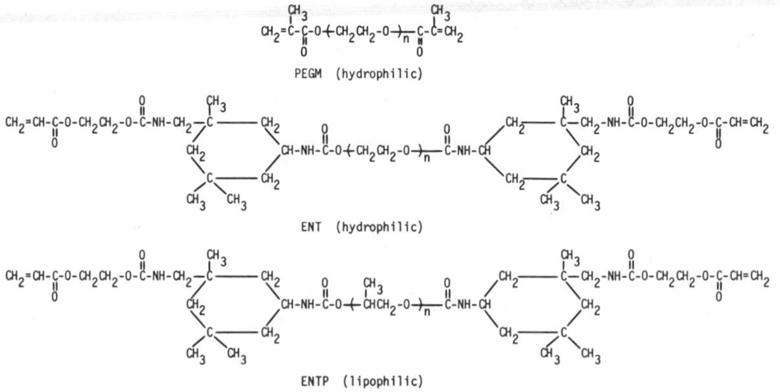

$$CH_2{=}C(CH_3)-C(=O)-O-(CH_2CH_2-O)_n-C(=O)-C(CH_3){=}CH_2$$

PEGM (hydrophilic)

ENT (hydrophilic)

ENTP (lipophilic)

Figure 1 Structures of typical photo-cross-linkable resin prepolymers. PEGM, ENT, and ENTP are trade names of prepolymers (ENT shows entrapment).

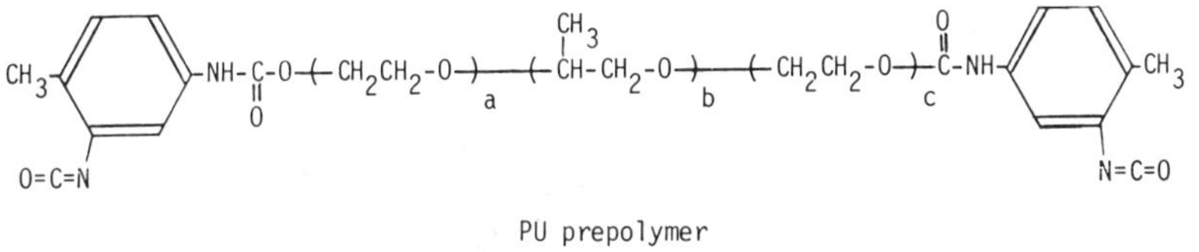

$$CH_3\text{-}C_6H_3(N{=}C{=}O)\text{-}NH\text{-}\underset{\underset{O}{\|}}{C}\text{-}O\text{-}(CH_2CH_2\text{-}O)_a\text{-}(\overset{\overset{CH_3}{|}}{CH}\text{-}CH_2\text{-}O)_b\text{-}(CH_2CH_2\text{-}O)_c\text{-}\overset{\overset{O}{\|}}{C}\text{-}NH\text{-}C_6H_3(N{=}C{=}O)\text{-}CH_3$$

PU prepolymer

Prepolmer	$\bar{Mw}$ of diol	NCO content (%)	Ethylene glycol content (%)
PU-3	2529	4.2	57
PU-6	2627	4.0	91
PU-9	2616	4.0	100

Figure 2 **Structure and properties of urethane prepolymers. PU-3 gives hydrophobic gels, whereas PU-6 and PU-9 give hydrophilic gels.**

nesium ion, and NAD, the immobilized cells converted over 75% of 20 mM adenosine to ATP. The system was stable for about 55 hr at 28°C (3). *Saccharomyces cerevisiae* cells encapsulated with ethylcellulose were packed in a column reactor, and the conversion of AMP to ATP was carried out at 35°C with a supply of glucose, phosphate, and magnesium ion at a space velocity of 0.2 hr^{-1}. At the beginning of the reaction, the conversion of 43 mM AMP to ATP was more than 70%. Although the enzyme activity decreased gradually, a conversion of more than 50% was maintained even after 10 days of operation (95). A more stable ATP-generating system was prepared by entrapping *S. cerevisiae* cells in polyacrylamide gel (76). Under the optimal conditions, the ATP-generating activity of the immobilized cells was 7.0 μmol/hr/ml of gel. The half-life of the column packed with these immobilized cells was 19 days at a space velocity of 0.3 hr^{-1} at 30°C.

These ATP-generating systems can be combined with the ATP-consuming systems in the same microbial cells or in other cells. When CMP and choline were added to the ATP-generating system of *Hansenula jadinii,* CMP was phosphorylated to CDP and CTP and the concomitant formation of CDP-choline was observed (45, 46). In this case, about 50% of CMP was converted to CDP-choline, and the addition of ATP was not necessary in the reused immobilized cell system. CDP-choline was also produced by *S. cerevisiae* encapsulated with ethylcellulose and chitosan (95). The substrate solution composed of glucose, CMP (20 mM), choline, phosphate, and magnesium ion was passed through a column reactor at a space velocity of

0.25 hr^{-1} and at 35°C. During the first 8 days the conversion ratio of CMP to CDP-choline was maintained at 50–60%. The turnover number of ATP in this system was calculated to be about 1400 cycles/hr.

Intact cells of *S. cerevisiae* entrapped in polyacrylamide gel were applied to the continuous production of glutathione from L-glutamate, L-cysteine, and glycine using the glycolytic system in the same organism as energy source. For the continuous operation, a supply of a small amount of NAD was necessary, as in the case of ATP and CDP-choline production. The column reactor was stable and its half-life was about 25 days (72). This system of *S. cerevisiae* (75) was found to be superior to the co-immobilized system of *E. coli* (glutathione synthesizing) and *S. cerevisiae* (ATP generating) (73), though the activity was lower than that of the immobilized *E. coli* system using acetylphosphate as phosphate donor for ATP-generation (74).

A co-immobilized system was also constructed for the production of CoA from pantothenate and cysteine. *Brevibacterium ammoniagenes* (CoA synthesizing) and *S. cerevisiae* (ATP generating) co-entrapped in polyacrylamide gels showed a higher activity of CoA synthesis using glucose as energy source than immobilized *B. ammoniagenes* alone, supplemented with ATP (95).

Nicotinamide adenine dinucleotide (NAD) phosphate (NADP) was produced by the phosphorylation of NAD with *B. ammoniagenes* having NAD kinase activity and *S. cerevisiae* having glycolytic activity. Both the co-entrapped system of these two microorganisms and the mixed system of immobilized cells of each microorganism were stable over 10 days, keeping the activity of NADP formation of about 0.6 μmol/hr/ml gel at a space velocity of 0.3 hr^{-1} at 30°C (76). A mixed immobilized cell system prepared by encapsulation with cellulose-acetate-butylate was also stable in batch reactions showing the conversion ratio of 50–55% with 6 g/liter of NAD (2).

In addition to the enzyme systems, many efforts have been made to generate ATP by photophosphorylation and oxidative phosphorylation. The photosynthetic generation of ATP by intact blue-green algae is especially interesting (97). However, as mentioned above, the yeast glycolytic system seems to be the most convenient, cheap, and stable ATP-generating system at present.

BIOCONVERSION OF HIGHLY LIPOPHILIC AND WATER-INSOLUBLE COMPOUNDS BY IMMOBILIZED CELLS

Enzymatic and fermentative reactions, even in cases where water-insoluble, lipophilic compounds are substrates, usually have been carried out in aqueous systems, because enzymes and microbial cells were believed to be unstable in organic solvent systems. Immobilization often gives stability for biocatalysts against the denaturation by organic solvents.Thus, bioconversions of water-insoluble compounds, such as steroids, have been carried out with immobilized microbial cells in organic solvent systems. Hydrophobicity of gels entrapping microbial cells sometimes seriously affects the efficiency of such conversions.

For the conversions of such lipophilic compounds, several solvent systems are applicable: water-water-miscible organic solvent homogeneous systems, water-water-immiscible organic solvent two-phase systems, and organic solvent systems. Water-miscible organic co-solvents, such as methanol, are often useful in constructing homogeneous reaction mixtures with immobilized microbial cells. For example, dehydrogenation of Reichstein's compound S by *Pseudomonas testosteroni* (133) and of hydrocortisone by *Arthrobacter simplex* (107, 108) was carried out successfully in aqueous systems containing 10% methanol. Methanol was also utilized for the bioconversions of hydrocortisone (58, 83) and of Reichstein's compound S (82) by immobilized growing cells. Thus, water-water-miscible organic solvent systems are useful in many cases, but the concentration of substrates to be dosed is limited by their solubilities in such systems.

A few reports have been published about the application of water-organic solvent two-phase systems and organic solvent systems to bioconversions of water-insoluble compounds by free microbial cells. However, free microbial cells seem to be unstable in the latter systems, and the former systems are not applicable to homogeneous and continuous bioconversion systems.

Nocardia rhodocrous cells entrapped with photo-cross-linkable resin prepolymers or urethane prepolymers were found to be active in catalyzing dehydrogenation of steroids at several positions, even in organic solvents (Table 4). Several water-saturated organic solvents were used, depending on the solubilities of substrates and products and on the stability of enzymes —benzene : *n*-heptane, 1:1 by volume (Solvent A); chloroform: *n*-heptane, 1:1 by volume (Solvent B); benzene:*n*-heptane, 4 : 1 by volume (solvent C). In the conversion of cholesterol (a highly hydrophobic compound) to cholestenone, hydrophobicity of gels entrapping the cells seriously affected the activity of the cells (87). That is, hydrophobic gel-entrapped cells showed a high activity, whereas hydrophilic gel-entrapped cells did not show any activity of the conversion in solvent A. This result coincided with the partition of cholesterol between the cell-entrapping gels and the external reaction solvent. Highly hydrophobic cholesterol could not penetrate into the hydrophilic gels when nonpolar solvent A was used as reaction solvent. The permeation of the substrate was improved in a more polar solvent, solvent B (89). However, polar solvents sometimes inactivate the enzymes in the cells. A less hydrophobic substrate, dehydroepiandrosterone, was transformed to 4-androstene-3,17-dione by both hydrophobic and hydrophilic gel-entrapped cells, the former cells revealing a higher activity than the latter. Thus, the advantage of hydrophobic gels was clarified in the bioconversions of hydrophobic compounds in nonpolar solvent systems. The prepolymer methods mentioned previously are very convenient for preparing gels of different hydrophobicities or hydrophilicities. In the case of steroid Δ^1-dehydrogenation, an appropriate electron acceptor, such as water soluble phenazine methosulfate (PMS), is necessary for the reaction (16, 132). Permeation of steroids is limited in the hydrophilic gels, whereas that of PMS is limited in the hydrophobic gels. This difference in permeation of PMS into gels has a dramatic effect on the diverse transformation of testosterone in solvent C. That is, when the hydrophilic gel-entrapped cells were used, the Δ^1-dehydrogenation product, Δ^1-dehydrotestosterone, was accumulated. This phenomenon can be explained by an abundant supply of PMS, and inhibition of 17β-hydroxysteroid dehydrogenase by Δ^1-dehydrotestosterone accumulated inside gel matrices. On the other hand, the hydrophobic gel-entrapped cells converted testosterone to 4-androstene-3,17-dione by the action of 17β-hydroxysteroid dehydrogenase, which did not essentially require PMS and was activated by a low concentration of PMS. The product could not be dehydrogenated at Δ^1-position because of a low concentration of PMS in the hydrophobic gels (16). It is very difficult to obtain such intermediate products by the free cell systems because both intermediates are converted to 1,4-androstadiene-3,17-dione in the presence of PMS. This concept would be applied to control diverse bioconversions of various hydrophobic compounds and to selective transformations of specific compounds in mixtures.

Stereoselective hydrolysis of *dl*-menthyl ester to yield *l*-menthol has been carried out in water-saturated *n*-heptane by *Rhodotorula minuta* entrapped with urethane prepolymers or a photo-cross-linkable resin prepolymer (88). The entrapped cells were very stable in the organic solvent, showing 55–63 days of half-life, whereas the half-life of the free cells was only 2 days.

Although many organic compounds have been excluded from enzymatic reactions by their low solubilities in aqueous solution, the examples mentioned above may encourage transformations of various hydrophobic compounds by using immobilized biocatalysts in organic solvent systems.

IMMOBILIZED GROWING CELLS

In addition to immobilized treated cells, immobilized living cells, especially immobilized growing cells, are attracting worldwide attention because of the superior stability that results from the self-regenerating nature of catalytic systems. Immobilized growing cells can catalyze not only single conversions and salvage synthesis, but also synthesis of various complex

Table 4 Steroid conversions in organic solvents catalyzed by immobilized *Nocardia rhodocrous* cells

Substrate		Product
Steroid Δ^1-dehydrogenation		
4-Androstene-3,17-dione	→	1,4-Androstadiene-3,17-dione
Testosterone	→	Δ^1-Dehydrotestosterone
17β-0-Acetyltestosterone	→	17β-0-Acetyl-Δ^1-dehydrotestosterone
3β-Hydroxy-Δ^5-steroid dehydrogenation		
Cholesterol	→	Cholestenone
β-Sitosterol	→	β-Sitostenone
Stigmasterol	→	Stigmastenone
Dehydroepiandrosterone	→	4-Androstene-3,17-dione
Pregnenolone	→	Progesterone
17β-Hydroxysteroid dehydrogenation		
Testosterone	→	4-Androstene-3,17-dione
Δ^1-Dehydrotestosterone	→	1,4-Androstadiene-3,17-dione
β-Estradiol	→	Estrone

compounds produced at present by conventional fermentation techniques.

When living microbial cells were immobilized and cultivated with a supply of adequate nutrients, the immobilized cells were found to form a dense layer near the gel surface (123, 124). This result indicates that cells grow at the place where nutritive balance is optimal for the growth. Such findings are also applicable to the preparation of effective immobilized biocatalysts. That is, after immobilized cells are grown and specific enzymes are induced in situ, the grown cells, whether living or dead, can be utilized as excellent catalysts. Immobilized *Kluyvera citrophila* cells were shown to have high penicillin acylase activity by this procedure and were used to prepare ampicillin after treatment with alkali (70).

Production of L-glutamate de novo from glucose by *Corynebacterium glutamicum* is the first example of laboratory application of immobilized growing cells (103). The immobilized cells produced more L-glutamate in successive batch reactions than did the free cells. Larsson and his coworkers (58, 83, 84) reported an interesting technique: activation of entrapped cells by nutrients and a specific inducer. *Arthrobacter simplex* cells entrapped in polyacrylamide gels were incubated to convert hydrocortisone to prednisolone. In the presence of peptone and hydrocortisone the transformation activity of the cells increased with successive batch reactions, until it was about 10 times higher than that of the original cells. The activated cells were stable over a long period. Similar activation of immobilized cells by suitable nutrients was also reported in the degradation of benzene by *Pseudomonas putida* (104), of phenol by *Candida tropicalis* (47), and of side-chain of cholesterol by *Mycobacterium phlei* (4), and in the dehydrogenation of hydrocortisone by *Arthrobacter globiformis* (52).

Recent topics in the field of immobilized biocatalysts are the production of biofuels. Production of ethanol from glucose was attempted by Kierstan & Bucke (44), who used resting cells of *S. cerevisiae* entrapped in calcium alginate gels. With a residence time of 10 hr, 10% of glucose was converted to ethanol with 90% of the theoretical yield. Half-life of the entrapped cells was calculated to be 250 hr.

Chibata and his co-workers (6, 8) have extensively investigated the production of ethanol by κ-carrageenan-entrapped *S. carlsbergensis.* When the gels containing about 3.5×10^6 cells per ml of gel were incubated in a nutrient medium at 30°C for 60 hr, the number of living cells reached a constant level of about 5.4×10^9 cells per ml of gel. A typical example of the cell growth inside the κ-carrageenan gel is shown in Figure 3. Under these conditions, the immobilized cells produced continuously about 100 mg per ml (12.7% by volume) of ethanol from 200 mg per ml of glucose at a retention time of 2.5 hr (8). This system was also successfully applied to the production of ethanol from molasses. Continuous production of ethanol higher than 110 mg per ml at a retention time of 2.8 hr was maintained over 6 months (6). *S. cerevisiae* cells entrapped with κ-carrageenan also produced a similar concentration of ethanol (125).

Ghose & Bandyopadhyay (23) also reported the production of ethanol from molasses by immobilized *S. cerevisiae.* The immobilized cells produced 71 mg per ml of ethanol at a residence time of 2.86 hr over 75 days. The productivity was 24.9 g per liter per hr. In the fluidized bed reactor packed with aluminium alginate gel-entrapped *S. cerevisiae,* 80 mg per ml of ethanol was produced continuously at a residence time of 4.2 hr over 17 days (20). Ethanol production from glucose was carried out by using *S. cerevisiae* adsorbed on gelatin-coated Raschig rings. The productivity (7.4 g per liter per hr) in the immobilized cell reactor was 4.2 times better than the stirred reactor with free cells (102).

Advantages of the immobilized cell systems are that the reactor volumes are smaller, because active cellular concentrations are higher than in fermentors employing free cells, and that the inhibition by both substrate and product can be reduced in the continuous systems.

Hydrogen gas and methane are also important biofuels. *Clostridium butyricum* entrapped in polyacrylamide gels (38) or in agar gels (109) produced hydrogen gas from glucose or wastewater, respectively. The immobilized cells were far more stable than the free cells and produced hydrogen gas over 20 days at a rate of 6 ml per min per kg of wet gels (corresponding to 100 g of wet cells). The resting cells of *Anabaena cylindrica,* adsorbed on glass beads, produced hydrogen gas by photolysis of water. About 2 ml of the gas per mg of algae was accumulated after 30 days of incubation (57). Agar gel-entrapped *Anabaena* cells were used to construct a photochemical fuel cell system. The hydrogen productivity of the immobilized algae (0.52 μmol/hr/g of wet gel) was three times higher than that of the free algae. The system composed of the immobilized *Anabaena* reactor equipped with an oxygen-removing reactor and the hydrogen-oxygen fuel cell continuously produced a photo-current of 15–20 mA for 7 days (41). A mixed culture of *Rhodospirillum rubrum* and *Klebsiella pneumoniae* entrapped in agar gels also produced hydrogen gas from glucose substrate (128). The operational half-life of the immobilized preparation was 1000 hr.

Methane was produced by agar gel-entrapped methanogenic population from wastewater. The rate of methane production was initially 1.8 μmol per hr per g of gel (corresponding to 20 mg of wet cells) and was increased to a constant level of 4.5 μmol per hr per g of gel after 25 days of incubation. This activity was maintained over a 90-day period (35).

In addition to the products mentioned above, a variety of compounds have been produced by immobilized growing cells. These include L-isoleucine (126), bacitracin (69), 2,3-butanediol (10), α-amylase (50), protease (51), etc (see Table 3). Vitamin B_{12} (S. Fukui, B. Yongsmith, K. Sonomoto, A. Tanaka, unpublished observation) is also an interesting product. Fermentation techniques with immobilized microbial cells will be applied extensively in future.

ENTRAPMENT OF SPORES

Immobilization of living microbial cells inside gel matrices has been mainly limited to bacteria and yeasts. It is very difficult to entrap homogeneously living fungal mycelia, and the homogenization of mycelia and subsequent immobilization often result in the loss of enzyme activities. Furthermore, enzymes leak from mycelia-entrapped gels when the gels are cut into small pieces. However, fungi are important microorganisms that are used in industry to produce and convert many compounds. Thus, the entrapment of fungal spores has been attempted very recently.

As in the case of *Fusarium* sp. (5), the spores themselves hydrolyze penicillin V to 6-aminopenicillanic acid. However, germination of spores is necessary in many cases to reveal metabolic activity. Spores of *Curvularia lunata* were immobilized in calcium alginate beads. Entrapment in polyacrylamide gels resulted in less germination and far lower enzyme activity. When the entrapped spores were incubated in a nutrient medium, the

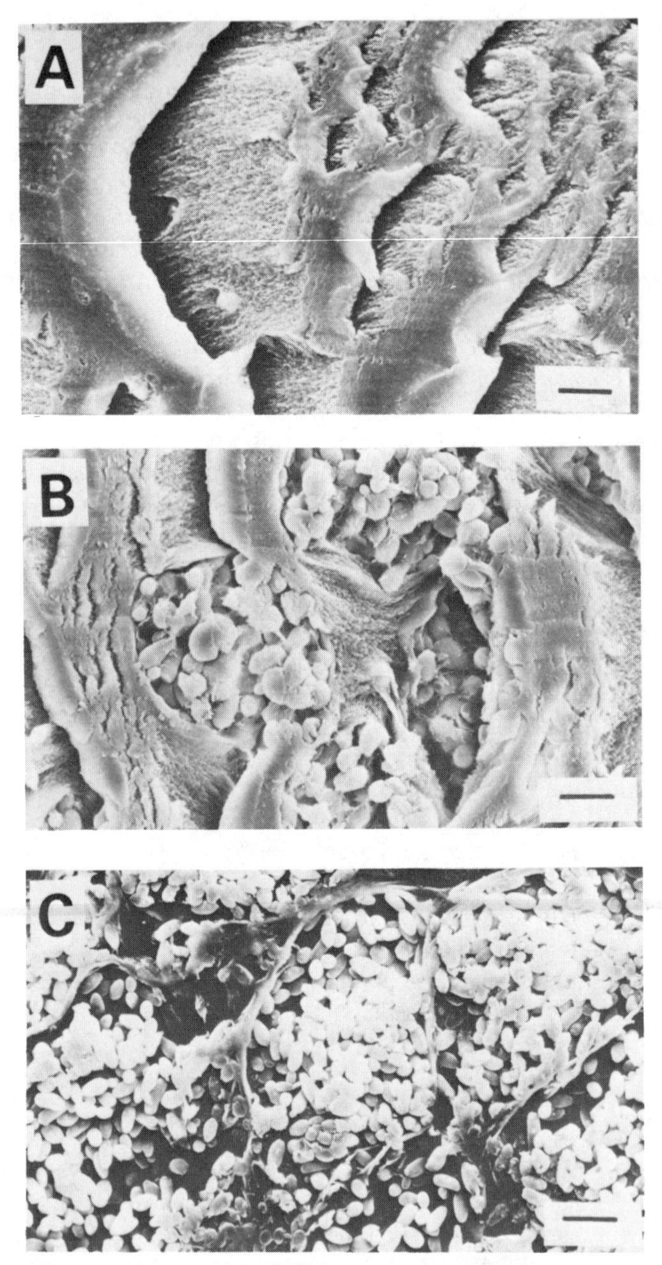

Figure 3 Growth of *Saccharomyces* sp. cells in κ-carrageenan gel. The yeast cells entrapped with κ-carrageenan (*A*) were incubated for 16 (*B*) and 120 hr (*C*) in a nutrient medium. Bar, 10 μm. (Courtesy of I. Chibata and T. Tosa.)

activity of steroid 11β-hydroxylation converting Reichstein's compound S to hydrocortisone was enhanced with the germination of spores (82). Among various entrapment methods, entrapment of *C. lunata* using a photo-cross-linkable resin prepolymer with a suitable chain-length was found to be the best for the development of mycelia, and, subsequently, for the activity of the resulting gel-entrapped mycelia to hydroxylate Reichstein's compound S to hydrocortisone (105). The size of gel matrices seems to have an important effect on the development of mycelia in situ. The entrapped mycelia were found to be far more stable than the free mycelia and could be reactivated by incubation in a nutrient broth (106). Figure 4 shows photographs of photo-cross-linked gels in which growth of *C. lunata* inside gel matrices can be seen clearly. Production of patulin was also reported with *Penicillium urticae* (13). These techniques will be applied widely to bioconversions or fermentations by fungi.

MICROBIAL ELECTRODES

Enzymes, whether free or immobilized, have been widely applied to the assay of various compounds because of their high specificity to substrates

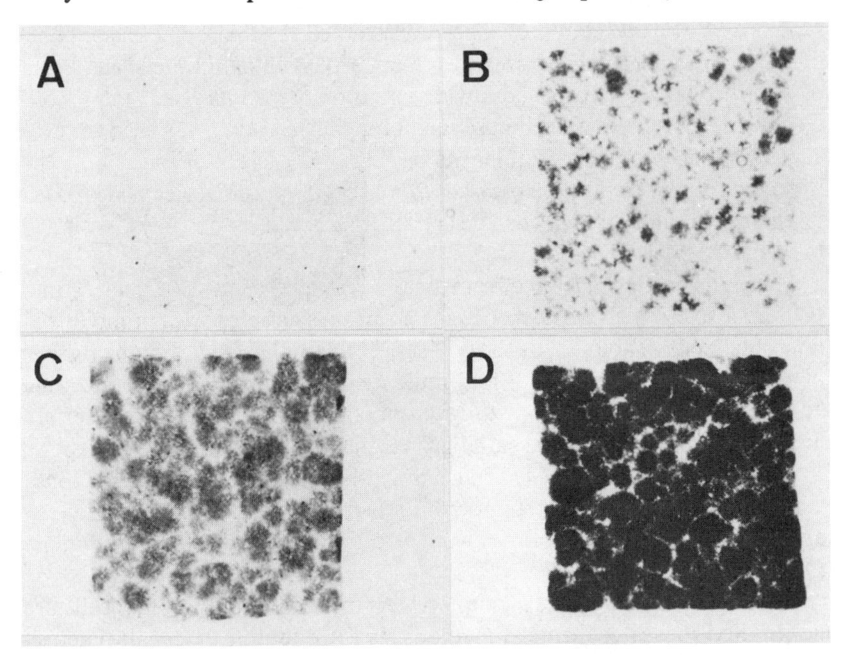

Figure 4 Growth of *Curvularia lunata* inside gel matrices. Spores of *C. lunata* were entrapped with a photo-cross-linkable resin prepolymer (*A*) and were incubated for 28 (*B*), 48 (*C*), and 96 hr (*D*) in a nutrient medium. Size of each gel, 5 X 5 mm.

and their high sensitivity. Microbial cells can also be utilized as excellent biocatalysts in these assay systems if their specificities are retained.

Several types of electrochemical assay systems have been developed with devices equipped with microbial electrodes or microbial sensors (26–30, 36, 37, 39, 40, 66, 67, 86). These systems are composed of devices to measure respiratory activities of immobilized microbial cells coupled to an oxygen electrode or to measure metabolic products of immobilized cells coupled to different electrodes. Application of these systems to on-line measurement in fermentation industries has been tried.

For example, *Trichosporon brassicae* cells adsorbed on porous acetyl cellulose membrane were attached on an oxygen electrode and were applied to assays of ethanol and acetic acid (40). Above pH 6, the yeast oxidized only ethanol among several lower alcohols, lower fatty acids, and sugars tested. The measurable range of ethanol was below 22 mg per liter, and the ethanol electrode was stable for more than 3 weeks and 2100 assays (27). Assay of acetic acid by this system must be carried out below the pK value for acetic acid (4.75 at 30°C), because acetate ion cannot pass through the gas-permeable membrane. The electrode was also found to be stable, but the substrate specificity was not high (26). An interesting microbial sensor was reported by Karube et al (36). They constructed the microbial sensor composed of two microbial electrodes by using a recombination-deficient strain of *Bacillus subtilis* (Rec^-) and a wide strain of *B. subtilis* (Rec^+) immobilized with a porous acetylcellulose membrane. The sensor was used to measure the concentration of mutagens based on the inhibitory action of the mutagens on the respiration of *B. subtilis* (Rec^-). Microbial electrodes based on their respiratory activity can be used in a wide variety of assays, as shown in Table 5.

Microbial products derived from assay substrates, such as CO_2, NH_3, and H^+, are also assayed by the respective electrodes coupled with immobilized microbial cells (Table 5). In these cases, both single-step reactions (assays of glutamate, glutamine, etc) and multistep reactions (assays of nicotinic acid, nitrate, etc) can be appled. Microorganisms, which require a certain vitamin or amino acid as essential growth factor, are used for the purpose.

To construct microbial electrodes, microbial cells need not be immobilized on or in membrane supports. Free microbial cells surrounded by appropriate dialysis membranes were also used in the assays of glutamate (93), arginine (92), etc.

The measurement systems coupled with immobilized microbial cells are not limited to electrochemical methods. Spectrophotometric, calorimetric, and other methods are also applicable as detection devices associated with metabolism of immobilized microbial cells.

Table 5 Application of immobilized microbial cells for assay

Assay substance	Microorganism	Measured substance	Method	Ref.
Glucose	*Pseudomonas fluorescens*	O_2	Amperometry (O_2 electrode)	39
Sugars	*Brevibacterium lactofermentum*	O_2	Amperometry (O_2 electrode)	29
Acetic acid	*Trichosporon brassicae*	O_2	Amperometry (O_2 electrode)	40
Ethanol	*T. brassicae*	O_2	Amperometry (O_2 electrode)	40
Methanol	Unidentified bacterium	O_2	Amperometry (O_2 electrode)	40
Glutamic acid	*Escherichia coli*	CO_2	Potentiometry (CO_2 electrode)	28
Glutamine	*Sarcina flava*	NH_3	Potentiometry (NH_3 electrode)	93
Serine	*Clostridium acidiurici*	NH_3	Potentiometry (NH_3 electrode)	14
Histidine	*Pseudomonas* sp.	NH_3	Potentiometry (NH_3 electrode)	127
Arginine	*Streptococcus faecium*	NH_3	Potentiometry (NH_3 electrode)	92
Cephalosporins	*Citrobacter freundii*	H^+	Potentiometry (pH electrode)	66
Nystatin	*Saccharomyces cerevisiae*	O_2	Amperometry (O_2 electrode)	37
Nicotinic acid	*Lactobacillus arabinosus*	H^+	Potentiometry (pH electrode)	67
Nitrate	*Azotobacter vinelandii*	NH_3	Potentiometry (NH_3 electrode)	49
Ammonia	Nitrifying bacterium	O_2	Amperometry (O_2 electrode)	40
Phenol	*Trichosporon cutaneum*	O_2	Amperometry (O_2 electrode)	79
Methane	*Methanomonas* sp.	O_2	Amperometry (O_2 electrode)	86
BOD	*T. cutaneum*	O_2	Amperometry (O_2 electrode)	30

MISCELLANEOUS

In addition to the topics mentioned above, several interesting reactions have been carried out by using immobilized microbial cells.

Erwinia herbicola cells having β-tyrosinase activity were entrapped with collagen, cross-linked by dialdehyde starch, and lyophilized. The immobilized cells synthesized L-tyrosine from phenol, pyruvate, and ammonia in a flow system with a conversion ratio of 65–70% (130). Similar reactions to produce L-tryptophan from indole, pyruvate, and ammonia were carried out by polyacrylamide gel-entrapped *E. coli* cells that had tryptophanase activity. Productivity of 2.0 mmol per hr and specific activity of 0.66 mmol per hr per g of bacteria were attained in a plug flow reactor (65).

The D-form of phenylglycine-related amino acids is an important starting material for the synthesis of penicillin and cephalosporin derivatives. To obtain these amino acids, Yamada and his co-workers (129) have developed an interesting process combining synthetic and enzymatic reactions. That is, racemic hydantoins synthesized from aldehydes, HCN, and $(NH_4)_2CO_3$ were hydrolyzed stereoselectively, giving *N*-carbamyl-D-amino acids, which were further hydrolyzed chemically to D-amino acids. Alkalophilic *Bacillus* sp. entrapped in polyacrylamide gels produced *N*-carbamyl-D-phenylglycine and *N*-carbamyl-D-thienylglycine with a conversion ratio of about 100% and *N*-carbamyl-D-*p*-hydroxyphenylglycine with a lower conversion ratio. Under alkalic conditions, L-hydantoins remained racemized, giving a conversion ratio as high as 100%. This method is very important for the production of various types of phenylglycine derivatives.

To establish an economical procedure of NADP production, immobilized cell systems having NAD kinase activity were investigated (25, 122). *Achromobacter aceris* entrapped in polyacrylamide gels and incubated at pH 4.0 to inactivate ATP-degrading activity showed an operational half-life of 20 days, and 76% of NAD in the substrate solution was converted to NADP in a column reactor at a space velocity of 0.1 hr^{-1} and at 37°C. Although the advantage of this system over the enzymatic system was mentioned, addition of ATP as co-substrate was not economical. Therefore, *Brevibacterium ammoniagenes* with polyphosphate NAD-kinase activity was selected to improve the process (71). The cells entrapped in polyacrylamide gels and activated by treatment with acetone phosphorylated NAD with metaphosphate as the phosphate donor. The half-life of the column packed with the immobilized cells was about 8 days at a space velocity of 0.1 hr^{-1} and at 36.5°C. At the initial phase of the reaction, 100% of NAD (2.1 mM) was converted to NADP. Immobilized *B. ammoniagenes* coupled with the immobilized yeast cells that had ATP-generating activity via glycolysis was mentioned previously.

Production of an anti-viral reagent, adenine arabinoside, from uracil arabinoside and adenine is also of interest and importance. *Enterobacter aerogenes* entrapped with a photo-cross-linkable resin prepolymer or a urethane prepolymer showed a stable transglycosylation activity over 30 days at 60°C in a buffer containing 40% dimethyl sulfoxide as the co-solvent. This organic co-solvent was necessary to dissolve the substrate (adenine) and the product (adenine arabinoside) for the construction of a homogeneous reaction system with a high concentration of substrate (134).

Mayer et al (67a) have applied genetic engineering techniques to obtain *E. coli* cells with significantly higher penicillin G-acylase levels. The gene of the enzyme of *E. coli* ATCC 11105 was cloned using initially cosmid packaging to isolate the gene directly from the total bacterial genome and then was subcloned on pBR322 and pOP203-3 plasmids. A new hybrid strain obtained by the procedures, *E. coli* 5K (pHM12), had constitutively the enzyme at a markedly enhanced level compared with the wild strain. The hybrid strain cells were entrapped in gels of various types (calcium alginate, polymethacrylamide, polyurethane, and epoxide resin) and the kinetic activity as well as mechanical properties were compared. The cells entrapped in the epoxide beads proved to be most excellent in the continuous production of 6-aminopenicillanic acid not only from penicillin G but also from penicillin V. The immobilized cells were much more stable (half-life, 40 days) than free cells (half-life, 1 day) (48a).

CONCLUSION

To date, industrialization of immobilized cell techniques has been limited to single-step reactions with treated cells. However, immobilized microbial cells, whether living or dead, have many advantages over immobilized enzymes. For example, it is not necessary to extract and purify intracellular and, in some cases, unstable enzymes from cells, and it is very easy to carry out multistep reactions to produce complex compounds. Several topics reviewed here concerned mainly multistep reactions mostly carried out by immobilized living cells. These processes seem to be more economical than conventional fermentations, because the immobilized systems are expected to have higher productivities per unit volume of reactor when they are in continuous operation.

In the future, genetic engineering combined with enzyme engineering holds even greater promise for continuous advances in the development and application of immobilized cell systems. The enhancement of penicillin acylase in *E. coli* mentioned above is one such example.

Literature Cited

1. Abbott, B. J. 1977. *Ann. Rep. Ferment. Processes* 1:205–33
2. Ado, Y., Kimura, K., Samejima, H. 1980. *Enzyme Eng.* 5:295–304
3. Asada, M., Morimoto, K., Nakanishi, K., Matsuno, R., Tanaka, A., Kimura, A., Kamikubo, T. 1979. *Agric. Biol. Chem.* 43:1773–74
4. Atrat, P., Hüller, E., Hörhold, C. 1981. *Eur. J. Appl. Microbiol. Biotechnol.* 12: 157–60
5. Charles, M. 1980. *Abstr. 6th Int. Ferment. Symp.*, p. 120
6. Chibata, I. 1980. *Abstr. 2nd German-Japanese Workshop on Enzyme Technology*, p. 39
7. Chibata, I., Tosa, T. 1977. *Adv. Appl. Microbiol.* 22:1–27
8. Chibata, I., Tosa, T. 1980. *Trends in Biochem. Sci.* 5:88–90
9. Chibata, I., Tosa, T., Sato, T. 1974. *Appl. Microbiol.* 27:878–85
10. Chua, J. W., Erarslan, A., Kinoshita, S., Taguchi, H. 1980. *J. Ferment. Technol.* 58:123–27
11. Constantinides, A. 1980. *Biotechnol. Bioeng.* 22:119–36
12. Constantinides, A., Bhatia, D., Vieth, W. R. 1981. *Biotechnol. Bioeng.* 23: 899–916
13. Deo, Y. M., Costerton, J. W., Gaucher, G. M. 1980. *Abstr. 6th Int. Ferment. Symp.*, p. 122
14. Di Paolantonio, C. L., Arnold, M. A., Rechnitz, G. A. 1981. *Anal. Chim. Acta* 128:121–27
15. Dunnill, P. 1980. *Phil. Trans. R. Soc. London Ser. B* 290:409–20
16. Fukui, S., Ahmed, S. A., Omata, T., Tanaka, A. 1980. *Eur. J. Appl. Microbiol. Biotechnol.* 10:289–301
17. Fukui, S., Sonomoto, K., Itoh, N., Tanaka, A. 1980. *Biochimie* 62:381–86
18. Fukui, S., Tanaka, A., Gellf, G. 1978. *Enzyme Eng.* 4:299–306
19. Fukui, S., Tanaka, A., Iida, T., Hasegawa, E. 1976. *FEBS Lett.* 66: 179–82
20. Fukushima, S. 1980. *Abstr. 2nd Ger-*

man-Japanese Workshop on Enzyme Technology, p. 38
21. Fukushima, S., Nagai, T., Fujita, K., Tanaka, A., Fukui, S. 1978. *Biotechnol. Bioeng.* 20:1465–69
22. Garde, V. L., Thomasset, B., Tanaka, A., Gellf, G., Thomas, D. 1981. *Eur. J. Appl. Microbiol. Biotechnol.* 11:133–38
23. Ghose, T. K., Bandyopadhyay, K. K. 1980. *Biotechnol. Bioeng.* 22:1489–96
24. Gulaya, V. E., Turkova, J., Jirku, V., Frydrychova, A., Čoupek, J., Ananchenko, S. N. 1979. *Eur. J. Appl. Microbiol. Biotechnol.* 8:43–47
25. Hayashi, T., Tanaka, Y., Kawashima, K. 1979. *Biotechnol. Bioeng.* 21: 1019–30
26. Hikuma, M., Kubo, T., Yasuda, T., Karube, I., Suzuki, S. 1979. *Anal. Chim. Acta* 109:33–38
27. Hikuma, M., Kubo, T., Yasuda, T., Karube, I., Suzuki, S. 1979. *Biotechnol. Bioeng.* 21:1845–53
28. Hikuma, M., Obana, H., Yasuda, T., Karube, I., Suzuki, S. 1980. *Anal. Chim. Acta* 116:61–67
29. Hikuma, M., Obana, H., Yasuda, T., Karube, I., Suzuki, S. 1980. *Enzyme Microb. Technol.* 2:234–38
30. Hikuma, M., Suzuki, H., Yasuda, T., Karube, I., Suzuki, S. 1979. *Eur. J. Appl. Microbiol. Biotechnol.* 8:289–97
31. Hirano, K., Karube, I., Suzuki, S. 1977. *Biotechnol. Bioeng.* 19:311–21
32. Itoh, N., Hagi, N., Iida, T., Tanaka, A., Fukui, S. 1979. *J. Appl. Biochem.* 1: 291–300
33. Jack, T. R., Zajic, J. E. 1977. *Biotechnol. Bioeng.* 19:631–48
34. Jirku, V., Turkova, J., Kuchynkova, A., Krumphanzl, V. 1979. *Eur. J. Appl. Microbiol. Biotechnol.* 6:217–22
35. Karube, I., Kuriyama, S., Matsunaga, T., Suzuki, S. 1980. *Biotechnol. Bioeng.* 22:847–57
36. Karube, I., Matsunaga, T., Nakahara, T., Suzuki, S. 1981. *Anal. Chem.* 53: 1024–26
37. Karube, I., Matsunaga, T., Suzuki, S. 1979. *Anal. Chim. Acta* 109:39–44
38. Karube, I., Matsunaga, T., Tsuru, S., Suzuki, S. 1976. *Biochim. Biophys. Acta* 444:338–43
39. Karube, I., Mitsuda, S., Suzuki, S. 1979. *Eur. J. Appl. Microbiol. Biotechnol.* 7: 343–50
40. Karube, I., Suzuki, S., Okada, T., Hikuma, M. 1980. *Biochimie* 62:567–73
41. Kayano, H., Karube, I., Matsunaga, T., Suzuki, S., Nakayama, O. 1981. *Eur. J. Appl. Microbiol. Biotechnol.* 12:1–5
42. Kennedy, J. F. 1978. *Enzyme Eng.* 4:323–28
43. Kennedy, J. F., Barker, S. A., Humphreys, J. D. 1976. *Nature* 261:242–44
44. Kierstan, M., Bucke, C. 1977. *Biotechnol. Bioeng.* 19:387–97
45. Kimura, A., Tatsutomi, Y., Matsuno, R., Tanaka, A., Fukuda, H. 1981. *Eur. J. Appl. Microbiol. Biotechnol.* 11:78–80
46. Kimura, A., Tatsutomi, Y., Mizushima, N., Tanaka, A., Matsuno, R., Fukuda, H. 1978. *Eur. J. Appl. Microbiol. Biotechnol.* 5:13–16
47. Klein, J., Hackel, U., Schara, P., Washausen, P., Wagner, F., Martin, C. K. A. 1978. *Enzyme Eng.* 4:339–41
48. Klein, J., Kluge, M. 1981. *Biotechnol. Lett.* 3:65–70
48a. Klein, J., Wagner, F. 1980. *Enzyme Eng.* 5:335–45
49. Kobos, R. K., Rice, D. J., Flournoy, D. S. 1979. *Anal. Chem.* 51:1122–25
50. Kokubu, T., Karube, I., Suzuki, S. 1978. *Eur. J. Appl. Microbiol. Biotechnol.* 5:233–40
51. Kokubu, T., Karube, I., Suzuki, S. 1981. *Biotechnol. Bioeng.* 23:29–39
52. Koshcheenko, K. A., Sukhodolskaya, G. V., Tyurin, V. S., Skryabin, G. K. 1981. *Eur. J. Appl. Microbiol. Biotechnol.* 12:161–69
53. Kumakura, M., Yoshida, M., Kaetsu, I. 1978. *Eur. J. Appl. Microbiol. Biotechnol.* 6:13–22
54. Kumakura, M., Yoshida, M., Kaetsu, I. 1978. *J. Solid-Phase Biochem.* 3:175–83
55. Kumakura, M., Yoshida, M., Kaetsu, I. 1979. *Appl. Environ. Microbiol.* 37: 310–15
56. Kumakura, M., Yoshida, M., Kaetsu, I. 1979. *Biotechnol. Bioeng.* 21:679–88
57. Lambert, G. R., Daday, A., Smith, G. D. 1979. *FEBS Lett.* 101:125–28
58. Larsson, P. O., Ohlson, S., Mosbach, K. 1976. *Nature* 263:796–97
59. Linko, P. 1980. *Abstr. 6th Int. Ferment. Symp.,* p. 128
60. Linko, P., Poutanen, K., Weckstrom, L., Linko, Y.-Y. 1980. *Biochimie* 62:387–94
61. Linko, Y.-Y., Pohjola, L., Linko, P. 1977. *Process Biochem.* 12(7/8):14–16 & 32
62. Linko, Y.-Y., Poutanen, K., Weckström, L., Linko, P. 1979. *Enzyme Microb. Technol.* 1:26–30
63. Maddox, I. S., Dunnill, P., Lilly, M. D. 1981. *Biotechnol. Bioeng.* 23:345–54
64. Makiguchi, N., Arita, M., Asai, Y. 1980. *J. Ferment. Technol.* 58:17–21
65. Marechal, P. D.-Le, Calderon-Seguin, R., Vandecasteele, J. P., Azerad, R.

1979. *Eur. J. Appl. Microbiol. Biotechnol.* 7:33–44
66. Matsumoto, K., Seijo, H., Watanabe, T., Karube, I., Satoh, I., Suzuki, S. 1979. *Anal. Chim. Acta* 105:429–32
67. Matsunaga, T., Karube, I., Suzuki, S. 1978. *Anal. Chim. Acta* 99:233–39
67a. Mayer, H., Collins, J., Wagner, F. 1980. *Enzyme Eng.* 5:61–69
68. Morikawa, Y., Karube, I., Suzuki, S. 1979. *Biotechnol. Bioeng.* 21:261–70
69. Morikawa, Y., Karube, I., Suzuki, S. 1980. *Biotechnol. Bioeng.* 22:1015–23
70. Morikawa, Y., Karube, I., Suzuki, S. 1980. *Eur. J. Appl. Microbiol. Biotechnol.* 10:23–30
71. Murata, K., Kato, J., Chibata, I. 1979. *Biotechnol. Bioeng.* 21:887–95
72. Murata, K., Tani, K., Kato, J., Chibata, I. 1978. *Eur. J. Appl. Microbiol. Biotechnol.* 6:23–27
73. Murata, K., Tani, K., Kato, J., Chibata, I. 1980. *Biochimie* 62:347–52
74. Murata, K., Tani, K., Kato, J., Chibata, I. 1980. *Eur. J. Appl. Microbiol. Biotechnol.* 10:11–21
75. Murata, K., Tani, K., Kato, J., Chibata, I. 1981. *Eur. J. Appl. Microbiol. Biotechnol.* 11:72–77
76. Murata, K., Tani, K., Kato, J., Chibata, I. 1981. *Enzyme Microb. Technol.* 3: 233–42
77. Murata, K., Uchida, T., Tani, K., Kato, J., Chibata, I. 1979. *Eur. J. Appl. Microbiol. Biotechnol.* 7:45–51
78. Nabe, K., Izuo, N., Yamada, S., Chibata, I. 1979. *Appl. Environ. Microbiol.* 38:1056–60
79. Neujahr, H. Y., Kjellen, K. G. 1979. *Biotechnol. Bioeng.* 21:671–78
80. Nishida, Y., Sato, T., Tosa, T., Chibata, I. 1979. *Enzyme Microb. Technol.* 1: 95–99
81. Ochiai, H., Shibata, H., Sawa, Y., Katoh, T. 1980. *Proc. Natl. Acad. Sci. USA* 77:2442–44
82. Ohlson, S., Flygare, S., Larsson, P. O., Mosbach, K. 1980. *Eur. J. Appl. Microbiol. Biotechnol.* 10:1–9
83. Ohlson, S., Larsson, P. O., Mosbach, K. 1978. *Biotechnol. Bioeng.* 20:1267–84
84. Ohlson, S., Larsson, P. O., Mosbach, K. 1979. *Eur. J. Appl. Microbiol. Biotechnol.* 7:103–10
85. Ohmiya, K., Ohashi, H., Kobayashi, T., Shimizu, S. 1977. *Appl. Environ. Microbiol.* 33:137–46
86. Okada, T., Karube, I., Suzuki, S. 1981. *Eur. J. Appl. Microbiol. Biotechnol.* 12:102–106
87. Omata, T., Iida, T., Tanaka, A., Fukui, S. 1979. *Eur. J. Appl. Microbiol. Biotechnol.* 8:143–55
88. Omata, T., Iwamoto, N., Kimura, T., Tanaka, A., Fukui, S. 1981. *Eur. J. Appl. Microbiol. Biotechnol.* 11:199–204
89. Omata, T., Tanaka, A., Fukui, S. 1980. *J. Ferment. Technol.* 58:339–43
90. Omata, T., Tanaka, A., Yamane, T., Fukui, S. 1979. *Eur. J. Appl. Microbiol. Biotechnol.* 6:207–15
91. Petre, D., Noel, C., Thomas, D. 1978. *Biotechnol. Bioeng.* 20:127–34
92. Rechnitz, G. A., Kobos, R. K., Riechel, S. J., Gebauer, C. R. 1977. *Anal. Chim. Acta* 94:357–65
93. Rechnitz, G. A., Riechel, T. L., Kobos, R. K., Meyerhoff, M. E. 1978. *Science* 199:440–41
94. Saif, S. R., Tani, Y., Ogata, K. 1975. *J. Ferment. Technol.* 53:380–85
95. Samejima, H., Kimura, K., Ado, Y., Suzuki, Y., Tadokoro, T. 1978. *Enzyme Eng.* 4:237–44
96. Sato, T., Nishida, Y., Tosa, T., Chibata, I. 1979. *Biochim. Biophys. Acta* 570: 179–86
97. Sawa, Y., Kanayama, K., Ochiai, H. 1980. *Agric. Biol. Chem.* 44:1967–69
98. Scherer, P., Kluge, M., Klein, J., Sahm, H. 1981. *Biotechnol. Bioeng.* 23: 1057–65
99. Schnarr, G. W., Szarek, W. A., Jones, J. K. N. 1977. *Appl. Environ. Microbiol.* 33:732–34
100. Seyhan, E., Kirwan, D. J. 1979. *Biotechnol. Bioeng.* 21:271–81
101. Shimizu, S., Tani, Y., Yamada, H. 1979. *ACS Symp. Series* 106:87–100
102. Sitton, O. C., Gaddy, J. L. 1980. *Biotechnol. Bioeng.* 22:1735–48
103. Slowinski, W., Charm, S. E. 1973. *Biotechnol. Bioeng.* 15:973–79
104. Somerville, H. J., Mason, J. R., Ruffell, R. N. 1977. *Eur. J. Appl. Microbiol.* 4:75–85
105. Sonomoto, K., Hoq, M. M., Tanaka, A., Fukui, S. 1981. *J. Ferment. Technol.* 59:465–69
106. Sonomoto, K., Hoq, M. M., Tanaka, A., Fukui, S. 1982. *Appl. Environ. Microbiol.* In press
107. Sonomoto, K., Jin, I.-N., Tanaka, A., Fukui, S. 1980. *Agric. Biol. Chem.* 44:1119–26
108. Sonomoto, K., Tanaka, A., Omata, T., Yamane, T., Fukui, S. 1979. *Eur. J. Appl. Microbiol. Biotechnol.* 6:325–34
109. Suzuki, S., Karube, I., Matsunaga, T., Kuriyama, S., Suzuki, N., Shirogami, T., Takamura, T. 1980. *Biochimie* 62: 353–58

110. Takata, I., Tosa, T., Chibata, I. 1977. *J. Solid-Phase Biochem.* 2:225–36
111. Takata, I., Yamamoto, K., Tosa, T., Chibata, I. 1980. *Enzyme Microb. Technol.* 2:30–36
112. Tanaka, A., Hagi, N., Gellf, G., Fukui, S. 1980. *Agric. Biol. Chem.* 44:2399–2405
113. Tanaka, A., Hagi, N., Itoh, N., Fukui, S. 1980. *J. Ferment. Technol.* 58:391–94
114. Tanaka, A., Hagi, N., Yasuhara, S., Fukui, S. 1978. *J. Ferment. Technol.* 56:511–15
115. Tanaka, A., Jin, I.-N., Kawamoto, S., Fukui, S. 1979. *Eur. J. Appl. Microbiol. Biotechnol.* 7:351–54
116. Tanaka, A., Yasuhara, S., Fukui, S., Iida, T., Hasegawa, E. 1977. *J. Ferment. Technol.* 55:71–75
117. Tanaka, A., Yasuhara, S., Gellf, G., Osumi, M., Fukui, S. 1978. *Eur. J. Appl. Microbiol. Biotechnol.* 5:17–27
118. Tanaka, A., Yasuhara, S., Osumi, M., Fukui, S. 1977. *Eur. J. Biochem.* 80:193–97
119. Tischer, W., Tiemeyer, W., Simon, H. 1980. *Biochimie* 62:331–39
120. Tosa, T., Sato, T., Mori, T., Chibata, I. 1974. *Appl. Microbiol.* 27:886–89
121. Tosa, T., Sato, T., Mori, T., Yamamoto, K., Takata, I., Nishida, Y., Chibata, I. 1979. *Biotechnol. Bioeng.* 21:1697–1709
122. Uchida, T., Watanabe, T., Kato, J., Chibata, I. 1978. *Biotechnol. Bioeng.* 20:255–66
123. Wada, M., Kato, J., Chibata, I. 1979. *Eur. J. Appl. Microbiol. Biotechnol.* 8: 241–47
124. Wada, M., Kato, J., Chibata, I. 1980. *J. Ferment. Technol.* 58:327–31
125. Wada, M., Kato, J., Chibata, I. 1981. *Eur. J. Appl. Microbiol. Biotechnol.* 11: 67–71
126. Wada, M., Uchida, T., Kato, J., Chibata, I. 1980. *Biotechnol. Bioeng.* 22:1175–88
127. Walters, R. R., Moriarty, B. E., Buck, R. P. 1980. *Anal. Chem.* 52:1680–84
128. Weetall, H. H., Sharma, B. P., Detar, C. C. 1981. *Biotechnol. Bioeng.* 23:605–14
129. Yamada, H., Shimizu, S., Shimada, H., Tani, Y., Takahashi, S., Ohashi, T. 1980. *Biochimie* 62:395–99
130. Yamada, H., Yamada, K., Kumagai, H., Hino, T., Okamura, S. 1978. *Enzyme Eng.* 3:57–62
131. Yamamoto, K., Tosa, T., Chibata, I. 1980. *Biotechnol. Bioeng.* 22:2045–54
132. Yamane, T., Nakatani, H., Sada, E., Omata, T., Tanaka, A., Fukui, S. 1979. *Biotechnol. Bioeng.* 21:2133–45
133. Yang, H. S., Studebaker, J. F. 1978. *Biotechnol. Bioeng.* 20:17–25
134. Yokozeki, K., Yamanaka, S., Utagawa, T., Takinami, K., Hirose, Y., Tanaka, A., Sonomoto, K., Fukui, S. 1982. *Eur. J. Appl. Microbiol. Biotechnol.* In press

Ann. Rev. Microbiol. 1982. 36:173–98

PHYCOBILISOMES: STRUCTURE AND DYNAMICS

A. N. Glazer

Department of Microbiology and Immunology, University of California, Berkeley, California 94720

CONTENTS

INTRODUCTION

The process of photosynthesis is initiated by the absorption of light. Emerson & Arnold (26, 27) showed that the photochemical reactions were complete in less than 10^{-5} s, whereas about 0.02 s at 25°C was required for the completion of the dark reactions of photosynthesis. By exploiting this kinetic difference, they concluded from studies with *Chlorella pyrenoidosa* "that for every 2480 molecules of chlorophyll there is present in the cell one unit capable of reducing one molecule of carbon dioxide each time it is

0066-4227/82/1001-0173$02.00

suitably activated by light" (27). The actual parameter measured by Emerson & Arnold (26, 27) was the evolution of oxygen. Since it is now known that primary photochemistry involves the promotion of a single electron transfer per quantum absorbed, and that the in vivo requirement for the evolution of a molecule of oxygen is eight quanta (23), the actual size of the photosynthetic unit deduced from the data of Emerson & Arnold is 2480/8, i.e. 310 chlorophylls *a* and *b*.

In oxygen-evolving photosynthetic organisms, the photosynthetic unit contains two distinct reaction centers: photosystem II (P680), responsible for the photolysis of water; and photosystem I (P700), readily quantitated by measurement of photobleaching at 700 nm (64). Each reaction center is associated with an antenna of light-harvesting protein-pigment complexes (3, 61, 111). The photosynthetic unit size, expressed as the molar ratio of total light-harvesting chlorophyll per P700, ranges from about 150 for cyanobacteria (62, 83, 90, 117) to ~3000 for the diatom *Skeletonema costatum* and the chlorophyte *Dunaliella tertiolecta* (28).

Why are such large antennae necessary? The obvious nature of this requirement was first pointed out by Gaffron & Wohl (33). The frequency of absorption acts for a molecule is determined by the equation (95) $n = 4 \times 10^{-21} \alpha N_{h\nu}$, where $\alpha = 2.3\epsilon$, ϵ is the molar extinction coefficient of the molecule, and $N_{h\nu}$ is the light flux in quanta/(s•cm^2). In direct sunlight at noon the light flux is $\sim 10^{17}$ visible quanta/(s•cm^2). Assuming an average α for visible light for chlorophyll *a* of 3×10^4, one molecule would absorb about 12 quanta/s (95). Since the dark reactions require 8 quanta/0.02 s, even under such extraordinary intense light flux, photosynthesis would be light limited unless some 35 chlorophyll molecules were able to supply an isolated reaction center.

Although chlorophyll *a* is the universal reaction center pigment in oxygen-evolving organisms, the nature of the pigments of the antenna complexes varies widely (45). The phycobilisome is an important light-harvesting component of the photosynthetic apparatus of cyanobacteria and red algae (35, 36). Phycobilisomes are large multiprotein particles of intricate structure. They are found attached in regular arrays to the outer surface of the thylakoids of cyanobacterial cells and of red algal chloroplasts (35). The phycobilisome consists largely of biliproteins (phycobiliproteins). These intensely colored proteins absorb visible light in the range of 450-650 nm and transfer the energy to the protein-chlorophyll complexes of photosystems II and I in the photosynthetic lamellae. Phycobilisomes provide 30–50% of the total light-harvesting capacity of cyanobacterial and red algal cells (62) and distribute the energy to both photosystems (121).

Available purification procedures do not perturb the organization and spectroscopic properties of the phycobilisome. Consequently, this particle is a unique experimental object for the detailed analysis of a light-harvesting system. The individual biliproteins are highly water soluble, readily purified, and crystallizable. They are made up of polypeptide chains of low molecular weight whose amino acid sequence can be easily determined. Each biliprotein has distinctive spectroscopic properties that can be exploited to examine its structure, its interaction with other phycobilisome components, and its role in light energy absorption and transfer. Phycobilisomes vary in size and complexity. The goal of defining the structures of the simplest of these particles at high resolution appears to be attainable.

Numerous reviews dealing with bilin structure (5, 55, 85, 99–101, 112), phycobiliproteins (5, 6, 9, 21, 43–46, 100, 101, 112), phycobilisomes (35, 36, 106), and chromatic adaptation (9, 106) have appeared in the past 10 years. The emphasis in this review is on phycobilisome structure, assembly, and function.

COMPONENTS OF PHYCOBILISOMES

Purification of Phycobilisomes

The purification of phycobilisomes exploits their large size (5–20 $\times$ 10^6 daltons) and high density. Phycobilisomes are unstable in solutions of low ionic strength, but they can be isolated intact in concentrated solutions of certain salts (37). The most commonly used procedures for the preparation of phycobilisomes involve breakage of cells in concentrated NaK-phosphate at pH 7 (37, 40) or 8 (129), in the presence of Triton X-100 (37) or of zwitterionic detergents (53), removal of cell debris, and centrifugation on sucrose density gradients. Other procedures, which utilize polyethyleneglycol in place of high salt, have been described (97, 98). The common denominator between the various media used for phycobilisome isolation is a greatly diminished activity of water.

The integrity of phycobilisomes is assessed by examining their fluorescence emission spectra. For excitation wavelengths $\leqslant$650 nm, the emission of phycobilisomes consists of a strong 680-nm component and a weak 660-nm component (56, 129, 131). Breakdown of the particles leads to a decrease in these fluorescence emissions and the appearance of strong fluorescence from the major biliproteins, such as phycocyanin and phycoerythrin (39). These changes are diagnostic of the loss of energy transfer between the biliprotein components of the phycobilisome.

Biliproteins are the major components of the phycobilisome and represent ~85% of the mass of the structure. The balance of the phycobilisome is made up of several linker polypeptides that function in the assembly of the particle and its attachment to the thylakoid membrane (109). Phycobilisomes are believed to consist entirely of proteins.

Biliproteins

BILIN PROSTHETIC GROUPS The characteristic absorbance of the biliproteins in the visible part of the spectrum arises from open-chain tetrapyrrole prosthetic groups covalently attached to these proteins. The structures of the major "blue" chromophore, phycocyanobilin (PCB), and "red" chromophore, phycoerythrobilin (PEB), are well established (Figure 1) (e.g. 5, 55, 63, 85, 99, 100). Phycourobilin (PUB) is responsible for the 495- to 500-nm absorption peak of B- and R-phycoerythrins (51, 63, 87, 89, 116), and a bilin of undetermined structure is bound to the α subunit of phycoerythrocyanin (15).

BILIPROTEINS: PROPERTIES AND SOME SPECULATIONS ON THEIR BIOSYNTHESIS Numerous detailed reviews dealing with biliproteins have appeared during the past decade (e.g. 43, 44, 46, 85, 90, 100, 101, 137). Consequently, only a brief summary of their properties is presented here.

The subunit structure and bilin composition of biliproteins, purified to homogeneity, are presented in Figure 2. The basic building block is a monomer, $\alpha\beta$, consisting of two dissimilar subunits of molecular weights in the range of 16,000–22,000. Pure biliproteins exist in a variety of aggrega-

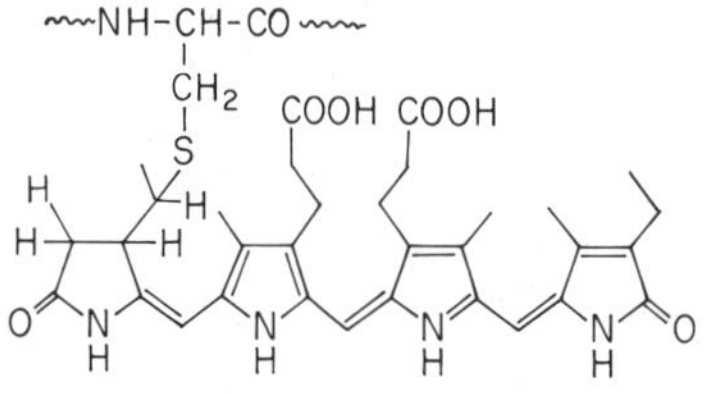

NH-CH-CO
CH_2 COOH COOH
S
(I)

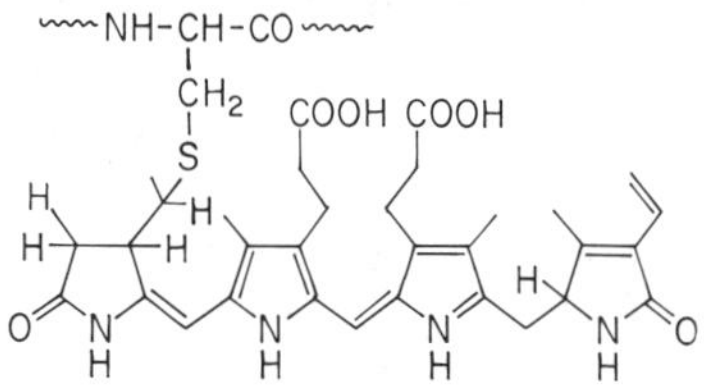

NH-CH-CO
CH_2 COOH COOH
S
(II)

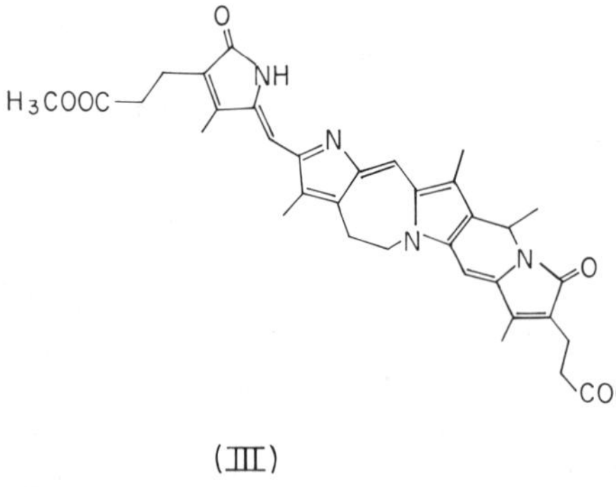

H_3COOC
$COOCH_3$
(III)

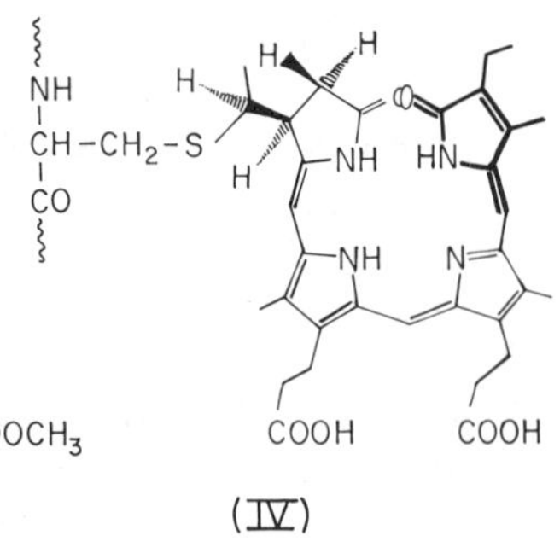

NH
CH-CH_2-S
CO
COOH COOH
(IV)

Figure 1 Structures of polypeptide-bound PCB (*I*) and PEB (*II*), isophorcabilin (*III*), and polypeptide-bound PCB (*IV*) in a cyclohelical conformation.

tion states, depending on the particular biliprotein, its concentration, the pH, and ionic strength. The amino acid sequences of several phycocyanin (30, 31, 113, 120, 124) and allophycocyanin (22, 105) subunits have been completely determined, and in addition substantial information exists on the amino-terminal sequences of numerous biliproteins (32, 45, 73, 115), and on sequences surrounding sites of bilin attachment (11, 17, 81). Biliproteins belonging to different spectroscopic classes show considerable sequence identity, and the accumulated data indicate these proteins arose by consecutive gene duplication of a single ancestral gene (43, 45, 105). It has also been suggested that a short insertion played a role in the evolution of the β subunit ancestral gene from that for the α subunit (105).

As illustrated in Figure 1, the bilins are linked to cysteinyl residues in the polypeptide chains (52). For PCB and PEB (11, 17, 66, 67, 69, 114, 124), there is substantial evidence that a single thioether bond attaches each prosthetic group to the polypeptide. For PUB, the question is not fully resolved. The structure of this bilin has not been rigorously determined (116). There is at least one thioether bond, and perhaps two, between this bilin and the polypeptide (52, 85, 134).

The following questions have yet to be answered concerning the biosynthesis of biliproteins. What determines which cysteine residues serve as sites of bilin attachment? Is the formation of the thioether linkage to the prosthetic group an enzyme-catalyzed reaction? What determines which bilin is present at a given position on the polypeptide chain of a biliprotein?

There are free sulfhydryl groups on numerous biliproteins in addition to those in thioether linkage (30, 105, 124). Cysteinyl residues within similar sequences on different biliproteins carry a PCB in one instance and a PEB in another (17). The presence of a cysteinyl residue may be the sole amino acid sequence requirement for bilin attachment. The conformation of the apopolypeptide may determine which cysteinyl residues are sterically accessible for bilin attachment. As established for cytochrome *c* (2), the formation of the thioether bond to the bilin is probably enzyme mediated. The addition of free bilins to thiol groups on proteins has been reported (86). However, such spontaneous addition is unlikely to show the required stereospecificity (101). The most difficult question to answer is what determines which bilin is present at a given position on a biliprotein. Very speculative hypotheses can be put forward to account for the observed bilin distribution. Binding of a bilin to a protein, e.g. bilirubin to serum albumin, preferentially stabilizes certain stereoisomers (8). One hypothesis would assume that a single type of bilin, PCB, is attached in an enzyme-catalyzed reaction to all available cysteinyl residues. The other bilins (PEB, PUB, etc) would then be formed by modification of polypeptide-bound PCB. Two alternative mechanisms for such transformations can be proposed. In the first it is assumed that different PCB conformers are stabilized by the protein folding around different sites of attachment. Isomerases specific for a given conformer would catalyze the proton transfer reactions needed to convert PCB to PEB or PUB. In an alternate mechanism, the conversion of different polypeptide-bound PCB stereoisomers to other bilins is envisaged to be driven by the folding energy of the conjugated protein, i.e. the isomerizations would be non-enzymatic (see 68). An observation relevant to the above suggestions comes from studies of the composition of R-phycoerythrin isolated from the red alga *Callithamnion roseum* grown at different light intensities. The protein from cells grown at $6 \mu E$ m^{-2} s^{-1} showed a ratio of PEB/PUB of 2.17, whereas that from cells grown at 37 μE m^{-2} s^{-1} was 3.06 (134). The total number of bilins per 240,000 daltons was unaltered. These data suggest that at certain positions on the polypeptide chains of *C. roseum* R-phycoerythrin, either a PEB or a PUB group

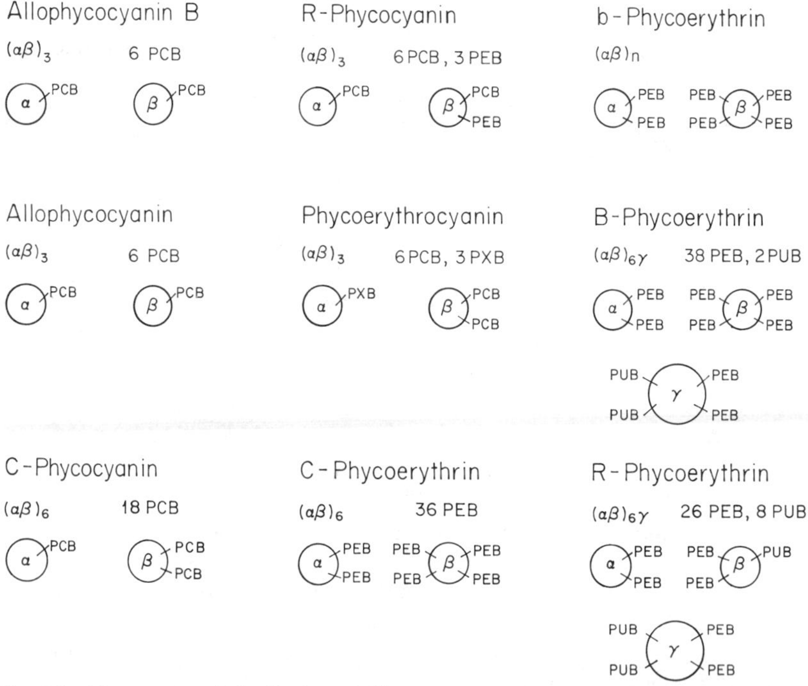

Figure 2 The nature and distribution of bilin prosthetic groups among the subunits of the various biliproteins, and the numbers present in higher aggregates of each protein. For example, R-phycocyanin,$(\alpha\beta)_3$, contains six PCB and three PEB groups per trimer; each α subunit carries one PCB, whereas each β subunit carries one PCB and one PEB. Certain cyanobacterial phycoerythrins carry two PEB on the α subunit and three on the β subunit (80). Phycoerythrin of the cyanobacterium *Gloeobacter violaceus* has an unusual bilin content: three PEB on the α subunit, three PEB and one PUB on the β subunit (14). Data are from multiple references (15, 48, 50, 51, 80, 81, 132, 134).

can be present. The simplest (but not exclusive) explanation of the data is that the ratio of PEB/PUB at these sites is dependent on the level and/or activity of an isomerase controlled by light (134).

Other hypotheses are not excluded by available experimental information, and the above speculations are advanced in the hope that they will stimulate research towards discovering the real answers.

SPECTROSCOPIC PROPERTIES OF THE BILIPROTEINS The manner in which the phycobilisome functions as a light-harvesting unit is determined by the absorption and fluorescence emission properties of its constituent biliproteins. Three factors govern the spectroscopic properties of a native biliprotein: (*a*) the type and number of bilin prosthetic groups; (*b*) the conformation and environment of the bilins within the subunit to which they are attached; and (*c*) the changes in that environment produced upon assembly of the subunit into structures of higher complexity. Such changes in the environment include conformational changes resulting from intersubunit interaction as well as electromagnetic coupling between bilins on adjacent subunits. For example, the monomer, $\alpha\beta$, of allophycocyanin has λ_{max} at ~ 615 nm, whereas the trimer absorbs maximally at 648 nm (76). The monomer of C-phycocyanin has a λ_{max} ~ 615 nm (ϵ_M ~ 2.3×10^5 M^{-1} cm^{-1}), whereas the hexamer, $(\alpha\beta)_6$, has a λ_{max} ~ 621 nm (ϵ_M 3.33×10^5 M^{-1} cm^{-1}) (49). Trimeric or hexameric complexes of phycocyanin with different linker polypeptides display distinctive absorption, fluorescence, and circular dichroism spectra (74, 132, 133, 135). By exploiting the discriminating power of the spectroscopic measurements, it has been possible to distinguish between those components obtained from dissociated phycobilisomes that have retained their native assembly features from those aggregates formed in the course of purification procedures.

The plasticity of the spectroscopic properties of the biliproteins is in large measure a consequence of inherent properties of the prosthetic groups. Different conformers of open-chain tetrapyrroles differ markedly in spectroscopic properties. In solution, free bilins can take up various configurations ranging from cyclic porphyrin-type structures to elongated polyene-type conformations (101). The coexistence of different conformers is responsible for the broad absorption bands of bile pigments in solution. The spectra of bile pigments show a low energy band at ~ 500–700 nm and a near-UV band between 300 and 380 nm (see 107, for example). The ratio of $\epsilon^{UV}/\epsilon^{vis}$ (where ϵ^{UV} and ϵ^{vis} are molar extinction coefficients at the wavelengths of maximal absorption in the near-UV and the visible) ranges from ~ 5 to ~ 1 for various bile pigments (107). In contrast, in biliproteins the visible absorption band is much stronger than the near-UV band. Upon denaturation, the absorption spectra of the biliproteins become more similar to those of the free pigments (86, 88). Molecular orbital calculations (e.g. 18, 20,

119), and studies of bile pigments structurally constrained to assume extended conformations (see Figure 1 *III*) (10), indicate that such conformations are associated with very strong absorbance in the visible and weak absorbance in the near-UV region of the spectrum, whereas bilins in cyclohelical conformations (Figure 1 *IV*) have strong near-UV absorbance and weak absorbance in the visible. Moreover, the fluorescence of a relatively rigid molecule, such as isophorcabilin (Figure 1 *III*), is stronger than that of bilins, which are less conformationally constrained (101). An important property of native biliproteins is their high fluorescence quantum yield, which is markedly diminished upon denaturation. This indicates that radiationless deactivation processes (such as isomerizations) are minimized within the native structure.

Linker Polypeptides

Biliproteins purified to homogeneity do not interact with one another (15). Their assembly into phycobilisomes is dependent on a group of polypeptides named linker polypeptides (74). Linker polypeptides were first described by Tandeau de Marsac & Cohen-Bazire (109). The number of these polypeptides varied among different phycobilisomes and was shown to be related to their biliprotein composition. For example, in chromatic adapters, the presence of certain linker polypeptides could be shown to be correlated with the presence of phycoerythrin. Tandeau de Marsac & Cohen-Bazire (109) classified the polypeptides of phycobilisomes on the basis of molecular weight: group I, 70,000–120,000; group II, 30,000–70,000; group III, 25,-000–30,000; and group IV, 15,000–25,000. Group IV polypeptides were the subunits of the biliproteins. Under appropriate conditions, group I polypeptides could be shown to remain with the thylakoid membrane and were therefore assumed to function in the attachment of the phycobilisome to the membrane. Group II and III polypeptides were associated solely with the biliproteins, and Tandeau de Marsac & Cohen-Bazire (109) proposed that these polypeptides were involved in assembly and positioning of the biliproteins. The general conclusions of the initial study have been amply supported by the results of numerous subsequent investigations.

Structural studies of linker polypeptides have been limited thus far to linkers of 75, 33, 30, and 27 kdaltons, purified from *Synechococcus* 6301 phycobilisomes (74). The linkers are very tightly associated with the biliproteins and their purification has involved chromatography under strongly denaturing conditions. The 75,000-dalton linker is a biliprotein that bears one phycocyanobilin group (75). Phycocyanobilin chromophores are also present on the 95,000-dalton linker of *Porphyridium cruentum* phycobilisomes (96), the 99,000-dalton linker of *Synechocystis* 6701 phycobilisomes (J. C. Gingrich, A. N. Glazer, unpublished observations), and the 120,000-dalton linker of *Anabaena variabilis* phycobilisomes (132). In each instance, it is the highest-molecular-weight linker of the phycobilisome that carries the phycocyanobilin. As noted above, that is also the polypeptide that functions in the attachment of the phycobilisome to the thylakoid membrane. *Synechococcus* 6301 33,000-, 30,000-, and 27,000-dalton linkers do not carry bilins (75). Each linker polypeptide yielded a unique tryptic peptide map (74). The linker polypeptides are basic, whereas the biliproteins are acidic (74). This suggests that electrostatic interactions represent an important component in the association of linkers with biliproteins.

As noted above, the amino acid sequences of the biliproteins have been shown to be highly conserved. One might suspect that the sequences of the linker polypeptides have likewise been highly conserved. Direct information on this point is lacking. However, polypeptides of roughly equal molecular weight are associated with the same biliprotein in the phycobilisomes of different organisms (13, 53, 123, 129, 135). Moreover, it has been shown that the 27,000- the 30,000-, and 33,000-dalton linker polypeptides, which interact with the phycocyanin of *Synechococcus* 6301 phycobilisomes, in vitro perform equivalent assembly functions with the phycocyanins of *Anabaena variabilis* and *Agmenellum quadruplicatum* PR6 (74).

PHYCOBILISOME STRUCTURE AND ASSEMBLY

Phycobilisome Morphology

Comprehensive discussions of phycobilisome morphology may be found in a recent excellent review by Gantt (35) and in a paper by Bryant et al (16). The ultrastructure and size of phycobilisomes vary widely. Three classes of structures have been described to date: (*a*) hemi-ellipsoidal phycobilisomes, thus far seen only in red algae; (*b*) hemi-discoidal phycobilisomes, found among red algae, cyanobacteria, and certain cyanelles; (*c*) "rod bundle" phycobilisomes, to date represented only by those of the thylakoid-less cyanobacterium *Gloeobacter violaceus* (58).

Hemi-ellipsoidal phycobilisomes have molecular weights of $\sim 20 \times 10^6$, whereas those of the other types range from $\sim 5–8 \times 10^6$ (35, 36). The hemi-discoidal phycobilisomes from *Rhodella violacea* (78) and from numerous cyanobacteria (16, 53, 98, 123) have been particularly amenable to study by electron microscopy. The most common type has a core of three contiguous objects, which appear disc-like in face-view projection, arrayed in an equilateral triangle, from which radiate in a hemi-discoidal array up to six rods composed of stacked discs. In the second type, of which *Synechococcus* 6301 phycobilisome is the sole known representative, the core consists of two contiguous disc-like objects surrounded by an arrangement of six rods similar to that described for the first type (53). These two types of phycobilisomes are represented schematically in Figure 3.

Synechococcus 6301 PHYCOBILISOME

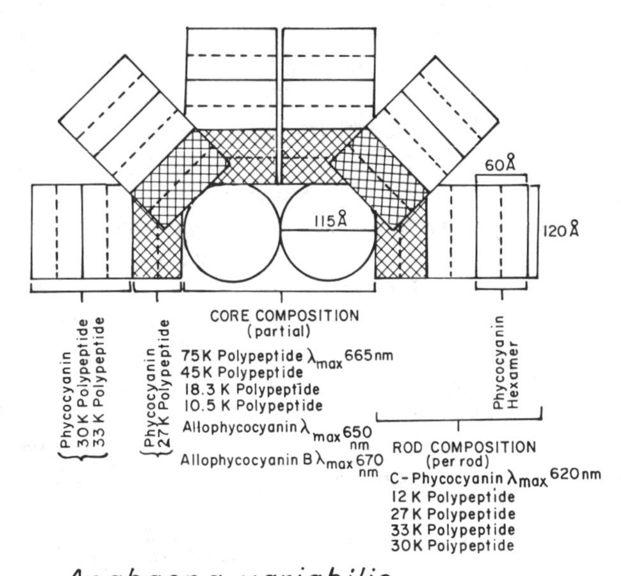

Anabaena variabilis
PHYCOBILISOME

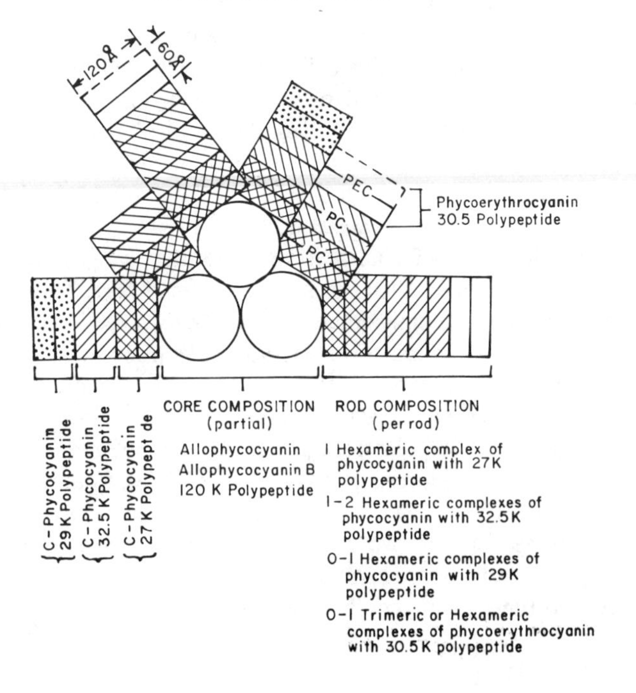

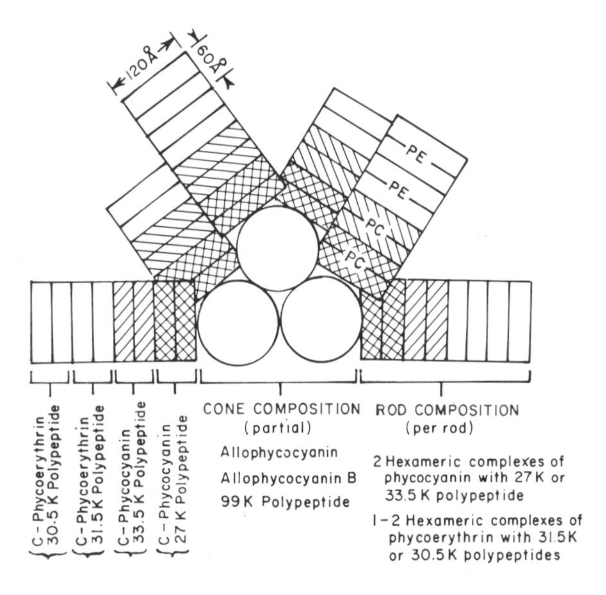

Figure 3 Diagrammatic representation of three different types of cyanobacterial phycobilisomes showing the location of the biliproteins and linker polypeptides. In *A. variabilis* phycobilisomes, it is not known whether phycoerythrocyanin is present as a trimeric or a hexameric aggregate (see text). PC and PE designate hexameric assemblies of phycocyanin and phycoerythrin, respectively.

As illustrated in Figure 3, the individual discs of the rod substructures are ~ 12 nm in diameter and ~ 6 nm thick (16, 53, 98, 123). Each disc is made up of two 3-nm-thick half-discs presumably arranged face-to-face (29). The dimensions of the disc, 6 X 12 nm, correspond to those determined by electron microscopy for $(\alpha\beta)_6$ aggregates of pure phycocyanin (7, 15) and $(\alpha\beta)_6\gamma$ aggregates of pure phycoerythrin (29, 34, 46). The half-discs, 3 X 12 nm, are inferred to represent trimeric assemblies, $(\alpha\beta)_3$. Each of the core units is a stack of four discs 3 X 11.5–12 nm (16, 130). Trimeric assemblies of allophycocyanin have been reported to have such dimensions (79).

Gross Localization of the Major Biliproteins Within Phycobilisomes

It appears certain that phycocyanin, phycoerythrin, and phycoerythrocyanin are contained within the rod substructures of the phycobilisome, whereas allophycocyanin is a major component of the core assembly.

The general location of the major biliproteins was first established for the hemi-ellipsoidal phycobilisomes of *Porphyridium cruentum* by Gantt and her co-workers from studies of the kinetics of biliprotein release upon dissociation of these particles (39) and from examination of partially dissociated thylakoid membrane vesicle-bound phycobilisomes by immunoelectron microscopy (38). The model derived from these studies placed phycoerythrin at the periphery of the structure, R-phycocyanin in the middle, and allophycocyanin in the core and proximal to the thylakoid membrane.

Complexes of phycoerythrin with phycocyanin, stable at low ionic strength, have been isolated from dissociated phycobilisomes of *Rhodella violacea* (65, 79), *Pseudoanabaena* 7409 (13), and *Porphyridium sordidum* (72). Phycoerythrocyanin-phycocyanin complexes have been found among the products of dissociation of *Mastigocladus laminosus* phycobilisomes (84). Such complexes show the stacked-disc morphology characteristic of the rod substructures. Moreover, excitation of phycoerythrin or phycoerythrocyanin leads to emission almost exclusively from phycocyanin, indicating that in these complexes radiationless energy transfer from phycoerythrin or phycoerythrocyanin to phycocyanin competes with the fluorescence decay of the former biliproteins with an efficiency comparable to that seen in intact phycobilisomes.

Direct evidence for the location of phycocyanin and phycoerythrin has come from correlation of biliprotein composition and phycobilisome ultrastructure in chromatic adapters. For example, *Synechocystis* 6701 grown in green light produces phycoerythrin, phycocyanin, and allophycocyanin in a molar ratio of ~ 2:2:1; when grown in red light, *Synechocystis* 6701 produces the biliproteins in a ratio of ~ 0.4:2:1 (123). The red- and green-light phycobilisomes have morphologically indistinguishable core assemblies. However, the rods of red-light phycobilisomes contain ~ 2 stacked discs per rod, whereas those of green-light particles contain 3–4 discs each (16, 123). In *Pseudoanabaena* 7409, where increase in phycocyanin in red-light compensates almost exactly for the decrease in phycoerythrin, and the reciprocal effect is seen in green light, the rod length remains the same independent of the ratio of phycoerythrin to phycocyanin (16) (see also 98). Such experiments show that both phycocyanin and phycoerythrin are contained within the rods, and that phycoerythrin makes up the portion of the rods distal from the core.

Mutants that make incomplete phycobilisomes have been particularly valuable in establishing the locations of the biliproteins. For example, phycobilisomes of wild-type *Synechococcus* 6301 contain phycocyanin and allophycocyanin in a molar ratio of ~ 4:1 (129), whereas those of mutant strain AN112 contain these biliproteins in a ratio of ~ 1.4:1 (130). The core assembly in the wild-type and mutant phycobilisomes appears the same, but the length of the rods is decreased from an average of three stacked discs per rod in the wild type to a single disc per rod in the mutant (130). *Synechocystis* 6701 mutants have been isolated that are totally lacking in phycoerythrin, independent of culture conditions. Phycobilisomes from such mutants have rod substructures only two stacked discs in length compared to four in the wild type (42). In other *Synechocystis* 6701 mutants that make greatly lowered amounts of phycoerythrin and phycocyanin, the core assembly is unaltered as judged by spectroscopic criteria, composition, and electron microscopy (123).

Thus, as might be expected on spectroscopic grounds, biliproteins such as phycoerythrin ($\lambda_{max} \sim 565$ nm) and phycoerythrocyanin ($\lambda_{max} = 568$ nm) that absorb green light from the distal portion of the rods, biliproteins such as R- and C-phycocyanin ($\lambda_{max} \sim 620$ nm) that absorb red light form the portions proximal to the core, and allophycocyanin ($\lambda_{max} = 650$ nm) is a component of the core.

Assembly of Rod Substructures

The assembly of the rod substructures of the phycobilisome is a consequence of a series of specific interactions of biliproteins with several linker polypeptides. Three different types of rod substructures have been examined in detail: those of *Synechococcus* 6301, which contain only phycocyanin; those of *Synechocystis* 6701, which contain phycocyanin and phycoerythrin in equal amounts; and those of *Anabaena variabilis,* which contain phycocyanin as the major and phycoerythrocyanin as the minor biliprotein (see Figure 3). Less extensive studies of the rod substructures of *Pseudoanabaena* 7409 (13) and *Porphyridium sordidum* (72), which contain phycocyanin and phycoerythrin, have also appeared.

From these studies, the following general conclusions have emerged. Each disc 6 X 12 nm in a rod substructure is a complex, $(\alpha\beta)_6X$, where $\alpha\beta$ is a biliprotein monomer and X is one copy of a linker polypeptide (42, 74, 135). Different linker polypeptides interact with the same biliprotein to give complexes with physical properties determined by the linker polypeptide (74, 133, 135). As discussed below, some complexes are involved in propagation of rod growth whereas others terminate rod growth (74, 135). In the in vitro assembly process (Figure 4), a given linker polypeptide binds tightly to a specific biliprotein to form an $(\alpha\beta)_3X$ complex (133). Under suitable conditions, $(\alpha\beta)_3X$ serves as a template for the formation of an ($\alpha\beta)_6X$ complex, probably formed by addition of monomeric biliprotein (74, 133). The two biliprotein trimers within $(\alpha\beta)_6X$ lie face-to-face (29). A large domain of X is buried within the trimer formed initially, leaving a segment of the linker polypeptide projecting from the face of that trimer (133). The projecting segment links the $(\alpha\beta)_6X$ complex onto another $(\alpha\beta)_6X$ that is already a part of the rod substructure. Examination of rod substructures, reconstituted in vitro, by electron microscopy shows that half-discs, which may correspond to either $(\alpha\beta)_3$ or $(\alpha\beta)_3X$, are frequently present at the ends of rods (74, 135). Native phycobilisomes rarely show rods that end in half-discs (53, 135). In the case of the disc proximal to the core, the projecting segment of the linker polypeptide specific to that disc is inferred to function in the attachment of that disc to the core (130).

What is the evidence in support of this assembly mechanism? The scheme detailed above is based on data gathered from in vitro reconstitution experiments, studies of aberrant phycobilisomes from mutants, and examination of phycobilisomes altered in vivo through manipulation of growth conditions.

The rod substructures of *Synechococcus* 6301 phycobilisomes contain phycocyanin as the only biliprotein. Three linker polypeptides, 27,000, 30,000, and 33,000 daltons, are also present within the rods, as well as a 12,000-dalton component of as yet unknown function (74, 128, 130). In vitro reconstitution experiments showed that mixtures of 30,000-dalton, or 33,000-dalton polypeptide and phycocyanin, in 0.6 M phosphate at pH 8.0, formed rods made up of stacked discs of phycocyanin hexamers, $[(\alpha\beta)_6 \cdot 30K]_n$ and $[(\alpha\beta)_6 \cdot 33K]_n$ (where K is 1000 daltons). In mixtures containing both the 30,000- and 33,000-dalton polypeptides and phycocyanin, the rods were on the average longer than those obtained with either polypeptide alone (74). In contrast, only single discs, $(\alpha\beta)_6 \cdot 27K$, were formed in mixtures of phycocyanin and the 27,000-dalton polypeptide (74). Moreover, when the 27,000- and 33,000-dalton (or 30,000) polypeptides were added together to phycocyanin, the length of the resulting rods decreased with increase in the 27,000/33,000 ratio, i.e. the $(\alpha\beta)_6 \cdot 27K$ disc functioned as a terminator of rod growth (74). Mutants of *Synechococcus* 6301 that produce phycobilisomes with reduced phycocyanin content have been isolated. These mutants fall into two classes: where the 30,000-dalton polypeptide is absent; where both the 30,000- and 33,000-dalton polypeptides are absent (128). In both classes of mutants, the amounts of the 75,000- and 27,000-dalton linker polypeptides, allophycocyanin, and allophycocyanin B per phycobilisome are the same as in wild type. Electron microscopy and spectroscopic measurements show no change in the core substructure of the mutant phycobilisomes (130). In the phycobilisomes that lack the 30,000-dalton polypeptide the rod substructures are shorter than in wild-type particles, and in phycobilisomes that lack both the 30,000- and 33,000-dalton polypeptides their length is reduced to a single hexameric disc (130).

Phycobilisomes obtained from *Synechococcus* 6301 cells in the early phase of nitrogen starvation are smaller than those of wild type, are partly depleted of phycocyanin, and have lost the 30,000-dalton polypeptide (126). These observations lead to a description of the rod substructure in *Synechococcus* 6301 phycobilisomes. The disc proximal to the core is an $(\alpha\beta)_6$•27K complex, the next one is an $(\alpha\beta)_6$•33K complex, and subsequent discs are either $(\alpha\beta)_6$•33K or $(\alpha\beta)_6$•30K discs (see Figure 3).

Synechocystis 6701 phycobilisomes are more complicated. Rod substructures from phycobilisomes from cells grown in white light contain roughly equimolar amounts of phycocyanin and phycoerythrin and four linker polypeptides, 33,500, 31,500, 30,500, and 27,000 daltons. Cells grown in red light produce only 20% of the white-light level of phycoerythrin, but the level of phycocyanin is unaltered (108). Phycobilisomes from such cells do not contain the 30,500-dalton polypeptide and the amount of 31,500-dalton polypeptide is greatly reduced (42). Although *Synechococcus* 6301 phycobilisomes may have rods that range from 1 to 7 discs in length on a single particle, *Synechocystis* 6701 phycobilisome rods are 4 ± 1 discs in length (16, 123). Extensive analysis of numerous mutants with altered rod substructures shows that the average rod is made up of four complexes (starting with the disc proximal to the core) (Figure 3): phycocyanin, $(\alpha\beta)_6$•27K and $(\alpha\beta)_6$•33.5K; phycoerythrin, $(\alpha\beta)_6$•31.5K and $(\alpha\beta)_6$•30.5K (42).

Studies of *Anabaena variabilis* phycobilisomes have been particularly useful in defining the assembly process. The rod substructures of these phycobilisomes contain the biliproteins phycocyanin and phycoerythrocyanin, and linker polypeptides of 27,000, 29,000, 32,500, and 30,500 daltons (132, 135). The amount of phycoerythrocyanin varies with culture conditions, but it does not exceed 20% of the biliprotein in the rod substructures (15). Extensive fractionation of the products of dissociation of *Anabaena variabilis* phycobilisomes in 0.05 M phosphate at pH 7.0 led to the isolation of three trimeric phycocyanin complexes, $(\alpha\beta)_3$•27K, $(\alpha\beta)_3$•32.5K, and $(\alpha\beta)_3$•29K, and a phycoerythrocyanin complex, $(\alpha\beta)_3$•30.5K (Figure 5) (132, 135). Reassociation experiments, performed in high salt, showed that only $(\alpha\beta)_3$•32.5K assembled into long rods, $[(\alpha\beta)_6\text{•}32.5\text{K}]_n$, the other two phycocyanin complexes formed single disc hexamers, and the aggregation state of the trimeric phycoerythrocyanin complex did not change (see Figure 5). These observations suggested that there were two different types of rod substructures in *Anabaena variabilis* phycobilisomes, those that terminated in a phycoerythrocyanin-containing disc and those that terminated in a phycocyanin $(\alpha\beta)_6$•29K disc. The arrangement of complexes in such rod substructures is shown in Figure 5. In a recent study of *Mastigo-*

cladus laminosus phycobilisomes, Nies & Wehrmeyer (84) reported the isolation of heterologous hexamers made up of one trimer of phycoerythrocyanin and one of phycocyanin. It is possible that such hybrid hexamers terminate the phycoerythrocyanin-containing rods (84).

How do the linker polypeptides mediate rod assembly? The availability of stable trimeric complexes of biliproteins with linker polypeptides offered an opportunity to explore this question. The results of limited tryptic degradation of the $(\alpha\beta)_3$•32.5K and $(\alpha\beta)_3$•27K phycocyanin complexes are presented in Figure 6. It is evident that a large domain of the linker polypeptide was protected from trypsin within each of these complexes. A 28,000-dalton-polypeptide domain of the 32,500-dalton polypeptide was protected within the $(\alpha\beta)_3$•32.5K trimer. Removal of the 4500-dalton-polypeptide domain did not prevent hexamer formation in high salt, but totally abolished the formation of rods (Figure 6) (133). The compelling conclusion is that the 4500-dalton-polypeptide domain attaches one $(\alpha\beta)_6$•32.5K disc to the disc immediately below it in the rod substructure (Figure 3). The conformation of the upper face of the tryptic degradation product, $(\alpha\beta)_3$ •28K, was unaffected by the cleavage of the linker polypeptide, and hence the hexameric $(\alpha\beta)_6$•28K complex was still able to form (133). Parallel studies with the $(\alpha\beta)_3$•27K complex, showed that a 21,000-dalton-polypeptide domain is protected from proteolysis within the trimeric complex. The $(\alpha\beta)_3$•21K degradation product was still able to form $(\alpha\beta)_6$•21K discs, which could copolymerize with $(\alpha\beta)_6$•32.5K discs to incorporate into rods. By analogy with the results obtained with phycobilisomes from other organisms (see above), the $(\alpha\beta)_6$•27K disc may be assigned to the position proximal to the core. Presumably the 6000-dalton linker polypeptide serves to attach the $(\alpha\beta)_6$•27K complex to the core substructure. On the basis of these results, the sequence of steps in the in vitro assembly pathway may be formulated as shown in Figure 4.

Structure of the Core

Studies of the ultrastructure of cyanobacterial phycobilisomes with "triangular" cores (16) and of those with "two-cylinder" cores (129) indicate that each of the core cylinders is made up of a stack of four discs $\sim 12 \times 3$ nm (see Figure 7). As noted above, these dimensions are characteristic of biliprotein trimers. The core contains allophycocyanin and the highest-molecular-weight polypeptide of the phycobilisome (13, 75). Various allophycocyanin components, obtained from partially dissociated phycobilisomes, have been described. Spectroscopically distinguishable allophycocyanins I, II, and III have been isolated from *Nostoc* sp. (19, 115, 136). Allophycocyanin I, a complex of trimeric allophycocyanin with a 37,000-dalton polypeptide, had a significantly lower molar absorbance at

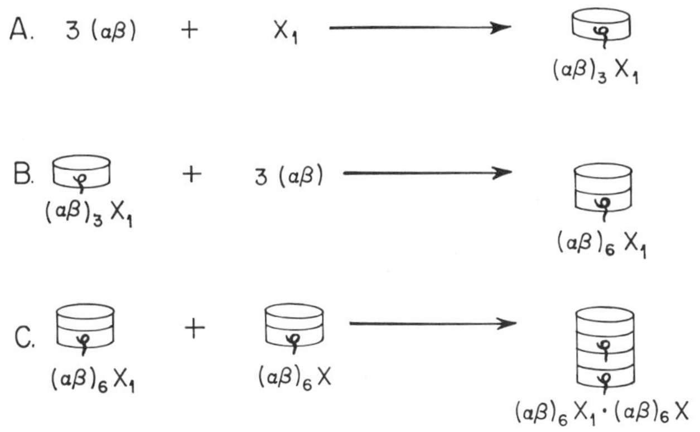

Figure 4 Pathway of in vitro assembly of the rod substructures of cyanobacterial phycobilisomes. Biliprotein monomer is represented by $\alpha\beta$. X_1 and X designate linker polypeptides, which may be the same or different.

650 nm relative to allophycocyanin III, and a fluorescence emission maximum red shifted to 680 nm (19, 136). The molecular basis of the difference between allophycocyanins II and III has yet to be established (115). Two allophycocyanins, from *Mastigocladus laminosus,* appeared to be different conformers of the same protein (59, 60). Since extensively purified allophycocyanin has a distinctive spectrum and a unique amino acid sequence, it is evident that partial dissociation of phycobilisomes leads to release of allophycocyanin in complexes and in conformations that retain in part the organizational features of this biliprotein exhibited in the intact core. These various observations hint at considerable complexity of organization of core components.

Spectroscopic analysis of purified biliproteins had led to the view that allophycocyanin B, with its 680-nm emission maximum, was a terminal energy acceptor of the phycobilisome (47, 71). Recently, it has been shown that allophycocyanin B shares a common β subunit with allophycocyanin, i.e. the monomers of these proteins can be represented as $\alpha^A\beta$ and $\alpha^B\beta$, where α^A and α^B are the subunits specific to allophycocyanin and allophycocyanin B, respectively. The subunits hybridize freely in vitro. Any trimer that contains an α^B subunit, e.g. $\alpha_2^A\alpha^B\beta_3$, fluoresces at 680 nm (73). From considerations of stoichiometry (see below), it appears that half of the allophycocyanin in the core is in the form of trimers that contain one α^B subunit and transfer energy to that subunit (73, 131).

The demonstration that the high-molecular-weight linker polypeptide associated with the core is likewise a 680-nm fluorescence emitter (75, 96) poses the question of defining the pathway(s) of energy flow from the core to the protein-chlorophyll complexes of the thylakoid membrane.

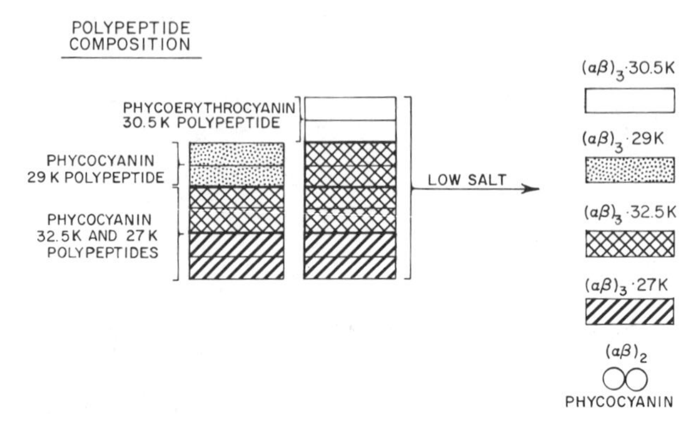

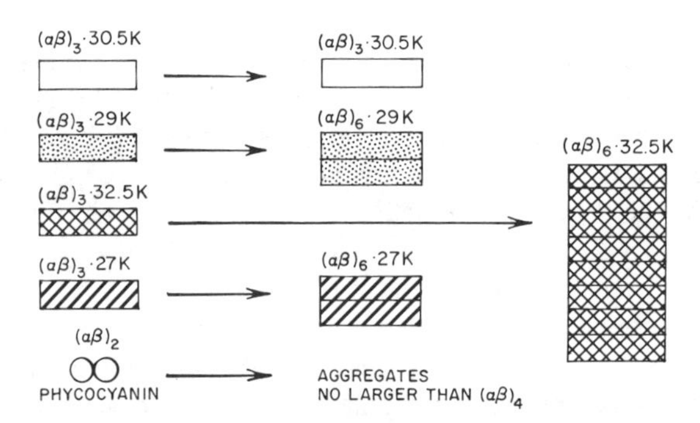

Figure 5 (*A*) Free phycocyanin and stable phycocyanin and phycoerythrocyanin complexes with linker polypeptides obtained upon dissociation of the rod substructures of *Anabaena variabilis* phycobilisomes in 0.05 M NaK-phosphate at pH 7.0. (*B*) Association behavior of these complexes in 0.6 M NaK-phosphate at pH 8.0. The structure of the intact rods, compatible with these observations, is shown in *A*. (From 132 and 135.)

Insight into the organization of the core has come from studies of a *Synechococcus* 6301 mutant, strain AN112 (130, 131) (Figure 7). The rods of the phycobilisomes of this mutant are only one disc in length, hence these phycobilisomes are greatly enriched in core components. The AN112 phycobilisome contains phycocyanin and allophycocyanin in a molar ratio of ~ 1.5:1. Its other components are allophycocyanin B, the 27,000- and 75,000-dalton linker polypeptides, an 18,300-dalton bilin-bearing polypeptide, and a 10,500-dalton polypeptide. Fractionation of AN112 phycobilisomes partially dissociated in 50 mM Tricine-5 mM $CaCl_2$-10% (wt/vol) glycerol, pH 7.8, by sucrose density gradient centrifugation, led to the separation of three fractions of distinctive composition. Fraction III contained a subassembly of unique composition, the 18S particle (Figure 7*C*).

This fraction contained all of the 75,000- and 18,300-dalton polypeptides, half of the allophycocyanin, and none of the allophycocyanin B of the phycobilisome. Fraction II was made up largely of the phycocyanin $(\alpha\beta)_6 \cdot 27K$ complex, whereas fraction I contained phycocyanin, allophycocyanin, allophycocyanin B, and most of the 10,500-dalton polypeptide. Extensive examination of the amount of each biliprotein in these fractions, as well as of the physical properties of these components, led to the model of the AN112 phycobilisome shown in Figure 7 (131). The most important firm conclusion from this study is that the two terminal energy acceptors, allophycocyanin B and the 75,000-dalton linker polypeptide are located within different domains of the core structure.

Structure of the Phycobilisome in Relation to Energy Transfer

The major features of energy transfer in intact cyanobacterial and red algal cells and in phycobilisomes have been established by both steady-state and

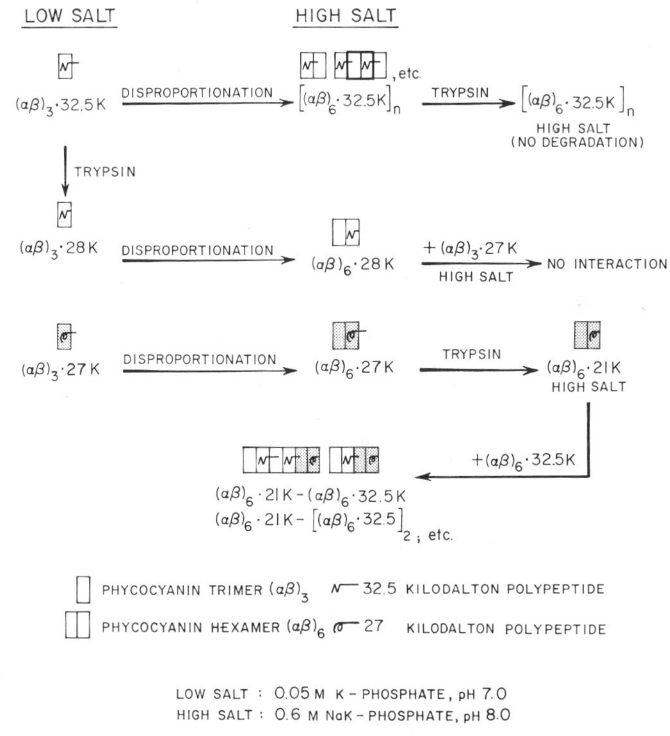

Figure 6 Tryptic degradation of *Anabaena variabilis* phycocyanin complexes with the 32,-500- and 27,000-dalton linker polypeptides and the aggregation behavior of the modified complexes. For a discussion, see text. (From 133 and 135.)

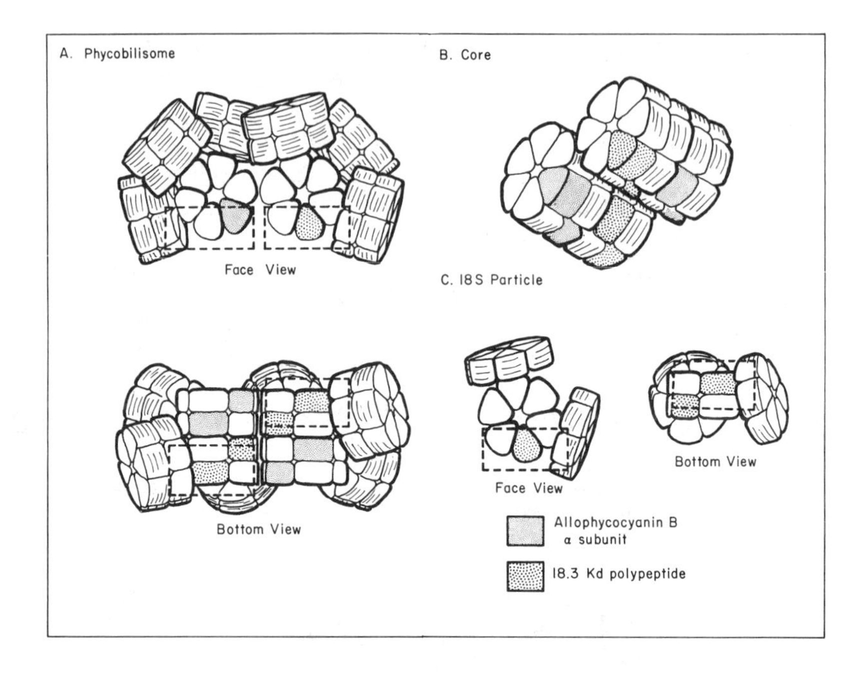

Figure 7 A schematic model of the *Synechococcus* 6301 strain AN112 phycobilisome (130, 131). (*A*) The intact phycobilisome. The face view shows the two-disc core domain parallel to its cylindrical axis, whereas the bottom view is that from the direction of the thylakoid membrane, perpendicular to this axis. In each view, the core is represented by two congruent cylindrical elements surrounded by six double-disc $(\alpha\beta)_6$·27K phycocyanin complexes. Each α or β biliprotein subunit is represented by a one-sixth sector of each single-disc trimer. In the bottom view, each core cylinder is shown as a stack of two non-equivalent double-disc hexamers consisting principally of allophycocyanin. One hexamer of each core cylinder is attached to two hexameric discs of phycocyanin, contains the 75,000-dalton linker polypeptide (represented by the dashed box) as an integral component, and upon partial dissociation of the phycobilisome is found in the 18S particle shown in *C*. Under the same conditions of dissociation, the other core hexamer (free allophycocyanin domain) dissociates primarily to two trimers of free biliprotein. In the intact phycobilisome, this latter hexamer interacts with only one hexamer of phycocyanin. In addition to the α and β subunits of allophycocyanin, two other polypeptides are present in the core cylinders: the α subunit of allophycocyanin B, found exclusively in the free allophycocyanin domain at a maximum of one copy per single disc trimer; and the 18,300-dalton polypeptide (see text), found only in the other core domain, also at a maximum of one copy per trimer. (*B*) The core is shown in an isometric view to emphasize the geometry of the two congruent cylindrical elements. (*C*) The 18S particle derived from AN112 phycobilisomes upon partial dissociation. Each 18S particle contains two trimeric discs of phycocyanin, each attached to the core domain through its interaction with the 27,000-dalton linker polypeptide. Unlike the phycocyanin hexamers that are attached to the core through the free allophycocyanin domains and are released as $(\alpha\beta)_6$·27K complexes upon partial dissociation of the phycobilisome, those interacting with the 18S core domains dissociate such that one phycocyanin trimer is released whereas the residual $(\alpha\beta)_3$·27K complex remains attached (131). The assumption that the 18,300-dalton polypeptide occupies the position of an allophycocyanin α subunit is not established. The number of copies of the

time-resolved measurements of fluorescence emission. The pathway of energy transfer is as follows (23, 36, 57, 93, 94, 102):

Phycoerythrin λ_{max} 565					Allophycocyanin B λ_{max} 670	
or	→	Phycocyanin λ_{max} 620–38	→	Allophycocyanin λ_{max} 650	→	→ Chlorophyll *a*
Phycoerythrocyanin λ_{max} 568					High Mol Wt Polypeptide λ_{max} 665	

The details of the pathway between allophycocyanin and chlorophyll *a* remain to be established.

Although the observed fluorescence lifetimes for the individual biliproteins are on the order of 2–4 ns (1 ns = 10^{-9} s) (56, 125), the induced resonance energy transfer in the above pathway from donor to acceptor is on a <100-ps time scale (1 ps = 10^{-12} s) (94). In intact cells, the overall efficiency of energy transfer from the phycobilisome to chlorophyll *a* exceeds 90% (94). What are the features of phycobilisome structure that contribute to the high efficiency of this process?

A major contribution comes from the increase in the oscillator strength of the long wavelength absorption band in the spectra of individual biliprotein upon formation of the hexameric complexes characteristic of the phycobilisome (74, 135). This change is paralleled by a red shift in the fluorescence emission spectrum and an increase in the quantum yield of the biliprotein (74, 135).

Two features of the rod substructures appear important. Electron microscopy and model building suggest that the rods are not in close contact with each other (e.g. 16). Consequently, radiationless energy transfer is limited to pathways within single rods. Second, the arrangement of discs in the rods is such that energy transfer is polar, proceeding preferentially from disc to disc towards the core, independent of the point of entry of the excitation quantum. This is not only because of obvious factors such as the placement of biliproteins absorbing in the green, such as phycoerythrin, at the periphery of the rods and of phycocyanin in the proximal portion. In rods made up solely of phycocyanin, such as those of *Synechococcus* 6301 phycobilisomes, the properties of the individual discs are modulated in such a manner by interactions with linker polypeptides as to ensure polar energy transfer

←

18,300-dalton polypeptide and of the *α* subunit of allophycocyanin B is 2–4 per phycobilisome. The exact numbers remain to be determined. A 10,500-dalton polypeptide also appears to be associated with the core (128).

towards the core. The order of the disc complexes in the rod substructures in *Synechococcus* 6301, from the end proximal to the core, is $(\alpha\beta)_6$•27K, $(\alpha\beta)_6$•33K, and $(\alpha\beta)_6$•30K (74, 128). This is paralleled by their fluorescence emission maxima at 652, 648, and 643 nm, respectively (74). These structural features are designed to minimize "random walk" of the excitation quantum. The desirability of polar energy flow is evident both from theoretical calculations (e.g. 103, 104) and from experimental data. *Synechococcus* 6301 mutant AN112 has phycobilisomes with rod substructures that consist of a single hexameric phycocyanin-27K complex (130). Comparison of the fluorescence emission spectrum of the phycobilisomes of this mutant with those of wild type, where the rod substructures are on the average three hexamers in length, demonstrated that some 5% of the energy absorbed by the distal hexamers in the wild-type particles was lost as fluorescence (130). There is as yet little detailed information on the topology of the energy transfer chain within the core.

It is interesting to note that uphill energy transfer from chlorophyll *a* to allophycocyanin has been detected in cyanobacterial cells (82, 122). The existence of such transfer opens up the possibility, entirely hypothetical at present, that the phycobilisome may serve as a bridge for energy flow between photosystems II and I in cyanobacteria and red algae, and that spillover between these photosystems proceeds through the base of the phycobilisome.

DYNAMICS OF PHYCOBILISOME STRUCTURE

Chromatic Adaptation

In certain of the cyanobacteria that synthesize the red biliproteins, C-phycoerythrin ($\lambda_{max} = 565$ nm) and phycoerythrocyanin ($\lambda_{max} = 568$ nm), the synthesis of these proteins is controlled by the wavelength of the incident light (9, 32, 41, 110, 118). Phycocyanin synthesis is enhanced in red light, phycoerythrin synthesis is enhanced in green light. Tandeau de Marsac (108) showed that phycoerythrin-producing cyanobacteria fall into three groups: group I, those that do not adapt chromatically; group II, those that modulate only phycoerythrin synthesis; and group III, those that modulate both phycoerythrin and phycocyanin synthesis. Chromatic adaptation provides a convenient way of varying in vivo the biliprotein composition of a cyanobacterial cell. How does such variation influence phycobilisome structure?

A key observation is that the level of allophycocyanin per phycobilisome does not change in chromatic adapters and that the ultrastructure of the core is unaltered (16, 123). In group II chromatic adapters, the length of the rod substructures decreases in proportion to the decrease in phycoerythrin. In group III organisms, an additional phycocyanin (PC-2) is produced in red light (12). The amount of the second phycocyanin is such that it compensates on a molar basis for the decrease in phycoerythrin (12). The length of the rods in the phycobilisomes of group III organisms does not change with light quality (16, 98). Pre-existing phycoerythrin is not destroyed on chromatic adaptation to red light, but is diluted by cell multiplication (4). It can be inferred that new phycobilisomes produced in red light have phycocyanin PC-2 hexamers in place of phycoerythrin (12). As might be expected, the linker polypeptides associated with phycoerythrin disappear when this biliprotein is absent (42, 109, 123).

These observations show that chromatic adaptation is achieved by a limited modification of a complex particle. Only the distal portion of the rod substructures are affected. The phycobilisome can be regarded as a modular structure in which certain modules at the periphery can be removed, replaced, or increased in number, without affecting the organization of the remainder of the components. Such modules are the hexameric biliprotein complexes of the rods.

Other Physiological Adaptations

The amount of biliprotein in cyanobacterial cells is affected by a variety of parameters. For example, in *Synechococcus* 6301 high levels of phycocyanin are obtained under conditions of high temperature, light intensity, and CO_2 concentration (25, 54, 91). In contrast, nitrogen starvation under high light intensity and CO_2 concentration leads to decrease in the biliprotein level (1, 70, 92, 127). Such fluctuations are brought about by two mechanisms: change in the number of phycobilisomes and change in their composition (126, 127). For *Synechococcus* 6301, conditions favoring increased phycocyanin synthesis resulted in the production of phycobilisomes with longer rod substructures, but with unaltered cores (127). Under conditions of nitrogen starvation, an early response is a decrease in the length of the rods on pre-existing phycobilisomes (127). Hence, diverse physiological adaptations, such as those that lead to increased phycocyanin synthesis, or those in response to nitrogen starvation, employ a molecular mechanism equivalent to that seen in chromatic adaptation—addition or removal of the distal segments of the rod substructures.

Acknowledgments

Sections of this article have presented concepts developed over several years in collaboration with Gregory Yamanaka, Daniel J. Lundell, and Robley C. Williams. The valuable contributions of Myeong-Hee Yu and Jeffrey C. Gingrich are also gratefully acknowledged. The work from our laboratory was supported by a grant (PCM 79-10996) from the National Science Foundation.

Literature Cited

1. Allen, M. M., Smith, A. J. 1969. *Arch. Mikrobiol.* 69:114–20
2. Basile, G., DiBello, C., Taniuchi, H. 1980. *J. Biol. Chem.* 255:7181–91
3. Bearden, A. J., Malkin, R. 1975. *Q. Rev. Biophys.* 7:131–77
4. Bennett, A., Bogorad, L. 1973. *J. Cell Biol.* 58:419–35
5. Bennett, A., Siegelman, H. W. 1979. *The Porphyrins,* ed. D. Dolphin, 6(A):493–520. New York: Academic. 932 pp.
6. Berns, D. S. 1971. *Biol. Macromol.* 5(A):105–48
7. Berns, D. S., Edwards, M. R. 1965. *Arch. Biochem. Biophys.* 110:511–16
8. Blauer, G., Wagnière, G. 1975. *J. Am. Chem. Soc.* 97:1949–54
9. Bogorad, L. 1975. *Ann. Rev. Plant Physiol.* 26:369–401
10. Bois-Choussy, M., Barbier, M. 1978. *Heterocycles* 9:677–90
11. Brown, A. S., Offner, G. D., Ehrhardt, M. M., Troxler, R. F. 1979. *J. Biol. Chem.* 254:7803–11
12. Bryant, D. A. 1981. *Eur. J. Biochem.* 119:425–29
13. Bryant, D. A., Cohen-Bazire, G. 1981. *Eur. J. Biochem.* 119:415–24
14. Bryant, D. A., Cohen-Bazire, G., Glazer, A. N. 1981. *Arch. Microbiol.* 129:190–98
15. Bryant, D. A., Glazer, A. N., Eiserling, F. A. 1976. *Arch. Microbiol.* 110:61–75
16. Bryant, D. A., Guglielmi, G., Tandeau de Marsac, N., Castets, A.-M., Cohen-Bazire, G. 1979. *Arch. Microbiol.* 123:113–27
17. Bryant, D. A., Hixson, C. S., Glazer, A. N. 1978. *J. Biol. Chem.* 253:220–25
18. Burke, M. J., Pratt, D. C., Moscowitz, A. 1972. *Biochemistry* 11:4025–31
19. Canaani, O. D., Gantt, E. 1980. *Biochemistry* 19:2950–56
20. Chae, Q., Song, P. S. 1975. *J. Am. Chem. Soc.* 97:4176–79
21. Chapman, D. J. 1973. In *The Biology of Blue-Green Algae,* ed. N. G. Carr, B. A. Whitton, pp. 162–85. Berkeley: Univ. Calif. 676 pp.
22. DeLange, R. J., Williams, L. C., Glazer, A. N. 1981. *J. Biol. Chem.* 256:9558–66
23. Duysens, L. N. M. 1964. *Prog. Biophys. Mol. Biol.* 14:3–104
24. Eiserling, F. A., Glazer, A. N. 1974. *J. Ultrastruc. Res.* 47:16–25
25. Eley, J. H. 1971. *Plant Cell Physiol.* 12:311–16
26. Emerson, R., Arnold, W. 1932. *J. Gen. Physiol.* 15:391–420
27. Emerson, R., Arnold, W. 1932. *J. Gen. Physiol.* 16:191–205
28. Falkowski, P. G., Owens, T. G., Ley, A. C., Mauzerall, D. C. 1981. *Plant Physiol.* 68:969–73
29. Fisher, R. G., Woods, N. E., Fuchs, H. E., Sweet, R. M. 1980. *J. Biol. Chem.* 255:5082–89
30. Frank, G., Sidler, W., Widmer, H., Zuber, H. 1978. *Hoppe-Seyler's Z. Physiol. Chem.* 359:1491–507
31. Freidenreich, P., Apell, G. S., Glazer, A. N. 1978. *J. Biol. Chem.* 253:212–19
32. Füglistaller, P., Widmer, H., Sidler, W., Frank, G., Zuber, H. 1981. *Arch. Microbiol.* 129:268–74
33. Gaffron, H., Wohl, K. 1936. *Naturwissenschaften* 24:81–90
34. Gantt, E. 1969. *Plant Physiol.* 44: 1629–38
35. Gantt, E. 1980. *Int. Rev. Cytol.* 66: 45–80
36. Gantt, E. 1981. *Ann. Rev. Plant Physiol.* 32:327–47
37. Gantt, E., Lipschultz, C. A. 1972. *J. Cell Biol.* 54:313–24
38. Gantt, E., Lipschultz, C. A. 1977 *J. Phycol.* 13:185–92
39. Gantt, E., Lipschultz, C. A., Zilinskas, B. 1976. *Biochim. Biophys. Acta* 430: 375–88
40. Gantt, E., Lipschultz, C. A., Grabowski, J., Zimmerman, B. K. 1979. *Plant Physiol.* 63:615–20
41. Gendel, S., Ohad, I., Bogorad, L. 1979. *Plant Physiol.* 64:786–90
42. Gingrich, J. C., Blaha, L. K., Glazer, A. N. 1981. *J. Cell Biol.* 92:261–68
43. Glazer, A. N. 1976. *Photochem. Photobiol. Rev.* 1:71–115
44. Glazer, A. N. 1977. *Mol. Cell. Biochem.* 18:125–40
45. Glazer, A. N. 1980. In *The Evolution of Protein Structure and Function,* ed. D. S. Sigman, M. A. B. Brazier, pp. 221–44. New York: Academic. 350 pp.
46. Glazer, A. N. 1981. In *The Biochemistry of Plants,* ed. M. D. Hatch, N. K. Boardman, 8:51–96. New York: Academic. 521 pp.
47. Glazer, A. N., Bryant, D. A. 1975. *Arch. Microbiol.* 104:15–22
48. Glazer, A. N., Fang, S. 1973. *J. Biol. Chem.* 248:663–71
49. Glazer, A. N., Fang, S., Brown, D. M. 1973. *J. Biol. Chem.* 248:5679–85
50. Glazer, A. N., Hixson, C. S. 1975. *J. Biol. Chem.* 250:5487–95
51. Glazer, A. N., Hixson, C. S. 1977. *J. Biol. Chem.* 252:32–42

52. Glazer, A. N., Hixson, C. S., DeLange, R. J. 1979. *Anal. Biochem.* 92:489–96
53. Glazer, A. N., Williams, R. C., Yamanaka, G., Schachman, H. K. 1979. *Proc. Natl. Acad. Sci. USA* 76:6162–66
54. Goedheer, J. C. 1976. *Photosynthetica* 10:411–22
55. Gossauer, A., Plieninger, H. 1979. *The Porphyrins,* ed. D. Dolphin, 6(A):585–650. New York: Academic. 932 pp.
56. Grabowski, J., Gantt, E. 1978. *Photochem. Photobiol.* 28:39–45
57. Grabowski, J., Gantt, E. 1978. *Photochem. Photobiol.* 28:47–54
58. Guglielmi, G., Cohen-Bazire, G., Bryant, D. A. 1981. *Arch. Microbiol.* 129:181–89
59. Gysi, J., Zuber, H. 1974. *FEBS Lett.* 48:209–13
60. Gysi, J., Zuber, H. 1976. *FEBS Lett.* 68:49–54
61. Hiller, R. G., Goodchild, D. J. 1981. In *The Biochemistry of Plants,* ed. M. D. Hatch, N. K. Boardman, 8:1–49. New York: Academic. 521 pp.
62. Kawamura, M., Mimuro, M., Fujita, Y. 1979. *Plant Cell Physiol.* 20:697–705
63. Killilea, S. D., O'Carra, P., Murphy, R. F. 1980. *Biochem. J.* 187:311–20
64. Kok, B. 1956. *Biochim. Biophys. Acta* 22:399–401
65. Koller, K.-P., Wehrmeyer, W., Mörschel, E. 1978. *Eur. J. Biochem.* 91:57–63
66. Köst-Reyes, E., Köst, H.-P. 1979. *Eur. J. Biochem.* 102:83–91
67. Köst-Reyes, E., Köst, H.-P., Rüdiger, W. 1975. *Justus Liebigs Ann. Chem.* 1975:1594–600
68. Kufer, W., Scheer, H. 1979. *Hoppe-Seyler's Z. Physiol. Chem.* 360:935–56
69. Lagarias, J. C., Glazer, A. N., Rapoport, H. 1979. *J. Am. Chem. Soc.* 101:5030–37
70. Lau, R. H., Mackenzie, M. M., Doolittle, W. F. 1977. *J. Bacteriol.* 132:771–78
71. Ley, A. C., Butler, W. L., Bryant, D. A., Glazer, A. N. 1977. *Plant Physiol.* 59:974–80
72. Lipschultz, C. A., Gantt, E. 1981. *Biochemistry* 20:3371–76
73. Lundell, D. J., Glazer, A. N. 1981. *J. Biol. Chem.* 256:12600–6
74. Lundell, D. J., Williams, R. C., Glazer, A. N. 1981. *J. Biol. Chem.* 256:3580–92
75. Lundell, D. J., Yamanaka, G., Glazer, A. N. 1981. *J. Cell Biol.* 91:315–19
76. MacColl, R., Csatorday, K., Berns, D., Traeger, E. 1980. *Biochemistry* 19:2817–20
77. Mörschel, E., Koller, K.-P., Wehrmeyer, W. 1980. *Arch. Microbiol.* 125:43–51
78. Mörschel, E., Koller, K.-P., Wehrmeyer, W., Schneider, H. 1977. *Cytobiologie* 16:118–29
79. Mörschel, E., Wehrmeyer, W., Koller, K.-P. 1980. *Eur. J. Cell Biol.* 21:319–27
80. Muckle, G., Rüdiger, W. 1977. *Z. Naturforsch. Teil C* 32:957–62
81. Muckle, G., Otto, J., Rüdiger, W. 1978. *Hoppe-Seyler's Z. Physiol. Chem.* 359:345–55
82. Murata, N. 1977. In *Photosynthetic Organelles,* ed. S. Miyachi, S. Katoh, Y. Fujita, K. Shibata, pp. 9–13. Tokyo: Jpn. Soc. Plant Physiol.
83. Myers, J., Graham, J. R., Wang, R. T. 1980. *Plant Physiol.* 66:1144–49
84. Nies, M., Wehrmeyer, W. 1981. *Arch. Microbiol.* 129:374–79
85. O'Carra, P., O'hEocha, C. 1976. In *Chemistry and Biology of Plant Pigments,* ed. T. W. Goodwin, 2nd ed., pp. 328–76. New York: Academic. 870 pp.
86. O'Carra, P., O'hEocha, C., Carroll, D. M. 1964. *Biochemistry* 3:1343–50
87. O'Carra, P., Murphy, R. F., Killilea, S. D. 1980. *Biochem. J.* 187:303–9
88. O'hEocha, C. 1963. *Biochemistry* 2:375–82
89. O'hEocha, C., O'Carra, P. 1961. *J. Am. Chem. Soc.* 83:1091–93
90. Öquist, G. 1974. *Physiol. Plant.* 30:38–44
91. Öquist, G. 1974. *Physiol. Plant.* 30:45–48
92. Paone, D. A. M., Stevens, S. E. Jr. 1981. *Plant Physiol.* 67:1097–100
93. Pellegrino, F., Wong, D., Alfano, R. R., Zilinskas, B. A. 1981. *Photochem. Photobiol.* 34:691–96
94. Porter, G., Tredwell, C. J., Searle, G. F. W., Barber, J. 1978. *Biochim. Biophys. Acta* 501:232–45
95. Rabinowitch, E. I. 1951. *Photosynthesis and Related Processes.* 2(I):838. New York: Intersci. Publ.
96. Redlinger, T. E., Gantt, E. 1981. *Plant Physiol.* (Suppl.) 67:64
97. Rigbi, M., Rosinski, J., Siegelman, H. W., Sutherland, J. C. 1980. *Proc. Natl. Acad. Sci. USA* 77:1961–65
98. Rosinski, J., Hainfeld, J. F., Rigbi, M., Siegelman, H. W. 1981. *Ann. Bot.* 47:1–12
99. Rüdiger, W. 1971. *Fortschr. Chem. Organ. Naturst.* 29:59–139
100. Rüdiger, W. 1975. *Ber. Dtsch. Bot. Ges.* 88:125–39
101. Scheer, H. 1981. *Angew. Chem. Int. Ed. Engl.* 20:241–61

102. Searle, G. F. W., Barber, J., Porter, G., Tredwell, C. J. 1978. *Biochim. Biophys. Acta* 501:246–56
103. Seely, G. R. 1973. *J. Theor. Biol.* 40:173–87
104. Seely, G. R. 1973. *J. Theor. Biol.* 40:189–99
105. Sidler, W., Gysi, J., Isker, E., Zuber, H. 1981. *Hoppe-Seyler's Z. Physiol. Chem.* 362:611–28
106. Stanier, R. Y., Cohen-Bazire, G. 1977. *Ann. Rev. Microbiol.* 31:225–74
107. Stoll, M. S., Gray, C. H. 1977. *Biochem. J.* 163:59–101
108. Tandeau de Marsac, N. 1977. *J. Bacteriol.* 130:82–91
109. Tandeau de Marsac, N., Cohen-Bazire, G. 1977. *Proc. Natl. Acad. Sci. USA* 74:1635–39
110. Tandeau de Marsac, N., Castets, A.-M., Cohen-Bazire, G. 1980. *J. Bacteriol.* 142:310–14
111. Thornber, J. P. 1975. *Ann. Rev. Plant Physiol.* 26:127–58
112. Troxler, R. F. 1977. In *Chemistry and Physiology of Bile Pigments,* ed. P. D. Berk, N. J. Berlin, pp. 431–54. Washington: DHEW Publ. No. (NIH) 77–100. 539 pp.
113. Troxler, R. F., Brown, A. S. 1979. *Fed. Proc.* 38:510
114. Troxler, R. F., Brown, A. S., Köst, H.-P. 1978. *Eur. J. Biochem.* 87:181–89
115. Troxler, R. F., Greenwald, L. S., Zilinskas, B. A. 1980. *J. Biol. Chem.* 255:9380–87
116. Vaughan, M. H. Jr. 1964. *Structural and comparative studies of the algal protein phycoerythrin.* PhD dissertation. Mass. Inst. Technol., Cambridge. 458 pp.
117. Vierling, E., Alberte, R. S. 1980. *Physiol. Plant.* 50:93–98
118. Vogelmann, T. C., Scheibe, J. 1978. *Planta* 143:233–39
119. Wagnière, G., Blauer, G. 1976. *J. Am. Chem. Soc.* 98:7806–10
120. Walsh, R. G., Wingfield, P., Glazer, A. N., DeLange, R. J. 1980. *Fed. Proc.* 39:2060
121. Wang, R. T., Stevens, C. L. R., Myers, J. 1977. *Photochem. Photobiol.* 25: 103–8
122. Wang, R. T., Myers, J. See Ref. 82, pp. 3–7
123. Williams, R. C., Gingrich, J. C., Glazer, A. N. 1980. *J. Cell Biol.* 85:558–66
124. Williams, V. P., Glazer, A. N. 1978. *J. Biol. Chem.* 253:202–11
125. Wong, D., Pellegrino, F., Alfano, R. R., Zilinskas, B. A. 1981. *Photochem. Photobiol.* 33:651–62
126. Yamanaka, G. 1981. *Structure and dynamics of cyanobacterial phycobilisomes.* PhD thesis. Univ. Calif., Berkeley. 344 pp.
127. Yamanaka, G., Glazer, A. N. 1980. *Arch. Microbiol.* 124:39–47
128. Yamanaka, G., Glazer, A. N. 1981. *Arch. Microbiol.* 130:23–30
129. Yamanaka, G., Glazer, A. N., Williams, R. C. 1978. *J. Biol. Chem.* 253:8303–10
130. Yamanaka, G., Glazer, A. N., Williams, R. C. 1980. *J. Biol. Chem.* 255:11004–10
131. Yamanaka, G., Lundell, D. J., Glazer, A. N. 1982. *J. Biol. Chem.* In press
132. Yu, M.-H. 1981. *Biliproteins of cyanobacteria and rhodophyta: studies of structure and assembly.* PhD thesis. Univ. Calif., Berkeley. 241 pp.
133. Yu, M.-H., Glazer, A. N. 1982. *J. Biol. Chem.* In press
134. Yu, M.-H., Glazer, A. N., Spencer, K. G., West, J. A. 1981. *Plant Physiol.* 68:482–88
135. Yu, M.-H., Glazer, A. N., Williams, R. C. 1981. *J. Biol. Chem.* 256:13130–36
136. Zilinskas, B. A., Zimmerman, B. K., Gantt, E. 1978. *Photochem. Photobiol.* 27:587–95
137. Zuber, H. 1978. *Ber. Dtsch. Bot. Ges.* 91:459–75

Ann. Rev. Microbiol. 1982. 36:199–216

INFECTIONS DUE TO *HAEMOPHILUS* SPECIES OTHER THAN *H. INFLUENZAE*

W. L. Albritton

Department of Medical Microbiology, University of Manitoba, Winnipeg, Manitoba R3E OW3, Canada

CONTENTS

INTRODUCTION

The 8th edition of *Bergey's Manual of Determinative Bacteriology* (14) includes 14 species in the genus *Haemophilus* with an additional 5 species incertae sedis. Ten of these species other than *Haemophilus influenzae* have been associated with human infections and this review covers selected clinical and microbiological aspects related to these species. One species listed

0066-4227/82/1001-0199$02.00

under the name *Haemophilus vaginalis* was considered not to belong in the genus *Haemophilus.* Recent studies support inclusion of this organism in a separate genus, *Gardnerella,* and it is not discussed in this review (103).

MICROBIOLOGY

The genus *Haemophilus* was created by the American Committee on Classification and Nomenclature in 1920 to include a number of hemophilic organisms (140). Presently, however, the genus is restricted to gram-negative, facultatively anaerobic, rod-shaped bacteria requiring hemin or certain other porphyrins (X-factor) and/or nicotinamide adenine dinucleotide (NAD) or certain definable coenzyme-like substances (V-factor) (142). The type species, *H. influenzae,* however, was isolated and described by Pfeiffer in 1892 in association with a clinical syndrome, and it was originally regarded to be the etiologic agent of influenza (102). Although the etiologic agent of acute contagious conjunctivitis (*H. aegyptius*) had been described by Koch (78) in 1883 and by Weeks (138) in 1886, and the etiologic agent of chancroid or soft chancre (*H. ducreyi*) had been described by Ducrey (28) in 1889, there was no confusion in the early literature regarding the identification of these organisms because of their association with clinically distinct syndromes. Further studies of haemophilus-like organisms from a wide variety of human and animal sources has resulted in the introduction into the literature of an increasing number of taxonomic species, some of which have questionable clinical validity. Seven species of *Haemophilus* that can be isolated from the human respiratory tract and that have been introduced into the literature since the description of Pfeiffer's bacillus in 1892 are listed in Table 1. Hemolytic strains of the influenza bacillus (Bacillus X) were reported in 1919 by Pritchett & Stillman (108). These strains were further characterized by Stillman & Bourn (126) in 1920 and subsequently were designated *H. haemolyticus* in the 1st edition of *Bergey's Manual.* Studies on the influenza bacillus following the 1918 pandemic showed that true influenza bacilli (*H. influenzae*) required two accessory growth factors present in blood, one of which was heat stable, and that certain influenza-like bacilli (*H. parainfluenzae*) required only the heat-labile factor (111). Thjötta & Avery (131) introduced the terms X-factor and V-factor to refer, respectively, to the heat-stable substance subsequently identified as hemin and the heat-labile vitamin-like substance subsequently identified as NAD. Fildes in 1924 described hemolytic strains that required both X- and V-factors, as well as strains that required only V-factor (35). In 1927 Valentine & Rivers proposed a classification of hemophilic bacilli based on requirement for accessory growth factors (132). This classification, however, did not assign separate species status to the hemolytic and non-hemolytic variants of *H. influenzae* and *H. parainfluenzae,* which were designated by their accessory growth factor requirements, respectively, for X- and V-factor and V-factor only. The growth factor requirements of *H. ducreyi* were not known until Lwoff & Pirosky established a requirement for X- but not V-factor in 1937 (87), and the growth factor requirements of *H. aegyptius* were established by Knorr in 1924 (77). Since the hemolytic strains referred to as Bacillus X by Pritchett & Stillman (108) had been given species status as *H. haemolyticus* without regard to requirement for accessory growth factors, there was considerable uncertainty in clinical reports of infections as to which organisms were being described, until 1953 when Pittman recommended the creation of two separate species within the hemolytic strains, *H. haemolyticus* and *H. parahaemolyticus,* in accordance with their accessory growth factor requirements and in conformity with the species designation of nonhemolytic strains (105). Thus, from a single clinically relevant species, the influenza bacillus, four taxonomic species had been created based only on accessory growth factor requirements and hemolysis. All four species were isolated from the same clinical sources and it is interesting to speculate that the fractionation of the species "influenza bacillus" was due to the continuing controversy regarding the true etiology of the clinical syndrome influenza, which was not established until 1933. Since most invasive strains of *H. influenzae* are serotype *b,* had serological studies such as those by Pittman (104) been carried out in the early 1900s rather than nutritional studies, one wonders if an entirely different basis for designation of species would not have occurred. By 1933, however, all three species (*H. influenzae, H. parainfluenzae,* and *H. haemolyticus*) had been associated with clinical infections and there was already precedent for assigning species status based on limited distinguishing characteristics. Thus *H. aphrophilus, H. paraphrophilus,* and *H. paraphrohaemolyticus* were described in 1940, 1968, and 1971, respectively, based primarily on a requirement for increased CO_2 for primary isolation and only limited biochemical characteristics other than hemolysis and the requirement for X- or V-factors for growth. As is discussed later, the clinical relevance of some of these species is questionable since they demonstrate no unique clinical, ecological, or pathological properties. On the other hand, *H. aegyptius,* a perfectly valid clinical species, epidemiologically associated with acute contagious conjunctivitis by Koch in 1883 (78), isolated by Weeks in 1886 (138), and proposed as a new species by Pittman & Davis in 1950 (106), does not appear to warrant species status based on current taxonomy of the genus and has recently been proposed as a hemagglutinating variety of *H. influenzae* (71).

Several taxonomic studies of the genus *Haemophilus* have been reported since 1971 that fail to support specific status for several species included in

Table 1. Particularly useful was the study by Kilian in 1976 (68) in which differential characteristics other than hemolysis, CO_2 requirement, and X- and V-factor requirement were considered. Differential characteristics of the genus *Haemophilus* as proposed by Kilian (68) and discussed in a recent critical review by Zinnemann (143) are given in Table 2. Confusing discrepancies in biochemical characteristics occur between the original published descriptions of the species, various taxonomic studies, and the description of the species in the 8th edition of *Bergey's Manual* and these should be considered briefly.

The oxidase test was reported negative by Sneath & Johnson (123) for the *H. influenzae* group in their numerical taxonomic study of *Haemophilus* but positive by Kilian (68), who used the same test, with variable results for biotype IV strains. Variable results with the oxidase test are to be expected with the use of different tests, substrates, and growth media. Nevertheless, current studies identify strains as belonging to a species based on tests that differ from the originally described strains. For example, in the original description of the species *H. paraphrohaemolyticus* by Zinnemann et al (145) the species is said to be oxidase negative, yet the study by Kilian (68) indicated six of eight strains grouped in this species were oxidase positive and this was included in the critical review by Zinnemann (143) as a species characteristic. Similar confusion exists with regard to description of the oxidase test for *H. paraphrophilus* (68, 144) and *H. ducreyi* (68, 124). As is shown in Table 2 only a few characteristics other than CO_2 requirement distinguish *H. paraphrohaemolyticus* from *H. parahaemolyticus* and the variability of these tests does not help matters. *H. paraphrohaemolyticus* is said to be variably positive for indole production in the 8th edition of *Bergey's Manual* and in the description of the species by Zinneman et al (145), but it was reported negative by Kilian (68) and by Zinnemann (143)

Table 1 Description of selected species of the genus *Haemophilus*

Species	Description	Reference
H. aegyptius	1883[a]	78
H. ducreyi	1889	28
H. influenzae	1892	102
H. haemolyticus	1919	108
H. parainfluenzae	1922	111
H. aphrophilus	1940	65
H. parahaemolyticus	1953	105
H. paraphrophilus	1968	144
H. paraphrohaemolyticus	1971	145

[a] The Koch-Weeks bacillus was described as *H. aegyptius* by Pittman & Davis in 1950 (106).

Table 2 Differential characteristics of the genus *Haemophilus*[a]

Species	X-factor requirement	V-vactor requirement	Indole	Urease	Ornithine decarboxylase	Oxidase	Catalase	ONPG	Glucose, acid	Sucrose, acid	Lactose, acid
Classical											
H. influenzae	+	+	V[b]	V	V	+	+	–	+	–	–
H. aegyptius	+	+	–	+	+	+	–	–	+	–	–
H. parainfluenzae	–	+	–	V	V	+	V	V	+	+	–
H. ducreyi	+	–	–	–	–	–	–	–	–	–	–
Hemolytic											
H. haemolyticus	+	+	V	+	–	+	+	–	+	–	–
H. parahaemolyticus	–	+	–	+	V	+	+	V	+	+	–
H. paraphrohaemolyticus	–	+	–	+	–	V	+	V	+	+	–
CO_2 requiring											
H. aphrophilus	+	–	–	–	–	–	–	+	+	+	+
H. paraphrophilus	–	+	–	–	–	+	–	+	+	+	+
Miscellaneous											
H. haemoglobinophilus	+	–	+	–	–	+	+	V	+	+	–
H. influenzae-murium	+	–	–	–	NR[c]	–	+	V	+	+	–

[a] All members of the genus are gram-negative, facultatively anaerobic rod-shaped bacilli that reduce nitrate to nitrite.
[b] V, Variable.
[c] NR, Not reported.

in his recent critical review. Kilian (68) described non-hemolytic, non-CO_2-requiring, V-factor-dependent strains that had urease activity, whereas Zinnemann (143) and Sneath & Johnson (123) reported *H. parainfluenzae* strains to be urease negative. Also, the catalase test is reported to be positive for *H. parainfluenzae* in the 8th edition of *Bergey's Manual* and by Sneath & Johnson, but it is reported to be only variably positive by Kilian (68) and Zinnemann (143).

Such discrepancies are compounded in reviewing the clinical literature where species designations are usually based on minimal characteristics. Many clinical laboratories assign species status based on growth factor requirement only and do not routinely test for hemolysis or CO_2 requirement, which in the absence of other differential characteristics would be necessary for even preliminary assignment of species. This presumably occurs because of the lack of perceived clinical relevance to speciating non-*H. influenzae* strains (5, 43). However, even the determination of growth factor requirements by the usual satellite phenomenon with factor discs presents difficulties. Misidentification based on this test alone is frequent (69), and satellite growth around catalase-positive organisms such as staphylococci is easily misinterpreted. All of these factors make the interpretation of clinical literature regarding infections due to *Haemophilus* species other than *H. influenzae* difficult.

INFECTIOUS DISEASES

In this review of the literature of infections due to *Haemophilus* species other than *H. influenzae*, it is important to remember the following: (*a*) No attempt was made to independently verify claims of first reports; (*b*) the reported identification of organisms rarely contained sufficient data to verify species designation; (*c*) the majority of the literature reviewed was derived from English language publications contained in Index Medicus 1960–1980.

Classical Species

H. PARAINFLUENZAE Russell & Fildes (115) first reported disease in man due to *H. parainfluenzae* in 1928, some 6 years after description of the species. The organism, a non-hemolytic, V-factor-requiring strain, was isolated from an 18-year-old female with endocarditis. Postmortem studies clearly determined the pathogenicity of the organism.

As shown in Table 3, endocarditis represents the most common reported infection due to *H. parainfluenzae*, with 44 cases reported since 1970 (3, 9, 10, 17, 18, 23, 30, 32, 41, 45, 48, 53, 60, 61, 63, 88, 96, 121, 133). Seven of the isolates were said to have been confirmed in a reference laboratory, but criteria for confirmation were not provided. An additional seven isolates were simply stated to have been confirmed by biochemical and fermentation reactions. Seven isolates were said to conform to Kilian's biotypes of *H. parainfluenzae*, which would not have distinguished *H. parahaemolyticus* and *H. paraphrohaemolyticus*. One isolate was said to ferment lactose but not sucrose, and another isolate produced indole and failed to reduce nitrate to nitrite. These isolates do not conform to previous descriptions of the species. The remainder of the isolates were identified by growth factor requirements and hemolysis only or the microbiology was not stated. Thus, it is clear that insufficient data are provided for cases since 1970 to distinguish clearly between the V-factor-requiring species.

All *Haemophilus* species account for less than 1% of cases of endocarditis, and only nine cases of *H. parainfluenzae* (hemolytic and non-hemolytic strains) endocarditis were included by Craven et al (20) in their review in 1940. Clinical features of *H. parainfluenzae* endocarditis are well reviewed in recent publications (17, 30, 60, 88). Its occurrence in relatively young adults without underlying valvular heart disease and a high frequency of major arterial embolization appear to be distinguishing features.

Gullekson & Dumoff (47) first reported neonatal meningitis due to *H. parainfluenzae* in 1966, and Smith & Berger (122) reported a case in an adult in 1974. The organism described by Gulleksen & Dumoff, however, was identified only on the basis of satellite growth around *Staphylococcus aureus* on nutrient agar medium. Difficulties with using the satellite phenomenon in differentiation of *Haemophilus* species was discussed by Klein & Blazevic (76), and even with the use of factor discs Kilian has reported erroneous results in nearly 20% of tests (69). For this reason recent studies have indicated the porphyrin test modified from Biberstein et al (7) by Kilian (66) and evaluated by Lund & Blazevic (86) is preferable for determining X-factor requirement. Twenty two cases of *H. parainfluenzae* meningitis have been reported since 1966 (2, 4, 16, 47, 48, 55, 80, 89, 122, 141). The series by Hable et al (48) included 7 patients with *H. parainfluenzae* meningitis from a group of 10 patients with *Haemophilus* meningitis compared to only 3 of 56 patients in the series reported by Wort (141). No tests other than growth factor requirements and hemolysis were reported in either series and neither used the porphyrin test. One of three isolates reported by Holt et al (55) produced indole. As noted by Michaels & Phillips (94), the lack of confirmatory tests and the lack of distinguishing epidemiologic or clinical features raises real questions regarding the validity of many reported cases of *H. parainfluenzae* meningitis.

Seven cases of pneumonia (19, 48, 96, 125) have been reported, including thoracic empyema as well as cases of septic arthritis (97, 109, 136) and epiglottitis (16, 112). As with previous cases of meningitis it is possible some of these organisms represent misidentification of *H. influenzae,* which is commonly associated with these syndromes. Unlike the patients with meningitis, however, these patients tend to be older. Hollin et al (54) reported the first case of brain abscess due to *H. parainfluenzae* in 1967 and two subsequent cases have been reported (48, 89). *H. paraphrophilus,* also associated with brain abscess, is easily misidentified as *H. parainfluenzae* and infections due to these organisms are clinically indistinguishable. Finally *H. parainfluenzae* may be isolated from the genitourinary tract as well as the respiratory tract and may be an infrequent pathogen in these sites (11, 114).

Miscellaneous infections due to *H. parainfluenzae* other than those reported in Table 3 include soft tissue abscess (119), peritonitis associated with an intrauterine contraceptive device (39), and puerperal bacteremia and neonatal sepsis (146).

H. DUCREYI Chancroid, or soft chancre (ulcus molle), was first differentiated from syphilis by Bassereau in 1852. The causative organism was first described by Ducrey in 1889 (28), although he was unable to grow the

Table 3 Infections due to *H. parainfluenzae*

Clinical syndrome	No. of cases	Age[a]	Reference
Endocarditis	44[b]	33	115
Meningitis	22	4	47
Pneumonia	7	42	48
Septic arthritis	4	28	97
Epiglottitis	2	42	112
Brain abscess	3	16	54
Urinary tract infection	2	35	11

[a] Mean age if more than one case.
[b] Reported cases since 1970.

organism in vitro. Sullivan (128), in his review of chancroid in 1940, credited Lenglet in 1898 with first culturing *H. ducreyi,* athough Himmel (52) in 1903 had credited Petersen in 1895 and Istoamanov and Akopiantz in 1897 with the first successful culture. Lwoff & Pirosky (87) in 1937 first demonstrated hemin requirement and this was confirmed more recently by Kilian (68) and Hammond et al (49). *H. ducreyi* appears to be associated clinically only with genital ulcer disease or direct inoculation infections (46, 128). Clinical features of an urban outbreak of chancroid in North America have recently been published (51). *H. ducreyi* is the only human *Haemophilus* species with a requirement for X-factor only as determined by the porphyrin test (49, 68). Thus, there would appear to be no difficulty in accepting *H. ducreyi* as a legitimate clinical and taxonomic species.

H. AEGYPTIUS Since the isolation of the bacillus of acute catarrhal conjunctivitis by Weeks in 1886 (138), there has been little doubt that certain strains of *Haemophilus* are ocular pathogens. Knorr (77) in 1924 showed these organisms to require both X- and V-factor, but it was not until 1950, when Davis et al (25) found that strains of the Koch-Weeks bacillus had a hemagglutinin and *H. influenzae* did not and Pittman & Davis (106) reported their work on the identification of the Koch-Weeks bacillus, that definite criteria were available for differentiating *H. aegyptius* from nonserotypable *H. influenzae.* These criteria included the following: (*a*) *H. aegyptius* tended to be slender and longer morphologically and grew more slowly; (*b*) it was indole negative; (*c*) it did not ferment xylose; (*d*) it was serologically distinguishable; and (*e*) it agglutinated human red blood cells. In addition, Orfila & Courden (98) differentiated *H. influenzae* and *H. aegyptius* isolated from ocular diseases by bacteremia curves after intravenous injection in mice. They found a high correlation between in vitro hemagglutination and sustained bacteremia in mice, but this correlation did not hold for xylose fermentation. Davis & Pittman (24) clearly recognized that strains of either *H. influenzae* or *H. aegyptius* could be recovered from cases of acute conjunctivitis. Not unexpectedly, several studies reported isolation of *Haemophilus* species from normal and infected conjunctiva that were indistinguishable from *H. influenzae,* a common respiratory tract commensal (58, 101, 135). Thus many clinical laboratories reported all X- and V-factor-requiring strains of *Haemophilus* isolated from the conjunctiva as *H. aegyptius* or *H. influenzae* without differentiating between the species since hemagglutination tests were difficult to perform and some *H. influenzae* strains could be shown to agglutinate human erythrocytes (59). On the basis of this study Ivler et al (59) suggested that hemagglutination was not a valid criteria for separating *H. influenzae* and *H. aegyptius.* In 1976 Kilian et al (71) reported a taxonomic study of *Haemophilus* conjunctival isolates and suggested that *H. aegyptius* was a hemagglutinating variety of *H. influenzae* biotype III. They never suggested, however, that all biotype III strains of *H. influenzae* were *H. aegyptius.* Unfortunately many laboratories adopting Kilian's method of biotyping have abbreviated the biochemical characteristics to indole production, urease, and ornithine decarboxylase activities. As such there is a recent report of pneumonia due to *H. aegyptius* (91) based on identification of a biotype III strain of *H. influenzae* by the abbreviated biochemical characteristics. It is unfortunate that a clinically relevant species associated with a clearly defined syndrome, acute catarrhal conjunctivitis, and for nearly 100 years never associated with systemic disease in man, loses species status because it lacks easily definable phenetic characteristics. *H. influenzae* and *H. aegyptius* appear to be genetically related (82, 83) and are both isolated from the conjunctiva of patients with ocular disease (24, 42) just as serotype b and nonserotype strains of *H. influenzae* are isolated from the respiratory tract of patients with systemic disease. The clinical relevance of distinguishing serotype b strains from nonserotype strains of *H. influenzae* is clear, but additional studies are needed to determine the clinical relevance of distinguishing *H. influenzae* from hemagglutinating X- and V-factor-requiring strains of *Haemophilus* associated with acute conjunctivitis.

Hemolytic Species

H. HAEMOLYTICUS Although a case of subacute bacterial endocarditis due to a hemolytic hemophilic bacillus was described by Miller & Branch in 1923 (95), only 4 years after the description of hemolytic strains by Pritchett & Stillman (108), it must be remembered that early descriptions of hemolytic organisms did not include growth factor requirements. It was not until the work of Rivers in 1922 (110) and Fildes in 1924 (35) that growth factor requirements were reported and the vast majority of hemolytic strains required only the V-factor, including strains resembling the original Bacillus X of Pritchett & Stillman (108). Thus, in the reported case of endocarditis due to hemolytic para-influenza bacillus and the review by Lichty in 1937 (85), there appear to be no clearly documented cases of systemic disease in man due to hemolytic X- and V-factor-requiring strains of *Haemophilus* (*H. haemolyticus*) and no cases have been reported since factor requirements have been routinely determined.

H. PARAHAEMOLYTICUS Although *H. parahaemolyticus* can be isolated frequently from the human respiratory tract (13, 120), it has been associated only rarely with disease, as shown in Table 4. As indicated above, Lichty (85) identified eight cases of probable *H. parahaemolyticus* endocarditis in his review in 1937. Since that time, however, a number of reports of *H. parainfluenzae* endocarditis have not mentioned hemolytic strains where this property has been assessed. Two isolated cases of gallbladder empyema (100) and periorbital cellulitis (48) have been reported. Branson (13) reported the isolation of *H. parahaemolyticus* from the pharynx of patients with pharyngitis, but the clinical significance of such observations is doubtful since such strains are also recovered from normal throats and no clinical sequelae have been reported.

In the 8th edition of *Bergey's Manual* the species of *H. parahaemolyticus* included hemolytic V-factor-requiring strains of human and animal origin. Two recent studies support separate species designation of strains from human and porcine sources (67, 72). Although there is disagreement in the literature (143), the taxonomic studies of Sneath & Johnson (123) and Kilian (68) do not support separate species status for *H. parahaemolyticus* and little clinical relevance appears for the species (5, 43).

H. PARAPHROHAEMOLYTICUS The original description of the species by Zinnemann et al in 1971 (145) included 11 clinical isolates that grew in

Table 4 Infections due to hemolytic *Haemophilus* species

Clinical syndrome	No. of cases	Age[a]	Reference
H. haemolyticus	0	—	—
H. parahaemolyticus	10		
Endocarditis	(8)	20	95
Gallbladder empyema	(1)	40	100
Periorbital cellulitis	(1)	1	48
H. paraphrohaemolyticus	1		
Liver abscess	(1)	73	27

[a] Mean age if more than one case.

pure or almost pure culture and were presumed pathogens. All sources, however, were superficial cultures associated with respiratory infections or in two cases urethral discharge. One case report is of a liver abscess due to *H. paraphrohaemolyticus* (27). Although a reference laboratory agreed with the identification, it was reported to be ornithine decarboxylase positive. Zinnemann (143) clearly states that *H. paraphrohaemolyticus* is ornithine decarboxylase negative, and as indicated in Table 2, this organism could have been *H. parahaemolyticus,* which, as noted, has previously been associated with gallbladder empyema. As indicated in Table 2, *H. paraphrohaemolyticus* appears taxonomically more related to the hemolytic species than the CO_2-requiring species and this is compatible with clinical observations as well. Both *H. parahaemolyticus* and *H. paraphrohaemolyticus* have been included as hemolytic variants of *H. parainfluenzae* by Kilian (68) and clinical findings support this conclusion.

CO_2-Requiring Species

H. APHROPHILUS Khairat (65) in 1940 described a hemophilic organism that required X-factor and increased CO_2 for isolation. Subsequent taxonomic studies support separate speciation of this group of organisms (123). The first isolate of this species was from the blood of a 28-year-old female with endocarditis. Since the original description of the species over 100 infections have been reported, with 90 cases included in a recent review by Bieger et al (8). Approximately half of all reported infections due to *H. aphrophilus* have been associated with endocarditis and 25 cases have been reported since 1970 (Table 5), including 7 cases not included in the review by Bieger et al (1, 29, 30, 93, 134, 137). Cases have been reported in pregnancy (1), in childhood (93), and in association with prosthetic heart valves (44) and malignancy (33). One case report is of a dual infection with *H. aphrophilus* and *H. paraphrophilus* (41). Clinical features associated with *H. aphrophilus* endocarditis may be found in several recent reports (6, 8, 30, 31, 137).

Brain abscess is the second most common source of *H. aphrophilus* since first described by Fager in 1961 (34). Nineteen cases were included in the review by Bieger et al (8) and four additional cases have been reported (40, 107, 127). Brain abscess appears to be more common than endocarditis in children (139), probably because of the lower incidence of endocarditis in general in the pediatric population. Kilian & Schiott (73) isolated *H. aphrophilus* from dental plaque and Kraut et al (79) recovered *H. aphrophilus* as part of the normal oral flora. Presumably oral flora are the source of organisms in both endocarditis and brain abscess. Nevertheless, CO_2-requiring strains of *Haemophilus,* although not particularly common causes of brain abscess (118), appear to be more frequently isolated from this source than are the hemolytic strains (*H. parahaemolyticus* and *H. paraphrohaemolyticus*) and *H. parainfluenzae,* all of which are part of the normal oral flora (73). As indicated in Table 5, cases of other clinical syndromes have been reported, including pneumonia/empyema (15), fasciitis (21), cholecystitis (57), and sinusitis (99). Soft tissue or wound infections (99, 139) are particularly notable in that they are uncommonly reported with other *Haemophilus* species.

The reviews by Sutter & Finegold (129) and Bieger et al (8) are excellent summaries of published clinical and microbiological studies of *H. aphrophilus.*

Although *H. aphrophilus* was originally included in the genus *Haemophilus* because of its requirement for X-factor, this requirement may be lost on subculture (12), and *H. aphrophilus,* unlike other hemin-requiring *Haemophilus,* has the biosynthetic pathway for hemin synthesis as indicated by a positive porphyrin test (66). In addition the G+C content of *H. aphrophilus* is closer to non-hemin-requiring species of the genus (68). Members of the genus *Haemophilus* have a close phenetic relationship to the *Actinobacillus* and *Pasteurella* genera and there appears to be a very close relationship of *H. aphrophilus* to *Actinobacillus actinomycetemcomitans,* both taxonomically and clinically (68, 75, 90, 116, 123). *Eikenella corrodens,* as well, has an obligate requirement for hemin when grown aerobically and is stimulated by CO_2 (113). As shown in Table 2, *H. aphrophilus* can be differentiated from the clinically related species *A. ac-*

Table 5 Infections due to CO_2-requiring *Haemophilus* species

Clinical syndrome	No. of cases	Age[a]	Reference
H. aphrophilus			
Endocarditis	25[b]	40	65
Brain abscess	7[b]	16	34
Pneumonia/empyema	4	57	15
Fasciitis	1	35	21
Cholecystitis	1	60	57
Sinusitis	3	61	99
Soft tissue/wound	17	36	99
H. paraphrophilus			
Endocarditis	2	23	26
Epiglottitis	1	6 mo.	62
Pneumonia	1	6 mo.	56
Brain abscess	2	–[c]	144
Osteomyelitis	1	–	144

[a] Mean age if more than one case.
[b] Reported cases since 1970.
[c] Age not reported.

tinomycetemcomitans (positive catalase, negative fermentation of sucrose and lactose), *E. corrodens* (positive ornithine decarboxylase and lysine decarboxylase, as well as negative fermentation of sucrose and lactose), and *Cardiobacterium hominis* (indole positive, nitrate reductase negative, and negative fermentation of lactose). Further work with this group of organisms is clearly needed.

H. PARAPHROPHILUS The original 24 strains that served as the basis for the description of the species by Zinnemann et al (144) included two patients with brain abscess and one patient with osteomyelitis (Table 5). Since that time several clinical infections due to this organism have been reported. de Silva et al (26) reported the first case of endocarditis due to this organism and it should be noted that the isolate was identified by the laboratory and in one reference laboratory as *H. parainfluenzae.* Only the Centers for Disease Control in Atlanta, Ga., identified the organism as *H. paraphrophilus.* Additional cases of endocarditis have been published by Hammond et al (50), Csukás & Bán (22), and Ellner et al (30). Isolated case reports of epiglottitis (62) and pneumonia (56) have appeared as well. In every case confusion with *H. parainfluenzae* was apparent, and other than the absence of hemolysis, the isolate by Csukás & Bán (22) would appear to be *H. paraphrohaemolyticus* since it was urease and catalase positive and failed to ferment lactose (Table 2). As noted in the report by Jones et al (62) of *H. paraphrophilus* epiglottitis, 49 of 59 *H. parainfluenzae* strains sent to the Center for Disease Control when examined were *H. paraphrophilus* according to the CO_2-dependence criteria of Zinnemann et al (144). Nevertheless interpretation of this test is difficult, as noted by Frazer et al (37). *H. paraphrophilus* is taxonomically more related to *H. aphrophilus* than to *H. parainfluenzae,* and it can be distinguished from other V-factor-requiring strains by several phenetic traits (Table 2). Clinically there appear to be few, if any, distinguishing characteristics to infections due to *H. paraphrophilus* as compared to infections due to *H. aphrophilus* or as compared to other V-factor-requiring strains of *Haemophilus.* Because of the apparent widespread misidentification of *H. paraphrophilus* as *H. parainfluenzae,* however, this is not surprising.

Miscellaneous Species

With the designation of *H. parahaemolyticus* strains of porcine origin as *H. pleuropneumoniae,* only two species isolated from animal sources have been associated with human infections.

H. HAEMOGLOBINOPHILUS Originally described by Friedberger in 1903 (38) and isolated from the prepuce of dogs, *H. haemoglobinophilus* has been reported in association with two human infections: A 10-year-old girl with severe hypogammaglobulinemia developed otitis media and from the discharge a pure culture was obtained (36); and a 12-year-old boy developed osteomyelitis of the right radius after receiving a superficial dog bit (81).

H. INFLUENZAE-MURIUM *H. influenzae-murium* was first described by Kairies & Schwartzer in 1935 and was associated with an epizootic respiratory infection of mice (64). There is a single report of the possible isolation of this organism from a perineal abscess of a 45-year-old woman (117). The clinical significance of this isolate is doubtful.

Two other species are at least worth mentioning. Taylor et al (130) reported the presence of agglutinins in human non-gonococcal urethritis patients to an equine organism referred to as *H. equigenitalis.* This organism requires neither hemin nor NAD and currently it cannot be considered a member of the genus. Kilian & Theilade proposed a new species *H. segnis* for certain V-factor-requiring strains isolated primarily from human dental plaque and characterized by a relatively high G+C content and inert biochemical activity (74). Although only one human infection due to this proposed species has been described (14a), these strains will almost certainly be reported by most laboratories as *H. parainfluenzae* and if accepted will create additional confusion, as previously described for clinical infections due to the V-factor-requiring strains of *Haemophilus* (*H. parainfluenzae, H. paraphrophilus, H. parahaemolyticus, H. paraphrohaemolyticus*).

CONCLUDING REMARKS

The importance of *Haemophilus* species other than *H. influenzae* in human infections has been increasingly recognized in recent years. However, a number of species have been created by taxonomists based on limited phenotypic traits of uncertain significance without regard to clinical relevance. Classification of prokaryotes suffers when compared to biological classification methods for higher eukaryotes in relying almost exclusively on numerical phenetics and serology. As discussed recently by Mayr (92), such an approach fails to undertake a careful analysis of the phenotypic traits that almost always would identify certain traits that have greater weight in establishing relationships. If, as indicated by Mayr, biological classifications serve as the basis of biological generalizations and serve as the key to an information storage system, then clearly the present speciation of *Haemophilus* fails in both respects. As examples, it is virtually impossible to retrieve from the literature cases of *H. paraphrophilus* endocarditis and distinguish them from cases of *H. parainfluenzae* endocarditis, and the biological significance of the type *b* capsule of *H. influenzae* is clear, yet hemolysis, which does not identify any biological generalizations, is the basis for three species of *Haemophilus.* It is easy, therefore, to agree with Kilian (68) that *H. parahaemolyticus* and *H. paraphrohaemolyticus* do not merit species status because they lack sufficient distinguishing phenetic characteristics, and differentiation from *H. parainfluenzae* appears, at present, to be of little clinical relevance. Likewise there is no clinical relevance to the species *H. haemolyticus* and there appears to be no clinical reason to differentiate *H. paraphrophilus* from either *H. aphrophilus* or *H. parainfluenzae.* If studies of the taxonomy of these organisms indicate a closer relationship of *H. paraphrophilus* to *H. aphrophilus* than to *H. parainfluenzae* then phenetic traits other than V-factor requirement would be more relevant to the clinical laboratory. On the other hand it is not so easy to agree with Kilian et al (71) that *H. aegyptius* does not merit species status because it lacks sufficient distinguishing phenetic characteristics when historically there has been little confusion in the clinical differentiation of infections due to *H. aegyptius* and *H. influenzae.* As we learn more of the genetic relatedness and presumed evolutionary history of these organisms and better understand the basis of pathogenesis, perhaps classification systems will emerge that satisfy both the clinician and the taxonomist.

Literature Cited

1. Anand, C. M., Mackay, A. D., Evans, J. I. 1976. *J. Clin. Pathol.* 29:812–14
2. Bachman, D. W. 1975. *Pediatrics* 55: 526–30
3. Bamrah, V. S., Williams, G. W., Hughes, C. V., Rose, H. D., Tristani, F. E. 1979. *Am. J. Med.* 66:543–46
4. Barnshaw, J. A., Phillips, C. F. 1970. *Pediatrics* 45:856–57
5. Bartlett, R. C. 1974. *Am. J. Clin. Pathol.* 61:867–72
6. Bauer, C. L., Walker, W. J. 1966. *Calif. Med.* 104:475–79
7. Biberstein, E. L., Mini, P. D., Gills, M. G. 1963. *J. Bacteriol.* 86:814–19
8. Bieger, R. C., Brewer, N. S., Washington, J. A. II. 1978. *Medicine* 57:345–55
9. Blair, D. C., Walker, W., Sodeman, T., Pagano, T. 1977. *Chest* 71:146–49
10. Blair, D. C., Weiner, L. B. 1979. *Am. J. Dis. Child.* 133:617–18
11. Blaylock, B. L., Baber, S. 1980. *Am. J. Clin. Pathol.* 73:285–87
12. Boyce, J. M. H., Frazer, J., Zinnemann, K. 1969. *J. Med. Microbiol.* 2:55–62
13. Branson, D. 1968. *Appl. Microbiol.* 16:256–59
14. Buchanan, R. E., Gibbons, N. E., eds. 1974. *Bergey's Manual of Determinative Bacteriology.* Baltimore: Williams & Wilkins. 8th ed. 1268 pp.

14a. Bullock, D. W., Devitt, P. G. 1981. *J. Infect. Dis.* 3:82–85

15. Capelli, J. P., Savacool, J. W., Randall, E. L. 1965. *Ann. Intern. Med.* 62: 771–77
16. Chow, A. W., Bushkell, L. L., Yoshikawa, T. T., Guze, L. B. 1974. *Am. J. Med. Sci.* 267:365–68
17. Chunn, C. J., Jones, S. R., McCutchan, J. A., Young, E. J., Gilbert, D. N. 1977. *Medicine* 56:99–113
18. Cole, R. A., Winickoff, R. N. 1979. *South. Med. J.* 72:516–18
19. Cooney, T. G., Harwood, B. R., Meisner, D. J. 1981. *Arch. Intern. Med.* 141:940–41
20. Craven, E. B. Jr., Lexington, N. C., Poston, M. A., Orgain, E. S. 1940. *Am. Heart J.* 19:434–52
21. Crawford, S. A., Evans, J. A., Crawford, G. E. 1978. *Arch. Intern. Med.* 138:1714–15
22. Csukás, Z., Bán, E. 1978. *Acta Microbiol. Acad. Sci. Hung.* 25:179–83
23. Dahlgren, J., Tally, F. P., Brothers, G., Ruskin, J. 1974. *Am. J. Clin. Pathol.* 62:607–11
24. Davis, D. J., Pittman, M. 1950. *Am. J. Dis. Child.* 79:211–22
25. Davis, D. J., Pittman, M., Griffitts, J. J. 1950. *J. Bacteriol.* 59:427–31

26. de Silva, M., Rubin, S. J., Lyons, R. W., Liss, J. P., Rotatori, E. S. 1976. *Am. J. Clin. Pathol.* 66:922–26
27. Douglas, G. W., Buck, L. L., Rosen, C. 1979. *J. Clin. Microbiol.* 9:299–300
28. Ducrey, A. 1889. *Monatsh. Prakt. Dermatol.* 9:387–405
29. Dusheiko, G. M., Cassel, R. 1978. *S. Afr. Med. J.* 54:245–47
30. Ellner, J. J., Rosenthal, M. S., Lerner, P. I., McHenry, M. C. 1979. *Medicine* 58:145–58
31. Elster, S. K., Mattes, L. M., Meyers, B. R., Jurado, R. A. 1975. *Am. J. Cardiol.* 35:72–79
32. Emmerson, A. M., Perinpanayagam, R. M., Barnado, D. E. 1981. *Postgrad. Med. J.* 57:117–19
33. Enck, R. E., Bennett, J. M. 1976. *J. Clin. Microbiol.* 4:194–95
34. Fager, C. A., 1961. *Lahey Clin. Bull.* 12:108–12
35. Fildes, P. 1924. *Br. J. Exp. Pathol.* 5:69–74
36. Frazer, J., Rogers, K. B. 1972. *J. Clin. Pathol.* 25:179–80
37. Frazer, J., Zinnemann, K., Boyce, J. M. H. 1969. *J. Med. Microbiol.* 2:563–66
38. Friedberger, E. 1903. *Zentralbl. Bakteriol. Parasitenkd. Infektionskr. Hyg. Abt. I Orig.* 33:401–6
39. Gallant, T. E., Malinak, L. R., Gump, D. W., Mead, P. B. 1977. *Am. J. Obstet. Gynecol.* 129:702–3
40. Garner, J. G. 1979. *NZ Med. J.* 89:384–86
41. Geraci, J. E., Wilkowske, C. J., Wilson, W. R., Washington, J. A. II. 1977. *Mayo Clinic Proc.* 52:209–15
42. Gigliotti, F., Williams, W. T., Hayden, F. G., Hendley, H. O. 1981. *J. Pediatr.* 98:531–36
43. Goldberg, R., Washington, J. A., II. 1978. *Am. J. Clin. Pathol.* 70:899–904
44. Goldsweig, H. G., Matsen, J. M., Castaneda, A. R. 1972. *J. Thorac. Cardiovasc. Surg.* 63:408–11
45. Gordon, A. M., Love, W. C. 1970. *J. Med. Microbiol.* 3:550–54
46. Gregory, J. E., Henderson, R. W., Smith, R. 1980. *Br. J. Vener. Dis.* 56:414–15
47. Gullekson, E. H., Dumoff, M. 1966. *J. Am. Med. Assoc.* 198:1221
48. Hable, K. A., Logan, G. B., Washington, J. A. II. 1971. *Am. J. Dis. Child.* 121:35–37
49. Hammond, G. W., Lian, C. J., Wilt, J. C. Albritton, W. L., Ronald, A. R. 1978. *J. Clin. Microbiol.* 7:243–46
50. Hammond, G. W., Richardson, H., Lian, C. J., Ronald, A. R. 1978. *Am. J. Med.* 65:537–41
51. Hammond, G. W., Slutchuk, M., Scatliff, J., Sherman, E., Wilt, J. C., Ronald, A. R. 1980. *Rev. Infect. Dis.* 2:867–79
52. Himmel, J. 1901. *Am. Inst. Pasteur* 15:928–40
53. Hodge, J. L. R., Bremner, D. A. 1974. *NZ Med. J.* 79:824–25
54. Hollin, S. A., Hayashi, H., Gross, S. W. 1967. *Oral Surg. Oral Med. Oral Pathol.* 23:277–93
55. Holt, R. N., Taylor, C. D., Schneider, H. J., Hallock, J. A. 1974. *Clin. Pediatr.* 13:666–68
56. Howard, J. 1981. *Canad. J. Med. Technol.* 43:174–75
57. Huck, W., Britt, M. R. 1978. *Am. J. Clin. Pathol.* 69:361–63
58. Ingham, H. R., Turk, D. C. 1969. *J. Clin. Pathol.* 22:258–62
59. Ivler, D., Preston, H. M., Portnoy, B. 1963. *Proc. Soc. Exp. Biol. Med.* 114: 232–34
60. Jemsek, J. G., Greenberg, S. B., Gentry, L. O., Welton, D. E., Mattox, K. L. 1979. *Am. J. Med.* 66:51–57
61. Johnson, R. H., Kennedy, R. P., Marton, K. I., Thornsberry, C. 1977. *South. Med. J.* 70:1098–102
62. Jones, R. N., Slepack, J., Bigelow, J. 1976. *J. Clin. Microbiol.* 4:405–7
63. Julander, I., Lindberg, A. A., Svanbom, M. 1980. *Scand. J. Infect. Dis.* 12:85–89
64. Kairies, A., Schwartzer, K. 1936. *Zentralbl. Bakteriol. Parasitenkd. Infektionskr. Hyg. Abt. I Orig.* 137:351–59
65. Khairat, O. 1940. *J. Pathol. Bacteriol.* 50:497–505
66. Kilian, M. 1974. *Acta Pathol. Microbiol. Scand. Sect. B* 82:835–42
67. Kilian, M. 1976. *Acta Pathol. Microbiol. Scand. Sect. B* 84:339–41
68. Kilian, M. 1976. *J. Gen. Microbiol.* 93:9–62
69. Kilian, M. 1980. See Ref. 84, pp. 330–36
70. Kilian, M., Heine-Jensen, J., Bülow, P. 1972. *Acta Pathol. Microbiol. Scand. Sect. B* 80:571–78
71. Kilian, M., Mordhorst, C.-H., Dawson, C. R., Lautrop, H. 1976. *Acta Pathol. Microbiol. Scand. Sect. B* 84:132–38
72. Kilian, M., Nicolet, J., Biberstein, E. L. 1978. *Int. J. Syst. Bacteriol.* 28:20–26
73. Kilian, M., Schiott, C. R. 1975. *Arch. Oral Biol.* 20:791–96
74. Kilian, M., Theilade, J. 1978. *Int. J. Syst. Bacteriol.* 28:411–15
75. King, E. O., Tatum, H. W. 1962. *J. Infect. Dis.* 111:85–94

76. Klein, M., Blazevic, D. J. 1969. *Am. J. Med. Technol.* 35:695–701
77. Knorr, M. 1924. *Zentralbl. Bakteriol. Parasitenkd. Infektionskr.* 93:385–92
78. Koch, R. 1883. *Wien. Med. Wochenschr.* 33:1548–51
79. Kraut, M. S., Attebery, H. R., Finegold, S. M., Sutter, V. L. 1972. *J. Infect. Dis.* 126:189–92
80. Krishnaswami, R., Schwartz, J., Boodish, W. 1972. *Pediatrics* 50:498–99
81. Lavine, L. S., Isenberg, H. D., Rubins, W., Berkman, J. I. 1974. *Clin. Orthop.* 98:251–53
82. Leidy, G., Hahn, E., Alexander, H. E. 1959. *Proc. Soc. Exp. Biol. Med.* 102:86–88
83. Leidy, G., Jaffee, I., Alexander, H. E. 1965. *Proc. Soc. Exp. Biol. Med.* 118:671–79
84. Lennette, E., ed. 1980. *Manual of Clinical Microbiology.* Washington, DC: ASM. 3rd ed. 1044 pp.
85. Lichty, J. A. Jr. 1937. *Am. J. Dis. Child.* 54:1311–19
86. Lund, M. E., Blazevic, D. J. 1977. *J. Clin. Microbiol.* 5:142–44
87. Lwoff, A., Pirosky, I. 1937. *CR Soc. Biol.* 124:1169–71
88. Lynn, D. J., Kane, J. G., Parker, R. H. 1977. *Medicine* 56:115–27
89. Maller, R., Ansèhn, S., Frydén, A. 1977. *Scand. J. Infect. Dis.* 9:241–42
90. Mannheim, W., Pohl, S., Holländer, R. 1980. *Zentralbl. Bakteriol. Parasitenkd. Infektionskr. Abt. I Orig. Reihe A* 246:512–40
91. Marraro, R. V., McCleskey, F. K., Mitchell, J. L. 1977. *J. Clin. Microbiol.* 6:172–73
92. Mayr, E. 1981. *Science* 214:510–16
93. Mesko, Z. G., Bauza, J., Vinas, C. 1976. *J. Pediatr.* 89:1031–32
94. Michaels, R. H., Phillips, D. M. 1972. *Am. J. Dis. Child.* 124:788–89
95. Miller, C. D., Branch, A. 1923. *Arch. Intern. Med.* 32:911–26
96. Oill, P. A., Chow, A. W., Guze, L. B. 1979. *Arch. Intern. Med.* 139:985–88
97. Okubadejo, O. A., Fayinka, O. A. 1967. *West Afr. Med. J.* 16:79–80
98. Orfila, J., Courden, B. 1961. *Ann. Inst. Pasteur* 100:252–56
99. Page, M. L., King, E. O. 1966. *N. Engl. J. Med.* 275:181–88
100. Parsons, M., Faris, I. 1973. *J. Clin. Pathol.* 26:604–5
101. Perkins, R. E., Knudsin, R. B., Pratt, M. V., Abrahamsen, I., Leibowitz, H. M. 1975. *J. Clin. Microbiol.* 1:147–49
102. Pfeiffer, R. 1892. *Dtsch. Med. Wochenschr.* 18:28
103. Piot, P., van Dyck, E., Goodfellow, M., Falkow, S. 1980. *J. Gen. Microbiol.* 119:373–96
104. Pittman, M. 1931. *J. Exp. Med.* 53: 471–92
105. Pittman, M. 1953. *J. Bacteriol.* 65: 750–51
106. Pittman, M., Davis, D. J. 1950. *J. Bacteriol.* 59:413–26
107. Plotkin, G. R. 1979. *Am. J. Med. Sci.* 278:195–200
108. Pritchett, I. W., Stillman, E. G. 1919. *J. Exp. Med.* 29:259–66
109. Renne, J. W., Tanowitz, H. B., Chulay, J. D. 1976. *Pediatrics* 57:573–74
110. Rivers, T. M. 1922. *Bull. Johns Hopkins Hosp.* 33:149–51
111. Rivers, T. M. 1922. *Bull. Johns Hopkins Hosp.* 33:429–31
112. Robineau, M., Veyssier, P., Boussougant, Y., Pelisse, J. M., Ghanassia, J.-P., Modai, J., Domart, A. 1973. *Scand. J. Infect. Dis.* 5:229–31
113. Rubin, S. J., Granato, P. A., Wasilauskas, B. L. See Ref. 84, pp. 263–87
114. Ruhen, R. G., Genat, R. J. 1977. *Med. J. Aust.* 1:756
115. Russell, D. S., Fildes, P. 1928. *J. Pathol. Bacteriol.* 31:651–56
116. Russell, J. P. 1965. *Am. J. Clin. Pathol.* 44:86–93
117. Ryan, W. J. 1968. *J. Gen. Microbiol.* 52:275–86
118. Samson, D. S., Clark, K. 1973. *Am. J. Med.* 54:201–10
119. Sanders, D. Y., Russell, D. A., Gilliam, C. F. 1968. *Pediatrics* 42:683–84
120. Sims, W. 1970. *J. Med. Microbiol.* 3:615–25
121. Smith, P. W., Chambers, W. A., Walker, C. A. 1979. *Am. J. Med. Sci.* 278:173–76
122. Smith, W. K., Berger, H. W. 1974. *J. Med. NY* 41:543–48
123. Sneath, P. H. A., Johnson, R. 1973. *Int. J. Syst. Bacteriol.* 23:405–18
124. Sottnek, F. T., Biddle, J. W., Kraus, S. J., Weaver, R. E., Stewart, J. A. 1980. *J. Clin. Microbiol.* 12:170–74
125. Spernoga, J. F. 1980. *Arch. Intern. Med.* 140:864
126. Stillman, E. G., Bourn, J. M. 1920. *J. Exp. Med.* 32:665–82
127. Sugarman, B., Pesanti, E. L. 1979. *Arch. Neurol.* 36:859
128. Sullivan, M. 1940. *Am. J. Syphilis* 24:482–521
129. Sutter, V. L., Finegold, S. M. 1970. *Ann. NY Acad. Sci.* 174:468–87
130. Taylor, C. E. D., Rosenthal, R. O., Taylor-Robinson, D. 1977. *Lancet* 1:700–1

131. Thjötta, T., Avery, O. T. 1921. *J. Exp. Med.* 34:97–114
132. Valentine, F. C. O., Rivers, T. M. 1927. *J. Exp. Med.* 45:993–1002
133. Vandenbroucke, J. J., Verhaegen, J., Vandepitte, J. 1980. *Acta Clin. Belg.* 35:166–71
134. Varghese, R., Melo, J. C., Barnwell, P., Chun, C. H., Raff, M. J. 1977. *Chest* 72:680–82
135. Vastine, D. W., Dawson, C. R., Hoshiwara, I., Yoneda, C., Daghfous, T., Messadi, M. 1974. *Appl. Microbiol.* 28:688–90
136. Warman, S. T., Rewinitz, E., Klein, R. S. 1981. *J. Am. Med. Assoc.* 246:868–69
137. Wauters, J. P., Ferguson, R. K., Michaud, P. A., Knobel, P., Fontolliet, C. H. 1978. *Clin. Nephrol.* 9:73–76
138. Weeks, J. E. 1886. *Arch. Ophthalmol.* 15:441–51
139. White, C. B., Lampe, R. M., Copeland, R. L., Morrison, R. E. 1981. *Pediatrics* 67:434–35
140. Winslow, C. E. A., Broadhurst, J., Buchanan, R. E., Krumwiede, C., Rogers, L. A., Smith, G. H. 1920. *J. Bacteriol.* 5:191–229
141. Wort, A. J. 1975. *Can. Med. J.* 112: 606–7
142. Zinnemann, K. 1967. *Int. J. Syst. Bacteriol.* 17:165–66
143. Zinnemann, K. 1980. *Zentralbl. Bakteriol. Parasitenkd. Infektionskr. Hyg. Abt. I Orig. Reihe A* 247:248–58
144. Zinnemann, K., Rogers, K. B., Frazer, J., Boyce, J. M. H. 1968. *J. Pathol. Bacteriol.* 96:413–19
145. Zinnemann, K., Rogers, K. B., Frazer, J., Devaraj, S. K. 1971. *J. Med. Microbiol.* 4:139–43
146. Zinner, S. H., McCormack, W. M., Lee, Y.-H., Zuckerstatter, M. H., Daly, A. K. 1972. *Pediatrics* 49:612–14

Ann. Rev Microbiol. 1982. 36:217–38

MAGNETOTACTIC BACTERIA

Richard P. Blakemore

University of New Hampshire, Durham, New Hampshire 03824

CONTENTS

INTRODUCTION

Many organisms have been known for a long time to sense the earth's magnetic field. Certain bacteria are also geomagnetically sensitive. The way these bacteria interact with the geomagnetic field has proven useful in thinking about ways higher organisms may also do so. In fact, the ferromagnetic detection system in bacteria has some interesting parallels in geomagnetically responsive animals. Thus, just as prokaryotes have contributed to such fields as genetics and biochemistry, they are now valuable in newly emerging fields.

Just as importantly, the magnetic property of these bacteria has permitted us to identify, collect, and study them and to begin to evaluate their niche. The magnetotactic bacteria at first appear so unique as to deserve

0066-4227/82/1001-0217$02.00

special taxonomic status. Indeed, all of those examined share several traits. They are Gram negative, motile (by means of flagella), microaerophilic, and aquatic, and they synthesize intracellular enveloped magnetic grains termed magnetosomes. However, despite these and perhaps other similarities, the diversity of cell morphology within the group (see Figures 1–7) and the fact that one isolate is similar to members of *Aquaspirillum* argues for rather widespread taxonomic distribution of the trait. Thus, "magnetotactic bacteria" is a descriptor similar to "gliding bacteria" or "hydrogen-oxidizing bacteria" and does not have taxonomic meaning. "Magnetotaxis" denotes cell motility directed by a magnetic field. "Taxis" implies that the magnetic field influences the swimming direction but not the absolute velocity of magnetotactic cells.

A review of current knowledge of magnetotactic bacteria might be considered premature, especially in view of the paucity of research papers available. However, my purpose is not only to critically discuss available information, but also, hopefully, to spark in some readers interest and activity in this new area of research.

The subject is a rich one that inherently crosses traditional disciplinary boundaries. Studies of magnetotactic bacteria have proven of interest and value to microbiologists, evolutionary biologists, biochemists, physicists, geochemists, paleogeologists, and animal behaviorists. These organisms also offer intriguing possibilities for biotechnology. Progress in the field has been greatly stimulated by interdisciplinary collaborations and discussions.

Discovery of Magnetotactic Bacteria

In the early 1970s as a graduate student in the laboratory of E. Canale-Parola, I was interested in the fascinating microbial populations undergoing natural enrichment within sulfide-rich mud samples stored on a laboratory shelf. Some of these he had collected from a marshy area adjacent to the Eel Pond in Woods Hole, Mass. They were of particular interest because they became enriched in the elusive *Spirochaeta plicatilis* as well as *Thiospira* and large *Beggiatoa* species (6), none of which have been cultured axenically. I casually observed that these natural enrichments also contained highly motile bacteria that migrated nearly unidirectionally across the microscope field of view. These latter microorganisms became dislodged from sediment particles in the course of making preparations for phase contrast microscopy. They then persistently swam toward and accumulated at one edge of drops of sediment transferred to depression slides. They swam in the same geographic direction even when the microscope was turned around, moved to another location, or covered with a pasteboard box. Thus, it was evident their swimming direction was influenced not by light, as I had at first supposed, but by some pervasive stimulus. That the cells were, in fact, magnetically responsive was vividly demonstrated when a magnet was brought near the microscope. To my astonishment, the hundreds of swimming cells instantly turned and rushed away from the end of the magnet! They were always attracted by the end that also attracted the North-seeking end of a compass needle and they were repelled by its opposite end. Their swimming speed was very fast, on the order of 100 μm/s, and the entire population consisting of hundreds of freely and independently swimming cells swerved in unison as the magnet was moved about nearby (4).

I wish to emphasize that this was a completely unexpected finding. A research proposal requesting support to search for geomagnetically sensitive bacteria would then have been met by peer review with exactly the same degree of attention as one submitted today proposing to detect sound production by bacteria. Yet, as far as I can determine, this simple serendipitous observation in a drop of muddy water appears to have attracted rather broad interest. For example, it seems to have unlocked thinking in at least one other field, that concerned with animal homing and migration, by unequivocally demonstrating that some organisms respond to geomagnetism by means of a ferromagnetic mechanism. It also suggests a possible origin of some of the magnetic remanence contained in the paleogeologic record.

BIOPHYSICS OF MAGNETOTAXIS IN BACTERIA

Magnetotactic bacteria each possess intracellular iron grains visible by means of electron microscopy (4, 5, 36; see Figures 1–8). Before the chemical state of the iron in these organisms was known, Kalmijn & Blakemore (27), at the suggestion of E. M. Purcell, proved that each cell contained a permanent magnetic dipole moment. Bacteria from mud enrichments were exposed to a brief (1 μs) monophasic magnetic pulse of several hundred gauss (G) delivered antiparallel to their swimming direction (27, 34). At sufficiently high pulse intensities (ca 200–800 G), the cells instantly became remagnetized by the treatment, indicating that each was intrinsically permanently magnetic. Remagnetized cells instantly made U-turns and thereafter swam opposite to their original course. Remagnetization was also effected by exposing the bacteria to Alternating Current demagnetizing fields (8, 17, 35). However, by this method a maximum of 50% of the population became repolarized. Thus, each cell behaved as though it contained a single magnetic dipole in that it could not be demagnetized. Some cells present in a population were also repolarized by being made to swim toward a finely pointed magnetic needle (8) or by encountering at close range the pole of a strong magnet. This latter observation should guide the

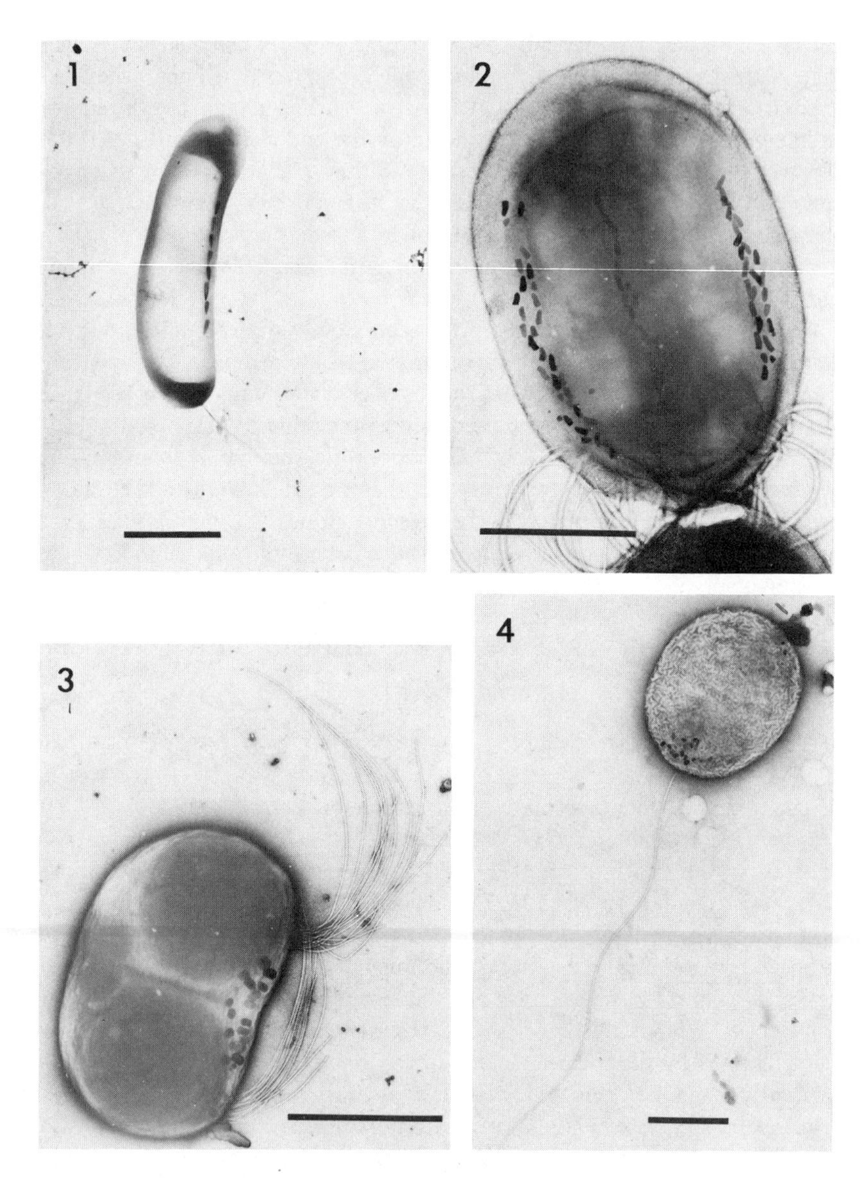

Figures 1–4 Magnetotactic bacteria recovered from sediments by application of a magnetic field. Cells lightly stained with uranyl acetate and viewed by TEM. Bars each represent 1 μm. (*1*) Curved, rod-shaped cell with single polar flagellum and single chain of tooth-shaped magnetosomes. Recovered from Little Styx River, South Island, New Zealand. (*2*) Ovoid, rod-shaped cell with multiple, sheathed flagella at one pole. Cells of this type were recovered from a New Hampshire bog. Each possesses one or more chains of tooth-shaped magnetosomes situated in each of three lateral cell positions. (*3*) Coccoid cell with two bundles of flagella

student of magnetotaxis away from the use of inhomogeneous magnetic fields (i.e. bar magnets) in favor of steady, uniform magnetic fields created with electromagnetic coil systems such as Helmholtz coil pairs (see 7).

Cell Behavior in Uniform Magnetic Fields

The local geomagnetic field is fairly uniform in configuration and intensity over most of the earth, with a value of approximately 0.5 G. It acts on the bacterial magnetic dipole in the same way it interacts with a compass needle. A cell with the axis of its magnetic moment (M) positioned at an angle (Θ) with respect to the direction of the ambient magnetic field (H) experiences a torque tending to align it in the field direction (see Figure 9 *A*). Once aligned, no further magnetic forces are exerted on the cell. Thus, magnetotactic bacteria are not pulled northward or southward by magnetic interactions, they are merely aligned in the geomagnetic field. Indeed, killed cells do the same, also being magnetic (see Figure 9*B*). Magnetotaxis results from passively oriented cells swimming along magnetic field lines (see Figure 9*C*). Conclusive evidence of their geomagnetic sensitivity was obtained with portable Helmholtz coils operated in the field, far from the magnetic noise associated with most laboratory buildings, power transmission lines, and the like (26).

Interaction of the Cell Magnetic Dipole with the Geomagnetic Field

How well adapted are these bacteria for interacting with earth's magnetic field? Basically, to assess this we must evaluate the cell magnetic moment. Its interactive energy with the magnetic field, which tends to produce cell orientation, may then be compared with kT, the thermal forces associated with Brownian motion that tend to randomize cell orientation in water.

The average alignment of a population of noninteracting magnetic dipoles in a magnetic field is described by the Langevin function for classical paramagnetism. This relationship states that

$$\cos \Theta = L\ (MH/\mathrm{k}T),$$

---

positioned at the (concave) side of the cell. Cells of this type, which are similar to those studied by Moench & Konetzka (36), were recovered from Mill Pond, Durham, New Hampshire. The cell shows a developing division plane bisecting its cluster of parallelepiped shaped magnetosomes. The large spherical dark structures contain phosphorus and potassium (see text). (*4*) Coccoid cell with a single polar unsheathed flagellum found in sediments of a New Hampshire bog. Parallelepiped-shaped magnetosomes are clustered at each end of cells of this type. The cell surface has the characteristically wavy appearance of the Gram-negative outer cell wall.

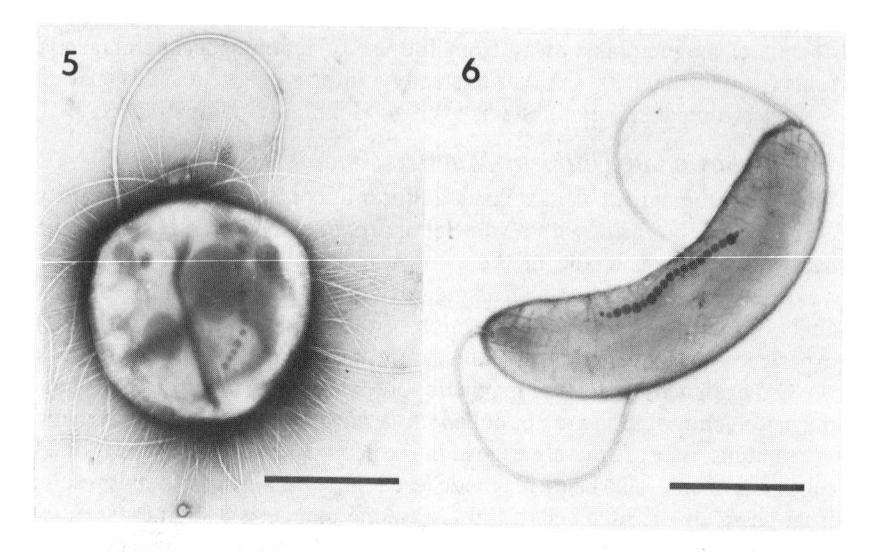

Figures 5 and 6 Magnetotactic bacteria with internal chains of cuboidal to octahedral magnetosomes. Cells stained lightly with uranyl acetate and viewed by TEM. Bars each represent 1 μm. (*5*) A coccoid cell recovered from Cedar Swamp, Woods Hole, Massachusetts. The cell has two lateral bundles of flagella and is covered with numerous, fine, pili. (*6*) A curved, rod-shaped cell with single bipolar flagella. Recovered from Durham, New Hampshire, water-purification plant.

where Θ is the angle between the direction of the cell magnetic moment (M) and the direction of the ambient field (H) (see Figure 9). Magnetotactic bacteria behave as noninteracting magnetic dipoles (14). Thus, their average alignment in an applied magnetic field is determined by the ratio of their interactive magnetic energy with the applied field (MH) to their thermal energy (kT). When the ratio MH/kT exceeds a value of about 10, the particles (here the bacteria) are fully aligned in the field direction (16).

At least three approaches have been used to evaluate the strength of the bacterial magnetic dipole moment. First, Frankel et al (18) showed that the magnetosomes of *Aquaspirillum magnetotacticum* strain MS-1, a magnetotactic spirillum isolated from a swamp (discussed below), consist of the magnetic iron oxide, magnetite (Fe_3O_4), also known as lodestone. Balkwill et al (2) determined the configuration, shape, size, and average number of magnetosomes per cell in this species grown under similar conditions. Each cell contained approximately 20 cuboidal-to-octahedral crystalline magnetosomes averaging 420 Å on a side and arranged in a chain (Figures 7, 8). The particles were within the single magnetic domain size range for magnetite, which is from 400–1000 Å (11, 29), although occasional smaller (superparamagnetic) particles were also present. It is not clear how

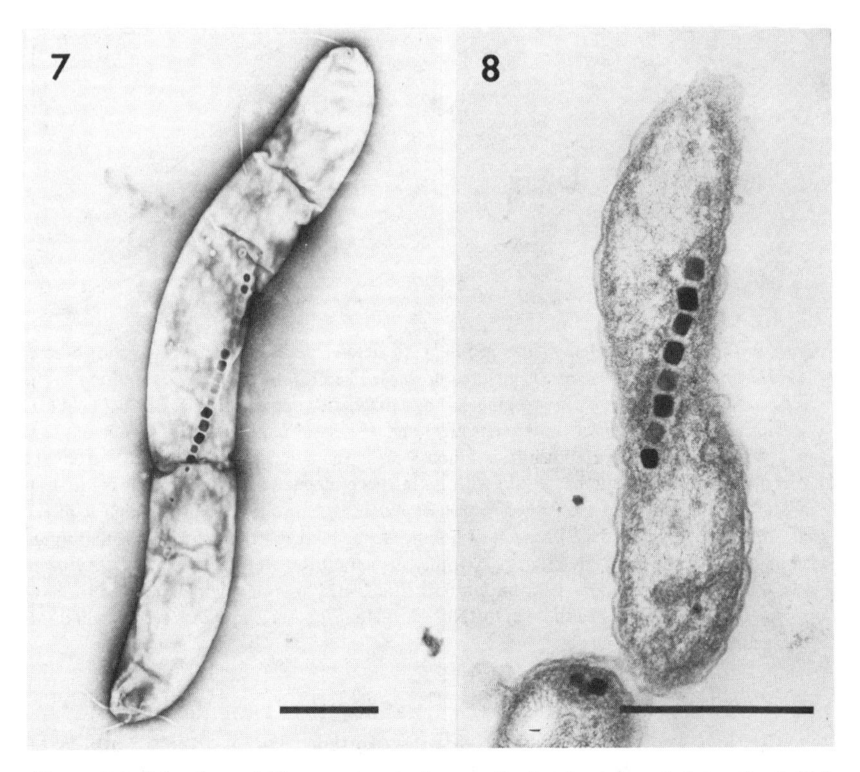

Figures 7 and 8 *Aquaspirillum magnetotacticum* cells showing internal chain of cuboidal magnetosomes. Bars each represent 0.5 μm. (*7*) Negatively stained cell showing single bipolar flagella. (*8*) Thin-sectioned cell showing Gram-negative wall type and portion of a chain of enveloped magnetosomes. Note the transversely sectioned cell revealing proximity of the magnetosomes to the cell boundary layers.

magnetotactic bacteria limit the dimensions of their magnetosomes to the magnetically efficient single-domain size, or why these structures exist in a chain, as is frequently the case (see Figures 1, 2, 5, 6, 7). However, it is tempting to speculate that constraint is somehow invoked by the envelope surrounding each magnetite particle and by the relationship of the envelope to the remaining cell structure. Single magnetic domains constrained to lie in a chain will lie with their magnetic axes parallel, head-to-tail (24). Thus, the total magnetic energy of the magnetosome chain is the sum of the individual particle magnetic moments. From these considerations, from the volume of magnetite per cell (then estimated to average 22 particles, each cuboidal and 500 Å on a side) and its saturation magnetization value (480 erg/G • cm^3), the calculated total cell magnetic energy was 1.3×10^{-12} erg/G. In the 0.5-G geomagnetic field, the magnetic energy of such a cell (6.6×10^{-13} erg) was found to be 16 times greater than thermal energy, kT

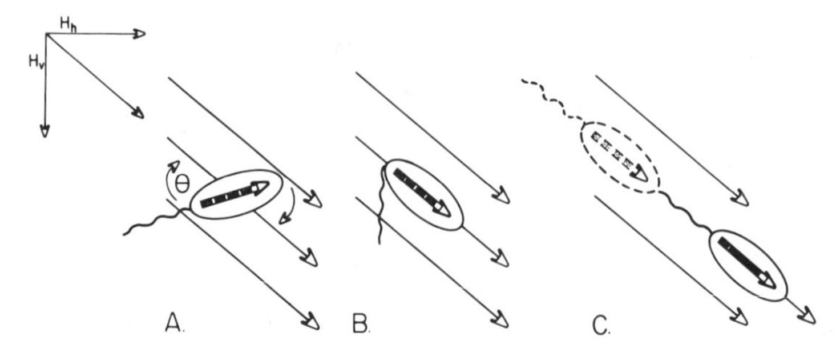

Figure 9 The Northern Hemisphere geomagnetic field is inclined downward as shown here; that of the Southern Hemisphere is inclined upward. The field direction, by convention, is the direction indicated by the North-seeking end of a magnetic compass needle. The net field (H) is comprised of a horizontal component (H_h) and a vertical component (H_v). The angle of inclination (dip) in New Hampshire is much steeper than that shown, being 73° from the horizontal. (*A*) An aquatic bacterium with its intrinsic magnetic dipole moment lying at an angle (Θ) with respect to the field direction experiences a torque tending to align it in the uniform geomagnetic field. (*B*) A dead or nonmotile magnetic bacterium with magnetosomes also aligns with the field direction. (*c*) Motile magnetic cells are magnetotactic; they swim preferentially along magnetic field lines. Southern Hemisphere magnetotactic bacteria have opposite magnetic polarity to those in the Northern Hemisphere. Thus, both types swim down along geomagnetic field lines.

(4.1×10^{-14} erg). From these considerations, it is clear the magnetic apparatus of *A. magnetotacticum* easily accounts for the organism's ability to align in the geomagnetic field. It serves as a ferromagnetic biocompass (16).

Recently, the average magnetic moment of a population of living *A. magnetotacticum* cells was determined from light scattering (39) as well as birefringence measurements (C. Rosenblatt, F. F. Torres de Araujo, R. B. Frankel, submitted for publication). Values of $2.2 \pm 0.2 \times 10^{-13}$ and $4.3 \pm 0.5 \times 10^{-13}$ erg/G were obtained for cells in each of two spirillum suspensions. These were within 15–20% of the estimated values for the two samples based upon the total average volume of magnetite per cell as determined by electron microscopy. The method permits direct evaluation of moments for cells suspended in liquid (culture) media and is insensitive to the presence of nonmagnetic organisms, dust, or other suspended particles that would contribute to light scattering. The light scattering method also provides rapid measurement of average cell length without the need for electron microscopy.

The cell magnetic moment has also been measured in an elegant, although tedious, way (25, 41). Kalmijn et al (25, 41) tracked single magnetotactic bacteria made to run back and forth across a ruled grid on a microscope slide situated within a Helmholtz coil pair. Thus, the swimming direction and translational velocity of individual bacteria was controlled by varying the amount and direction of current flow through the coils. A cell's migration rate (i.e. its translational velocity, not its absolute velocity) was measured as a function of the applied field strength. The migration rates of single cells plotted for various values of applied magnetic field fit a curve described by the Langevin function. Since their swimming velocity remained constant over the observation period, the average value of their angle of deviation from the field direction $\langle \cos \Theta \rangle$ was directly related to the cell migration rate. At low ambient field strengths cells deviated from the field direction more than at high field intensities. By measuring cell migration velocity for known values of H at constant temperature, values of M were obtained. The method provided values of 6.2×10^{-13} and 7.3×10^{-13} erg/G for each of two individual magnetotactic cocci obtained from a swamp-water enrichment (25). The estimated average magnetic moment of cells from the same enrichment based upon electron microscope evaluation was 8.6×10^{-13} erg/G.

The average value of the bacterial cell magnetic moment as determined by these independent methods appears to be surprisingly constant between different species. The results of Langevin analysis applied to spirilla in pure culture (16) or coccoid cells found in natural enrichments (25) indicated that magnetotactic bacteria synthesize sufficient but not excessive amounts of magnetite for their efficicnt interaction with the 0.5 G geomagnetic field. They are 80–90% fully aligned by a field of 0.5 G at 30°C. More than the amount of magnetite present within cells would not significantly improve the efficiency with which they interact with the geomagnetic field. Furthermore, larger cell magnetic moments than those observed might produce cell-to-cell magnetic interactions sufficiently great to override their electrostatic repulsive (zeta potential) forces, thereby causing them to clump. This may provide the selection pressure responsible for limiting the number of magnetosomes per cell in natural populations.

NATURAL ABUNDANCE, DISTRIBUTION, AND ECOLOGY OF MAGNETOTACTIC BACTERIA

Detection Methods

With the advantage of their magnetically directed motility, it has been a simple matter to detect magnetotactic bacteria in diverse natural environments (5, 8, 17, 36). The cells readily swim out of sediments when a steady field of 1–10 G is applied and can be made to accumulate in regions where they may be counted or collected for further study. Effective uniform magnetic fields are easily created with electromagnetic coils. The Helmholtz configuration (paired coils separated by a distance equal to the radius of each coil) (see 7) are preferred because of the uniformity of the resulting magnetic fields. Serially diluting homogenized sediment samples prior to magnetic cell separation has made it possible to directly enumerate these organisms in many natural environments, as well as in laboratory experiments with altered fields.

Distribution and Abundance

Commonly, as many as 10^3–10^4 cells/ml of sediment slurry occur in aquatic environments sampled throughout New England. Moench (35) detected, but did not enumerate, magnetotactic bacteria in approximately 37 of 41 temperate freshwater and marine environments sampled. Only those polluted with oil, chlorine, or copper sulfate and a site containing acid mine drainage contained no magnetotactic bacteria. Other habitats in which we have sought but have not found these organisms include a limestone cavern, a thermal spring, thawed sediments collected in Antarctica, and iron-rich seeps containing abundant *Gallionella* and *Leptothrix.* Settling basins of water purification plants, sewage-treatment oxidation ponds, and natural ponds with accumulated organic sediments are particularly good collection sites. Magnetotactic bacteria are abundant in sediments collected from beneath *Typha.* Moench (35) investigated a possible relationship between *Lemna* and a freshwater magnetotactic coccus. Such relationships, if they exist at all, have been difficult to establish. However, it is not unlikely that plant-derived phenolic compounds may serve to chelate iron in a form useful to these bacteria for magnetite synthesis. Ferric quinate is a satisfactory source of chelated iron in a chemically defined culture medium for the magnetotactic spirillum (9). It is also possible that organic acids produced by plant roots serve as satisfactory sources of carbon and energy for these bacteria.

Global Distribution and Cell Magnetic Polarity

Magnetotactic bacteria are ubiquitous. They have been found in the Arctic (L. Greenfield, personal communication), in Baltic Sea sediments (R. S. Wolfe, personal communication) in South America (17), in Australia and New Zealand (8, 28), and throughout North America (5, 7, 36). Without exception, those found in the Northern Hemisphere swim predominantly northward (5, 7, 36) and those of the Southern Hemisphere swim predominantly southward (7, 8, 28). This seeming curiosity is of considerable ecological significance: It involves selective survival of cells with one direction of magnetic polarity over those of opposite polarity. What may account for this segregation of polarity types in each global hemisphere?

As mentioned previously, magnetotactic bacteria swim in unison along magnetic field lines. The earth's magnetic field is directed northward over the surface of the globe. However, the local geomagnetic field also has a vertical component, which is directed upward in the Southern Hemisphere, decreasing to a value of zero at the geomagnetic equator, and is inclined downward in the Northern Hemisphere, with a value increasing with latitude (20). Because of this inclination, north-seeking bacteria found in the Northern Hemisphere are also directed downward there (Figure 9). South-seeking bacteria, which predominate in the Southern Hemisphere, are directed downward south of the geomagnetic equator (8, 28). Consequently, magnetotactic cells in either hemisphere are magnetically polarized relative to their principal swimming direction in such a way that they swim almost exclusively downward. This undoubtedly is an important factor determining their prevalence in sediments and absence from surface waters.

Recently, Frankel et al (17) observed that populations of magnetotactic bacteria at the geomagnetic equator consist of roughly equal numbers of cells of each magnetic polarity. Apparently cells there, when disturbed, migrate principally horizontally along the magnetic field lines, avoiding detrimental upward excursions. (The field is totally horizontal there.) However, since they exist in sediments at the equator as they do elsewhere, and not in the surface waters, it is likely that other tactic responses, to oxygen or to reduction-oxidation potential, for example, can play an important role in determining their vertical distribution in sediments.

Effects of Altered Magnetic Fields on Predominant Cell Polarity

Studies in my laboratory have been aimed at establishing the effect of altered magnetic fields on survival of magnetotactic bacteria (R. Blakemore, W. O'Brien, R. Caplan, manuscript in preparation). Loosely stoppered 125-ml vials containing 10 ml of sediment slurry and 90 ml of water from enrichment cultures containing predominantly northward and downward swimming cells were incubated within electromagnetic coils. Four coils (each consisting of 80 turns of 24-gauge magnet wire; coil length, 8 cm; coil diameter, 7.5 cm) were connected in series and were powered by an AC adapter (Archer, cat. no. 273–1435; 3-V DC; 100-mA output). With current flowing, the vertical component of the local geomagnetic field within each coil was inverted as determined with a gaussmeter. Cells in a drop of sediment from the enrichment, when placed within each coil, migrated upward and accumulated at the top of the drop when the current was on.

At approximately 4-day intervals for a period of up to 2 months, samples within each of the four coils as well as control vials placed within nonenergized coils were gently homogenized by swirling for 20 s. Direct cell counts were then made of magnetotactic bacteria present in serially diluted samples from each vial. Total population densities as well as the numbers of each polarity type present were determined by allowing the cells to migrate to their respective edges of carefully quantitated drops placed on the stage of a dissecting microscope also within an electromagnetic coil. The total population density within control and experimental vials decreased initially from about 10^6 cells/ml and after 2 weeks it stabilized at roughly 10^4 cells/ml. The number of south-seeking bacteria in the experimental vials became detectably more numerous within 3 days after supplying current to the coils. South-seeking bacteria progressively increased in number in the experimental vials, becoming predominant after 3 weeks. The ratio of South-seeking to total magnetotactic bacteria was 0.9 to 1.0 after 5–8 weeks. This trend was reversed when current to the coils was interrupted. No corresponding changes were noted in the control population. These results confirm that the global segregation of polarity types is a response to the direction of the vertical component of the geomagnetic field. They lend strong support to the notion that in natural environments, a very real consequence of magnetotaxis in bacteria is that it directs them downward away from surface waters and toward sediments.

The direction of cell magnetic polarity could be transmitted to progeny if at cell division at least some magnetosomes were partitioned to each daughter cell (see Figure 3). Nascent magnetosomes would then have the same direction of magnetization as the inherited particles. Occasionally, a division might produce a daughter cell without magnetosomes or with superparamagnetic magnetosomes (e.g. incipient areas of crystal formation occupied by particles too small to have stable magnetic remanence). After synthesizing single-domain-sized magnetosomes, these cells would have an equal probability of being of either polarity type. Any natural population of magnetotactic bacteria includes a few cells (less than 0.5% of the total population) of the "wrong" polarity. These undoubtedly comprised the progenitors of the predominantly South-seeking population that developed in the field-reversal experiments.

Mud samples similar to those described were also incubated under zero field conditions in a mu-metal enclosure (a chamber constructed of a metal with high magnetic permeability and therefore enclosing a field-free space). After 1 month they contained equal numbers of cells of each magnetic polarity; without a vertical magnetic field there was no selection for either polarity type. However, a similar experiment indicated that oxygen may help control the vertical distribution of the bacteria. When vials were sealed to prevent continued access of air, cells of each polarity type were soon prevalent in the oxygen-depleted water above the sediments, as well as in the sediments. If the seals were then removed from some of the vials, allowing air into them, they shortly thereafter contained cells of each polarity at the sediment/water interface but not in the surface waters.

These laboratory observations collectively point up the primary role of the vertical component of the geomagnetic field in selecting for downward-seeking cells among natural populations of magnetotactic bacteria. They also indicate the independent but consistent role of oxygen in effecting vertical migration and distribution of these bacteria in natural environments. Magnetotaxis would benefit these organisms in the absence of well-defined oxygen gradients. It might also make aerotaxis more efficient by promoting straight-line motion, thereby tending to minimize random non-vertial excursions.

ENRICHMENT, ISOLATION, AND CULTIVATION OF MAGNETOTACTIC BACTERIA

Despite considerable effort spent in several laboratories, magnetotactic bacteria have not readily yielded to isolation and axenic growth. Only one species, *Aquaspirillum magnetotacticum* strain MS-1, has been grown in pure culture (9). Consequently, much of what we know of growth conditions and metabolism of magnetotactic bacteria has been learned from studies of this organism. Many of the findings may not apply to other species.

The remarkably simple and prolific enrichment of magnetotactic organisms in stored muds (5, 36) contrasts with the difficulty in their isolation. With no special precautions for maintaining anaerobic conditions, sediments may be scooped up and placed in glass or plastic containers. When these are stored loosely covered in indirect light without being frequently mixed or disturbed, they undergo an ecological succession, the supernatant water clears, and the sediment stratifies. Orangey or lighter-colored surface bands generally overlie black layers just beneath the surface sediments. Magnetotactic bacteria are abundant in the surface sediments after 2 to 5 weeks. They remain abundant for periods of up to several years with replacement of water lost by evaporation. Flowing (open system) enrichments were exploited by Moench (35), but with less success than batch (closed system) enrichments in glass jars and tubes. In many natural environments containing magnetotactic bacteria, the elements P and N are limiting for bacterial growth. Our attempts to selectively enrich for them by using PO_4^- and/or NO_3^- or NH_4^+ resulted in rapid overgrowth of other bacteria. Addition of iron salts or iron chelates did not enhance their enrichment. This is consistent with findings in my laboratory that iron concentrations greater than those commonly measured in natural environments containing these organisms (ca 1 mg of Fe/liter) do not significantly increase the magnetite yield in cells of *A. magnetotacticum.* At present, the only parameter known to selectively affect cell numbers in enrichment culture is, as discussed previously, oxygen. Presumably, their successful enrichment after a period of several weeks corresponds to the time required to establish microaerobic zones and microhabitats compatible with their growth and long-term survival. This is interesting because it also suggests that as a group, these microaerophiles may have predominated earlier in geological history when the atmospheric oxygen content was substantially lower than at present.

After studying enrichment and survival conditions for magnetotactic bacteria in diluted mud samples, a trial isolation medium was constructed. The inoculum consisted of cells magnetically accumulated from an enrichment and washed free of most contaminating nonmagnetic forms. The medium in semisolid form supported growth of *A. magnetotacticum,* which was then cloned in microaerobic agar shake dilution tubes containing a medium of similar composition (9). A small rod-shaped magnetotactic organism also formed colonies in this medium, but it was subsequently lost upon subculture. Efforts to isolate coccoid magnetotactic bacteria abundant in marine or freshwater enrichments have met with uniformly negative results (5, 35, 36). The prolific enrichment of magnetotactic forms, coupled with the fact that they may be easily accumulated by means of a magnetic field (5, 8, 17, 36), greatly facilitates isolation efforts, however.

A. magnetotacticum cells are routinely cultured microaerobically in sealed serum vials containing liquid medium and in semisolid medium in screw-capped culture tubes (9). They do not readily or consistently initiate growth from small inocula if the dissolved oxygen is greater than approximately 3–5% of saturation (4–7 μmol of O_2/liter at 30°C in the culture medium used). However, with headspace oxygen as high as 0.21 atm (21 kPa) in sealed, undisturbed vials of liquid medium, cells often eventually grow after a variable lag period. In this case final cell yields may be even higher than those obtained at lower initial oxygen concentrations. However, cells are not magnetotactic and do not synthesize magnetosomes under aerobic conditions (2, 9).

PHYSIOLOGY AND ULTRASTRUCTURE OF MAGNETOTACTIC BACTERIA

Aquaspirillum magnetotacticum cells are microaerophilic and chemoheterotrophic. They metabolize diverse organic acids including fumaric, tartaric, and succinic as sole sources of carbon and, presumably, energy (9). They do not grow in the absence of an organic carbon source. Escalante-Semerena et al (15) studied nitrate dissimilation by growing cells of this organism. The stoichiometry obtained (two nitrate utilized per tartrate oxidized) indicated that nitrate served as the principal electron acceptor for cells growing microaerobically on tartrate. A trace of O_2 was also consumed (7 vs 463 μmol of nitrate utilized). Oxygen utilization increased under nitrate-limiting conditions despite lower cell yields. It probably also serves as a terminal electron acceptor in this organism under appropriate conditions. However, cells will not grow anaerobically with nitrate. Thus, oxygen may also be required as a substrate for oxygenases participating in biosynthesis of hemes or other cell constituents. A study of the electron transport system of this spirillum has revealed the presence of *b*- and *c*-type cytochromes. *a*-Type and/or *o*-type cytochromes are also present, depending upon growth conditions. Although whole cells give a negative or delayed positive oxidase test (32), a positive test has been obtained with a preparation of respiratory membrane particles (W. O'Brien, unpublished results). A magnetotactic coccoid bacterium studied by Moench was found to possess a *c*-type cytochrome (35). Products of nitrate metabolism in *A. magnetotacticum* include nitrous oxide and ammonia (9, 15). Nitrite and nitric oxide are not detectable. Nitrogen is also a presumed product since under nitrate-limited conditions nitrous oxide accumulates only transiently (3, 15). Thus, this organism is a denitrifyer that can concommitantly carry out assimilatory nitrate reduction. Under conditions of low oxygen tension (less than 0.2 kPa) or with nitrate and ammonia, the organism also reduces acetylene. Acetylene reduction activity was totally inhibited by 0.2 mM-NH_4^+ or NO_3^- (3).

In sealed culture vessels with the partial pressure of oxygen in excess of about 0.06 atm (6 kPa), cells of *A. magnetotacticum* grow to high final cell yields, but only after a prolonged, variable lag period. Moreover, as mentioned, they do not synthesize magnetosomes and are not magnetotactic at high oxygen values. Magnetite appears to be synthesized only in the presence of a suitable iron supply and when oxygen is limiting for growth. Neither cells of this species (32) nor magnetotactic bacteria accumulated from enrichment cultures (principally coccoid cells) (N. Blakemore, unpublished results) contain detectable catalase activity. This apparently explains their sensitivity to high oxygen at low cell densities because catalase, when supplied exogenously as a constituent of culture media, permitted aerobic growth of *A. magnetotacticum,* although cells were not magnetotactic and lacked magnetosomes (9). Cells of this organism are very sensitive to H_20_2, even more so than those of *Chromobacterium* spp. (N. Blakemore, unpublished results). It is well known that iron salts catalyze decomposition of H_20_2. Thus, extracted bacterial magnetosomes visibly decompose H_20_2 (D. Bazylinski, unpublished results.) Magnetite synthesis, especially if proceeding under microaerobic conditions in areas adjacent to cellular sites of H_20_2 formation, might well protect the cell from H_20_2 toxicity by preventing its accumulation. Nonmagnetotactic cells, even though they lack magnetosomes, also have high intracellular iron content (0.2% of cell dry weight), which might offer similar protection. Iron salts that promote aerotolerance in other microaerophiles (10, 23) do not appear to do so with this species, however (9). A role in destruction of toxic products of oxygen metabolism by high cellular Mn in *Lactobacillus plantarum* has been identified (1). In stored sediments, magnetotactic bacteria survive for long periods. In laboratory cultures of magnetotactic spirilla, which often consist of mixtures of magnetic and nonmagnetic cells, the magnetotactic cells remain actively motile and persist the longest (N. Blakemore, unpublished observation). Magnetite synthesis, or reactions leading to it, may promote long-term survival of these organisms (particularly if they occasionally experience superoptimal oxygen tensions) by helping to rid cells of toxic products of 0_2 metabolism. This might be of significance in many natural settings where they occur. Due to the generally depressed availability of combined forms of nitrogen (i.e. nitrate) in these environments, cells might use more oxygen in respiration, thereby generating inhibitory amounts of peroxides or other toxic intermediates of 0_2 metabolism. Superoxide dismutase activity by cells has not been measured. However, exogenously supplied bovine superoxide dismutase, known to enhance aerotolerance of the microaerophile *Campylobacter fetus* (22), had no detectable effect on cells of *Aquaspirillum magnetotacticum* (R. Blakemore, unpublished results).

Cells also appear equipped to avoid excess oxygen. They show evidence of aerotaxis in semisolid medium (9) and have been observed to localize in bands around the periphery of coverslips or air bubbles in wet mounts, as do other known aerotactic bacteria. Magnetotactic cells supplied with bog or marsh water (36) or iron quinate (9) as iron sources accumulated intracellular iron some 20,000- to 40,000-fold over its extracellular concentration. Siderophores elaborated by magnetotactic bacteria have not been detected under these conditions. Iron uptake probably occurs via nonspecific (37) transport because of the relatively high extracellular iron concentration of ca 1 mg/liter. At the values of pH and Eh' at which cells grow [pH 5–7; Eh' (+)40 to (+)400 mV), dissolved iron is likely to be in the ferrous form. Chelated ferric iron would likely be released intracellularly by a reductive process (19, 37). Could iron oxidation provide an energy source for these organisms?

One method of evaluating the possible contribution of Fe_30_4 synthesis to the overall energy budget of *A. magnetotacticum* is to consider the Gibbs free energy change for Fe_30_4 synthesis, where

$$\Delta G = \Delta G^\circ + \mathrm{RT} \ln a_{\mathrm{products}}/a_{\mathrm{reactants}}. \quad (1)$$

Since the pathway for bacterial magnetite synthesis is not yet known, this evaluation may be made from the standard free energies of formation of reactants and products, where

$$\Delta G^{\circ} = \Sigma \Delta G^{\circ} \,(_{\text{formation of products}}) - \Sigma \Delta G^{\circ} \,(_{\text{formation of reactants}}) \qquad (2)$$

and

$$\Delta G^{\circ} = -n\mathfrak{F}\epsilon^{0}. \qquad (3)$$

The reactions expressing formation of Fe_30_4 from standard state reactants and products based upon values of electrode potentials or standard free energies of formation are

Reaction	ϵ°; volt	ΔG°_f kcal/mol
$6e^- + 3Fe^{2+}_{(aq)} \rightarrow 2Fe_{(s)}$	-0.4(33)	+60.9
$3Fe_{(s)} + 20_2 \rightarrow Fe_30_4$		-242.4(30)
$3H_2O \rightarrow 3/20_2 + 6H^+ + 6e^-$	-1.23(33)	+170.1

$$3Fe^{2+} + 1/20_2 + 3H_2O \rightarrow Fe_3O_4 + 6H^+ \quad \Sigma\Delta G^{\circ}_f = -11.5 \text{ kcal/mol.}$$

Thus, the $\Sigma\Delta G^{\circ}_f = -11.5$ kcal/mol for the charge balanced overall reaction for magnetite synthesis. Substitution of this value into equation 1 with appropriate values for the concentration of products and reactants under conditions where magnetite is synthesized by cells (pH = 7.0; $Fe^{2+} = 2.9 \times 10^{-5}$ M; $O_2 = 7 \times 10^{-6}$ M) affords a ΔG value of -46.9 kcal/mol of magnetite produced. The negative value suggests that cells might actually derive energy from this reaction; it proceeds spontaneously. When cultured under the conditions cited, *A. magnetotacticum* attained a cell yield of approximately 30 mg (dry wt) per liter and the cells took up roughly 30% of the iron available in the medium (8 μmol of Fe/liter), thereby producing approximately 2.5 μmol of Fe_30_4 per liter. Assuming 7–12 kcal are utilized to form 1 mol of ATP, then the energy associated with magnetite synthesis is sufficient to yield only 12 μmol of ATP per liter of culture. Assuming a value of $Y_{ATP} = 10.5$ μg of cell dry weight per μmol of ATP, the expected cell yield from magnetite synthesis would be 126 μg/liter. This is less than 1% of the observed cell yield. Under laboratory growth conditions, magnetite synthesis by oxidation of ferrous iron would produce an insignificant fraction of the cells' total energy budget. This supports the conclusion arrived at by Moench & Konetzka (36) that magnetite within magnetotactic bacteria is not merely an accumulated by-product produced during energy transformation involving iron oxidation. However, studies in progress in my laboratory are directed at the possibility that cells may use Fe^{3+} during heme synthesis (12) or as a terminal electron acceptor under appropriate conditions.

The process of magnetite synthesis by bacteria is currently under study (W. O'Brien, R. Frankel, R. Blakemore, manuscript in preparation). It has as an interesting precedent in the studies of magnetite biomineralization in the marine chiton (Polyplacophora). The radular teeth of this mollusk are arranged in long parallel rows and undergo a progressive sequential mineralization that leads to a surface coating of magnetite on the most anterior denticles (31). The stages in magnetite formation in chitons include accumulation of ferritin within an organic unmineralized micellar framework, and transformation of ferrihydrite ($5\ Fe_2O_3 \cdot 9H_2O$) in the ferritin cores to magnetite. The latter stage involves reduction of one third of the ferric iron to ferrous iron, oxygen removal, dehydration, and finally a change in crystal form from hexagonal close packing to cubic inverse spinel (29, 42, 43). Mössbauer spectroscopic studies of *A. magnetotacticum* have resulted in some interesting parallels. Mössbauer spectra of this organism contain, in addition to the typical six-line spectrum of magnetite, a central pair of absorption lines that correspond to those obtained from ferritin (18). Until recently, however, ferritin was not known to be present in prokaryotic cells (21, 40). Consequently, its role(s) and distribution in these organisms have not been widely studied. Recent studies of ^{57}Fe-enriched magnetotactic bacteria at low temperature have confirmed the ferritin-like behavior of this material, suggesting that it may also be a magnetite precursor in the bacteria. If this process of magnetic synthesis is common to both organisms, the magnetosome sheath may then correspond functionally and structurally to the micellar boundary layer (29, 43) observed in the mineralizing chiton radular denticle.

The magnetotactic bacteria examined to date all appear to be Gram negative. They stain appropriately and most have the typical wavy outer cell wall characteristic of this group (2; see also Figures 3, 4, 6, 8). Many species accumulated from sediments appear quite different morphologically from described species of bacteria. One ultrastructural feature common to all, however, is the presence of intracellular magnetosomes. In those species in which they have been studied, the magnetosomes were shown by means of Mössbauer spectroscopy (18), X-ray diffraction (14), or electron optical analysis (44) to consist of magnetite, Fe_3O_4. However, the magnetosome composition of all species is not known and conceivably other magnetic minerals such as other oxides of iron or cubic iron sulfides similar to greigite (13) might be present in some. Extracted magnetosomes appear either to be contaminated with cell substances or contain substances in addition to Fe_3O_4. Despite different extraction procedures used in two separate studies, the magnetosome fraction consisted of from 14 (14) to 79% (44) magnetite by weight. Mössbauer spectroscopic measurements of extracted magnetosomes of *A. magnetotacticum* have revealed the presence of high-spin Fe^{3+} in oxygen coordination, which, as mentioned previously, behaves at low temperature as ferritin. This organism's magnetosomes, unlike those of a magnetotactic coccus studied by Towe & Moench (44), contain no detectable titanium and exist as cubic-to-octahedral grains (42 nm on a side) rather than hexagonal prisms (parallelepipeds approximately 100 X 62 nm). Magnetosomes consistently exist in chains situated at the cell boundary in the spirilla (2; Figure 8), although as indicated by Figures 3 and 4, they may occur in clumped or scattered arrangements in other bacteria (see also 35, 44). At cell division, the magnetosomes appear to be more or less equally partitioned to each daughter cell by the cell division plane (7; Figure 3).

Recent elemental mapping studies of magnetotactic bacteria using energy-dispersive X-ray analysis have indicated the presence of substantial amounts of phosphorus within cells. This appears to be a consistent feature of cell composition in the several species examined. However, in the coccoid types (Figures 3, 5), the cell phosphorus is localized in large spherical bodies along with potassium (R. Blakemore, unpublished results). Iron, sulfur, calcium, and magnesium also present within these bacteria in substantial amounts are absent from the phosphate bodies.

All magnetotactic bacteria examined to date possess flagella, usually positioned at one side of the cell or at the ends of cells. Some cells have bipolar flagella as does *A. magnetotacticum.* Bipolarly flagellated organisms swim in either direction along magnetic field lines. Presumably, the direction chosen under natural conditions is determined by factors such as concentration gradients of oxygen. Other magnetotactic bacteria appear to be more or less unidirectional in their motility: swimming persistently forward with a characteristic wobbling motion. These do not show the run-and-tumble motility pattern characteristic of many chemotactic flagellated bacteria. However, other tactic responses may override magnetotaxis, because, as mentioned previously, even these magnetotactic bacteria may be found in oxygen-depleted surface waters.

An interesting feature of the ultrastructure of these bacteria is the presence of surface appendages (pili) often observed on coccoid forms (Figure 5). Other forms (the spirilla) produce extracellular polysaccharide material that stains with ruthenium red (2). These aspects of the outer surfaces of cells are consistent with microscope observations that these organisms adhere to sediment particles. Perhaps it is only when disturbed or dislodged from surfaces to which they may attach themselves that magnetotaxis may affect their distribution and location.

CONCLUDING REMARKS

As expected of a newly described phenomenon, research opportunities with magnetotactic bacteria abound. This is a fascinating, diverse group of organisms with a single species in pure culture. The cells sequester iron, accumulating it intracellularly to a significant fraction (2–4%) of their dry weight as a magnetic iron oxide containing two thirds of its iron in the ferric form and one third in the ferrous state. Because of the unique size, shape, and configuration of its magnetic grains, a single cell cannot be demagnetized, and its magnetic moment interacts effectively with the geomagnetic field so as to produce a torque tending to align the organism with geomagnetic field lines. It is interesting that each cell's magnetic axis coincides well enough with its axis of motility to effect rather direct swimming along magnetic field lines. An ecological consequence is that in natural environments when a vertical component of the ambient magnetic field is present, cells' direction of magnetic polarity relative to their principal swimming direction guides them generally downward toward sediments. This also keeps them away from more aerobic surface waters. The possibility that bacterial intracellular magnetite is a product of some physiological process common to all magnetotactic forms is under investigation. An understanding of the manner in which magnetite is formed by these organisms will contribute to an appreciation of their role(s) in biogeochemical cycling. These bacteria illustrate how the environmental cycles of nitrogen, carbon, and iron, for instance, may be coupled together by microorganisms.

The deposition of biogenic magnetic fine particles in sediments due to death and lysis of bacteria should have obvious interest among paleomagnetists. A study to investigate the possible contribution of bacterial magnetite to magnetic remanence of sulfide-rich salt marsh sediments was unsuccessful (13), but that does not rule out the possibility that the fine-grain magnetic iron sulfides detected may be of microbial origin. They appeared similar to particles found within magnetotactic cells shown in Figures 5 and 6, which are of unknown composition. The possible involvement of bacteria is strengthened by recent findings (W. Ghiorse, personal communication) of particles with the shape and dimensions of bacterial magnetite grains both within and outside of cells found in sediments collected in an area of manganese nodule formation in Oneida Lake of New York State (see also 5).

Recently, large numbers of magnetic cells of *Tetrahymena* were successfully produced by feeding these protozoans intact cells of *Aquaspirillum magnetotacticum* (J. Rifkin, unpublished data). We have also observed magnetically responsive protozoans grazing on accumulated magnetotactic bacteria in enrichment cultures. Although these latter observations may provide a useful means of manipulating these organisms, they bear little relation to their natural activities. One may also ponder the ecological consequences and biological adaptiveness of magnetism in at least two other instances recently reported among microorganisms. A non-magnetotactic, magnetic, aerobic, motile, Gram-negative bacterium, strain 1173, was isolated by Rades-Rohkohl et al from a weathered rock surface (38). Neither it, nor several other iron-accumulating bacterial strains, cells of which were attracted to the poles of magnets placed near cultures, possessed magnetosomes, however.

Finally, a magnetotactic green alga belonging to the genus *Chlamydomonas* was discovered (H. Lins de Barros, et al, unpublished information) in a polluted brackish coastal lagoon of Brazil. These eukaryotic cells contain a permanent magnetic dipole moment estimated to be ten times greater than those of magnetotactic bacterial cells. The direction of their cell polarity is such that they are steered generally downward in the geomagnetic field. The presence of magnetosomes has not been substantiated, but it is inferred from demagnetization studies. The ecological (behavioral) significance of magnetotaxis in this alga is not clear. Above a certain light threshold, the organisms are negatively phototactic. Possibly cells with their direction of magnetic polarity orienting them so they swim up and localize in the highly illuminated surface waters might be selected against, whereas those appropriately magnetized to swim down would be consistently favored.

These recent findings all provide about as much wonder as they do information concerning the adaptive significance of magnetism in diverse microorganisms. They suggest that rewarding research, in addition to providing newly isolated strains now needed for comparative studies, will address the physiological implications of magnetite formation in microorganisms, studies that will require genetic and enzymatic techniques of analysis.

ACKNOWLEDGMENTS

I wish to thank graduate students and colleagues who contributed to this review. I thank N. A. Blakemore, D. Maratea, and D. L. Balkwill for providing material for figures used and J. A. Goodrich for typing. The thermodynamic considerations were developed through valued discussions with the late G. L. Klippenstein. N. A. Blakemore and R. B. Frankel, who critically read the manuscript, continue to be strong sources of encouragement, information, and inspiration. E. Canale-Parola and R. S. Wolfe set in motion many events leading to much of the work reported here. The studies carried out in my laboratory are currently supported by contract NOOO14–80–C–0029 from the Office of Naval Research and a grant PCM–7922224 from the National Science Foundation.

Literature Cited

1. Archibald, F. S., Fridovich, I. 1981. *J. Bacteriol.* 145:442–51
2. Balkwill, D. L., Maratea, D., Blakemore, R. P. 1980. *J. Bacteriol.* 141: 1399–408
3. Bazylinski, D.A., Blakemore, R. P. 1982. *Abstr. 82nd Annu. Meet. Am. Soc. Microbiol.*, I53, p. 103
4. Blakemore, R. P. 1975. *Abstr. 75th Annu. Meet. Am. Soc. Microbiol.*, I148, pg. 141
5. Blakemore, R. P. 1975. *Science* 190: 377–79
6. Blakemore, R. P., Canale-Parola, E. 1973. *Arch. Mikrobiol.* 89:273–89
7. Blakemore, R. P., Frankel, R. B. 1981. *Sci. Am.* 245:58–65
8. Blakemore, R. P., Frankel, R. B., Kalmijn, A. J. 1980. *Nature* 286:384–85
9. Blakemore, R. P., Maratea, D., Wolfe, R. S. 1979. *J. Bacteriol.* 140:720–29
10. Bowdre, J. H., Krieg, N. R., Hoffman, P. S., Smibert, R. M. 1976. *Appl. Environ. Microbiol.* 31:127–33
11. Butler, R. F., Banerjee, S. K. 1975. *J. Geophys. Res.* 80:4049–58
12. Dailey, H. A. Jr., Lascelles, J. 1977. *J. Bacteriol.* 129:815–20
13. Demitrack, A. 1981. *Eos Trans. Am. Geophys. Union* 62:850
14. Denham, C. R., Blakemore, R. P., Frankel, R. B. 1980. *I.E.E.E. Trans. Magnet.* Mag-16:1006–7
15. Escalante-Semerena, J. C., Blakemore, R. P., Wolfe, R. S. 1980. *Appl. Environ. Microbiol.* 40:429–30
16. Frankel, R. B., Blakemore, R. P. 1980. *J. Mag. Magn. Mtls.* 15–18:1562–64
17. Frankel, R. B., Blakemore, R. P., Torres de Araujo, F. F., Esquivel, D.M.S., Danon, J. 1981. *Science* 212: 1269–70
18. Frankel, R. B., Blakemore, R. P., Wolfe, R. S. 1979. *Science* 203:1355–56
19. Gaines, C. G., Lodge, J. S., Arceneaux, J. E. L., Byers, B. R. 1981. *J. Bacteriol.* 148:527–33
20. Gilbert, W., 1600. *De Magnete Magneticisque Corporibus et de Magne Magnete Tellure.* London: Petrus Short. 240 pp.
21. Harrison, P. M. 1979. *Nature* 279:15–16
22. Hoffman, P. S., George, H. A., Krieg, N. R., Smibert, R. M., 1979. *Can. J. Microbiol.* 25:8–16
23. Hoffman, P. S., Krieg, N. R., Smibert, R. M. 1979. *Can. J. Microbiol.* 25:1–7
24. Jacobs, I. S., Bean, C. P. 1955. *Phys. Rev.* 100:1060–67
25. Kalmijn, A. J., 1981. *I.E.E.E. Trans. Magnet.* Mag-17:1113–23
26. Kalmijn, A. J., Blakemore, R. P. 1977. *Proc. Int. Union Physiol. Sci.* 13:364
27. Kalmijn, A. J., Blakemore, R. P. 1978. In *Animal Migration, Navigation and Homing,* ed. K. Schmidt-Koenig, W. T. Keeton, pp. 354–55. New York: Springer
28. Kirschvink, J. L. 1980. *J. Exp. Biol.* 86:345–47
29. Kirschvink, J. L., Lowenstam, H. A. 1979. *Earth. Plan. Sci. Lett.* 44:193–204
30. Laitenen, H. 1960. *Chem. Anal.* New York: McGraw-Hill. 611 pp.
31. Lowenstam, H. A. 1972. *Geol. Soc. Am. Bull.* 73:435–38
32. Maratea, D., Blakemore, R. P. 1981. *Int. J. Syst. Bacteriol.* 31:452–55
33. Maron, S. H., Prutton, C. F. 1969. *Principles of Physical Chemistry.* London: Macmillan. 886 pp.
34. Microbial magnets. 1978. *Sci. Am.* 238:73–74
35. Moench, T. T. 1978. *Distribution, isolation and characterization of a magnetotactic bacterium.* PhD thesis. Indiana Univ., Ind. 110 pp.
36. Moench, T. T., Konetzka, W. A. 1978. *Arch. Microbiol.* 119:203–12
37. Neilands, J. B. 1977. *Adv. Chem.* 162: 3–32
38. Rades-Rohkohl, E., Fränzle, O., Hirsch, P. 1981. *Abstr. 5th Int. Sympos. Environ. Biogeochem, Stockholm* 1–5.6, p. 43
39. Rosenblatt, C., Torres de Araujo, F. F., Frankel, R. B. 1982. *J. Appl. Phys.* 53: 2727–29
40. Stiefel, E. I., Watt, G. D. 1979. *Nature* 279:81–3
41. Teague, B. D., Gilson, M., Kalmijn, A. J. 1979. *Biol. Bull.* 157:399
42. Towe, K. M., Bradley, W. F. 1967. *J. Colloid. Interface Sci.* 24:382–92
43. Towe, K. M., Lowenstam, H. A. 1967. *J. Ultrastr. Res.* 17:1–13
44. Towe, K. M., Moench, T. T. 1981. *Earth Plan. Sci. Lett.* 52:213–20

Ann. Rev. Microbiol. 1982. 36:239–58

VIROIDS AND THEIR INTERACTIONS WITH HOST CELLS

T. O. Diener

Plant Virology Laboratory, Plant Protection Institute, ARS, U.S. Department of Agriculture, Beltsville Agricultural Research Center, Beltsville, Maryland 20705

CONTENTS

INTRODUCTION

Viroids are low-molecular-weight nucleic acids ($1.1–1.3 \times 10^5$) of unique structure that can be isolated from certain higher plant species that suffer from specific diseases. They are not detectable in healthy individuals of the same species, but when introduced into such individuals, they are replicated autonomously in spite of their small size and cause the appearance of the characteristic disease syndrome. Viroids thus are the causative agents of the diseases in question. Unlike viral nucleic acids, viroids are not encapsidated, i.e. no virion-like particles can be isolated from infected tissue. All viroids so far identified are composed of RNA and all have been isolated from higher plants. Viroids constitute a novel class of subviral pathogens; they are the smallest known agents of infectious disease.

0066-4227/82/1001-0239$02.00

The viroid concept was advanced in 1971 (14) on the basis of newly established properties of the infectious agent responsible for the potato spindle tuber disease. These properties were found to differ basically from those of conventional viruses in at least five important respects: (*a*) The pathogen exists in vivo as an unencapsidated RNA; (*b*) virion-like particles are not detectable in infected tissue; (*c*) the infectious RNA is of low molecular weight; (*d*) despite its small size, the infectious RNA is replicated autonomously in susceptible cells, i.e. no helper virus is required; and, (*e*) the infectious RNA consists of one molecular species only. Viroids have been reviewed once before in this series, in 1974 (17). At that time the viroid concept was rather new and, understandably, much skepticism as to its correctness prevailed. Thus, much of the space allotted was taken up by the then available evidence indicating that viroids do exist and that they are distinct biological entities. Although, fortunately, this review is relieved of this obligation, it faces another problem. Ideally, it should represent an overall update of the earlier review, but because of the steadily increasing attention given to viroids and viroid diseases in recent years, the pertinent literature has grown to such an extent that an overall review would be impractical.

Fortunately, a number of up-to-date reviews exist and it appears more productive to concentrate here on a critical examination of some very recent developments in the viroid field that may be of particular interest to the microbiologically oriented reader. For other aspects of the subject, the reader will be referred to appropriate review articles.

CLASSIFICATION

Whether or not viroids should be classified among viruses depends on the particular definition of "virus" one accepts. Thus, the definition advanced in the latest (3rd) edition of the textbook by Luria et al (47) considers "the possession of an extracellular infective phase represented by specialized objects or *virions,* which are produced under the genetic control of the virus" an essential quality of viruses. Clearly, viroids do not fall within the scope of this definition.

Lwoff (48), on the other hand, considers viruses to be "potentially infectious and potentially pathogenic absolute parasites" and defines "absolute parasites" as those that utilize their host cells' enzyme systems for energy production (Lipmann system) and protein synthesis, including their host cells' tRNA's and ribosomes. According to this definition viroids evidently are viruses. Recognizing the pronounced differences between conventional viruses and viroids, Lwoff nevertheless proposes to divide the kingdom of viruses into two groups, the true viruses, or Euviruses, and the Viroids (48).

Lwoff's scheme, however, based as it is on the degree and nature of parasitic dependence, does not consider the profound difference in this regard between viruses and viroids. If viruses are absolute parasites in Lwoff's sense, then viroids might be regarded as ultra-absolute parasites because they depend to a far greater extent even than viruses on their host cells' biosynthetic machinery. Although viruses invariably contain genetic information that with the help of the host's protein synthesizing system is translated (at one stage or other of the virus' replication cycle) into one or more virus-specific proteins, viroids apparently do not possess this capacity (see below). Therefore, they must depend for their replication entirely on enzymes normally present in the cells of their hosts. Viroids thus are the only known pathogens that do not code for any proteins.

The extreme parasitic dependence of viroids as compared with that of viruses may, by itself, invalidate Lwoff's proposal to classify viroids as viruses, but a still more convincing argument against the proposal emerges when one considers the unique molecular structure of viroids (see below), which has no counterpart among viruses and which places in doubt any evolutionary relationship between the two types of pathogens.

In view of these and other considerations (see below), it might be best to defer a decision regarding the proper position of viroids in a classification scheme until more is known of their mode of replication, mechanism of pathogenesis, and evolutionary origin. In Table 1, some distinguishing properties of parasitic microorganisms, viruses, and viroids are listed.

VIROID DISEASES

The following naturally occurring diseases of higher plants are now known to be the result of infection with viroids: potato spindle tuber (14), citrus exocortis (70, 76), chrysanthemum stunt (23), chrysanthemum chlorotic

Table 1 Distinctions among parasitic agents

Property	Parasitic microorganisms	Viruses	Viroids
Growth	yes	no	no
Division	yes	no	no
Absolute parasitism[a]	no	yes	yes
Type nucleic acid	DNA and RNA	DNA or RNA	RNA (or DNA?)
M_r of genomic nucleic acid	$\geqslant 3 \times 10^8$	10^6–3×10^8	1.1–1.3×10^5
Nucleic acid encapsidated	NA[b]	yes[c]	no
Nucleic acid codes for protein(s)	yes	yes	no

[a] See Ref 48.
[b] NA, Not applicable.
[c] Except maturation-defective mutants.

mottle (69), cucumber pale fruit (80), hop stunt (72), tomato bunchy-top (2), tomato planta macho (28), and tomato apical stunt (81). Whether the three tomato diseases are caused by one and the same or by different viroids has not been determined. Identity appears possible between the tomato bunchy-top and the tomato apical stunt viroids, because they occur on the same continent, the former in South Africa, the latter in the Ivory Coast (tomato planta macho disease is known to occur only in Mexico). Viroid etiology of coconut cadang-cadang (66) and advocado sunblotch (62) has not yet been demonstrated rigorously, but it is virtually certain because viroid-like RNAs are detectable in tissue from plants with both diseases, but not in tissue from healthy plants. Furthermore, a viroid has been isolated from apparently healthy *Columnea erytrophae* plants that causes symptoms similar to, if not identical with, those of the potato spindle tuber viroid (PSTV) in tomato or potato, yet is distinct from PSTV in its primary structure (60).

A detailed discussion of viroid diseases is beyond the scope of this review. For more detail the reader is referred to Diener (19).

MOLECULAR STRUCTURE

Viroids are single-stranded, covalently closed-circular and linear RNA molecules (53, 71) that occur, because of extensive regions of intramolecular complementarity, in the form of collapsed circles and hairpin structures with the appearance in the electron microscope of double-stranded rods (77). The complete nucleotide sequence of PSTV has been determined (33), and on the basis of this sequence, as well as of that of quantitative thermodynamic and kinetic studies of its thermal denaturation (42, 44), a model for the secondary structure of PSTV has been proposed (33). According to this model, viroids exist in their native configuration as extended rod-like structures characterized by a series of double-helical sections and internal loops. Thus, the rigid, rod-like structure of the native viroid is based on a defective rather than a homogeneous RNA helix (26), in confirmation of conclusions arrived at earlier by different techniques (15, 77).

For further discussion of the molecular structure of viroids, the reader is referred to two recently published, excellent reviews (35, 43).

Recently, the complete nucleotide sequence of the chrysanthemum stunt viroid (CSV) has been established (41). In contrast to PSTV, which consists of 359 nucleotides, CSV is a covalently closed-circular, single-stranded RNA that consists of 356 nucleotides (41). Of the CSV sequence, 69% is contained in that of PSTV and the two viroids can form similar secondary structures (41).

Also, a mild strain of PSTV has been sequenced (34) and the resulting primary structure was compared with that of the previously determined severe strain. The mild strain was shown to differ from the severe strain by only three nucleotide exchanges at different sites of the molecule, namely AA to U in positions 120–121, A to U in position 310, and an insertion of U between positions 312 and 313 (34).

These results are consistent with the conclusion that viroids constitute genetic systems whose properties are encoded in the nucleotide sequences of their RNAs. Thus, the nucleotide sequences of individual "species" of viroids differ conspicuously from each other, whereas the sequences of strains of one "species" differ only slightly. It is interesting to note that the small differences in the sequences of the two PSTV strains examined, namely three nucleotide exchanges (34), have such profound biological consequences.

VIROID-HOST CELL INTERACTIONS

When viroids are introduced into susceptible cells, they are replicated autonomously, i.e. without the requirement of a helper virus (14). This basic biological fact raises a number of intriguing questions. Foremost among these are the following.

1. By what mechanisms are viroids replicated? Because viroids have been demonstrated to be distinct species of low-molecular-weight RNA that introduce only a very limited amount of genetic information into host cells, it appears a priori that pre-existing host enzymes are largely or entirely responsible for viroid replication.
2. By what mechanisms do viroids incite disease in certain hosts, yet replicate in other susceptible plant species without discernible damage to the host?
3. How did viroids originate?

Subcellular Location

Bioassays of subcellular fractions from PSTV-infected tomato leaves demonstrated that only the tissue debris and nuclear fractions contain appreciable infectivity (13). Chloroplast, mitochondrial, ribosomal, and soluble fractions contain only traces of infectivity. Most infectivity is chromatin associated and can be extracted as free RNA with phosphate buffer (13). The citrus exocortis viroid (CEV) is also located primarily in the nuclear fraction in close association with chromatin (70), but with CEV in *Gynura aurantiaca,* a significant portion of the viroid has been reported to be associated with a plasma-membrane-like component of the endomembrane system (75).

The fact that infectious PSTV is located primarily in the nuclei of infected cells does not prove it is synthesized there. However, experiments with an in vitro RNA synthesizing system, in which purified cell nuclei from infected tomato leaves were used as an enzyme source, suggested this is the case (79). It appears, therefore, that the infecting viroid migrates to the nucleus (by a mechanism as yet unknown) and is replicated there. The absence of significant amounts of PSTV in the cytoplasmic fraction of infected cells suggests that most of the progeny viroid remains in the nucleus.

Question of Viroid Translation

Viroids are of sufficient chain length to code for a polypeptide of about 10,000 daltons, although with circular PSTV the uneven number of nucleotides theoretically permits three rounds of translation with a frame shift each time.

Testing for in vitro messenger function of PSTV and CEV in a variety of cell-free protein-synthesizing systems indicated that neither viroid functions in this capacity (11, 40). Also, CEV is not translated in *Xenopus laevis* oocytes, even after polyadenylylation in vitro, and it does not interfere with the translation of endogenous messenger RNAs (73). Lack of mRNA activity of PSTV is not surprising in view of the fact that no AUG initiation codon is present in its nucleotide sequence (there are, however, several GUG triplets that possibly might act as initiators).

Although viroids do not act as mRNA's in these systems, they might be translated in vivo from a complementary RNA strand synthesized by preexisting host enzymes with the infecting viroid serving as a template. RNA sequences complementary to viroids have been identified in infected tissue (31, 57), and it has been suggested that these might act as mRNA's (52). With PSTV, the complementary strand (cPSTV), as constructed from the nucleotide sequence of the viroid (33), could theoretically serve as an mRNA (52). Although cPSTV also does not contain any AUG initiation triplets, it contains four GUG triplets and six possible termination triplets that could theoretically result in four polypeptides containing 108, 79, 43, or 28 amino acids (52). Whether or not viroid-complementary RNAs act as mRNA's is not known, but if they do, novel, viroid-specific proteins should be detectable in protein preparations from infected host tissue.

Comparisons of protein species in healthy and PSTV-infected tomato (86) and healthy and CEV-infected *Gynura aurantiaca* (8), however, did not reveal qualitative differences between healthy and infected plants. In both studies, synthesis of at least two proteins was enhanced in infected as compared with healthy tissue, but recent studies indicate that these proteins are host and not viroid specific (7a). Also, two-dimensional gel electrophoretic analyses of proteins synthesized in uninfected or PSTV-infected tomato cells derived from suspension cultures revealed neither quantitative nor qualitative changes resulting from maintenance of the viroid in the cell line (88).

Although more sensitive methods of analysis may yet disclose the presence of viroid-specified polypeptides in infected cells, in light of present knowledge, one must conclude that viroids do not act as mRNA's. If so, the complementary RNA sequences found in infected tissue must be synthesized entirely by pre-existing (but possibly activated) host enzymes.

Mechanism of Replication

Theoretically, viroid replication could involve transcription from either RNA or DNA templates. An RNA-directed mechanism requires the presence of RNA sequences complementary to the entire viroid in infected tissue, as well as a pre-existing host enzyme with the specificity of an RNA-directed RNA polymerase.

A DNA-directed mechanism requires the presence of DNA sequences complementary to the entire viroid. These DNA sequences might already be present, in repressed form, in uninfected hosts or they might be synthesized as a consequence of infection with viroids. In the latter case, a pre-existing host enzyme with the specificity of an RNA-directed DNA polymerase (reverse transcriptase) would also be required.

RNA-OR DNA-DIRECTED REPLICATION? Although data favoring both DNA- and RNA-directed replication have been reported, it is now evident that viroid replication occurs from RNA, not DNA, templates.

To distinguish between RNA- or DNA-directed replication, the effects of certain antibiotic compounds on viroid replication have been investigated. In an in vivo system, in which leaf strips from healthy and PSTV-infected plants were treated with water or actinomycin D, viroid replication was found to be sensitive to this inhibitor of DNA-directed RNA synthesis (24). Similar results were obtained with an in vitro RNA synthesizing system, in which purified cell nuclei from healthy or PSTV-infected tomato leaves were used as an enzyme source (79).

Sensitivity of viroid replication to actinomycin D has been confirmed in a study of cucumber pale fruit viroid (CPFV) synthesis in protoplasts isolated from tomato leaves (54). In the same study, the effect of α-amanitine on viroid synthesis has been investigated. Intracellular concentrations (10^{-8} M) of α-amanitine sufficient to inhibit tomato DNA-directed RNA polymerase II but not RNA polymerase III inhibited CPFV replication. In contrast to the studies with actinomycin D, in which the effect of the compound on viroid synthesis could conceivably have been the result, not of a specific inhibition of DNA-directed RNA synthesis but of a general toxic effect on cellular metabolism, the inhibitory effect of α-amanitine is not likely to be due to nonspecific, secondary effects of the compound on cell metabolism. This conclusion is strengthened by the demonstration that at an intracellular α-amanitine concentration sufficient to inhibit viroid replication by about 75% the biosynthesis of tobacco mosaic virus RNA or that of prominent cellular RNA species (tRNA, 5S RNA, 7S RNA, and rRNA) was not appreciably affected (54).

All of these studies seem to suggest that DNA-directed RNA synthesis is involved in viroid replication. This is most clearly evident in the experiments with α-amanitine, the results of which not only confirm those obtained with actinomycin D, but in addition implicate one specific enzyme, namely normally DNA-dependent RNA polymerase II.

Contradictory results, however, have been reported. Grill & Semancik (32), on the basis of infiltration of CEV-infected *Gynura aurantiaca* foliar tissue or of PSTV-infected potato tuber sprouts with varying concentrations of actinomycin D, concluded that the antibiotic had no specific inhibitory effect on viroid replication and that the inhibitory effects previously reported were a result of a general toxic effect of actinomycin D on cell metabolism. As discussed earlier (21), however, the evidence presented is somewhat equivocal.

ENZYME(S) INVOLVED Recently Rackwitz et al (65) have shown that purified DNA-directed RNA polymerases II from wheat germ or from callus cells or green leaves of a wild tomato species (*Lycopersicon peruvianum*) are capable of transcribing in vitro several synthetic and natural RNA templates, although at an efficiency about two orders of magnitude lower than with DNA templates. The authors showed that of all natural RNA templates tested, viroids are transcribed with the highest efficiency by either enzyme (65). Gel electrophoretic analysis under denaturing conditions of the in vitro transcription products with purified viroid as template revealed, aside from several smaller viroid-complementary molecules, full-length linear molecules (65).

Taken together, these findings strongly suggest that viroids are replicated by a novel mechanism in which the infectious RNA molecules are copied entirely by a pre-existing host enzyme and that this enzyme is normally DNA-directed RNA polymerase II. It thus appears that the enzyme responsible for the synthesis of pre-mRNA, under certain circumstances, can function as an RNA-directed RNA polymerase or replicase. It is tempting to speculate that it is the quasi-double-stranded DNA-like native structure of viroids that permits the enzyme to function relatively well in this capacity. Indeed, it has been shown that viroids readily form binary complexes with RNA polymerase II, that they compete with DNA for the template binding sites on the enzyme, and that they strongly inhibit DNA-directed RNA synthesis (65). In this view, the infecting viroid molecule commandeers nuclear RNA polymerase II for its own replication, and viroids may be regarded, as has been suggested previously (20), as selfish RNAs, in the sense introduced for noncoding DNA (reviewed in 46).

MOLECULAR PROBES Another approach to the study of viroid replication consists in the development of viroid-specific molecular probes and identification of viroid-related RNA or DNA sequences in nucleic acid extracts from plants by molecular hybridization. Three types of molecular probes have been used: (*a*) purified viroids labeled in vitro with ^{125}I; (*b*) in vitro prepared, single-stranded, viroid-complementary DNA (cDNA); and (*c*) double-stranded, viroid-related DNA obtained by recombinant DNA technology.

Initially, use of ^{125}I-labeled viroids as probes in hybridization experiments led to confusing results. On the basis of such experiments, two groups have reported the presence of viroid-complementary sequences in the DNA of viroid-infected (74) and even in that of uninfected host plants (38). Later work, however, clearly showed that these reports were in error. In one case (74), the reported DNA sequences have been identified subsequently as viroid-complementary RNA, not DNA (32), and in the other case (38), several investigators have clearly demonstrated that neither DNA from uninfected nor DNA from PSTV-infected tomato plants contains sequences complementary to major portions of the viroid. Thus, no viroid-complementary regions could be identified in DNA from either uninfected or viroid-infected tomato plants by conventional solution and filter hybridization techniques (87), or by Southern blot hybridization, using either ^{125} I-labeled PSTV (4) or fully defined ^{32}P-labeled, cloned double-stranded PSTVcDNA (see below) (36) as probes.

Because in the latter two studies sensitivity of the experimental methods was demonstrated to be adequate for the detection of less than one copy of viroid complement per haploid genome, these experiments rule out the presence of even a single complete and contiguous complement of PSTV in host DNA. It appears, therefore, that the supposed PSTV-DNA hybrids reported earlier (38) in fact were complexes between host DNA and ^{125}I-labeled cellular RNA contaminants in the labeled viroid preparations.

Similarly, in molecular hybridization experiments between ^{32}P cDNA to the chrysanthemum stunt viroid (CSV) and the genomes of either healthy or CSV-infected hosts (chrysanthemum and *Gynura aurantiaca*), no sequence homologies could be detected (61).

However, none of these experiments rule out the possibility that viroid-related sequences might be randomly located on host chromosomes or that host DNA contains only a small portion of the PSTV genome. In the latter case, it is conceivable that short, viroid-complementary DNA sequences might serve as recognition sites and might be involved in viroid pathogenesis. It is evident, however, that any such viroid-related DNA sequences could not act as templates for the synthesis of progeny viroids. It follows that viroids must be replicated from RNA templates.

This conclusion is supported by observations indicating that the primary structure of a viroid is faithfully maintained, regardless of the host in which the viroid is replicated (12, 56, 60)—as would be expected if the incoming viroid, and not host DNA, serves as the template for progeny synthesis.

REPLICATION INTERMEDIATES The most convincing evidence for an RNA-directed mechanism of viroid replication consists in the finding, by several groups of investigators, of viroid-complementary RNA molecules in nucleic acid extracts from infected plants and their absence in extracts from uninfected plants. Evidently, such molecules could represent intermediates in the process of viroid replication.

Viroid-complementary RNA sequences were first detected in extracts from CEV-infected tomato and *Gynura aurantiaca* leaves by solution hybridization with a ^{125}I-labeled viroid probe (31). Some viroid-complementary RNAs were detected in the LiCl supernatant fraction, but most were in the LiCl precipitable fraction (31). In a later publication (30) the authors attempted to characterize the molecular structures and sizes of the CEV-complementary molecules and suggested a variety of possible interpretations of their results. These included the association of viroid-complementary RNA with host DNA, RF (replicative form)— and RI (replicative intermediate)-like molecules, molecules enriched for single-stranded (ss) regions and others with significant lengths of both double-stranded (ds) and ss regions, and monomeric as well as higher-molecular-weight multimers or aggregates of viroid-complementary molecules. On the basis of their data, however, the authors could not choose among these possibilities (30).

If viroid-complementary RNAs are to serve as templates from which progeny viroids are transcribed, they evidently must contain a full complement of the viroid sequence, that is, they must be of equal or greater length than the viroid. Because in the investigations with CEV (30) RNAs were not denatured before analysis and were separated by gel electrophoresis under nondenaturing conditions, no conclusions as to the size(s) of the detected viroid-complementary molecules are possible.

The first convincing evidence for the existence in infected plants of full-length RNA molecules complementary to a viroid was obtained by Owens & Cress (57) in blot hybridization experiments in which a viroid-specific recombinant ds DNA was used as a probe.

This probe was synthesized by incubating polyadenylylated PSTV with reverse transcriptase, followed by removal of the PSTV template by heating. The resulting ss DNA (PSTVcDNA) was converted into S1 nuclease-resistant ds cDNA in a reaction using *Escherichia coli* DNA polymerase I. The ds PSTVcDNA was then inserted in the Pst I endonuclease site of plasmid pBR 322 by using the oligo(dC) oligo(dG)-tailing procedure. Tetracycline-resistant, ampicillin-sensitive transformants contained sequences complementary to PSTV[^{32}P]cDNA, and one recombinant clone (pDC-29) contained a 460-base-pair insert. This cloned ds PSTV cDNA contains the cleavage sites for six restriction endonucleases predicted by the published primary sequence of PSTV (57). Results of these and other experiments indicated that almost the entire sequence of PSTV had been cloned.[1] Labeled hybridization probes specific for RNA molecules that have the same polarity as PSTV or the opposite polarity were prepared by labeling the 5' termini of Bam HI-cleaved pDC-29 with ^{32}P and cleaving the labeled DNA with Pst I. Purification of the resulting products yielded a 110-base-pair fragment that specifically hybridizes with cPSTV (57).

Hybridization experiments with this probe demonstrated the presence in extracts from infected cells of RNA molecules of the same mobility (and presumably molecular weight) as linear PSTV but of opposite polarity (57). PSTV-complementary RNA molecules of this size were found after treatment of the nucleic acid extracts with RNase, denaturation of the RNAs by heating for 2 min at 100°C in 50% formamide, quenching, and analysis by gel electrophoresis at 55°C in the presence of 8 M urea. Thus, contrary to what has been stated later by others (5), RNAs were denatured prior to analysis, and gel analysis was performed under conditions known to prevent major reannealing of RNA.

In contrast to the results with CEV (31), cPSTV was found almost exclusively in the LiCl supernatant (low molecular weight and ds) fraction (57). The results also demonstrated that most, if not all, of cPSTV was present in nucleic acid extracts in the form of RNase-resistant duplex molecules, that is, base paired with PSTV. This is indicated by the observation that the yield of cPSTV was not significantly reduced if the RNA•RNA annealing prior to RNase treatment was omitted (57).

Recent evidence by several investigators indicates that viroid-infected cells contain, in addition to full-length viroid complements, viroid-specific molecules longer than unit length.

A first suggestion that such viroid-related RNAs may exist was obtained in blot hybridization experiments with nucleic acid extracts from PSTV-infected plants, in which two RNA species containing cPSTV were observed that migrated more slowly than PSTV (37).

PSTV-specific RNA molecules with electrophoretic mobilities slower than those of circular or linear PSTV were observed in another study, in

[1]From a practical agricultural standpoint, cloned, viroid-specific recombinant DNAs are of considerable importance, because they permit the design of rapid and sensitive diagnostic tests for the presence of viroids in cultivated plants (58).

which nucleic acid extracts from PSTV-infected plants were separated by gel electrophoresis and in which viroid-specific molecules were identified by Northern blot hybridization with either ^{125}I PSTV or ^{32}P cDNA probes (68). Seven cPSTV species were observed, six of which migrated more slowly than circular PSTV and one with about the same mobility as linear PSTV (68). Interpretation of the results presented in these studies (37, 68), however, is somewhat uncertain. Although the conditions used for electrophoresis are adequate to prevent major renaturation of previously denatured RNAs, they are not adequate to denature ds RNA. Because the authors (68) did not state whether or not their RNA preparations were denatured prior to electrophoresis, it is possible that some of the seemingly larger-than-unit-length molecules could have been complexes of unit-length PSTV and cPSTV, which, in the system used, would have migrated to the approximate position shown (68).

Convincing evidence for the existence of longer-than-unit-length cPSTV molecules has been obtained in a similar blot hybridization study, in which, however, two fully denaturing gel systems were utilized (5). Four discrete bands of cPSTV molecules were identified. Reference to DNA markers of known molecular weight indicated these bands contained molecules approximately 700, 1050, 1500, and 1800 nucleotides long, suggesting that they represent multimers of PSTV, which would contain 718 (dimer), 1077 (trimer), 1436 (tetramer), and 1795 (pentamer) bases (5). No unit-length cPSTV strands were detected, probably because of interference with hybridization by the significant quantities of unlabeled PSTV present in RNA from infected plants moving to the same position in the gel as cPSTV monomers (5). Enzymatic studies indicated that the cPSTV strands are composed exclusively of RNA and, as extracted, are present in complexes containing extensive ds regions (5). Because a well-defined band of cPSTV about 40 nucleotides longer than unit-length PSTV was observed after gentle treatment with RNase T_1, the authors suggested that a least some RNA species containing cPSTV appear to be composed of unit-length ds regions flanked by ss regions, the latter being composed of RNase T_1 -resistant sequences.

Based on their results, the authors hypothesized that the longer-than-unit-length cPSTV strands play a role in viroid replication, that the cPSTV complexes containing ds regions of the length of PSTV represent replication intermediates composed of cPSTV strands with approximately viroid-length tandem repeats, and that the PSTV strands present in these complexes are of unit length (5). These postulates, as well as the suggestion of a rolling-circle-type viroid replication mechanism were hypothetical, because (*a*) the kinetics of appearance during viroid infection of the putative replication intermediates were not determined, (*b*) no hybridization experiments with a probe capable of detecting molecules with the polarity of PSTV were reported, and (*c*) the gel electrophoretic systems used did not permit separation of circular from linear PSTV.

In still another investigation that is also based on gel electrophoretic separation of RNAs, followed by blot hybridization, the molecular structures of viroid-specific molecules have been further clarified; and some evidence has been presented that these structures indeed may represent intermediates in viroid replication (59). Hybridization probes specific for either PSTV or cPSTV were prepared as described above, except in the more recent studies the cloned insert consisted (as shown by direct DNA sequencing) of ds DNA representing the complete 359 nucleotide sequence of PSTV (9).

In agreement with the investigations already discussed, blot hybridization experiments using these recombinant DNA probes revealed the presence in RNA extracts from infected tissue of viroid-related mostly ds RNA species that migrate in gels more slowly than unit-length PSTV (59). The two most prominent of these slowly migrating RNAs were separated from one another by cellulose chromatography and preparative gel electrophoresis, and the composition, size, and configuration of the PSTV and cPSTV components of each were analyzed in a gel system that does not denature ds RNA, but prevents major reannealing of previously denatured RNAs. This gel system has the further advantage of separating circular from linear PSTV molecules (57). Untreated and RNase-treated separated ds components were analyzed in this system with or without prior denaturation (59).

The results of these analyses demonstrated that the two major, slowly migrating, viroid-specific ds RNAs are structurally related and that they are composed of unit-length circular and linear strands with the polarity of PSTV complexed with longer-than-unit-length RNA strands of opposite polarity. In Figure 1, a schematic summary of the results, as well as the authors' interpretation of the molecular structures of the identified components, is shown. At least five discrete zones of viroid-related RNAs are resolved (Figure 1, *left*). Two zones contain ss RNAs, the circular and linear forms of PSTV and cPSTV. Only traces of the latter, however, could be detected. At least three zones contain ds RNA, as shown by RNase treatment. The most rapidly migrating ds RNA species, a linear duplex containing unit-length linear PSTV and linear cPSTV, was only detected if the RNA had been treated with RNase before analysis. Therefore, it was not considered a potential PSTV replication intermediate. The two more slowly migrating ds PSTV RNAs, however, were shown to possess several of the characteristics expected for such intermediates. The more rapidly migrating one of these two ds RNAs was shown to contain unit-length linear or circular PSTV, but complementary strands somewhat longer than unit length. The more slowly migrating ds RNA was shown to also contain unit-length linear and circular PSTV and longer-than-unit-length cPSTV. In this component, however, the lengths of cPSTV ranged from unit length to at least twice that. The formation of these structures can be explained most readily if one assumes that cPSTV is synthesized on a circular PSTV template and that this synthesis continues past the origin of replication, leading to the synthesis of linear dimers and higher multimers of cPSTV. Such a scheme resembles, in some respects, the rolling circle model previously advanced to explain replication of certain viral RNAs (6). Results further suggest that unit-length linear PSTV synthesized from the cPSTV template may be circularized while still complexed to the template (59).

In this view the more slowly migrating ds RNA zones would contain dimers and higher multimers of cPSTV, whereas the more rapidly migrating ds RNA zone would contain circular ds molecules of unit length with ss cPSTV tails of varying lengths (Figure 1, *right*). That these structures probably are fragments of a larger PSTV replicative intermediate complex is indicated by the observation that synchronous synthesis of PSTV is accompanied by simultaneous synthesis of ds PSTV (59).

In summary, it appears reasonable to state that much of the mystery that in the past has surrounded the mechanism of viroid replication has now been dispelled. Although results of the discussed recent studies differ in detail, they nevertheless complement one another and evidently converge toward a unified concept of the molecular mechanisms involved. This concept includes the following postulates.

1. Viroids are transcribed from complementary RNA, not DNA templates.
2. These templates, as well as progeny viroids, are synthesized by a pre-

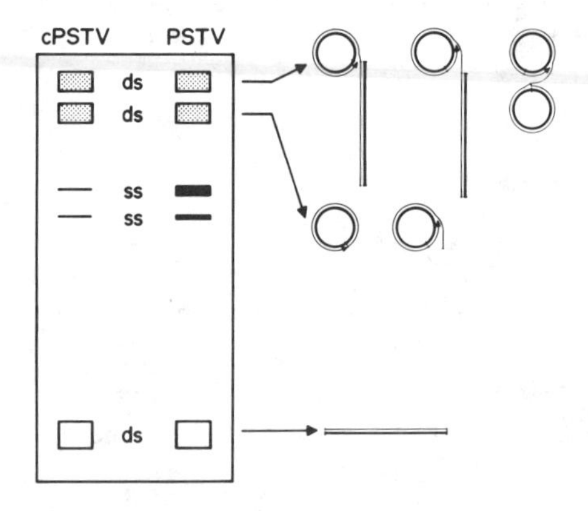

Figure 1 Major viroid-related RNA species identified (*left*) and tentative structures for dsPSTV RNA (*right*). For explanation, see text. (From 59.)

existing host enzyme, most likely RNA polymerase II functioning as an RNA-directed RNA polymerase (replicase).
3. Viroid-complementary RNA is transcribed from circular viroid molecules by a rolling-circle-type mechanism that results in the formation of multimeric templates and large replicative intermediate-like complexes.

POSSIBLE MECHANISMS OF PATHOGENESIS

By what mechanisms do viroids incite diseases in certain hosts, yet replicate in other susceptible species without inflicting discernible damage? The nuclear location of viroids and their apparent inability to act as mRNA's suggest that viroid-induced disease symptoms may be caused by direct interaction of the viroid with the host genome, that is, by interference with gene regulation in the infected cells. If so, viroids might be regarded as abnormal regulatory molecules (14). Alternatively, if viroids did originate, as is suggested below, from introns, their detrimental effects on host cell functions may be a result of interference with mRNA maturation processes (22). A third model of viroid pathogenesis posits that the commandeering of nuclear DNA-dependent RNA polymerase II by the infecting viroid molecule to perform its selfish replication inhibits or represses the synthesis of genomic mRNA's of the host cell and thus disturbs normal differentiation (65).

It should be stressed that all of these explanations are purely speculative and that for any model of pathogenesis to be plausible it must be able not only to account for the observed pathological consequences of viroid infection, but also for the fact that, as stated above, in certain plant species viroids are replicated efficiently without detectable damage to the host (19).

POSSIBLE ORIGIN

At the time when the viroid concept was advanced (14), viroids could reasonably be regarded as relatives of conventional viruses, being either very primitive or else degenerate representatives of the latter. Knowledge accumulated since then has rendered this concept increasingly less likely. The apparent lack of messenger function of viroids (or their complements) and their novel molecular structure—that has no counterpart among viruses—imply a far greater phylogenetic distance from viruses than could be imagined previously.

With the discovery in eukaryotic organisms of split genes and RNA splicing (reviewed in 10), it has been suggested (10, 18) that viroids might have originated by circularization of spliced-out intervening sequences (introns). One might speculate that if such excised sequences would permit extensive intramolecular base pairing (as do viroids) and if they would become circularized (as are viroids), they might become stabilized and thus escape degradation. Circularization of introns has been observed recently (3, 29, 39), including some with the approximate size of viroids (3). It is conceivable that if such introns would comprise appropriate recognition sequences, they might be transcribed by a host enzyme capable of functioning as an RNA-directed RNA polymerase and thus escape the control mechanisms of the host cell.

Small nuclear RNAs (snRNA's) associated with ribonucleoprotein particles are believed to be involved in the processing of the primary transcription products of split genes (45, 55, 67). The 5' end of one such RNA, U1, has been shown to exhibit complementarity with the ends of introns (1, 45, 67), and it is believed this affords a mechanism that ensures correct excision of the intron sequences and accurate joining of the coding sequences.

Although the primary structures of higher plant snRNA's are unknown, the recent demonstration of a split gene in a higher plant species (78) and the similarities of its intron-exon boundary sequences with those of other eukaryotes suggest that an snRNA homologous to U1 RNA exists in higher plants and that its 5'-end sequence resembles that of the latter. If so, the intron theory of viroid origin predicts that a specific nucleotide sequence on viroids exhibits complementarity to the 5' end of this putative plant snRNA, as well as to that of U1 RNA.

In view of these similarities, it was of interest to determine whether or not the nucleotide sequence of PSTV contains stretches of significant complementarity with the 5' end of U1 RNA, but a search for such sequences failed to reveal the possibility of stable complexes between the two RNAs.

However, because PSTV appears to be transcribed from an RNA template, the complementary strand, and not the viroid itself, might represent a stabilized intron and exhibit complementarity with U1 RNA.

Figure 2 shows that a complex of considerable stability is indeed possible between the 5' end of U1 RNA and nucleotides 257 to 279 of the PSTV complement (22). The hypothetical splice junction is located between nucleotides 262 and 263. The proposed intron deviates from a suggested consensus sequence (45) by ending with GG instead of AG. At least one mammalian intron, however, is known to end with GG and not with AG (45).

Although the striking complementarity possible between the 5' end of U1 RNA and the PSTV complement may be a fortuitous coincidence, the high stability of the complex, as compared with those of genuine splicing sites, rather tends to support a functional role of this nucleotide sequence of the PSTV complement.

The model suggests a mechanism of PSTV pathogenesis. Because PSTV contains a sequence homologous to the 5' end of U1 RNA, the viroid may interfere with the splicing process mediated by the latter's plant equivalent, possibly triggering incorrect excision of introns and thereby perturbing normal RNA processing (22).

QUESTION OF ANIMAL VIROIDS

Although viroids are definitely known to occur only in higher plants, similar agents may exist in other forms of life. It appears reasonable to search for viroids in the many instances in which viral etiology of an infectious disease has been assumed, but in which no causative agent has been identified.

One case in point is a group of animal and human diseases, the subacute spongiform encephalopathies (27). On the basis of comparisons of known properties of PSTV with those of the agent of one of these, scrapie, the hypothesis has been advanced that the latter may be a viroid (16). Efforts to isolate infectious nucleic acid from brain preparations of scrapie-infected animals, however, were fruitless (51, 82), and claims of a low-molecular-weight DNA component essential for the expression of scrapie infectivity (49, 50) have not been confirmed (63). On the other hand, convincing evidence for the presence in the scrapie agent of a hydrophobic protein essential for the expression of infectivity has been reported (64). The same investigators were unable to demonstrate a requirement for nucleic acid; and thus it appears that the viroid model does not apply to the agents of the subacute spongiform encephalopathies.

Warren and colleagues have reported the isolation of an infectious agent from synovial tissues and fluids of rheumatoid arthritis patients and the consistent induction by this agent of a transmissible acute and chronic polyarthritis in mice and rats (85) and of a crooked-toe syndrome in chicks

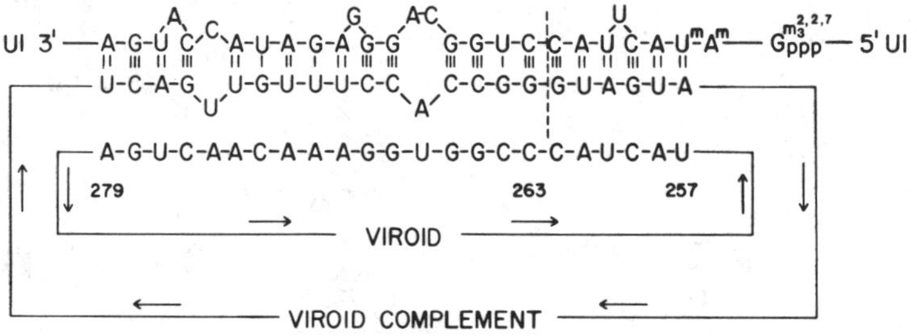

Figure 2 Possible base-pairing interactions between the PSTV complement and the 5' end of U1 RNA. ----, Hypothetical splice junction. (From 22.)

(83). Study of the physical-chemical properties of this putative agent led to the conclusion that it was a free RNA akin to a viroid (84). Efforts to confirm these reports showed that a factor present in synovial fluids from some rheumatoid arthritis patients, when injected into embryonated chicken eggs, indeed caused the crooked-toe syndrome, but that this factor was not a replicating agent but a heat-stable, dialyzable, polar lipid that could not be serially transmitted (25).

Finally, an unusual filterable oncogenic agent has been isolated from horizontally transmitted Syrian hamster lymphomas, and on the basis of its physical-chemical properties, this agent was considered to represent a mammalian viroid (7). This claim, however, appears somewhat premature in view of the fact that because of technical difficulties, properties of the putative viroid have not been extensively elucidated. Data indicating low molecular weight of the DNA are particularly fragmentary.

No convincing evidence exists at present for the existence of viroid-incited diseases in life forms other than higher plants. Evidently, only future work will determine whether viroids indeed are a peculiar feature of plants or whether, contrary to present indications, similar entities exist in other kinds of organisms.

Literature Cited

1. Avvedimento, V. E., Vogeli, G., Yamada, Y., Maizel, J. V., Pastan, I., de Crombrugghe, B. 1980. *Cell* 21:689–96
2. Benson, A. P., Raymer, W. B., Smith, W., Jones, E., Munro, J. 1965. *Potato Handb.* 10:32–36
3. Borst, P., Grivell, L. A. 1981. *Nature* 289:439–40
4. Branch, A. D., Dickson, E. 1980. *Virology* 104:10–26
5. Branch, A. D., Robertson, H. D., Dickson, E. 1981. *Proc. Natl. Acad. Sci. USA* 78:6381–85
6. Brown, F., Martin, S. J. 1965. *Nature* 208:861–63
7. Coggin, J. H. Jr., Oakes, J. E., Huebner, R. J., Gilden, R. 1981. *Nature* 290:336–38

7a. Conejero, V., Picazo, I., Segado, P. 1979. *Virology* 97:454–56

8. Conejero, V., Semancik, J. S. 1977. *Virology* 77:221–32
9. Cress, D. E., Owens, R. A. 1981. *Recombinant DNA* 1:90
10. Crick, F. 1979. *Science* 204:264–71
11. Davies, J. W., Kaesberg, P., Diener, T. O. 1974. *Virology* 61:281–86
12. Dickson, E., Diener, T. O., Robertson, H. D. 1978. *Proc. Natl. Acad. Sci. USA* 75:951–54
13. Diener, T. O. 1971. *Virology* 43:75–89
14. Diener, T. O. 1971. *Virology* 45:411–28
15. Diener, T. O. 1972. *Virology* 50:606–9
16. Diener, T. O. 1972. *Nature New Biol.* 235:218–19
17. Diener, T. O. 1974. *Ann. Rev. Microbiol.* 28:23–39
18. Diener, T. O. 1979. *Science* 205:859–66
19. Diener, T. O. 1979. *Viroids and Viroid Diseases.* New York: Interscience. 252 pp.
20. Diener, T. O. 1980. In *Plant Disease Etiology, Abstr. Meet. Fed. Br. Plant Pathol./Soc. Gen. Microbiol., London, Dec. 1980.*
21. Diener, T. O. 1981. *Ann. Rev. Plant Physiol.* 32:313–25
22. Diener, T. O. 1981. *Proc. Natl. Acad. Sci. USA* 78:5014–15
23. Diener, T. O., Lawson, R. H. 1973. *Virology* 51:94–101
24. Diener, T. O., Smith, D. R. 1975. *Virology* 63:421–27
25. Diener, T. O., Smith, D. R. 1982. *J. Rheumatol.* In press
26. Domdey, H., Jank, P., Sänger, H. L., Gross, H. J. 1978. *Nucleic Acids Res.* 5:1221–36
27. Gajdusek, D. C. 1977. *Science* 197: 943–60
28. Galindo, J. A., Smith, D. R., Diener, T. O. 1982. *Phytopathology.* 72:49–54

29. Grabowski, P. J., Zaug, A. J., Cech, T. R. 1981. *Cell* 23:467–76
30. Grill, L. K., Negruk, V. I., Semancik, J. S. 1980. *Virology* 107:24–33
31. Grill, L. K., Semancik, J. S. 1978. *Proc. Natl. Acad. Sci. USA* 75:896–900
32. Grill, L. K., Semancik, J. S. 1980. *Nature* 283:399–400
33. Gross, H. J., Domdey, H., Lossow, C., Jank, P., Raba, M., Alberty, H., Sänger, H. L. 1978. *Nature* 273:203–8
34. Gross, H. J., Liebl, U., Alberty, H., Krupp, G., Domdey, H., Ramm, K., Sänger, H. L. 1981. *Biosci. Rep.* 1: 235–41
35. Gross, H. J., Riesner, D. 1980. *Angew. Chem.* 19:231–32
36. Hadidi, A., Cress, D. E., Diener, T. O. 1981. *Proc. Natl. Acad. Sci. USA* 78: 6932–35
37. Hadidi, A., Hashimoto, J. 1981. *Phytopathology* 71:222 (Abstr.)
38. Hadidi, A., Jones, D. M., Gillespie, D. H., Wong-Staal, F., Diener, T. O. 1976. *Proc. Natl. Acad. Sci. USA* 73:2453–57
39. Halbreich, A., Pajot, P., Foucher, M., Grandchamp, C., Slonimski, P. 1980. *Cell* 19:321–29
40. Hall, T. C., Wepprich, R. K., Davies, J. W., Weathers, L. G., Semancik, J. S. 1974. *Virology* 61:486–92
41. Haseloff, J., Symons, R. H. 1981. *Nucleic Acids Res.* 9:2741–52
42. Henco, K., Riesner, D., Sänger, H. L. 1977. *Nucleic Acids Res.* 4:177–94
43. Kleinschmidt, A. K., Klotz, G., Seliger, H. 1981. *Ann. Rev. Biophys. Bioeng.* 10:115–32
44. Langowski, J., Henco, K., Riesner, D., Sänger, H. L. 1978. *Nucleic Acids Res.* 5:1589–610
45. Lerner, M. R., Boyle, J. A., Mount, S. M., Wolin, S. L., Steitz, J. A. 1980. *Nature* 283:220–24
46. Lewin, R. 1981. *Science* 213:634–36
47. Luria, S. E., Darnell, J. E., Baltimore, D., Campbell, A. 1978. *General Virology.* New York: Wiley. 578 pp. 3rd ed.
48. Lwoff, A. 1981. *Ann. Virol.* 132E(2)-:121–34
49. Malone, T. G., Marsh, R. F., Hanson, R. P., Semancik, J. S. 1979. *Nature* 278:575–76
50. Marsh, R. F., Malone, T. G., Semancik, J. S., Lancaster, W. D., Hanson, R. P. 1978. *Nature* 275:146–47
51. Marsh, R. F., Semancik, J. S., Medappa, K. C., Hanson, R. P., Rueckert, R. R. 1974. *J. Virol.* 13:993–96
52. Matthews, R. E. F. 1978. *Nature* 276:850
53. McClements, W. L., Kaesberg, P. 1977. *Virology* 76:477–84
54. Mühlbach, H.-P., Sänger, H. L. 1979. *Nature* 278:185–88
55. Murray, V., Holliday, R. 1979. *FEBS Lett.* 106:5–7
56. Niblett, C. L., Dickson, E., Fernow, K. H., Horst, R. K., Zaitlin, M. 1978. *Virology* 91:198–203
57. Owens, R. A., Cress, D. E. 1980. *Proc. Natl. Acad. Sci. USA* 77:5302–6
58. Owens, R. A., Diener, T. O. 1981. *Science* 213:670–72
59. Owens, R. A., Diener, T. O. 1982. *Proc. Natl. Acad. Sci. USA* 79:113–17
60. Owens, R. A., Smith, D. R., Diener, T. O. 1978. *Virology* 89:388–94
61. Palukaitis, P. *Molecular biology of viroids.* PhD thesis. Univ. Adelaide, Australia. 138 pp.
62. Palukaitis, P., Hatta, T., Alexander, D. McE., Symons, R. H. 1979. *Virology* 99:145–51
63. Prusiner, S. B., Groth, D. F., Bildstein, C., Masiarz, F. R., McKinley, M. P., Cochran, S. P. 1980. *Proc. Natl. Acad. Sci. USA* 77:2984–88
64. Prusiner, S. B., McKinley, M. P., Groth, D. F., Bowman, K. A., Mock, N. I., Cochran, S. P., Masiarz, F. R. 1981. *Proc. Natl. Acad. Sci. USA* 78:6675–79
65. Rackwitz, H. R., Rohde, W., Sänger, H. L. 1981. *Nature* 291:297–301
66. Randles, J. W. 1975. *Phytopathology* 65:163–67
67. Rogers, J., Wall, R. 1980. *Proc. Natl. Acad. Sci. USA* 77:1877–79
68. Rohde, W., Sänger, H. L. 1981. *Biosci. Rep.* 1:327–36
69. Romaine, C. P., Horst, R. K. 1975. *Virology* 64:86–95
70. Sänger, H. L. 1972. *Adv. Biosci.* 8: 103–16
71. Sänger, H. L., Klotz, G., Riesner, D., Gross, H. J., Kleinschmidt, A. K. 1976. *Proc. Natl. Acad. Sci. USA* 73:3852–56
72. Sasaki, M., Shikata, E. 1977. *Proc. Jpn. Acad. Ser. B.* 53:109–12
73. Semancik, J. S., Conejero, V., Gerhart, J. 1977. *Virology* 80:218–21
74. Semancik, J. S., Geelen, J. L. M. C. 1975. *Nature* 256:753–56
75. Semancik, J. S., Tsuruda, D., Zaner, L., Geelen, J. L. M. C., Weathers, L. G. 1976. *Virology* 69:669–76
76. Semancik, J. S., Weathers, L. G. 1972. *Virology* 47:456–66
77. Sogo, J. M., Koller, T., Diener, T. O. 1973. *Virology* 55:70–80
78. Sun, S. M., Slightom, J. L., Hall, T. C. 1981. *Nature* 289:37–41

79. Takahashi, T., Diener, T. O. 1975. *Virology* 64:106–14
80. VanDorst, H. J. M., Peters, D. 1974. *Neth. J. Plant Pathol.* 80:85–96
81. Walter, B. 1982. *CR Acad. Sci. Paris* 292(III):537–42
82. Ward, R. L., Porter, D. D., Stevens, J. G. 1974. *J. Virol.* 14:1099–103
83. Warren, S. L., Marmor, L., Boak, R., Liebes, D. M. 1971. *Arch. Intern. Med.* 128:619–22
84. Warren, S. L., Marmor, L., Horner, H. E., Stanley, W. D., Stoeckel, T. A., Gerken, S. L., Patton, G. A. 1979. *J. Rheumatol.* 6:135–46
85. Warren, S. L., Marmor, L., Liebes, D. M., Rosenblatt, H. M. 1972. *Arch. Intern. Med.* 130:899–903
86. Zaitlin, M., Hariharasubramanian, V. 1972. *Virology* 47:296–305
87. Zaitlin, M., Niblett, C. L., Dickson, E., Goldberg, R. B. 1980. *Virology* 104:1–9
88. Zelcer, A., Van Adelsberg, J., Leonard, D. A., Zaitlin, M. 1981. *Virology* 109:314–22

Ann. Rev. Microbiol. 1982. 36:259–84

METABOLIC ACQUISITIONS THROUGH LABORATORY SELECTION

R. P. Mortlock

Cornell University, Ithaca, New York 14853

CONTENTS

006–4227/82/1001–0259$02.00

INTRODUCTION

Studies in the history of evolution have usually been based upon the fossile record. More recently, the examination and comparison of existing macromolecules has been made possible by the techniques of molecular biology. Certain laboratories, however, have attempted to employ microorganisms, with their high populations and rapid growth rates, to observe evolution occurring under laboratory conditions. Such experiments have usually consisted of placing microorganisms under conditions where they must develop new metabolic capabilities to grow. Cultures that are successful, in that mutants are selected that have acquired the ability to grow, can then be studied in an attempt to determine the biochemical and genetic changes that led to the establishment of the new metabolic capability. Often it is found that regulatory mutations have permitted enzymes of a different pathway to be utilized for a new function. A more detailed discussion of some of these experiments can be found in an excellent discussion by Clarke (16).

ACQUISITIONS INVOLVING OXIDOREDUCTASE ACTIVITY

Selection of Mutants of Klebsiella pneumoniae and Klebsiella aerogenes Which have Acquired the Ability to Utilize Xylitol, L-Arabitol, and 6-Deoxy L-Talitol as Growth Substrates

THE MUTATION ESTABLISHING GROWTH ON XYLITOL AND THE ORIGIN OF THE XYLITOL DEHYDROGENASE ACTIVITY The pathways of degradation of ribitol, xylitol, and L-arabitol by coliform bacteria are shown in Figure 1. In each case the pentitol is oxidized by a nicotinamide adenine dinucleotide-linked dehydrogenase to the 2-ketopentose. A kinase activity then converts the ketopentose to the 5-phosphate, and epimerase activities convert the ketopentose 5-phosphate into the substrates for the transketolase and transaldolase rearrangements (89).

These two enzymes, dehydrogenase and kinase, of the ribitol pathway are under negative control by the regulatory gene *rbt*B and the inducer of the operon is the intermediate in the pathway, D-ribulose (6). Growth on both xylitol and L-arabitol was found to result from a mutation in the regulatory gene, *rbt*B, which permitted constitutive synthesis of the enzymes of the ribitol pathway. Ribitol dehydrogenase (EC 1.1.1.56) was found to possess slight activity for the oxidation of xylitol to D-xylulose. However, wild-type cells could not utilize the enzyme for growth on xylitol, simply because xylitol did not cause induction of the enzyme. Regulatory mutants constitutive for the enzyme could utilize the dehydrogenase activity to oxidize xylitol and were able to grow with xylitol as the substrate (52, 54, 55). Since

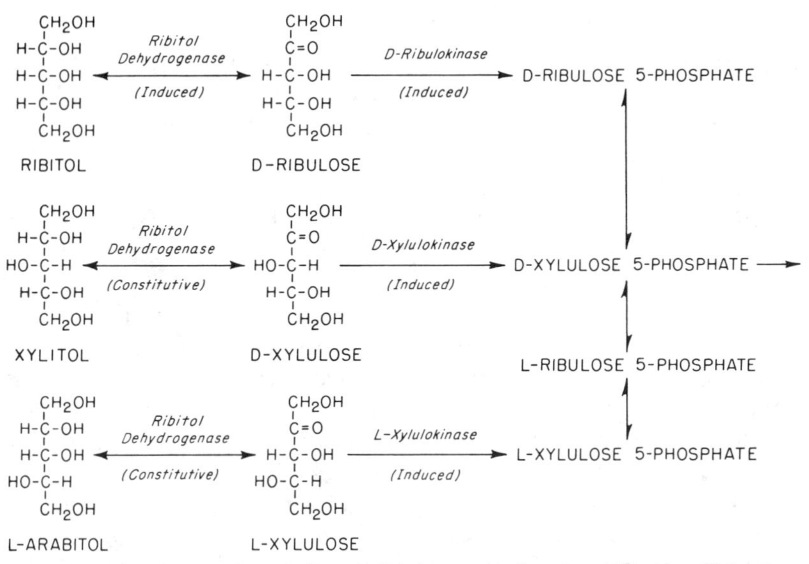

Figure 1 Pathways of catabolism of ribitol, D-arabitol, and L-arabitol by *Klebsiella.*

the ribitol operon was under negative control, any mutation that resulted in the loss of activity of the *rbt*B regulatory gene resulted in constitutive synthesis of the enzymes of the operon and growth on xylitol (15, 54).

MUTATIONS INCREASING THE GROWTH RATE WITH XYLITOL AS THE CARBON AND ENERGY SOURCE Lin and his co-workers (52, 92) obtained a mutant of *Klebsiella aerogenes* 1033 that had gained the ability to utilize xylitol as a growth substrate and was constitutive for the synthesis of ribitol dehydrogenase. Cells were treated with a mutagenic agent to increase the frequency of mutation and then were cultured again on xylitol, which resulted in the selection of a mutant that had a doubling time on xylitol medium reduced to only 2 hr (52, 92). This new mutant produced a ribitol dehydrogenase with improved activity for xylitol, resulting from mutations in the dehydrogenase structural gene (12). The activity of the mutant enzyme for ribitol had not significantly changed, but the binding affinity for xylitol had more than doubled, as had the maximum velocity with xylitol as a substrate.

Continued selection by this latter method led to the isolation of another new mutant, which had improved ability for the uptake of xylitol from the medium and which was constitutive for the synthesis of the enzymes of the D-arabitol catabolic pathway.

The transport system for D-arabitol was employed for the transport of xylitol into the cell, and constitutive synthesis of the transport system resulted in improved growth on xylitol, especially under conditions where the xylitol concentration was low. In this latter mutant all of the activities of the new xylitol pathway were under constitutive synthesis (92), with the interesting situation that the structural genes involved in the pathway were in different but adjacent operons (14, 15).

Mortlock & Wood (58) reported that cultivation of a xylitol-positive mutant of *Klebsiella pneumoniae* PRL-R3 in a chemotstat with xylitol as the limiting carbon source resulted in the selection of strains that produced very high levels of an apparently unaltered enzyme, and they suggested that the dehydrogenase structural gene had been duplicated. Hartley and his co-workers conducted more extensive chemostat experiments and concluded that the increased levels of ribitol dehydrogenase they observed were due to multiple copies of the dehydrogenase structural gene (39, 66). The first mutants isolated from the chemostat contained approximately a fivefold increase in the level of dehydrogenase activity and were relatively stable. Continued selection in the chemostat led to the isolation of mutants containing higher activities for ribitol dehydrogenase, some elevated 15-fold over the wild type, and were the result of duplications in the dehydrogenase structural gene. Later experiments by Hartley and his co-workers yielded mutants with an altered enzyme that had increased activity for xylitol as a substrate (12, 67, 80).

Inderlied & Mortlock reported chemostat experiments with *K. aerogenes* W-70 and xylitol as a substrate, resulting in duplication of the ribitol dehydrogenase structural gene (41). They isolated mutants that were hyperproducing ribitol dehydrogenase, but they could not detect greatly elevated levels of D-arabitol dehydrogenase (EC 1.1.1.11), even though the structural gene for this enzyme was located on the adjacent dal operon. They could not find highly elevated levels of the enzyme coded by the other structural gene known to be located on the *rbt* operon, D-ribulokinase (EC 2.7.1.47), which indicates that only the structural gene for ribitol dehydrogenase had been duplicated in that mutant. Neuberger & Hartley (60) tested some of their ribitol dehydrogenase hyperproducing strains for *rbt* duplications by using cloned *rbt* DNA as a hybridization probe. For most strains the duplication included both the ribitol dehydrogenase and D-ribulokinase structural genes with elevated levels of both of these enzymes. For one strain, however, both the ribitol dehydrogenase and the adjacent D-arabitol dehydrogenase structural genes appeared to have been duplicated with elevated amounts of both of these enzymes, whereas the D-ribulokinase activity was still similar to that of the parent strain.

Thompson (81) has conducted chemostat experiments with *K. pneumoniae* and used low levels of ribitol and higher levels of xylitol as mixed substrates. In three experiments the chemostat selected regulatory mutants with constitutive synthesis of ribitol dehydrogenase. In a fourth experiment the chemostat selected a mutant with no apparent change in regulation, but with an altered ribitol dehydrogenase, showing improved activity for xylitol as a substrate. In the latter case the continued presence of low levels of ribitol in the medium apparently maintained the induction of the enzyme.

ACQUISITION OF THE ABILITY OF K. PNEUMONIAE PRL-R3 TO UTILIZE L-ARABITOL AS A GROWTH SUBSTRATE After the observation that ribitol dehydrogenase was capable of catalyzing the oxidation of L-arabitol as well as xylitol (56, 57), L-arabitol-positive mutants of *Klebsiella pneumoniae* PRL-R3 were examined and most were found to be constitutive for ribitol dehydrogenase. The enzyme was shown to have activity for the oxidation of L-arabitol to L-xylulose and mutants constitutive for the ribitol operon but which had never been previously exposed to L-arabitol were capable of growth on L-arabitol without further mutation. Apparently the oxidation of L-arabitol to L-xylulose resulted in the induction of all the other enzymes necessary for the complete catabolism of L -xylulose.

ACQUISITION OF THE ABILITY OF K. AEROGENES TO UTILIZE 6-DEOXY L-TALITOL AS A GROWTH SUBSTRATE *Klebsiella aerogenes* strain W-70 possesses an inducible enzyme pathway for the catabolism of L-fucose (Figure 2). Experiments have indicated that the inducer of the pathway is the intermediate L-fuculose l-phosphate (68).

A mutant was constructed that was blocked in the catabolism of L-fucose, but which was constitutive for ribitol dehydrogenase. When this mutant was incubated with L-fucose under reducing conditions it excreted the product of fuculose reduction into the medium. This product was isolated and identified as 6-deoxy-L-talitol (71). Since ribitol dehydrogenase was responsible for the synthesis of 6-deoxy L-talitol, it could be predicted that it would also catalyze the reverse reaction, the oxidation of 6-deoxy-L-talitol to L-fuculose, and this was shown to be the case. However, the wild-type strain could not grow with 6-deoxy-L-talitol as a substrate, because the metabolism of this compound did not lead to induction of ribitol dehydrogenase. Therefore, selection for growth on 6-deoxy-L-talitol led to the isolation of mutants constitutive for synthesis of the dehydrogenase (71).

Methods by Which Escherichia coli K-12 Can Acquire the Ability to Utilize Xylitol and D-Arabitol as Growth Substrates

TRANSDUCTION OF THE RIBITOL OPERON FROM E. COLI C OR K. AEROGENES Scangos & Reiner found some strains of *Escherichia coli* C that could use both ribitol and D-arabitol (74). The ribitol operon could be

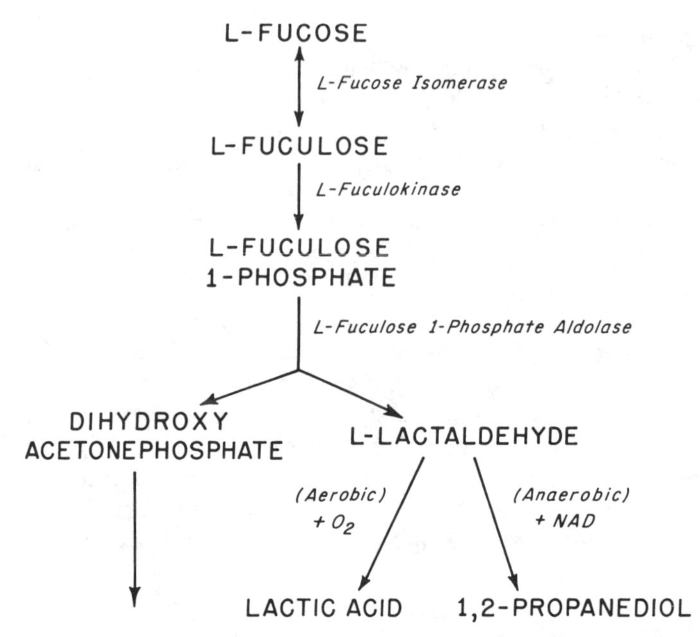

Figure 2 Enzymes involved in the metabolism of L-fucose by *E. coli* K-12.

transduced into *E. coli* K-12 so that the recipient strain could utilize ribitol and mutants could be selected from it that would grow on xylitol (65). The mutants that conferred constitutive synthesis of the ribitol operon and growth on xylitol were located in the regulator gene of the ribitol operon (74).

Rigby et al (67) were successful in transducing the pentitol genes from *Klebsiella aerogenes* to *E. coli* K-12. If the donor cells had been constitutive for the synthesis of ribitol dehydrogenase and had been able to grow on xylitol, the recipient cells showed the same phenotype.

UTILIZATION OF PROPANEDIOL DEHYDROGENASE FOR XYLITOL OXIDATION Wu (90) was able to construct a strain on *E. coli* K-12 that could grow on xylitol by using a dehydrogenase other than ribitol dehydrogenase. He had previously obtained a mutant of *E. coli* K-12 that had gained the ability to grow aerobically by using 1,2-propanediol as a substrate and he synthesized a propanediol dehydrogenase activity (EC 1.-1.1.55) constitutively. After it was observed that this dehydrogenase also possessed activity for the oxidation of xylitol to D-xylulose, Wu introduced a second mutation in the cells, which resulted in the constitutive synthesis of the enzymes of the D-xylose catabolic pathway (90). The constitutive D-xylose transport system permitted the entrance of xylitol into the cells, the propanediol dehydrogenase catalyzed the oxidation of xylitol to D-xylulose, and the D-xylulokinase (EC 2.7.1.17) of the D-xylose pathway phosphorylated the ketopentose to yield D-xylulose 5-phosphate. This strain, resulting from at least two regulatory mutations, could now grow slowly with xylitol as the sole source of carbon and energy with a doubling time of about 10 hr.

EXPERIMENTS TO IMPROVE THE GROWTH OF E. COLI K-12 ON XYLITOL Chemostat experiments using *E. coli* K-12 strains that had received the pentitol genes by transduction resulted in the isolation of mutants with improved activity for xylitol as a substrate, as well as mutants that were hyperproducing the wild-type dehydrogenase (39, 59, 66, 67). Neuberger & Hartley (58) constructed specialized transducing phage that carried the genes of the *rbt* operon, and they were able to show that when *E. coli* K-12 strains lysogenic for this phage were selected in chemostat experiments for faster growth on xylitol, ribitol dehydrogenase was hyperproduced because of the presence of multiple prophages within the cells.

UTILIZATION OF THE PROPANEDIOL TRANSPORT SYSTEM FOR D-ARABITOL TRANSPORT Wu was able to construct a mutant of *E. coli* K-12 that could grow by using D-arabitol as the growth substrate (91). The parent strain was a mutant that was constitutive for the synthesis of 1,2-propanediol dehydrogenase activity and was also constitutive for a transport system to bring propandiol into the cells. This same transport system was capable of causing the transport of D-arabitol into the cells. Selection on D-arabitol led to the isolation of cells able to use D-arabitol as a sole carbon and energy source. This new D-arabitol-positive mutant synthesized a dehydrogenase capable of catalyzing the oxidation of D-arabitol to D-xylulose. The structural gene for this new dehydrogenase was located at 68.5 min on the *E. coli* chromosome and possessed activity for D-galactose oxidation as well as for D-arabitol oxidation.

Acquisition of the Ability to Utilize Propanediol as an Aerobic Growth Substrate by E. coli K-12

Methyl pentoses such as L-fucose and L-rhamnose are degraded by many coliform bacteria by a route involving the isomerization of the methyl pentose phosphorylation to yield the keto sugar l-phosphate. In each case the phosphorylated intermediate is cleaved by an aldolase to dihydroxyacetone-phosphate and L-lactaldehyde (Figure 2). Under aerobic conditions the lactaldehyde is oxidized to lactic acid and then is further metabolized. Under anaerobic conditions, however, the sugars must be fermented and a lactaldehyde oxidoreductase (EC 1.1.1.55) must be synthesized to catalyze the reduction of lactaldehyde to 1,2-propanediol as the fermentation product.

E. coli is normally not able to grow with 1,2-propanediol as a growth substrate, but in 1969, Sridhara et al (75, 76) reported the isolation of a mutant capable of aerobic growth on this new substrate. Investigation showed that a mutation in the regulation of the lactaldehyde oxidoreductase, which was normally synthesized only under anaerobic conditions as a fermentation enzyme, permitted it to be synthesized constitutively even in the presence of oxygen. Since the enzyme was present aerobically, it could then be used as a propanediol dehydrogenase to catalyze the oxidation of propanediol to lactaldehyde (75, 76). Continued selection on propanediol led to increased growth rates on the substrate, but it also led to the complete loss of the ability to grow on L-fucose. These new mutants were found to possess constitutive L-fucose l-phosphate aldolase activity but were unable to synthesize the permease, isomerase, or kinase of the L-fucose catabolic pathway (17, 22–24).

Selected Modification in the Alcohol Dehydrogenase of S. cerevisiae

Yeast make several forms of alcohol dehydrogenase (EC 1.1.1.1). Located in the cytoplasm of the cell is an enzyme, ADH-l, formed under aerobic conditions and used for the oxidation of ethanol to acetaldehyde (85, 86). Another form of the enzyme (ADH-II) is synthesized constitutively and is used for ethanol formation during fermentation. Alcohol dehydrogenase will also catalyze the oxidation of allyl alcohol to acrolein. Although allyl alcohol is not toxic, acrolein is highly toxic and its formation will result in the death of the cells. Selection in the presence of allyl alcohol has been used for the isolation of mutants that have lost their alcohol dehydrogenase activity, since only such mutants are able to grow in the presence of allyl alcohol.

Mutant strains lacking cytoplasmic alcohol dehydrogenase are dependent upon the mitochondria to provide oxidized nicotinamide adenine dinucleotide (NAD) for the catabolism of sugars to continue. Petite strains (mitochondrial deficient) must retain alcohol dehydrogenase activity to survive. The only way a petite strain could grow in the presence of allyl alcohol would be to continue to reduce acetaldehyde to ethanol while preventing the oxidation of allyl alcohol to acrolein.

Wills & Phelps (88) isolated mutants of a petite strain of *Saccharomyces cerevisiae* that had become resistant to 120 mM concentrations of allyl alcohol. In contrast to expectations, kinetic analysis of alcohol dehydrogenase from these mutants did not show strong changes in the activity on allyl alcohol as a substrate. Studies (86) showed that it was the binding constants for both NAD and NADH (reduced NAD) that had been altered as a result of the mutation. The change in the kinetic constants for NAD and NADH resulted in an increase in the NADH-to-NAD ratio and thus a corresponding decrease in the acrolein to allyl alcohol ratio. The decrease in the production of acrolein permitted growth on the cells in the presence of allyl alcohol concentrations up to 1.6 mM, whereas the wild-type strains were killed by concentrations over 1 mM.

Wills & Jornvall (87) examined two of the mutant yeast alcohol dehydrogenases that had become resistant to allyl alcohol. For mutant strain S-AA-5 the change in structure of the enzyme was found to be a substitution of an arginine for histidine at position 44, whereas for mutant strain C-40 the change was a proline at position 316 becoming an arginine. For this latter mutant the K_m values for ethanol and acetaldehyde had been increased five- and two-fold, respectively. Many of the other mutant alcohol dehydrogenases isolated appeared to have amino acid substitutions at different positions in the enzyme.

ACQUISITIONS INVOLVING TRANSFERASE ACTIVITY

Ornithine Carbamoyltransferase and Aspartate Carbamoyltransferase in E. coli K-12

The reactions seen in Figure 3 are important biosynthetic reactions in the synthesis of arginine and pyrimidines by bacteria. The reaction involving the condensation of ornithine and carbamoylphosphate to form citrulline can be reversed by some bacteria and can be used for the degradation of arginine as a substrate (77). *Escherichia coli* K-12 is able to synthesize two separate enzymes, each a functional ornithinecarbamoyltransferase (EC 2.1.3.3.), to catalyze the biosynthetic reaction leading to citrulline. These two enzymes, produced from the structural genes *arg*F and *arg*I, are very similar in their molecular weights and kinetic properties, leading to the suggestion that they both originated from a common ancestral gene (49).

The biosynthetic intermediate carbamoylphosphate also serves as a precursor for the synthesis of pyrimidines such as uracil. Legrain et al (50, 51) investigated the possibility that citrulline and the reverse of the ornithine transcarbamoylase reaction might generate carbamoylphosphate, which could then serve for the synthesis of uracil. They started with a mutant that had lost activity for carbamoylphosphate synthetase and could not make that compound by the normal biosynthetic route. Since the carbamoylphosphate was not available to this mutant strain, both arginine and uracil had to be supplied to permit growth to occur.

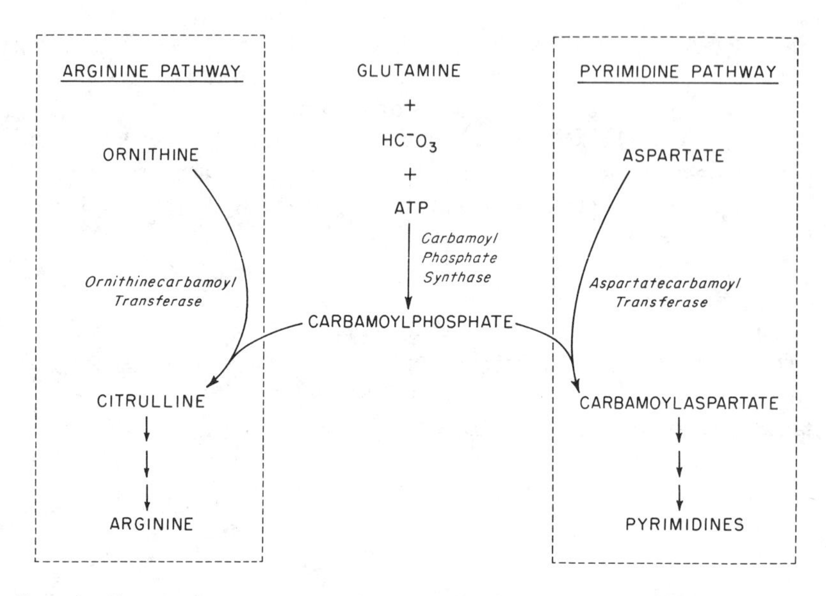

Figure 3 The role of carbamoylphosphate in the biosynthetic reactions leading to the synthesis of arginine and pyrimidines by *E. coli* K-12.

Cells of this strain were plated on a medium containing citrulline but not uracil as a growth factor. Under such circumstances arginine could be synthesized from the citrulline, but the only possible source of carbamoylphosphate for the synthesis of uracil would be from the breakdown of citrulline. Mutants were selected that had gained the ability to grow and synthesize uracil, and they were found to possess very high levels of ornithinetranscarbamoylase, some three to four times higher than those found in an *arg*R genetically derepressed mutant. Only the synthesis of the enzyme from the *arg*F genes was effected, and the *arg*I gene remained under repression. The authors suggested that the high levels of enzyme resulted from duplication of the *arg*F gene and that the extra copies of the gene that were being expressed were not under the normal repression control (50, 51).

The same workers tried a similar experiment, but in this case they began with a strain with a deletion in the *arg*F gene to determine if the *arg*I gene could be employed for the biosynthetic production of carbamoylphosphate from citrulline (51). Two types of mutants were isolated. The first type possessed high activity for ornithinetranscarbamoylase and the activity was only subject to weak repression by citrulline. The second type of mutant showed high activity for the enzyme, but the activity was repressed by the presence of arginine and uracil (51). In similar experiments the authors first selected a mutant capable of using carbamoylaspartate as a substrate for the synthesis of pyrimidines. By using this strain as the parent, they were next able to select a mutant that could use carbamoylaspartate as a source of carbamoylphosphate for the synthesis of citrulline and arginine (50, 51).

ACQUISITIONS INVOLVING HYDROLASE ACTIVITY

The Acquisition of the Ability to Utilize New Amides as Carbon or Nitrogen Sources for the Growth of Pseudomonas aeruginosa

The amidase (acylamide aminohydrolase, EC 3.5.1.4) produced by a strain of *Pseudomonas aeruginosa* and coded by structural gene *ami*E catalyzes the hydrolysis of ammonia from certain amides (16). Acetamide and propioniamide were substrates for the enzyme and both of these amides also served as good inducers (5). Mutations resulting in the constitutive synthesis of the enzyme took place in the regulator gene (*ami*R), which seemed to regulate the synthesis of the enzyme by a positive control mechanism. The enzyme was subject to catabolite repression and intermediates of the tricarboxylic acid cycle served as especially good catabolite repressors. In addition, certain inducer analogues also prevented induction of the enzyme and were even able to inhibit synthesis of the enzyme by some of constitutive mutants. Synthesis of the enzyme was repressed by butyramide, formamide, and other amides (8, 9), and this type of effect was termed amide analogue repression. Mutants resistant to this type of repression also mapped in the regulator gene. Mutants could also be found resistant to catabolyte repression by tricarboxylic acid cycle intermediates, such as succinate. Some of these mutants seemed to map close to the structural gene for the enzyme, in what might have been the promotor region.

Formamide showed very low activity as a substrate for the enzyme and was even a poorer inducer. In 1967, Brammar et al used a medium containing succinate as the carbon and energy source but formamide as the only nitrogen source to isolate mutants with improved efficiency for the hydrolysis of formamide to formic acid and ammonia (7). The mutants isolated showed a variety of phenotypes, with some constitutive for the synthesis of the enzyme and some that gained the ability to respond to formamide as an inducer of the enzyme. Since succinate served as a catabolite repressor of the synthesis of the enzymes, additional mutations were sometimes obtained that overcome this repression. In addition, formamide was an amide analogue repressor and additional regulatory mutations were sometimes isolated that overcame this particular type of repression of enzyme synthesis.

The wild-type organism was unable to use butyramide as a growth substrate, even though the amidase was able to hydrolyze butyramide at 2%

the rate of hydrolysis of acetamide. Brown et al (8) were able to isolate mutants that had gained the ability to use butyramide as a growth substrate. They used as parent strain for this experiment a spontaneous mutant that was constitutive for high levels of amidase production, but which was still repressed by succinate (catabolite repression) and butyramide (amide analogue repression). Studies with a butyramide-positive strain B_6 showed that it produced an altered enzyme, amidase B, that had improved activity for butyramide as a substrate. The rate of hydrolysis of butyramide by the wild-type or A-type of enzyme was only 2% of the rate of hydrolysis of acetamide, but the rate of hydrolysis of butyramide by the B enzyme was about 30% of the rate of acetamide hydrolysis (8). Later studies showed that the mutation resulting in the modified enzyme involved the substitution of a phenylalanine for serine at position seven from the N terminus of the enzyme. Since this mutation was a structural gene change and since the strain already possessed a mutation in the regulatory gene that permitted the constitutive synthesis of the enzyme, the genotype for this new strain was *ami*R_{11}, and *ami*E_{16}.

The evolutionary pathway resulting in the utilization of butyramide seemed to lead in a predictable manner through a regulatory mutation that permitted the synthesis of the enzyme in the absence of a normal substrate, followed by a structural gene mutation that improved the activity of the enzyme for the new substrate. This analysis was confused by the observation that some of the formamide-positive mutants were able to grow with butyramide as the substrate even though they continued to make the wild-type A amidase.

Analysis showed that the low activity of the wild-type enzyme for butyramide as a substrate was really sufficient to permit growth of the organism on butyrate, providing the cells were capable of synthesizing enough of the enzyme. Since butyramide was capable of causing amide analogue repression even in the constitutive mutant, those cells could not synthesize enough of the enzyme to permit growth with the wild-type activity for butyrate, and a mutation improving the activity of the enzyme for butyramide as a substrate was isolated.

Some of the other formamide-positive mutants isolated, in addition to becoming constitutive for synthesis of the enzyme, had also gained resistance to the formamide and butyramide repression of enzyme synthesis. With such alterations in regulation these mutants were able to synthesize large amounts of the enzyme in the presence of butyramide and to grow without requiring an improvement in the activity of the enzyme toward this substrate (9). Starting with the B_6 mutant as the parent strain, Brown et al isolated mutants capable of growth on the five-carbon amide, valeramide. These mutants proved to be a heterogenous group displaying different phenotypes (8). Examination of some the abilities of the valermide-positive strains for growth on acetamide gave variable results, with some of the valeramide-positive strains retaining the ability to grow on actamide and some losing that ability. Apparently there were different mutations within the structural gene for the amidase that could permit the utilization of valermide as a substrate.

Betz & Clark (5) reported the isolation of mutants that had gained the ability to use phenylacetamide as the substrate. Some 18 different mutants were isolated from various lines of decent. The parent and intermediate strains are shown in Figure 4. All of these 18 mutants that had gained the ability to use phenylacetamide as a growth substrate were still constitutive for enzyme synthesis, but none had retained the ability to grow on succinate-formamide agar. The 12 mutants isolated from the butyramide-positive strain were designated as the PhB group and they were similar in their growth abilities. None had retained the ability to use acetamide, but all could use as substrates butyramide, valeramide, hexonamide, oxtanomide, and phenylacetamide (5, 64). Three of the mutant types of phenylacetamide-using enzyme studied (PhV_1, PhB_3, and PhF_1), had low K_m values for phenylacetamide, ranging from 1-3 mM. The original constitutive *ami*E_{11} mutation was conserved through all the additional mutations.

In 1972, Brown & Clark isolated mutants that had gained the ability to use acetanilide as growth substrate (10). They started with a strain that had the genotype $amiR_{33}$, $amiE^+$, crp^{-7}, which was constitutive to amidase synthesis and resistant to amide analogue repression, and it was also insensitive to catabolite repression. Although acetanilide was a repressor of enzyme synthesis for the wild-type and gave the same effect for a constitutive mutant, it had no effect for repression of synthesis of the enzyme by the amide analogue-resistant strain PAC 142 (10). Analysis of the peptide fragments showed this modified enzyme to possess the substitution of an isoleucine residue for a threonine residue.

Tuberville & Clarke (82) made a genetic cross between two of their mutants, producing a new strain that produced a butyramide-utilizing amidase, but since it still had this wild-type regulator gene, only acetamide or propionamide would induce the amidase. Butyramide would not induce the amidase to a sufficient extent to permit growth of the cells with butyramide as the substrate. The authors subjected this new strain to mutagenic treatment and then plated cells with butyramide as the only carbon and energy source. Of over 500 mutants that had gained the ability to grow with butyramide as the substrate, all but one proved to have mutated to constitutive synthesis of the enzyme. This latter strain was inducible for the butyramide-type enzyme, with butyramide now acting as an inducer and increasing the synthesis of this enzyme 10-fold. Since the enzyme was not

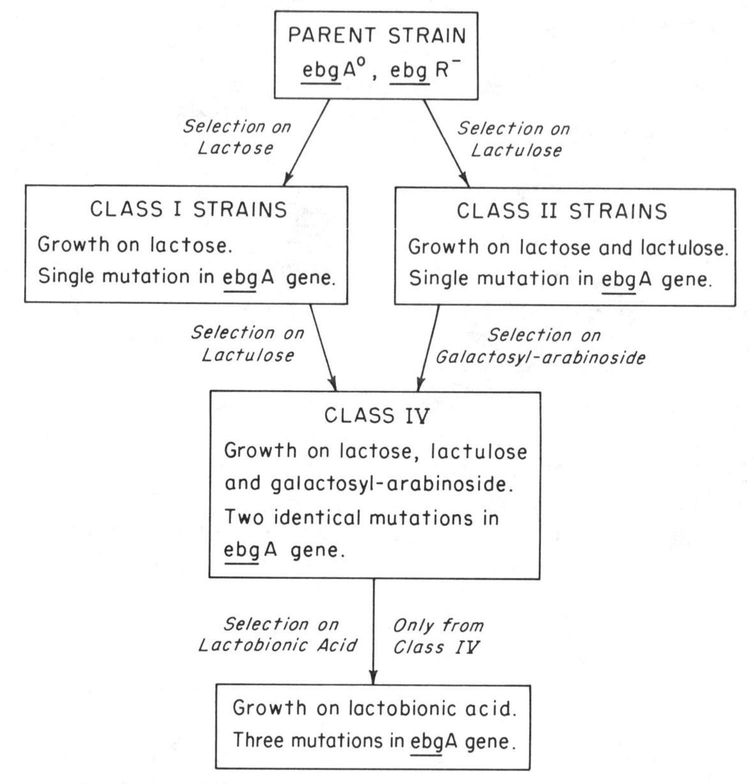

Figure 4 Selection of mutants leading to growth on lactobionic acid.

constitutive, the cells could not grow on formamide-succinate plates, but all could grow on acetamide or butyamide. The authors suggested that this type of mutation was rare because there were many different mutational events that could result in a constitutive phenotype, but perhaps only a few that could specifically alter the inducer specificity of the regulator gene.

Modifications in Acid Phosphatase Activity

In 1972, Francis & Hansche deliberately obtained modifications in the acid phosphatase of *Saccharomyces cerevisiae* (EC 3.1.3.2), an enzyme located in the yeast cell wall (19). In this procedure they used a chemostat with β-glycerophosphate as the only source of phosphate so that the organism was dependent upon phosphatase activity to catalyze the hydrolysis of inorganic phosphate. They set the pH of the chemostat at 6.0, higher than the pH for optimal activity of the acid phosphatase, to limit the activity of the enzyme so that the growth rate of the cells was dependent upon the activity of the phosphatase itself. The first change in growth rate resulted from a single mutational event termed M1, which caused a 35% increase in the efficiency for the metabolism of phosphate by the organism. A second mutation, M2, resulted in a shift in the pH optimum of the enzyme itself so that the enzyme was more efficient at the higher pH of 6.0. In further chemostat experiments the same authors obtained a mutant with increased growth rate (20), which was designated as mutant strain M4. Genetic analysis indicated that this strain resulted from a single mutational event in the region of the acid phosphatase structural gene. There was only a small shift in the K_m of the enzyme for its substrate, but the pH optimum had shifted from 4.2 to 5.2.

Uridine-5'-monophosphate is also a substrate for acid phosphatase and is hydrolyzed by the enzyme at half the rate of β-glycerophosphate. Another chemostat was employed by these authors with the M4 strain and the hydrolysis of uridine-5'-monophosphate as the only source of inorganic phosphate. Again, the pH was buffered at 6.0.

A strain was isolated from this chemostat where the in vivo activity of acid phosphatase was four times higher on uridine-5'-monophosphate than it had been in the parent strain. The activity for the hydrolysis of phosphate from β-glycerophosphate was increased four times over the M4 strain and eight times over the wild-type strain used at the beginning of these experiments.

About half of this increase in activity was due to two separate mutations, *wal* and *als,* which affected the cell wall structure and increased the amount of enzyme associated with the cell wall. An additional mutation observed in this strain was an apparently stable duplication of the previously evolved ACP-2 mutant acid phosphatase gene of strain M4, resulting in the synthesis of increased amounts of the enzyme (37, 38).

The ebg Operon of E. coli K-12; a New Operon for Lactose Utilization

At 67 min on the *Escherichia coli* chromosome, almost directly opposite from the *lac* operon, the *ebg* operon is located. This operon contains a gene, *ebg*A, that codes for an enzyme that can mutate to replace a missing β-galactosidase activity.

One way in which microorganisms can be utilized in studies of evolution is to eliminate a gene function and then challenge the organism to replace that function by employing another gene. In 1972, Warren obtained a mutant of *E. coli* K-12 that contained a deletion in the *lac*Z gene and was unable to grow on lactose because of the inability to synthesize β-galactosidase (83). Since this mutation was a deletion, true revertants would not be expected to occur. However, Warren was able to obtain cells that had regained the ability to grow on lactose because of a new mutation not linked to the *lac* operon. The new mutants synthesized a low level of a constitutive β-galactosidase activity. In 1973, Campbell et al (13) obtained a strain of *E. coli* that had a deletion that extended into the *lac* operon. By using this *lac*Z deletion strain as a parent strain, these workers were able to isolate mutants that could synthesize a new β-galactosidase and that had regained the ability to grow with lactose as the substrate. The new β-galactosidase differed in its physical properties from the original enzyme and mapped at a different location on the chromosome. The authors designated this gene *ebg,* or evolved β-galactosidase (4). Kinetic analysis showed that the new *ebg* β-galactosidase bound galactose more tightly than it bound lactose, and the authors suggested that the *ebg* enzyme may have evolved by modification of a gene normally involved with the metabolism of monosacchardide.

In 1974, Hall & Hartl reported similar findings with lactose-positive revertants of a *lac*Z deletion strain of *E. coli* (33). Selection was done on lactose as the growth substrate in the presence of isopropyl-l-thio-β-D-galactopyranoside, which would maintain continued induction of the lactose transport protein and permit lactose to enter the cells. The new enzyme activity for the hydrolysis of lactose was similar to the *ebg* enzyme described previously (13). The unmutated parent *ebg* gene was found to code for a protein that had galactosidase activity against O-Nitrophenyl-p-galactopyranoside, but virtually no activity against lactose. A single mutation within this gene changed it to *ebg*A^{+a} and permitted it to produce an enzyme with activity for lactose (25, 26, 34). The *ebg*A gene was under the negative control of a regulator gene termed *ebg*R (27, 32). They showed that for the growth on lactose to occur another mutation, increasing the amount of enzyme produced, was required in addition to the mutation increasing the activity of the enzyme for lactose. This mutation in the regulator gene was necessary to permit the synthesis of a larger quantity of enzyme. Thus, mutants that had gained the ability to grow on lactose had acquired two separate mutations, one in the structural gene producing *ebg* A^{+a} and one in the regulator gene *ebg*R, which permitted the synthesis of larger quantities of the mutant enzyme. Since any mutation resulting in the inactivation of the repressor should result in the constitutive synthesis of the enzyme, this type of regulation mutation would be expected to be isolated most frequently, but this was not the case. About 90% of the lactose-positive mutants isolated still had the enzyme under regulation, but with lactose as an improved inducer. This latter modified regulation mutation was designated *ebg*R^{+U}.

The *ebg* operon was found to contain a second structural gene designated *ebg*B (35, 36), producing a protein coordinately controlled with the *ebg*A gene. This latter protein has been designated protein B, but its function is unknown.

Further studies showed that the wild-type *ebg*A° product possessed very slight activity for the hydrolysis of methyl-B-D-galactopyranoside (MG), and mutants constitutive for the synthesis of the enzyme (*ebg*R⁻) could grow very slowly with MG as a substrate. Starting with a constitutive strain, selection on medium with only MG as the potential substrate led to the isolation of mutants with a faster growth rate and an improved activity of the enzyme for the new MG substrate. This illustrated how the *ebg* enzyme had different evolutionary routes it could follow depending upon which compounds were first presented to it as potential growth substrates (26).

A study of the various lactose-positive mutants showed that they could be placed into two different classes. The mutants in class I had gained, by a single point mutation in the *ebg*A structural gene, the ability to cleave lactose and to grow well with lactose as a substrate. The mutants placed in class II were the result of a point mutation at a separate location within the *ebg*A gene, although they were somewhat less efficient for the utilization of lactose; they had also gained the additional ability to use lactulose as a growth substrate. When the parent strain for those experiments was used for the selection of lactulose-positive mutants, the mutants isolated were identical to those previously placed in class II. Thus, class II mutants could be obtained by direct selection on lactulose and sometimes by selection on lactose itself (Figure 4).

When class I mutants were placed on lactulose to obtain mutants that had gained the ability to use this substrate through a second mutation, the strains obtained had not only acquired the ability to use lactulose, but the new mutation had also permitted the enzyme to use galactosyl-arabinoside as a substrate, something that the parent class I strain could barely accomplish. These mutants, capable of growth on lactose, lactulose, and galactosyl-arabinoside, were placed in a new class, class IV.

Surprisingly, when class II cells were placed on galactosyl-arabinoside to select for mutants that had directly acquired the ability to use this compound as a growth substrate, the mutants obtained were both phenotypically and genotypically identical to those previously placed into class IV. Although two different evolutionary routes had been used for the isolation of the mutants capable of using all three sugars, the genotypes of all class IV mutants were identical (28, 31, 36). Attempts to isolate mutants that could use lactobionic acid as a substrate were only successful when class IV mutants were used as the parent strain, indicating that three mutations in the structural gene were necessary to produce this phenotype.

During investigations on the regulation of the *ebg* operon, Hall (29) found that the *ebg*R gene would respond very slightly to lactulose as an inducer. With this knowledge he was able to select a mutant in the *ebg*R gene that permitted the genes repressor protein to recognize lactulose as a better inducer. These *ebg*R^{+L} mutants were found to be pleotrophic in their response in that they had increased response to lactose, lactulose, galactosyl-arabinoside, and even methyl-galactose as inducers (29), giving rise to the possibility of selecting for a range of individual mutants, each with increased response to one of these substrates.

Even mutants constitutive for synthesis of the improved *ebg*A enzyme could not grow with lactose as a substrate unless there existed some means for transporting lactose into the cell. All previous studies had been done with mutants either constitutive for the *lac*Y permease gene of the *lac* operon or that had this gene induced through the presence of a gratuitous inducer such as isopropyl-1-thio-β-D-galactopyranoside. The true inducer of the *lac* operon is allolactose, normally produced from lactose by a side reaction of the β-galactosidase of the *lac*Z gene. Since all of the strains used in the selection of the *ebg* mutants had been deleted for the *lac*Z gene, no allolactose was produced and the *lac* operon could not be naturally induced.

Starting with a constitutive strain (*ebg*R^{52}), which produced an improved enzyme (*ebg*A^{53}), a mutant was selected that grew more rapidly on lactose. This additional mutation within the *ebg*A gene had permitted it to convert some of the lactose into an inducer of the *lac*Y gene, resulting in more efficient transport of lactose into the cells (29).

In additional experiments, Hall (30) constructed merodiploid strains carrying both wild-type and evolved *ebg* genes; one strain possessed both wild-type and mutant *ebg*A genes and another strain possessed both wild-type and mutant *ebg*R genes. The strain possessing the *ebg*A mutation was still able to grow on lactose, since this gene was expressed and the gene product was able to hydrolyze lactose. The strain with the wild-type and mutant *ebg*R genes, however, was phenotypically lactose negative, since the operon was under negative control and the presence on the wild-type *ebg* R gene prevented the mutant *ebg*A gene from being expressed at a rate sufficient to permit growth.

ACQUISITIONS INVOLVING ISOMERASE ACTIVITY

Acquisition of the Ability of K. pneumoniae PRL-R3 to Use D-Arabinose as a Growth Substrate

The first reaction in the degradation of L-fucose by coliform bacteria normally is the isomerization of L-fucose to the 2-keto sugar (Figure 2). The synthesis of both isomerase (EC 5.3.1.4) and kinase (EC 2.7.1.51) of this pathway are under coordinate control by a regulator gene, *fuc*C.

Many *Klebsiella* species also possess an inducible pathway for the catabolism of the pentitol ribitol (Figure 1). The structural genes for the two enzymes of the ribitol catabolic pathway are located on an operon under the control of a regulator gene, *rbtB* (15). The inducer of the operon is the intermediate in the pathway of ribitol degradation, D-ribulose (6).

Mutants can be obtained from strains of *Klebsiella pneumoniae, Klebsiella aerogenes, Escherichia coli,* and *Salmonella typhimurium,* which have gained the ability to use D-arabinose as a growth substrate (14, 46, 47, 61, 69, 70). For each of the D-arabinose-positive mutants isolated, the initial reaction in the new D-arabinose pathway has been the isomerization of D-arabinose to D-ribulose, catalyzed by L-fucose isomerase (53). For mutants of the *K. pneumoniae* strain PRL-R3 studied, the mutation permitting growth on D-arabinose was found to be a mutation resulting in the constitutive synthesis of L-fucose isomerase. The constitutive isomerase catalyzed the isomerization of D-arabinose into D-ribulose, the natural inducer of the ribitol operon, resulting in the induction of D-ribulokinase, which catalyzed the second reaction for the degradation of L-fucose.

Continued cultivation on D-arabinose led to the selection and isolation of a mutant producing an altered isomerase that had more favorable kinetics for the isomerization of D-arabinose (62, 63), with an apparent decrease in the K_m for D-arabinose.

Acquisition of the Ability of E. coli K-12 to Utilize D-Arabinose as a Growth Substrate

Mutants of *E. coli* K-12 could be obtained that had gained the ability to grow on D-arabinose, and L-fucose isomerase was responsible for the isomerization of D-arabinose to D-ribulose in such mutants (46–48). However, *E. coli* K-12 did not possess a pathway for the degradation of ribitol. Sudies led to the realization that in these D-arabinose-positive mutants all three of the L-fucose pathway enzymes were being employed for the degradation of D-arabinose. L-Fucose isomerase catalyzed the isomerization of D-arabinose to D-ribulose and L-fuculokinase phosphorylated the ribulose at the 1-carbon position. The aldolase of the L-fucose pathway (EC 4.1.2.17) then catalyzed the cleavage of the D-ribulose 1-phosphate to dihydroxyacetone-phosphate and glycolaldehyde (47).

A further difference in the *E. coli* mutants was that the synthesis of the enzymes of the L-fucose pathway was not constitutive. Instead, incubation of the mutant cells with D-arabinose led to the induction of all three of these enzymes. Similar data were obtained for mutants of *S. typhimurium* that had gained the ability to grow on D-arabinose (61).

Acquisition of the Ability of K. aerogenes W-70 to Utilize D-Arabinose as a Growth Substrate

Both types of regulatory mutation were observed when D-arabinose-positive mutants of *K. aerogenes* W-70 were isolated and examined. The data for two such mutants, a constitutive strain fucC^6 and an inducible mutant

Table 1 L-Fucose isomerase activities of D-arabinose-positive mutants

Strain	Inducer	L-Fucose isomerase activity (μM/min per mg protein)	Strain phenotype
Parent	None[a]	< 0.07[b]	Cannot grow on D-Arabinose; isomerase activity not induced by D-Arabinose
	L-Fucose	0.60	
	D-Arabinose	< 0.07	
*fuc*C^6	None	0.488	Growth on D-Arabinose; isomerase constitutive
	D-Arabinose	0.355	
*fuc*C^{111}	None	< 0.003	Growth on D-Arabinose; presence of D-Arabinose results in isomerase induction
	D-Arabinose	0.595	

[a]Cells were grown on casein hydrolysate in the absence of any added carbohydrate

[b]Isomerase activity not detectable but less than the indicated value. Data from 69, 70.

fucC^{111}, are shown in Table 1. The inducer of the L-fucose pathway is the intermediate in the pathway L-fuculose-1-phosphate, and for the mutants that could grow on D-arabinose by an inducible phenotype, the mutant cells were able to recognize D-ribulose-1-phosphate as an alternate inducer. Thus, for strain fucC^{111}, both L-fuculose-1-phosphate and D-ribulose-1-phosphate could induce the L-fucose pathway enzymes (Figure 5). When this mutant strain was presented with D-arabinose as a potential growth substrate the following events took place: (*a*) basal, uninduced levels of 1-fucose isomerase and L-fuculokinase converted D-arabinose to D-ribulose-1-phosphate; (*b*) the latter induced higher levels of both the isomerase and kinase as well as the aldolase; (*c*) the aldolase cleaved some of the ribulose-1-phosphate to dihydroxyacetonephosphate and glycolaldehyde; (*d*) the formation of D-ribulose as an intermediate caused the induction of the ribitol operon; and (*e*) D-ribulokinase phosphorylated D-ribulose to the 5-phosphate that was epimerized and degraded by the enzymes of the cell.

Since D-ribulose was the natural substrate of D-ribulokinase most of the ribulose was to be phosphorylated to the 5-phosphate. If too little of the

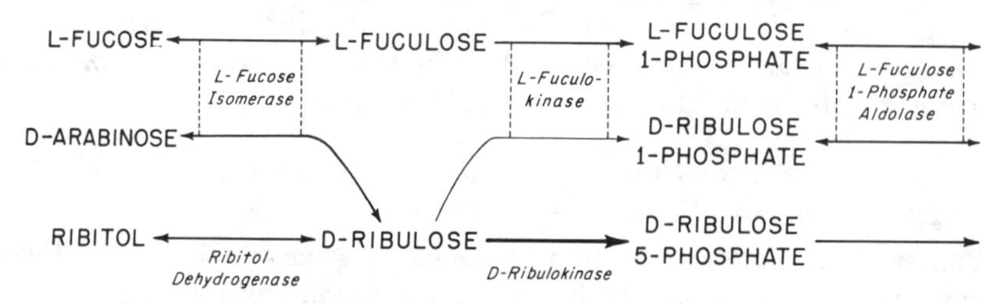

Figure 5 Pathway of D-arabinose utilization by *K. aerogenes* W-70 mutant *fuc* C^{111}.

1-phosphate was formed, however, the cells would be unable to maintain the induction of the isomerase. In that case the internal pool of D-ribulose would have to decrease, also decreasing the induced rate of synthesis of the enzymes of the ribitol operon and lowering the level of D-ribulokinase. In this case more D-ribulose would be available for conversion to the 1-phosphate (70, 71).

ACQUISITION OF THE ABILITY OF K. PNEUMONIAE PRL-R3 TO UTILIZE L-XYLOSE AS A GROWTH SUBSTRATE

Anderson & Wood (2, 3) reported that L-xylose was catabolized by *Klebsiella pneumoniae* strain PRL-R3 by a pathway involving isomerization to L-xylulose, and L-xylose-positive mutants were found to be constitutive for synthesis of high levels of L-fucose isomerase, which was used by the mutants to catalyze this reaction (14). The wild-type cells might also lack an efficient mechanism for the transport of L-xylose into the cells. Under such conditions the L-fucose isomerase constitutive mutants that could have been present in the initial inoculum would only grow exceedingly slowly. However, once the population of these mutants reached a sufficient level, there could be selection for a second mutation that permitted the more efficient transport of L-xylose into the cells, with growth then complete within a few days (56, 57).

Acquisition of the Ability of K. pneumoniae PRL-R3 to Utilize D-Lyxose as a Growth Substrate

Anderson & Allison (1) studied mutants of *Klebsiella pneumoniae* PRL-R3 that had acquired the ability to use D-lyxose as a growth substrate. These mutant cells were found to possess an isomerase activity capable of converting D-lyxose to D-xylulose, which could be further degraded by enzymes of either the D-xylose or D-arabitol inducible catabolic pathways. The new isomerase activity for D-lyxose was not present in the wild-type strain and could not be induced in the wild type by incubation of the cells with D-lyxose. The purified enzyme showed activity for D-mannose as a substrate in addition to D-lyxose, but the activity on mannose was actually poorer than the activity on D-lyxose, suggesting that mannose was not the natural substrate of the enzyme (1).

Acquisition of the Ability of E. coli K-12 to Use D-Lyxose as a Growth Substrate

Stevens & Wu (79) isolated mutants of *Escherichia coli* strain K-12 that were capable of growth on D-lyxose as a substrate. These new D-lyxose-positive mutants possessed a constitutive isomerase activity that could convert D-lyxose to D-xylulose. This enzyme was not found in the parent strain and could not be induced in the parent strain by the incubation of cells with either D-lyxose or D-mannose. Continued selection on D-lyxose led to the isolation of an additional mutation that produced twice as high a concentration of constitutive isomerase as the previous strain (78). The enzyme isomerized D-mannose as well as D-lyxose. The K_m of the enzyme for D-lyxose as a substrate was 300 mM, whereas the K_m for D-mannose was 80 mM and the ratio of D-lyxose to D-mannose activity was about 0.25. The isomerase mapped at about 85 min on *E. coli* chromosome, distant to the location of any other known isomerase genes (79).

Apparently basal levels of D-xylose isomerase (EC 5.3.1.5) converted some of the D-xylulose formed by the isomerization of D-lyxose to D-xylose, resulting in the induction of the D-xylose pathway enzymes. The D-xylose permease could then permit entrance of D-lyxose into the cells, and following isomerization to D-xylulose, the kinase of the D-xylose pathway could catalyze the phosphorylation to form D-xylulose-5-phosphate (40).

Stevens & Wu (79) suggested that the origin of the enzyme was from an isomerase that was not functional under normal conditions. They have speculated that a regulatory mutation permitted the synthesis of an enzyme from a structural gene left from a previous pathway for D-mannose catabolism, a pathway that had evolved before the presence of the phosphotransferase system for the transport of the D-mannose into the cells.

Selection for a New Isomerase for the Leucine Biosynthetic Pathway

The second enzyme specific for the synthesis of leucine in the leucine biosynthetic pathway in *Salmonella typhimurium* is *a*-isopropylmalate isomerase. This enzyme is composed of two polypeptides, each coded by separate genes, the *leu*C and the *leu*D genes (44). With mutants deleted for the *leu*D gene, revertants could be obtained that were no longer auxotrophic for leucine. This supressor mutation, *sup*Q, mapped at a distance from the *leu* operon near the genes for *pro*A and *pro*B. Mutations in the *sup*Q gene made available to the cells a product of another gene, the *new* D gene, which could replace the missing *leu*D gene product. Insertions, point mutations, and even deletions in the *sup*Q gene made the product of the *new*D gene available to combine with the product of the *leu*C gene, causing the formation of an active isomerase. Some of the *sup*Q deletions extended into the *pro*A gene and resulted in mutants now auxotrophic for proline (42).

Kemper (42, 43) has suggested that the normal product of the *sup*Q gene binds with the product of the *new*D gene. Mutation in the *sup*Q gene, especially deletion mutations, cause the release of the *new*D gene product and permit it to replace the missing product of the *leu*D gene in *leu*D$^-$ mutants (42, 43).

Fultz et al (21) used specialized transducing phage to integrate these genes into an F' episome. The F' factor carried the wild-type *new* gene, but it contained a *sup*Q with a deletion (*sup*Q394). It also carried the prolein A and B genes. When it was introduced into a *S. typhimurium* strain that possessed a deletion in the *leu*D gene (*leu*D798), required leucine for growth, and was also deficient for prolein biosynthesis, the recipient strain regained the ability to synthesize both prolein and leucine. This strain was a merodiploid in that it contained both the original genes, *sup*Q, *new*D and the *sup*Q394, *new*D genes introduced on the plasmid. The leucine-positive phenotype indicated that the *sup*Q394, *new*D$^+$ was dominant over the *sup* Q$^+$, *new*D$^+$ genotype. The introduction of an F' plasmid that carried the *sup*Q394, *new*D$^+$ genes into an *E. coli* strain that carried a *leu*D deletion resulted in a restoration of the cell's ability to synthesize leucine, showing that the *new*D gene could replace the *leu*D gene in *E. coli* as well as in *Salmonella* (45).

The wild-type α-isopropylmalate isomerase possessed a K_m for its substrate (isopropylmalate) of 3×10^{-4} M. The K_m values for either of the hybrid enzymes of the *leu*C-*new*D isomerase from either *Salmonella* or *Escherichia* for the same substrate were an order of 100-fold higher. This low activity of the new isomerase for its substrate appeared to be rate limiting. When a strain was constructed possessing two copies of the *new* D gene, the rate of synthesis of leucine no longer limited the growth rate of the cells (21). Such data leads to the speculation that continued culture of the strain possessing one copy of the *leu*C-*new*D gene in a chemostat where the growth of the cells was limited by the ability of the hybrid enzyme to synthesize leucine would lead either to the selection of a mutant of the *new*D gene with improved activity for the *leu*C-*new*D hybrid enzyme, or to selection for the duplication of the *leu*C-*new*D genes.

ACQUISITIONS INVOLVING SYNTHETASE ACTIVITY

Growth of E. coli on Fatty Acids

The first enzyme of the β-oxidation pathway for the degradation of fatty acids by *Escherichia coli* K-12 is acyl CoA synthetase (EC 6.2.1.3). It and the other enzymes of the β-oxidation pathway are regulated and synthesized when the cells are presented with an appropriate substrate such as oleate or palmitate. An *E. coli* strain constitutive for the synthesis of the enzymes of the glyoxylate bypass could grow with fatty acids of chain length of 3, 12, 14, 15, 16, and 18 carbons, but it could not grow with fatty acids of from 4 through 11 carbons (84).

The organism was unable to use decanoate or laurate as growth substrates even though it was found to have transport activity to bring decanoate into the cells (84). Weeks et al (84) were able to isolate mutants that had gained the ability to grow on decanoate and found such mutants to be constitutive for the synthesis of the enzymes of the β-oxidation pathway, suggesting that decanoate did not serve as an inducer of these enzymes and the regulatory mutation was required to establish growth.

Salanitro & Wegener (72) selected mutants of *E. coli* that would grow on decanoate and nonanoate. The nonanoate-positive mutants had also gained the ability to grow on all of the fatty acids, whereas the decanoate-positive mutants would grow on all with the exception of the 4-carbon and 5-carbon fatty acids, butyrate and valerate. The decanoate-positive strain was used as a parent to select for additional mutants permitting growth on either of these two fatty acids. A valerate-positive strain was obtained as well as a butyrate-positive strain. The valerate-positive strain would not grow on butyrate, but in contrast, the butyrate-positive strain had also acquired the ability to grow with valerate as a substrate. These latter mutations apparently involved the transport of the fatty acids through the cell membrane (72, 73).

SUMMARY

These studies represent preliminary investigations into the means by which bacteria present on the Earth today are able to modify their metabolic abilities to cope with minor changes in their environment. Since these experiments employ microorganisms that have already evolved precise control mechanisms for regulation of the synthesis and activity of their metabolic pathways, it should not be surprising that mutations in regulation are normally required to permit the utilization of an existing enzyme for a different metabolic pathway. If an organism is capable of using a present enzymatic activity to establish growth under new conditions, a mutation in the regulation of that enzyme may be required to permit the enzyme to be synthesized. Even if the enzymatic activity can only be established by mutation of a structural gene, a mutation in regulation permitting the altered enzyme to be expressed normally will also be required.

In some cases the origin of new enzymatic activity can be identified as being derived from a previously known activity for a common substrate, but in other cases the previous function of the gene that now expresses a new activity is really unknown, despite speculation. Future studies may reveal if these borrowed genes actually represent a useful but unknown enzymatic capability of the cell or if they had drifted into "silent genes" whose past function is no longer important but which could be evolved to satisfy new requirements of the organism.

ACKNOWLEDGMENT

I wish to thank B. G. Hall for providing me with information concerning some of his unpublished research investigations. My own research has been supported by Public Health Service Grant AI15328 from the National Institute of Allergy and Infectious Diseases.

Literature Cited

1. Anderson, R. L., Allison, D. P. 1965. *J. Biochem. Tokyo* 240:2367–72
2. Anderson, R. L., Wood, W. A. 1960. *J. Biol. Chem.* 237:269–303
3. Anderson, R. L., Wood, W. A. 1962. *J. Biol. Chem.* 237:1029–52
4. Arraj, J. A., Campbell, J. H. 1975. *J. Bacteriol.* 124:849–56
5. Betz, J. L., Clarke, P. H. 1972. *J. Gen. Microbiol.* 73:161–74
6. Bisson, T. M., Oliver, E. J., Mortlock, R. P. 1968. *J. Bacteriol.* 95:932–36
7. Brammer, W. J., Clarke, P. H., Skinner, A. J. 1967. *J. Gen. Microbiol.* 47:87–102
8. Brown, J. E., Brown, P. R., Clarke, P. H. 1969. *J. Gen. Microbiol.* 57:273–95
9. Brown, J. E., Clarke, P. H. 1970. *J. Gen. Microbiol.* 64:329–42
10. Brown, P. R., Clarke, P. H. 1972. *J. Gen. Microbiol.* 70:287–98
11. Burleigh, B. D., Rigby, P. W. J., Harley, B. S. 1974. *Biochem. J.* 143:341–52
12. Campbell, J. H., Lengyel, J. A., Langridge, J. 1973. *Proc. Natl. Acad. Sci. USA* 70:1841–45
13. Camre, K. P., Mortlock, R. P. 1965. *J. Bacteriol.* 90:1157–58
14. Charnetzky, W. T., Mortlock, R. P. 1974. *J. Bacteriol.* 119:162–69
15. Charnetzky, W. T., Mortlock, R. P. 1974. *J. Bacteriol.* 119:176–82
16. Clarke, P. H. 1978. *The Bacteria,* pp. 137–218. New York: Academic. 603 pp.
17. Cocks, G. T., Aguilar, J., Lin, E. C. C. 1974. *J. Bacteriol.* 118:83–88
18. Fossitt, D. D., Mortlock, R. P., Anderson, R. L., Wood, W. A. 1964. *J. Biol. Chem.* 239:2110–15
19. Francis, J. C., Hansche, P. E. 1972. *Genetics* 70:59–73
20. Francis, J. C., Hansche, P. E. 1973. *Genetics* 74:259–65
21. Fultz, P. N., Kwoh, D. Y., Kemper, J. 1979. *J. Bacteriol.* 137:1253–62
22. Hacking, A. J., Aguilar, J., Lin, E. C. C. 1978. *J. Bacteriol.* 136:522–30
23. Hacking, A. J., Lin, E. C. C. 1976. *J. Bacteriol.* 126:1166–72
24. Hacking, A. J., Lin, E. C. C. 1977. *J. Bacteriol.* 130:832–38
25. Hall, B. G. 1976. *J. Bacteriol.* 126:536–38
26. Hall, B. G. 1976. *J. Mol. Biol.* 107:71–84
27. Hall, B. G. 1977. *J. Bacteriol.* 129:540–43
28. Hall, B. G. 1978. *Genetics* 89:453–65
29. Hall, B. G. 1978. *Genetics* 90:673–81
30. Hall, B. G. 1980. *Genetics* 96:1007–17
31. Hall, B. G. 1981. *Biochemistry* 20:4042–49
32. Hall, B. G., Clarke, N. D. 1977. *Genetics* 85:193–201
33. Hall, B. G., Hartl, D. L. 1974. *Genetics* 76:391–400
34. Hall, B. G., Hartl, D. L. 1975. *Genetics* 81:427–58
35. Hall, B. G., Zuzel, T. 1980. *J. Bacteriol.* 144:1208–11
36. Hall, B. G., Zuzel, T. 1980. *Proc. Natl. Acad. Sci. USA* 77:3529–33
37. Hansche, P. E. 1974. *Genetics* 79:661–74
38. Hansche, P. E., Berres, V., Lange, P. 1978. *Genetics* 88:673–87
39. Hartley, B. S., Barleigh, B. D., Midwinter, G. G., Moore, C. M., Morris, H. R., Rigby, P. W. J., Smith, M. J., Taylor, S. S. 1973. *Proc. FEBS Meet., 8th, 1972,* 29:151–76
40. Henderson, S. K., Wu, T. T. 1977. *Fed. Proc.* 36:827
41. Inderlied, C. B., Mortlock, R. P. 1977. *J. Mol. Evol.* 9:181–90
42. Kemper, J. 1974. *J. Bacteriol.* 119:937–51
43. Kemper, J. 1974. *J. Bacteriol.* 120:1176–85

44. Kemper, J., Margolin, P. 1969. *Genetics* 63:263–79
45. Kowh, D. Y., Kemper, J. 1978. *J. Virol.* 27:535–50
46. LeBlanc, D. J., Mortlock, R. P. 1971. *J. Bacteriol.* 106:82–89
47. LeBlanc, D. J., Mortlock, R. P. 1971. *J. Bacteriol.* 106:90–96
48. LeBlanc, D. J., Mortlock, R. P. 1972. *Arch. Biochem. Biophys.* 150:774–81
49. Legrain, C., Halleux, P., Stalon, V., Glansdorff, N. 1972. *Eur. J. Biochem.* 27:93–102
50. Legrain, C., Stalon, V., Glansdorff, N. 1976. *J. Bacteriol.* 128:35–38
51. Legrain, C., Stalon, V., Glansdorff, N., Gigot, D., Pierard, A., Crabeel, M. 1976. *J. Bacteriol.* 128:39–48
52. Lerner, S. A., Wu, T. T., Lin, E. C. C. 1964. *Science* 146:1313–15
53. Mortlock, R. P. 1976. *Adv. Microb. Phys.* 13:1–53
54. Mortlock, R. P., Fossitt, D. D., Petering, D. H., Wood, W. A. 1965. *J. Bacteriol.* 89:129–53
55. Mortlock, R. P., Fossitt, D. D., Wood, W. A. 1964. *Bacteriol. Proc.*, p. 95
56. Mortlock, R. P., Wood, W. A. 1964. *J. Bacteriol.* 88:835–44
57. Mortlock, R. P., Wood, W. A. 1964. *J. Bacteriol.* 88:845–49
58. Mortlock, R. P., Wood, W. A. 1971. *Biochemical Response to Environmental Stress*, pp. 1–14. New York: Plenum
59. Neuberger, M. S., Hartley, B. S. 1979. *J. Mol. Biol.* 132:435–70
60. Neuberger, M. S., Hartley, B. S. 1981. *J. Gen. Microbiol.* 122:181–91
61. Old, D. C., Mortlock, R. P. 1977. *J. Gen. Microbiol.* 101:341–44
62. Oliver, E. J., Mortlock, R. P. 1971. *J. Bacteriol.* 108:287–92
63. Oliver, E. J., Mortlock, R. P. 1971. *J. Bacteriol.* 108:293–99
64. Paterson, A., Clarke, P. H. 1979. *J. Gen. Microbiol.* 114:75–84
65. Reiner, A. M. 1975. *J. Bacteriol.* 123: 530–36
66. Rigby, P. W. J., Burleigh, B. D. Jr., Hartley, B. S. 1974. *Nature* 251:200–4
67. Rigby, P. W. J., Gething, M. J., Harley, B. S. 1976. *J. Bacteriol.* 125:728–38
68. Rolseth, S. J., Fried, V. A., Hall, B. G. 1980. *J. Bacteriol.* 142:1036–39
69. St. Martin, E. J., Mortlock, R. P. 1976. *J. Bacteriol.* 127:91–97
70. St. Martin, E. J., Mortlock, R. P. 1977. *J. Mol. Evol.* 10:111–22
71. St. Martin, E. J., Mortlock, R. P., 1980. *J. Bacteriol.* 141:1157–62
72. Salanitro, J. P., Wegener, W. S. 1971. *J. Bacteriol.* 108:885–92
73. Salanitro, J. P., Wegener, W. S. 1971. *J. Bacteriol.* 108:893–901
74. Scangos, G. A., Reiner, A. M. 1978. *J. Bacteriol.* 134:492–500
75. Sridhara, S., Wu, T. T. 1969. *J. Biol. Chem.* 244:5233–38
76. Sridhara, S., Wu, T. T., Chusd, T. M., Lin, E. C. C. 1969. *J. Bacteriol.* 98: 87–95
77. Stalon, V., Legrain, C., Wiame, J. M. 1977. *Eur. J. Biochem.* 74:319–27
78. Stevens, F. J., Stevens, P. W., Hovis, J. G., Wu, T. T. 1981. *J. Gen. Microbiol.* 124:219–23
79. Stevens, F. J., Wu, T. T. 1976. *J. Gen. Microbiol.* 97:257–65
80. Taylor, S. S., Rigby, P. W. J., Harley, B. S. 1974. *Biochem. J.* 141:693–700
81. Thompson, L. K. W. 1981. *An Investigation of acquisitive evolution using ribitol dehydrogenease in Klebsiella pneumoniae.* PhD thesis. Lehigh Univ., Penn. pp. 182
82. Tuberville, C., Clarke, P. H. 1981. *FEBS Lett.* 10:87–90
83. Warren, R. A. 1972. *Can. J. Microbiol.* 18:19–44
84. Weeks, G., Shapiro, M., Burns, R. O., Wahil, S. J. 1969. *J. Bacteriol.* 97: 827–36
85. Wills, C. 1976. *Fed. Proc.* 8:2098–1
86. Wills, C. 1976. *Nature* 261:26–29
87. Wills, C., Jornvall, H. 1979. *Eur. J. Biochem.* 99:323–31
88. Wills, C., Phelps, J. 1975. *Arch. Biochem. Biophys.* 167:627–37
89. Wood, W. A., McDonough, M. J., Jacobs, L. B. 1961. *J. Biol. Chem.* 236: 2190–95
90. Wu, T. T. 1976. *Biochim. Biophys. Acta* 428:656–63
91. Wu, T. T. 1976. *J. Gen. Microbiol.* 94: 246–56
92. Wu, T. T., Lin, E. C. C., Tanaka, S. 1968. *J. Bacteriol.* 96:447–56

Ann. Rev. Microbiol. 1982. 36:285–309

MICROBIAL ENVELOPE PROTEINS RELATED TO IRON[1]

J. B. Neilands

Biochemistry Department, University of California, Berkeley, California 94720

CONTENTS

INTRODUCTION AND SCOPE

This review concerns both prokaryotic and eukaryotic microbial species, the envelope proteins of which have been shown to be affected by iron. Only a relatively small proportion of these proteins has as yet been assigned a biofunction. Most of the information available has been garnered with studies on bacteria, especially enteric species such as *Escherichia coli* and *Salmonella typhimurium.* This is a consequence of the more complete understanding of the genetics of the enterics as a class, as contrasted to

[1]This review is dedicated to Professor A. E. Braunstein, Institute of Molecular Biology, Moscow, on the occasion of his 80th birthday, May 26, 1982.

0066–4227/82/1001–0285$02.00

other bacterial species or fungi. In some cases an iron-affected envelope component has been shown to serve as an attachment site for noxious agents, such as phages, antibiotics, or bacteriocins, and in a few instances the true biofunction has been assigned as a role in iron transport. A conspicuous exception is the colicin Ia receptor, which although strongly induced at low iron growth, has not yet been dignified with a role in transport of any nutritious substance.

The main purpose of this article is to survey briefly the general contours of the siderophore (Greek: iron bearer) iron assimilation system, to review what is known about siderophore receptors, and to emphasize what still remains unknown about these interesting proteins.

The generic term siderophore, although not yet exactly a household word in biochemistry, was first proposed by Lankford (54) to denote the relatively low-molecular-weight carriers of Fe(III) commonly encountered in aerobic and facultative anaerobic microbial species. Lankford's (54) detailed review, still a classic, must be consulted for the earlier work on bacterial iron assimilation. For fungi, the reader is referred to the report by Emery (25). A comprehensive tome on microbial iron metabolism (75) appeared in 1974, the year in which siderophore receptors were discovered (116); unfortunately, it is devoid of any mention of the main topic of this review. The siderophores themselves have been reviewed recently (77) and hence here I eschew discussion of the coordination characteristics (96) of these ligands and present their formulae without further comment. However, it should be noted that Eng-Wilmot & vander Helm (26) and Llinas (57) have made major contributions to the X-ray crystallography and solution conformation, respectively, in the siderophore series. A recent review deals with the general topic of microbial iron assimilation (78), the genetics (bacterial) of which have been recorded by Kadner & Bassford (47).

The designation siderophore supercedes earlier trivial names, such as siderochrome, for the entire class of virtually Fe(III)-specific ligands that is over-produced under low iron stress. By analogy with the ionophore, the label siderophore denotes the iron-free molecule, whereas the iron-laden form is referred to as the ferri-siderophore. Thus, the correct terminology is ferrichrome and deferriferrichrome; ferric ferrichrome is redundant. The chromosomally programmed siderophore common to many enteric bacteria, the cyclic trimer of 2,3-dihydroxybenzoylserine, was isolated in 1970 from *E. coli* and *S. typhimurium* and was named enterochelin (82) and enterobactin (93), respectively. I prefer the latter as a more suitable designation for a metabolic product derived from a major endosymbiont of man.

In regard to surface structures binding specific siderophores, in those cases where the biofunction is known it seems preferable to so name the receptor. Thus the particular protein of *E. coli* acting as the attachment site for phages T1, T5, and $\phi 80$ and for colicin M and albomycin is designated properly as the ferrichrome receptor. When referring to this receptor there is an unfortunate tendency to equate it with the 78,000-dalton protein visible on sodium dodecyl sulfate polyacrylamide gel (SDS-PAGE) analysis. However, this protein is more precisely described as a component of the ferrichrome receptor complex. This qualification applies to all of the receptors treated in this review. Even though the receptors for ferrichrome (62) and ferric enterobactin (41) have been shown to bind their respective ferrisiderophores in vitro, the participation in vivo of other essential components, such as lipopolysaccharide, cannot be ruled out. In iron-starved cells of *E. coli* K-12, an SDS-PAGE gel analysis of the outer membrane reveals the presence of at least three prominent bands with molecular weights corresponding to 83,000, 81,000, and 74,000. Following Braun et al (14) it has become common practice to designate these as 83K, 81K, and 74K, but because of the vagaries of electrophoresis on polyacrylamide gels, these numbers may depart substantially from the true molecular weights.

In the four decades since Waring & Werkman (114) performed their elegant experiments on the iron nutrition of bacteria, much progress has been made in understanding how microorganisms assimilate this mundane but biologically precious metal. The problem with iron is not one of abundance, since it ranks fourth among all elements on the surface of the earth, but of availability in aerobic environments. This is a consequence of the near quantitative insolubility of Fe(III) at biological pH (K_{sol} $Fe(OH)_3 \simeq 10^{-38}$ M) and of the tendency to form oxyhydroxide polymers of the composition FeOOH. This property of Fe(III) limits its solubility at neutral pH to ca 10^{-18} M, a value too low to support anything approaching maximum growth rates. It seems that following the advent of O_2-evolving photosynthesis, most aerobic and facultative anaerobic microorganisms have evolved (74) a high-affinity pathway for assimilation of Fe(III), which consists of the low-molecular-weight carriers, the siderophores, and the cognate membrane receptors.

The requirement for a receptor follows logically from the observation that water-soluble compounds exceeding ca 500–600 daltons cannot permeate the small, water-filled pores in the sealed outer membrane of Gram-negative bacteria (81). The actual identification of a siderophore receptor, first achieved for ferrichrome (117), owes much to prior work on vitamin B_{12} transport in *E. coli* and the demonstration, in this case, that the vitamin and colicin E share a common binding site (13). An essential prelude was the discovery of a series of *sid* mutants of *S. typhimurium* resistant to an antibiotic analogue of ferrichrome, albomycin (61). The mutants were shown incapable of ferrichrome transport and were mapped at the *pan* locus on the chromosome. The fact that the *pan* locus in *E. coli* coincides with a classic genetic lesion, *ton*A, was instrumental in generating the thought that the biochemical function of its gene product must be that of a receptor for ferrichrome (116, 117). Here we have a situation in which *E. coli,* even though it does not synthesize ferrichrome, maintains a receptor for a siderophore that is, nonetheless, the product of all *Penicillium* spp. and of many other fungal species. Once the *ton*A protein had been identified as a component of the ferrichrome receptor, it was logical to assume that receptors for other siderophores should exist and that these in many instances might also serve as binding sites for specific phages and bacteriocins. Thus, the protective effect of ferric enterobactin against colicin B, which was first shown by Guterman (35), could be ascribed to competition for a common binding site in the outer membrane.

The essential features of the high-affinity iron assimilation system are depected in Figure 1. At low levels of available iron in the growth medium, which generally means less than about micromolar, the biosynthetic machinery for both siderophore and receptor is derepressed. In *E. coli* the genes for ferrichrome transport are clustered at about 3 min on the chromosome; for ferric enterobactin the corresponding operon, which includes in this instance biosynthetic genes, occurs at around min 13. The recent work of Williams (120) has disclosed the existence of iron transport plasmids in clinical isolates of *E. coli.* These extra chromosomal elements program the synthesis of a siderophore, aerobactin, and its transport system (115). In addition, several additional genetic loci located outside of the chromosomal operon or plasmid are needed for ferric siderophore utilization.

Although as chemical entities the siderophores display considerable structural variation, most can be classed as either hydroxamates or catechols. The formation constant for Fe(III) lies in the range of $10^{20} - 10^{50}$ M (96), and as chelating agents they are virtually specific for Fe(III). Although these chemical and physical properties would suggest a role in iron metabolism, an ultimate proof would seem to have been furnished by the observation that *ent* mutants of *S. typhimurium* fail to grow in citrate medium in the absence of either excess iron or trace amounts of an appropriate siderophore (92).

The function of the receptors, the principal topic of this review, is to bring the ferri-siderophore to, or through, the envelope where the metal ion undergoes a reductive separation from the ligand. As depicted in Figure 1, this may or may not require processing of the complex. The released iron regulates the uptake system by some process or processes, the precise details of which are currently unknown. When one contemplates the fact that *E. coli* K-12 can synthesize and transport ferric enterobactin, can transport exogenous siderophores such as ferrichrome and ferric rhodotorulate, can induce a system for transport of ferric citrate, may harbor plasmids for synthesis and transport of ferric aerobactin, and, additionally, may assimi-

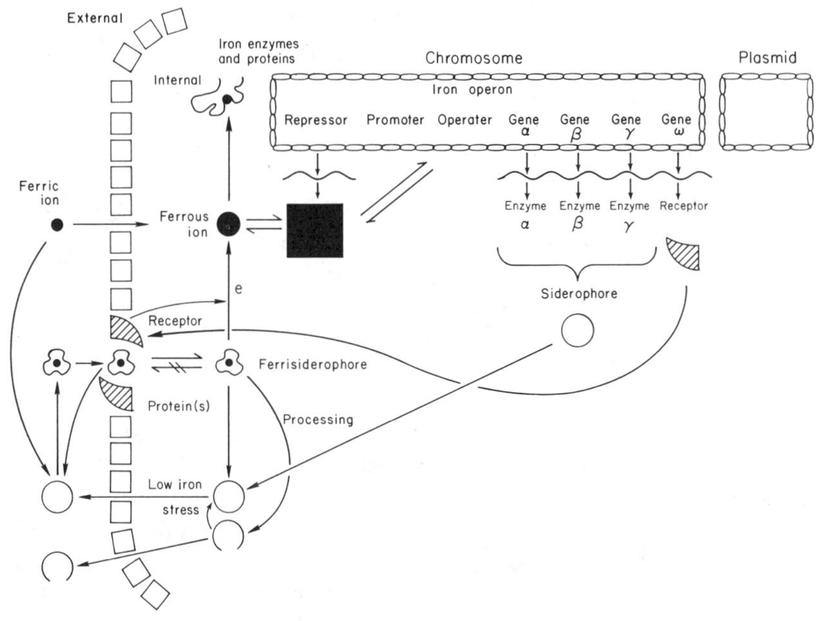

Figure 1 Schematic representation of low and high affinity iron assimilation in aerobic and facultative anaerobic microorganisms. In the former, environmental Fe(III) crosses the cell envelope not coordinated to specific ligands and without mediation by membrane-associated receptors. The high affinity pathway is comprised of Fe(III)-specific carriers (circles), termed siderophores, and the cognate receptors. The receptors for ferric enterobactin, a siderophore common to enteric bacteria, and ferrichrome, a typical fungal siderophore, have been prepared from the outer membrane of *Escherichia coli.* Four mechanisms of siderophore-mediated iron uptake are possible. Only the iron of the ferrisiderophore may be deposited in the envelope, or conversely, the intact complex may be incorporated. If the latter, the iron may be removed without processing of the ligand; conversely, the ligand may be either reversibly or irreversibly (broken circles) processed. Recent evidence indicates that plasmids, as well as the chromosome, are involved in specifying elements of the ferric aerobaction transport system in enteric bacteria (see the text). Although siderophore synthesis has been known for almost three decades to be controlled by iron, the molecular mechanism of regulation has not yet been established.

late iron via a relatively inefficient low-affinity pathway, the biochemist is confronted with an assimilatory system of stupefying complexity. Evidently, iron is an element that *E. coli* has determined must never be eliminated from its dietary regime.

THE VITAMIN B_{12} CONNECTION

It is instructive to compare the transport of siderophores with that of cobalamin, particularly since a colicin-shared receptor model had first been discovered for the latter nutrilite (13). *Escherichia coli* does not synthesize or normally require B_{12}, although it maintains both an uptake system for the vitamin and at least two enzyme proteins that use it as cofactor, a methyl tetra-hydrofolate-homocysteine methyl transferase and an ethanolamine ammonia lyase. Gene *btu*B codes for a 60,000-dalton outer membrane receptor that requires a divalent cation and lipopolysaccharide for maximum activity. The K_d for cyanocobalamin is 0.3×10^{-9} M. A few hundred molecules of receptor are normally present per cell, although this number can be reduced by an order of magnitude by growth in the presence of high levels of cobalamin. Modification of the *btu*B cistron via the *btu*A mutation impairs B_{12} transport without affecting sensitivity to the E colicins or phage BF23, which leads to the conclusion that different binding sites occur in the receptor. The product of the *btu*C gene is believed to be an inner membrane permease. In addition, a specific B_{12} binding protein is present in the periplasmic space, although mutants for this element have not yet been isolated. Finally, an intact *ton*B gene is required for B_{12} uptake, whereas this is not the case for successful attack by phage BF23 or the E group colicins.

The lambda receptor, the product of the *lam*B gene, has the biochemical function of transport of maltose and maltodextrins. It apparently forms aqueous channels in the outer membrane that display specificity for maltose transport (60). However, it has not been possible to measure any binding of maltose to the *lam*B protein. Nonetheless, the B_{12} system of uptake remains as the model that most closely parallels siderophore transport, including the participation of the elusive *ton*B gene function.

THE ENVELOPE OF GRAM-NEGATIVE BACTERIA

Most of the published work on siderophore receptors has derived from studies on Gram-negative bacteria. Hence, it is instructive to recall what is understood about the envelope of such species.

Systematic genetic and biochemical analyses of the cell envelopes of Gram-negative bacteria, in particular those of *Escherichia coli* and *Salmonella typhimurium,* have revealed substantial detail regarding the morphology, composition, and function of the triple-layered shell characteristic of these organisms (43). The inner or cytoplasmic membrane is a phospholipid bilayer containing a large number of peripheral and integral proteins. The major functions of these are the active transport of substrates, including iron, and energy transduction (101). In the periplasmic space between the inner and outer membranes lies a peptidoglycan or murein network that imparts shape and rigidity to the cell (38). In this periplasmic space are found proteins involved in binding or processing of transported solutes (22); none of these as yet has been involved in iron transport. The outer membrane, in which most of the iron-related proteins have been detected, is an asymmetric bilayer with chains of lipopolysaccharide protruding from the outer leaf (63). The molecular weights of the iron-related proteins, in the range of ca 80,000 daltons by SDS-PAGE analysis, are such that they can easily span the periplasmic space and contact the inner membrane. Possibly such contacts occur at the Bayer (9) adhesion zones visible in electron microscope examination of suitably prepared cells. The ferrichrome (previously T5) receptor is located at such a zone (8).

FERRICHROME RECEPTOR

The structure of ferrichrome is shown in Figure 2. Although it has been found so far only in fungi, the receptor for this prototypical iron hydroxymate siderophore has been studied *in extenso* in bacteria, where it appears to be widely distributed.

The background to the identification of the ferrichrome receptor in *Escherichia coli* has already been recounted. A similar protein apparently occurs in the outer membrane of *Salmonella typhimurium* LT-2, where it serves as a site for adsorption of phages ES18 and ES18•hl (59). The former attacks smooth and rough strains not lysogenic for Fels2, whereas ES18•h1 is a host range mutant capable of growth on strains carrying Fels2. In *E. coli* and *S. typhimurium* it is obvious that the availability of both suitable phages and the illicitly transported antibiotic albomycin aided the identification of the ferrichrome receptor.

The structural gene for the ferrichrome receptor is *ton*A (for T-one). It has been suggested it be renamed *fhu*A, for ferric hydroxamate uptake, a designation that would imply little ligand specificity in the gene product (48). Hence, it seems desirable to retain the original designation until the full range of hydroxamate-type siderophores transported by *E. coli* can be charted. A second phenotype, FhuB, has been defined as inability to use any type of hydroxamate siderophore. The gene order at 3.5 min on the chromosome was determined to be *pan-ton*A-*fhu*B-*met*D. It was proposed that *fhu*B codes for a cytoplasmic membrane component required for transport of all hydroxamate-type siderophores, since mutations at this site did not cause detectable alteration in the outer membrane protein profile. Mutants defective at the *ton*A locus acquired capacity to transport ferrichrome following treatment with pronase, whereas *fhu*B mutants remained transport negative. This supports an inner membrane location for *fhu*B (126).

When enteric bacteria are cultured at low iron levels the TonA protein is overexpressed, but not to the extent seen for the ferric enterobactin receptor (36). This suggests a distinct mode of regulation in the *ton*A gene. The biogenesis of the TonA protein has been studied in a plasmid clone carrying the *ton*A region. In a coupled transcription-translation system in vitro the TonA protein was detected as a precursor some 2000 daltons larger than the final product; processing to the mature form was achieved with inverted inner membrane vesicles (91). The kinetics of synthesis, as well as the yield, are unique for TonA since following an iron starvation insult its maximum rate of synthesis is reached well before that of the other low-iron-induced membrane proteins (49).

The TonA protein is located in the outer membrane of enteric bacteria, but as yet there is little information about its state of aggregation or association with other constituents of the bilayer. The molecular weight is 78,000 daltons in SDS-PAGE (14).

In both *S. typhimurium* (27) and *E. coli* (36), mutants constitutive for the synthesis of the iron-repressible outer membrane proteins have been isolated or constructed, but the genes, designated *fur,* have not been mapped.

The TonA protein is at least part of the receptor complex for phages T1, T5 and ϕ80, for colicin M, and for the antibiotic analogue of ferrichrome, albomycin. In addition to ferrichrome, siderophores of closely related structure, such as ferrichrysin and ferricrocin, use the TonA protein (62). More distantly related siderophores, such as rhodotorulic acid and aerobactin, apparently required separate receptors. Although ferrichrome is highly potent in protection versus lethal agents both in vivo and in vitro, it has not been established that all of these substances utilize an identical site in the receptor complex. Little purification work has been reported on the TonA protein and nothing is known about its chemical composition and quantitative affinity for ligands.

In *ent* mutants of *E. coli* and *S. typhimurium,* the iron label of [^{55}Fe-^{3}H] ferrichrome is taken up in preference to the label of the ligand, indicating the latter is recycled (55). It has been suggested (103) that the ligand may be acetylated on one of the hydroxylamino oxygens following reductive separation of the Fe(III), but a mutation affecting ferrichrome utilization that could be ascribed to this event has not been reported (37). In general, O-acylated hydroxamic acids are unstable. The fact that the kinetically inert Λ-*cis*-chromic-[^{3}H]deferriferrichrome was accumulated indicated that the siderophore must permeate the cytoplasmic membrane (55), although disuccinyl ferrichrysin, which would not be expected to be transported, displays substantial activity as a siderophore (37). It may be the case that the exchange inert chromic complex accumulates in the envelope and is leaked to the cytoplasm intact, whereas in the natural iron chelate, the Fe(III) suffers instantaneous reductive separation from the carrier in the membrane.

Inner membrane vesicles of iron-starved *E. coli* cells of an *ent* mutant were shown capable of transport of [^{3}H]ferrichrome with K_m of 0.2

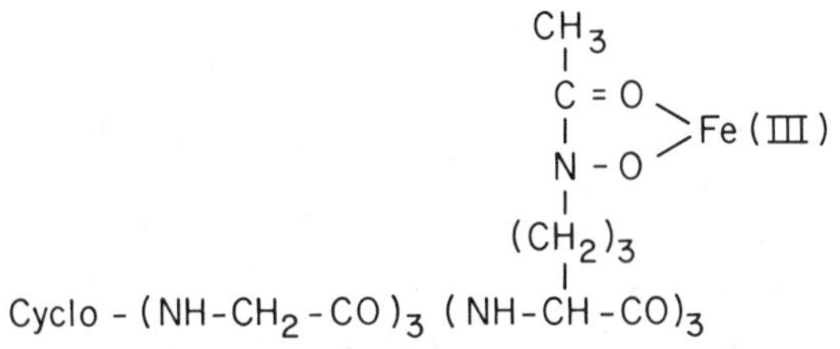

$$\begin{array}{c} CH_3 \\ | \\ C=O \\ | \qquad \rangle Fe(III) \\ N-O \\ | \\ (CH_2)_3 \\ | \\ \text{Cyclo}-(NH-CH_2-CO)_3\ (NH-CH-CO)_3 \end{array}$$

Figure 2 Ferrichrome is commonly found in a large number of fungi and acts as a siderophore in these species as well as in bacteria, although the latter have not yet been shown to synthesize this particular cyclohexapeptide ferric trihydroxamate.

μM (73). Essentially similar rates of transport were found for *ton*A and *ton*A$^+$ vesicles, which suggests the presence of a discrete transport system in the inner membrane. The use of metabolic inhibitors pointed to a crucial need for the energized membrane, and some evidence has been adduced for a symport transport mechanism possibly involving divalent cations (39).

In *E. coli* K-12 AB2847, the iron of synthetic Δ-*cis*-enantioferrichrome was accumulated at about half of the rate seen with natural Λ-*cis*-ferrichrome (124), whereas in *Penicillium parvum* the synthetic antipode was totally excluded. In other fungi, enantioferrichrome was observed to have some activity (123).

FERRIC ENTEROBACTIN RECEPTOR

Following identification of the ferrichrome receptor (116, 117), a ready explanation was forthcoming for the protective effect of ferric enterobactin (Figure 3) versus colicin B, a phenomenon first described by Guterman (35). Thanks to the classic work of Fredericq (29), colicins and phages had been known to share common receptors. The prediction (117) that the colicin B receptor acts for specific binding and uptake of ferric enterobactin was immediately confirmed by results from several laboratories [for review of genetics, see Kadner & Bassford (47)].

In 1975 Uemura & Mizushima (110) reported iron to be responsible for regulation of certain of the outer membrane proteins of *Escherichia coli,* and in the following year McIntosh & Earhart (65) and a number of other laboratories confirmed this observation and correlated the 81,000-dalton band seen on SDS-PAGE with colicin B binding and ferric enterobactin transport. Mutations *fep*A and *fep*B in the iron operon at min 13 of the *E. coli* chromosome were distinguished through studies with spheroplasts. However, as complementation data are lacking (97), the existence of a *fep* B gene remains to be proven. The *fep*A gene product is readily observed at 81,000 daltons in SDS-PAGE analysis of outer membranes of iron-depleted cultures and is variously designated as the FepA, Cbr, or FeuB protein; here it is simply Fep.

McIntosh & Earhart (66) correlated the production of catechol compounds with three outer membrane proteins, 74,000, 81,000, and 83,000 daltons, and they concluded the genes for all three were coordinately regulated by iron. This problem was reinvestigated by Klebba et al (49) through examination of the kinetics of induction and repression of the iron-regulated membrane proteins. The latter workers substituted biological chelators, such as deferriferrichrome A, transferrin, and ovotransferrin, for other methods of deferration, such as filtration, extraction, or the use of synthetic chelators in situ. Thus dipyridyl, although commonly added to media for iron complexation, is a general coordination agent for divalent metal ions, and furthermore, it leads to nonspecific permeability changes in *E. coli* cells (M. A. McIntosh, personal communication). Klebba et al (49) concluded that the 83,000-dalton components, Fep and Cir, are coordinately regulated, whereas other iron-related membrane proteins are under different regulation.

Laird et al (51, 52) have used phage Mu for construction of a partial order of nine genes in the ferric enterobactin biosynthesis and transport cluster. A combination of Tn5 mutagenesis, cloning of restriction fragments, and restriction mapping was employed to establish the gene order as *ent*D, *fes*, *ent*F, *fep*, and *ent*CAGBE across 26 kbases of DNA.

Enterobactin synthetase, an enzyme catalyzing formation of the siderophore from 2,3-dihydroxybenzoic acid and L-serine, is comprised of at least four possibly physically associated subunits designated D, E, F, and G. Evidence was found that each component is repressible by iron and that negative feedback controls the biosynthetic pathway (32, 33).

The work of Wettstein & Stent (119) established a correlation between iron deprivation and the maturation of certain specific tRNA's. The particular alteration is the presence of isopentenyl adenosine in place of the usual methylthioisopentenyl adenosine adjacent to the 3' side of the anticodon sequence. Recently, it was shown that such undermodified tRNA's for *phe*, *tyr*, and *trp* stimulate aromatic amino acid transport in *E. coli* and hence may serve to replete pools of aromatic amino acids diverted to the synthesis of enterobactin (15).

Enterobactin as well as a number of analogues have been prepared by chemical synthesis. Both a carbocyclic analogue consisting of 1,5,9-triaminocyclododecane fully acylated with 2,3-dihydroxybenzoic acid (41) and an aromatic analogue similarly derived from 1,3,5-tri-aminomethylbenzene (112) supported the growth of *ent* mutants of *E. coli*, apparently without degradation to 2,3-dihydroxybenzoic acid and resynthesis to enterobactin. Enantio-enterobactin, chemically synthesized from D-serine, was found incapable of either supplying iron to *ent* mutants of *E. coli* or competing with synthetic or natural enterobactin for the Fep protein (39). Enterobactin forms complexes with a number of group III and transition metal ions, with the complexes of Sc(III) and In(III) displaying antibacterial action versus *Klebsiella pneumoniae* (100). Agrobactin and parabactin, which form Λ coordination isomers with iron, fail to replace enterobactin, whereas their A forms, containing an opened oxazoline ring, which yield Δ isomers, display significant activity (80). These data indicate that the Fep protein recognizes the metal center of the siderophore, and that the platform to which it is attached is relatively less important as a determinant of specificity. Beyond this is the possibility that the specific chirality of the metal ion coordination compound may be one of the structural parameters evolved by microorganisms in their struggle to encapsulate iron in a form recognizable only to their individual transport systems (78).

The fact that enterobactin analogues that lack ester bonds act as siderophores in *ent* mutants of *E. coli* calls into question the nature of the *fes* mutation. This gene is believed to code for an esterase that degrades enterobactin to 2,3-dihydroxybenzoylserine and its linear conjugates, the iron complexes of which may have electron-accepting potentials within the range of natural reductants. In *Bacillus subtilis* WB2802, enterobactin, the aromatic analogue, and the sulfonated derivative of the aromatic analogue all served as sources of iron for growth of the organism. Extracts of the cells were found capable of reducing the Fe(III) in all three complexes in the presence of NADPH, FMN, and Mg^{2+} (58). Glutathione competes effectively for enterobactin iron at pH values less than 6.0 (35a).

As the Fep protein is very easily observed in outer membrane preparations from iron-depleted cultures, some biochemical work has been done with this component of the ferric enterobactin transport system. Outer membrane preparations prepared in tris(hydroxymethyl)aminomethane (Tris)-ethylenediaminetetraacetate buffer containing Triton X-100 retained ferric enterobactin binding activity (41). Subsequently, it was noted that either whole membranes or outer membranes upon incubation at 37°C experienced loss in ferric enterobactin binding potency, an event that could be correlated with processing of the Fep protein to a new component, 81K*, which behaved on SDS-PAGE analyses as though it had lost a ca 6000-dalton polypeptide (40). The modification activity could be adsorbed to a column of *p*-aminobenzamidine-//-sepharose and eluted with free benzamidine (28). It appears to be identical with protein *a*, a major outer membrane protein without previously designated function.

That protein *a* is responsible for the processing of Fep was confirmed by genetic analysis. Thus *E. coli* UT2300, which lacks Fep but is replete with respect to protein *a*, is able to process added Fep. Protein *a* has been mapped at 13 min (23). Mutants carrying large deletions in this area, such as strains UT 4400 and 5600, lack both protein *a* and Fep; Fep protein added to outer membrane preparations of these strains remains stable. Finally, UT4400 transformed with pGGC110, a plasmid containing the structural gene for protein *a,* acquired the ability to process Fep added to solubilized outer membrane preparations (E. H. Fiss, J. B. Neilands, unpublished observations).

As ferric enterobactin is a trivalent anion, an assay for Fep could be devised based on the passage of [^{55}Fe]enterobactin through a column of diethylaminoethyl cellulose (41). Purified Fep, prepared by standard fractionation methods and the use of affinity columns derived from the molybdate complex of bis-2,3-dihydroxybenzoyllysine, has afforded enough protein for limited chemical characterization work (E. H. Fiss, J. B. Neilands, unpublished observations). Neither Fep nor 81K* appears to be glycoproteins. The N-termini have been determined to be *phe* and *ala,* respectively. A survey of *E. coli* B and B/r and *Salmonella typhimurium* LT2 revealed no processing activity of the indigenous Fep protein in outer membrane preparations from these strains (E. H. Fiss, P. Samuels, J. B. Neilands, unpublished observations). This study revealed that *E. coli* B and B/r both convert Fep, added from strain K-12, to 81K*, thus indicating the presence of a functional protein *a* in the former strains. No processing of the indigenous Fep protein in outer membrane preparations of *S. typhimurium* could be seen and the K-12 Fep added to such preparations was not attacked. Like K-12, *E. coli* B and B/r were sensitive to colicin B. Not surprisingly, *S. typhimurium* was insensitive to this bacteriocin. All of this suggests there may be subtle differences between Fep as it occurs in the various enterics. It also suggests that the protease activity displayed by protein *a* is not required for insertion of Fep in the outer membrane or for normal operation of the ferric enterobactin system, which is presumably intact in all of the bacteria just described. Protease IV of the outer membrane of *E. coli* appears different from protein *a* in that, unlike the latter, it could be inhibited by Triton X-100 or ethylenediaminetetraacetate (98).

The 81K* protein is unable to bind either ferric enterobactin or colicin B, although the work of McIntosh et al (64) suggests that the two receptor functions can be separated by mutation.

Moore et al (69) reported evidence for the presence of enterobactin-specific immunoglobulins in normal human serum. The antibody, which is presumably of the A isotype, inhibited uptake of iron from ferric enterobactin but not from either ferric citrate or ferrichrome (68). Chromatographic evidence for the presence of enterobactin in the peritoneal washings of guinea pigs lethally infected with 10^8 cells of *E. coli* 0111K58H2 was announced by Griffiths & Humphreys (34). They concluded that host iron

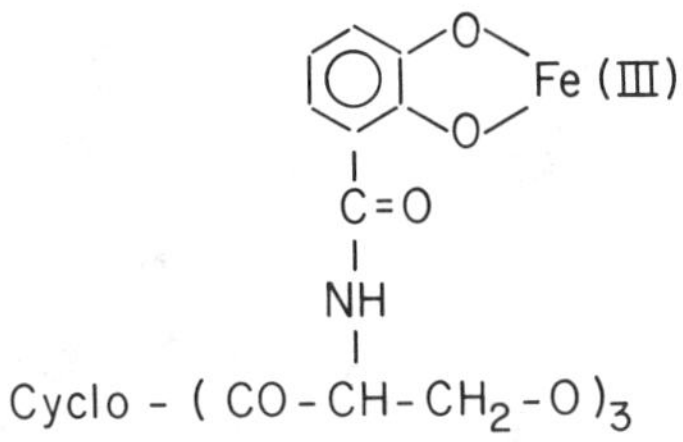

Fe (III)

C=O

NH

Cyclo - (CO - CH - CH_2 - O $)_3$

Figure 3 Ferric enterobactin, a siderophore commonly produced by enteric bacteria.

binding proteins do indeed influence the metabolism of the invading pathogen.

Pseudorevertants of *E. coli* K-12 mutants with multiple defects in the enterobactin system appear to excrete a molecule that can extract iron from synthetic chelators, but not from ferrichrome or ferric enterobactin (89). The pseudorevertants were affected at the *omp*B locus, the regulatory site for the porins.

FERRIC CITRATE RECEPTOR

No structure is written here for ferric citrate since the exact stoichiometry of the complex depends on the proportion of excess ligand, with a ratio of citrate/Fe(III) of about 20/1 required to achieve the mononuclear form (106).

Strains K-12 and B/r of *Escherichia coli* defective in the enterobactin transport system were shown to use iron citrate, whereas corresponding mutants of strain W were found to be inhibited by citrate on iron-poor media (102). This system is distinguished from the others described in this review in that it is induced by growth in the presence of citrate, a carbon source used by *Salmonella typhimurium* but not *E. coli.* Conversely, the former organism cannot utilize iron citrate (92). Woodrow et al (125) performed a genetic analysis of the iron citrate transport system in *E. coli* K-12 and located the *fec* genes at about min 6 on the chromosome. Subsequently, the *fec* gene locus was divided into *fec*A and *fec*B, with the former coding for the outer membrane protein FecA (42). The induction of this protein requires iron as well as citrate. Fluorocitrate and phosphocitrate induced citrate-dependent iron transport, whereas a number of other citrate analogues were inactive. The FecA protein is assigned a molecular weight of 80,500 on SDS-PAGE analysis (113).

CITRATE-HYDROXAMATE SIDEROPHORES

Three siderophores that may be classed chemically as citrate-hydroxamic acids have been described, namely, schizokinen, aerobactin, and arthrobactin. Nothing is known about the transport of the latter, which is identical with deferri-terregens factor, and it is not discussed further here. A structure for ferric aerobactin is given in Figure 4 and it is assumed on the basis of net charge that in all members of the series the Fe(III) is linked to the α-hydroxy acid function.

Mutants of *Bacillus megaterium* blocked in biosynthesis of schizokinen have been used to examine the mode of action of the siderophore. The intact Fe(III) complex is translocated across the cell envelope and the iron is removed by a reductive process (4, 17). The transport system does not utilize ferric aerobactin. Schizokinen occurs in the cyanobacterium *Anabaena* sp. ATCC 27898 (105), where it acts to transport iron against a concentration gradient (53). The K_m of transport was found to be 30 nM. Light-driven transport was blocked by ATPase inhibitors and uncouplers, and it was concluded that ATP serves as an energy source for uptake of the siderophore. Murphy et al (71) and Bailey & Taub (6) demonstrated inhibition, apparently by iron starvation, of algae by siderophores not utilized by the species under test.

In 1969, Gibson & McGrath (31) isolated aerobactin from *Aerobacter aerogenes* 62–1, an organism capable of enterobactin synthesis at least in its genetic wild state. Renewed interest in aerobactin was stimulated by a report from Payne (85) that this siderophore is synthesized and utilized by *Shigella flexneri.* A second *Shigella* species, *Shigella sonnei,* has been reported to form unidentified siderophores of both the catechol and the hydroxamate type (87). Other enteric bacteria, specifically *Salmonella* species *austin* and *memphis* and *Arizona hinshawii,* were shown to produce high yields of aerobactin following culture on Tris-succinate medium (10). Most of these bacteria could be shown to harbor plasmids.

Prior to the demonstration of rather widespread, but erratic, synthesis of aerobactin in enteric bacteria and its tentative correlation with plasmids, Williams (120) had discovered a novel plasmid-directed iron-harvesting system in colicin V (ColV)-bearing strains of *E. coli.* A substantial percentage of clinical isolates and invasive strains of *E. coli* harbor this plasmid, and its designation V, for virulence, seemed appropriate. The formation of ColV itself, however, is not essential for virulence (12, 95), and Williams

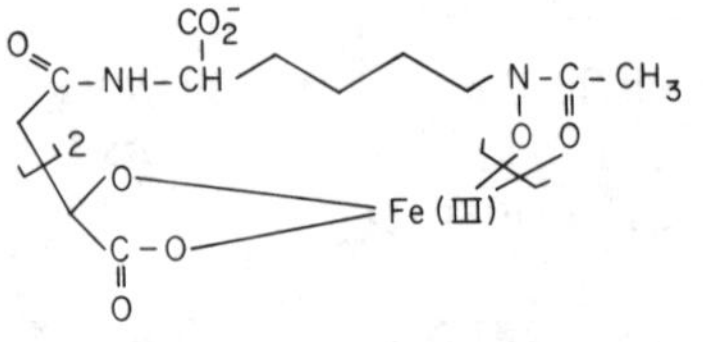

Figure 4 Ferric aerobactin, a siderophore produced by plasmid-bearing strains of enteric bacteria.

& Warner (121) were able to show that other determinants on the ColV plasmid enabled the cell to scavenge iron from transferrin and in environments low in iron. Two types of mutants were defined, *iuc* and *iut,* on the basis of failure to synthesize either chelator or transport system for iron uptake. The chelator was postulated to be of the hydroxamate type (107, 121), and the transport defect was suggested to be a membrane protein. Finally, the ColV-plasmid-specified siderophore was identified as aerobactin (115).

Of 42 clinical isolates of *E. coli,* bioassay revealed that 11 produce aerobactin (11). Study of the ferric aerobactin receptor is facilitated through use of the bacteriocin produced by strains of *Enterobacter cloacae,* cloacin DF13. At low iron growth, *E. cloacae,* like *A. aerogenes* and *A. hinshawii,* excretes both enterobactin and aerobactin. Ferric aerobactin protects *E. cloacae* and ColV strains of *E. coli* against cloacin DF13, presumably via binding at a common receptor site in the outer membrane (111, 111a). The cloacin attacks some, but not all, of the enteric species synthesizing aerobactin (A. Bindereif, personal communication). The list of organisms believed to form this siderophore is given in Table 1. Only in the case of the ColV strains of *E. coli* has it been proven unequivocally that synthesis is encoded or directed by a plasmid. Clearly, as pointed out by Williams & Warner (121), the correlation with ColV is purely fortuitous.

*TON*B

A functional locus, *ton*B, is required for utilization of all high-affinity iron carriers, including citrate, and by certain phages and colicins (for review see 76). Wookey & Rosenberg (127) failed to observe ferric enterobactin transport in spheroplasts from *ton*B cells and concluded the function was required for transport across the inner membrane. The problem was then

Table 1 Plasmids in enteric bacteria synthesizing aerobactin

Species	Criterion for aerobactin production	Plasmid(s)
Aerobacter aerogenes 62–1	Isolation	+
Shigella flexneri	Isolation	?
Salmonella austin	Isolation	?
Salmonella memphis	Isolation	?
Arizona hinshawii	Isolation	+
Escherichia coli, ColV	Isolation	+
E. coli, clinical isolates	Bioassay	+
E. coli, B	Utilization	?
Enterobacter cloacae	Isolation	?

taken up by Weaver & Konisky (118), who reported no transport of the siderophore in spheroplasts from wild-type cells. However, the latter workers did note a substantial rate of ferrichrome uptake in spheroplasts from *ton*B strains and concluded that the *ton*B gene product is required for physical association of inner and outer membrane, i.e. apposition sites as contrasted to Bayer adhesion sites. In the model proposed for cobalamin transport, the TonB protein would be located in the inner membrane, where in conjunction with the proton motive force it generates a diffusible substance, the function of which is to dislodge the transported substrate from the outer membrane receptor (99).

Postle & Reznikoff (94) cloned the *ton*B gene and demonstrated that it codes for a 36,000-dalton polypeptide, whereas Plastow & Holland (90) showed the inner membrane to be enriched with a 40,000-dalton protein specified by a λ-*ton*B transducing phage. In sum, the location and function of the TonB protein remain in doubt. However, it is apparent that its expression is not regulated by iron (36).

AGROBACTERIUM TUMEFACIENS AND *PARACOCCUS DENITRIFICANS*

When cultured at low iron, *Agrobacterium tumefaciens* and *Paracoccus denitrificans* excrete catechol-type siderophores derived from spermidine and named, respectively, agrobactin (84) and parabactin (88, 108) (Figure 5). The coordination characteristics of these siderophores parallel those of enterobactin, except that with the former the complexes are left-handed screws and the net charge is two minus (80). No work has been reported on transport of ferric parabactin in *P. denitrificans*, but agrobactin-

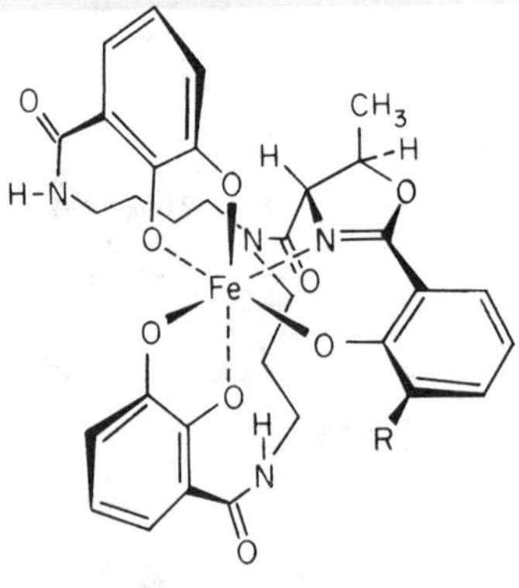

Figure 5 Ferric parabactin (R=H) and ferric agrobactin (R=OH), siderophores from *Paracoccus denitrificans* and *Agrobacterium tumefaciens*, respectively.

mediated uptake of iron in *A. tumefaciens* has been the subject of a preliminary investigation (56). Growth of the organism at low iron led to synthesis of several envelope proteins in the 80,000-dalton region. Mutant *agb*-3, unable to use agrobactin on agar plate diffusion tests, was found defective in uptake of the label from [^{55}Fe]agrobactin. The mutant lacked a band in the 80,000-dalton region in the SDS-PAGE profile, but it has not been proven that this is the ferric agrobactin receptor or a component thereof.

VIBRIO SPP.

Virulent strains of the fish pathogen *Vibrio anguillarum* have been found to contain the plasmid pJM1, and curing experiments established a link between the plasmid and iron uptake by the bacterium (20). Iron-starved cells of *V. anguillarum* show the presence of two new outer membrane proteins, one of which, designated OM2, has a molecular weight of 86,000 and is present only in the plasmid-bearing cells (21). In another study (16), plasmid pJM1 is said to code for an 80,000-dalton low-iron-induced outer membrane protein. No siderophore has as yet been identified from this species.

The membrane profile of *Vibrio cholerae* has been examined by SDS-PAGE analysis following growth of the organism at low iron (104). Although a number of proteins were shown to be induced under such conditions, none of these has been identified as a siderophore receptor.

PSEUDOMONAS SPP.

Although the various members of the *Pseudomonas* are known to form siderophores of diverse structure—ferribactin, various ferrioxamines, pyoverdine, pyochelin, and pseudobactin—little is understood about the mechanism of transport.

Pyochelin from *Pseudomonas aeruginosa* has just been characterized as a salicylic acid-substituted cysteinyl peptide (19). The structure proposed would be expected to display a rather weak chelate effect; nonetheless, it is reported to give a red color with Fe(III). Uptake of label from [^{55}Fe]pyochelin was shown to occur in two stages, an energy-independent step, presumably binding to the cell surface, followed by an energy-dependent process (18). The low iron induction of membrane proteins in *P. aeruginosa* and their possible correlation with the pyocin S2 has been described by Ohkawa et al (83).

Pseudobactin (Figure 6) from fluorescent *Pseudomonas* B10 is reported by Teintze et al (109) to be a linear hexapeptide bearing one each of the catechol, α-hydroxy acid, and hydroxamic acid Fe(III)-ligating functionali-

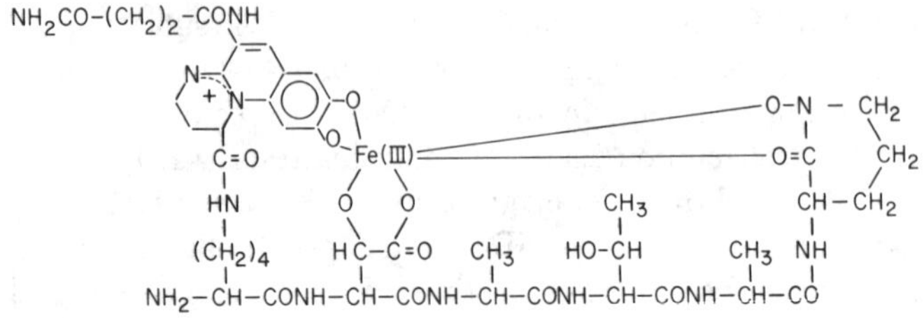

Figure 6 Ferric pseudobactin, a siderophore from *Pseudomonas* B10, is derived from the linear hexapeptide L-lys-D-*threo*-β-hydroxy-asp-L-ala-D-*allo*-thr-L-ala-D-N^{δ}-hydroxy-orn (109).

ties commonly found in siderophores. Pseudobactin is believed responsible for disease suppression in soil (50), possibly via a mechanism of species selection based on iron deprivation as in algal systems (71). No papers have appeared on the transport of ferric pseudobactin.

NEISSERIA SPP.

In a series of papers, Archibald & de Voe (1, 2) showed that in *Neisseria meningitidis* a variety of iron chelates (functional siderophores) and transferrin could supply iron to the cell. Although acquisition of iron was deemed essential both in vitro and in vivo, these and other (72) workers did not detect endogenous siderophore synthesis by the organism. In contrast, Payne & Finkelstein (86) produced tentative evidence for siderophore formation in both *N. meningitidis* and *Neisseria gonorrhoeae.* Subsequently, Yancey & Finkelstein (128, 129) partially purified meningobactin and gonobactin from the two neisseria and produced evidence that they belong to the hydroxamic acid series of siderophores. In the gonococcus, siderophore complexed iron was demonstrated to be important for virulence in the chick embryo model (130). Thus, although the neisseria do seem capable of siderophore synthesis, little information is available as yet concerning the membrane physiology and mechanism of transport of the Fe(III) complexes. The first order of business would seem to be the isolation of the siderophores and the demonstration that they are either new or known compounds.

EXPRESSION AND REGULATION

Minimal media are about micromolar in contaminating iron, and although there is considerable variation among microbial species, this is about the concentration at which derepression of iron assimilation systems can be seen. One order of magnitude above and below this figure can be called high and low iron, respectively. Apart from specific stimulation of growth rates by iron, additional evidence that the culture is in fact deficient in this element may come from a positive Arnow (3) test or ferric perchlorate reaction (5) for catechol- or hydroxamate-type siderophores, respectively, or from overproduction of certain membrane proteins, as revealed by SDS-PAGE.

Among the synthetic chelating agents that have found favor for limiting availability of iron are ethylenediamine-di-*o*-hydroxyphenyl-acetic acid, bipyridyl, or nitrilotriacetate. The natural counterparts of these reagents are deferriferrichrome A, conalbumin, or transferrin, and they are to be preferred to the synthetic compounds on the basis of their superior specificity for Fe(III) and their probable lack of toxic and undesirable effects. Dipyridyl, for example, leads to cell lysis in suspensions of *Escherichia coli* (M. A. McIntosh, personal communication).

For enteric bacteria, a medium high in Tris has been found to invoke iron starvation (65). Irvin et al (44, 45) showed that exposure of *E. coli* to pH 2.75 citrate buffer followed by suspension of the cells in pH 8.00 Tris led to shedding of the outer membrane. They also noted Tris-induced permeability changes in the outer membrane and concluded that "investigators utilizing Tris buffer are cautioned that Tris is not physiologically inert and that it may interact with the system under investigation" (44). Of equal importance for iron starvation is the carbon source, a tricarboxylic acid cycle intermediate invoking a higher demand for the element than, for example, glucose. In the former case, the substrate affords energy only by oxidation through the iron-containing cytochrome system.

Working with strains of *E. coli* K-12 defective in enterobactin synthesis, Klebba et al (49) developed a pulse label technique with which to measure the rate of synthesis of envelope proteins following sudden depletion and repletion of the culture in iron. The study indicated a common, coordinate mode of regulation for proteins of 83,000 daltons, Fep, and the colicin Ia receptor. The TonA protein was confirmed to be clearly under a different type of regulation. Two other proteins, with molecular weights of approximately 25,000 and 90,000, were also found to be under iron regulation; the latter is especially interesting since it appears only under conditions of iron repletion. Hantke (36) detected by SDS-PAGE analysis a protein corresponding in size to a ferritin subunit (19,000 daltons) that was repressed in the *lac* fusion mutants that overproduced the other iron-related membrane proteins. Physical evidence has recently appeared for the presence of ferritin-like iron in bacteria, including *E. coli* (7). The experiments of Klebba et al (49) demonstrate that the cell can sense instantly changes in the iron status of the environment, and this in turn is reflected in the rate of synthesis of membrane proteins. Enterobactin and heme were ruled out as participating in the regulatory process, a model for which is shown in Figure 7. A regulatory role for undermodified tRNA species has been demonstrated (15,

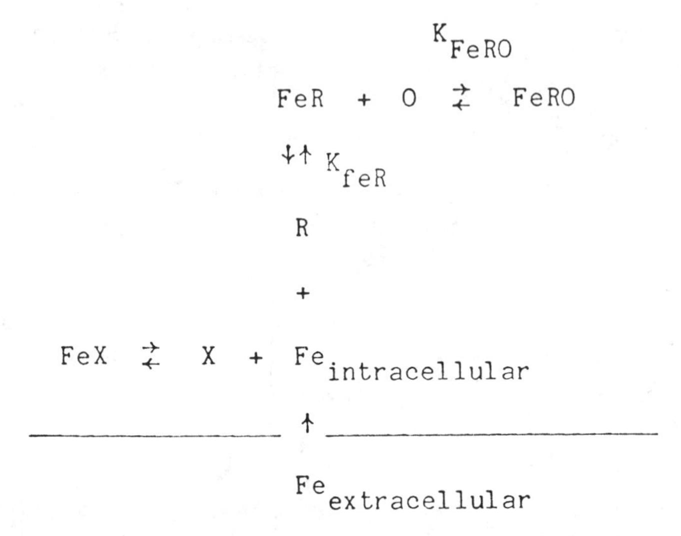

Figure 7 A regulatory model designed to explain the observed kinetics of repression and derepression of synthesis of membrane proteins in *Escherichia coli* K-12 following depletion and repletion of the medium in iron. R, Repressor; Fe, corepressor; O, operator; FeRO, nontranscribed repressor-corepressor-operator complex; X, iron storage protein (49).

67). Hence, it is tempting to speculate that iron assimilation is governed by a range of controls, with the actual number of switches closed and metabolic circuits operational at any time dependent on the severity of the deficit.

CONCLUSIONS

The changes in the membrane protein profile of microorganisms correlated with iron supply demonstrates in a most dramatic way how the cell has evolved mechanisms to sense and adapt to the external environment.

It remains now for molecular genetics to furnish the blueprint whereby the profound physiological effects of iron on its own assimilation become manifest. In enteric bacteria, cloning of the various iron operons would provide access to the DNA from which, in turn, insights into the regulatory mechanism and mode of action of the siderophore receptors might be forthcoming. The low dissociation constants of the siderophore-receptor complexes provides a rationale for the requirement of additional genetic elements, such as *ton*B, for the orderly assimilation of the iron. In spite of the quite negative potentials of siderophores, it does appear that cells have devised means for reductive separation of the ligand and iron. Unfortunately, the biochemistry of many of these events is still in a rudimentary state. A more detailed genetic analysis than is currently available will be required to liberate the regulatory mechanisms from the black box, illustrated in Figure 1, in which they have reposed these past three decades since synthesis of both hydroxamate- and catechol-type siderophores was first shown to depend on the iron level of the culture medium (30).

Several lines of evidence indicate that pathogenic species, when present in the tissues of the host, are derepressed with respect to iron and that the high-affinity siderophore-mediated pathway is functional. This observation, plus the fact that the siderophore mode of iron assimilation is widely distributed among all types of aerobic and facultative anaerobic microorganisms, affords new avenues of attack on agents of infectious disease, which range from conventional drug development to immunology.

One aspect of microbial virulence in dire need of further work is the role of heme, a chelate compound few pathogenic bacterial species seem capable of using. Janzen et al (46) have shown the *hem*A mutants of *Salmonella typhimurium* selected by growth on heme to be defective in envelope composition. Antibiosis by iron starvation has been suggested as a biofunction for the hemoglobin binding protein, haptoglobin (23a).

In Table 2 I have listed some of the envelope proteins related to iron that have been investigated in apparently plasmid-free strains of *Escherichia coli.* A conspicuous feature of this table is the substantial number of question marks in its fourth column. However, there must be a finite limit to the number of specific siderophore receptors in the outer membrane of this organism. Possibly some of the many proteins altered by iron are affected for relatively trivial reasons, such as the need to maintain some degree of uniformity in the protein content of the envelope.

To fulfill the promise made in the first sentence of this review, some remarks on siderophore transport in fungi are in order. Specific siderophore binding sites in the cell envelopes of *Ustilago sphaerogena* (24) and *Neurospora crassa* (122) have been proposed, and a recent study (70) in the latter organism has revealed the presence of such sites in both open membranes and closed vesicles. But the fungi, which have provided us with such a rich harvest of siderophores, have yet to yield up a single isolated receptor protein for complexed iron. By building on the sturdy foundation laid in bacteria, we can achieve rapid progress in our understanding of the mechanisms of siderophore transport in the eukaryotic microbial cell.

It is evident from the foregoing that nothing is yet known about the mode of translocation, insertion, and assembly of iron-related proteins in the microbial envelope. Based on analogy with other outer membrane proteins, it is anticipated that the siderophore receptors observed at the 80,000-dalton region in SDS-PAGE analyses are clustered in regular units of small whole numbers. The location of such clusters at Bayer (8) adhesion sites remains an attractive possibility, although one not yet proven by direct experimentation. The 80,000-dalton envelope proteins of *E. coli* believed to be formed only at discrete periods in the cell cycle and collectively designated d or D (for DNA) have been equated with the siderophore receptors (12a), the methodology employed in synchronizing the culture having precipitated an apparent iron-deficient induction of their synthesis.

Similarly, the molecular mechanism of action of the siderophore receptors has not been established. Only in the case of ferric enterobactin do we have some notion of the specificity range of the receptor. The functional interchange of siderophores among species provides a means with which to probe the kinship of high-affinity iron assimilation systems in different microbes (62a). At least some siderophores, such as ferrichrome, appear to enter the cell intact and to deliver the iron internally; in the case of other siderophores there is no evidence for permeation of the cytoplasmic membrane. Obviously, a mechanism in which penetration is coupled to a very efficient extrusion of the ligand is difficult to distinguish from the alternative, namely, a turnstile or surface exchange mechanism. Given that reduc-

Table 2 Envelope proteins related to iron in *Escherichia coli* K–12

Protein (SDS-page mol wt)	Mnemonic(s)	Map locus (min)	Function or designation[a]	Properties
90K[b]	—	—	?	Repressed by low iron
83K	—	—	?	Strongly induced by low iron; not in *E. coli* B
81K	*fep, cbr*	13	Ferric enterobactin receptor	Strongly induced by low iron; colicin B receptor
81K*	—	—	?	Larger fragment (~74K daltons) resulting from endolytic cleavage of 81K daltons by protein *a*
80.5K	*fec*A	6	Ferric citrate receptor	Induced by ferric citrate
78K	*ton*A, *fhu*A[c]	3	Ferrichrome receptor	Moderately induced by low iron; receptor for albomycin, T1, T5, ϕ80, colicin M
76K	—	—	?	Prominent in constitutively derepressed (*fur*) mutants
74K	*cir*	43	?	Strongly induced by low iron; colicin Ia receptor
40K	—	13	Protein *a*	Endoproteolytic cleavage of 81K daltons
36K	*ton*B	27	Transport of siderophores and vitamin B_{12}	Not yet detected in envelope preparations; not iron regulated
25K	—	—	?	Transient synthesis preceding low iron induction of 83K daltons, Fep, and Cir proteins
19K	—	—	?	Diminished in constitutively derepressed (*fur*) mutants

[a]The protein entity indicated may be only a component of a receptor complex. In addition, strains of *E. coli* carrying plasmids (121) responsible for synthesis and transport of aerobactin will express the common outer membrane receptor for cloacin DF13 and ferric aerobactin (111a).

[b]K = 1,000.

[c]Gene *fhu*B, also mapped at 3 min, is required for transport of hydroxamate type siderophores.

tion of the Fe(III) is the actual mode of separation of the metal ion from the siderophore, it is difficult to imagine how the extreme outer surface of the cell could protect the necessary reducing equivalents against atmospheric oxidation. The identification and characterization of specific permeases in the inner membrane would be evidence that ferric siderophores do indeed reach the cytoplasm.

The function of the iron-related membrane proteins in vivo is dependent on the remaining components of the iron transport system. For two common siderophores, ferric enterobactin and ferrichrome, the genetics of transport in even such well studied species as *E. coli* still requires further investigation. Research on genetic aspects of the receptor component of siderophore-mediated iron assimilation obviously will be facilitated in those cases where the genes occur naturally on plasmids or may be transferred to such extra chromosomal elements by the standard methods of manipulation of DNA.

Literature Cited

1. Archibald, F. S., DeVoe, I. W. 1979. *FEMS Microbiol. Lett* 6:159–62
2. Archibald, F. S., DeVoe, I. W. 1980. *Infect. Immun.* 27:322–34
3. Arnow, L. E. 1937. *J. Biol. Chem.* 118:531–37
4. Aswell, J. E., Haydon, A. H., Turner, H. R., Dawkins, C. A., Arceneaux, J. E. L., Byers, B. R. 1977. *J. Bacteriol.* 130:173–80
5. Atkin, C. L., Neilands, J. B., Phaff, H. 1970. *J. Bacteriol.* 103:722–33
6. Bailey, K. M., Taub, F. B. 1980. *J. Phycol.* 16:334–39
7. Bauminger, E. R., Cohen, S. G., Dickson, D. P. E., Levy, A., Ofer, S., Yariv, J. 1979. *J. Phys.* 40:523–25
8. Bayer, M. E. 1968. *J. Virol.* 2:346–56
9. Bayer, M. E. 1979. In *Bacterial Outer Membranes,* ed. M. Inouye, pp. 167–202. New York: Wiley
10. Bindereif, A., Garibaldi, J. A., Neilands, J. B. 1981. *Abstr. Annu. Meet. Am. Soc. Microbiol. Dallas,* p. 119
11. Bindereif, A., Montgomerie, J. Z., Neilands, J. B. 1982. *Abstr. Annu. Meet. Am. Soc. Microbiol., Atlanta,* Abstr. no. CCI, II6
12. Binns, M. M., Davies, D. L., Hardy, K. G. 1979. *Nature* 279:778–81

12a. Boyd, A., Holland, I. B. 1977. *FEBS Lett.* 76:20–24

13. Bradbeer, C. 1979. In *Vitamin B_{12},* ed. B. Zagalak, W. Friedrich, pp. 711–23. Berlin: de Gruyter
14. Braun, V., Hancock, R. E. W., Hantke, K., Hartmann, A. 1976. *J. Supramol. Struct.* 5:37–58
15. Buck, M., Griffiths, E. 1981. *Nucleic Acids Res.* 9:401–14
16. Buckley, J. T., Howard, P. S., Trust, T. J. 1981. *FEMS Microbiol. Lett.* 11: 41–46
17. Byers, B. R. 1974. In *Microbial Iron Metabolism,* ed. J. B. Neilands, pp. 83–105. New York: Academic
18. Cox, C. D. 1981. *J. Bacteriol.* 142: 581–87
19. Cox, C. D., Rinehart, K. L., Moore, M. L., Cook, J. C. 1981. *Proc. Natl. Acad. Sci. USA* 78:4256–60
20. Crosa, J. H. 1980. *Nature* 284:566–68
21. Crosa, J. H., Hodges, L. L. 1981. *Infect. Immun.* 31:223–27
22. Dills, S. S., Apperson, A., Schmidt, M. R., Saier, M. H. 1980. *Microbiol. Rev.* 44:385–418
23. Earhart, C. F., Lundrigan, C. L., Pickett, C. L., Pierce, J. R. 1979. *FEMS Microbiol. Lett.* 6:277–80

23a. Eaton, J. W., Brandt, P., Mahoney, J. R., Lee, J. L. 1982. *Science* 215:691–92

24. Emery, T. 1971 *Biochemistry* 10: 1483–88
25. Emery, T. 1978. In *Metal Ions in Biological Systems,* ed. H. Sigel, 7:77–126. New York: Dekker
26. Eng-Wilmot, D. L., vander Helm, D. 1980. *J. Am. Chem. Soc.* 102:7719–25
27. Ernst, J. F., Bennett, R., Rothfield, L. I. 1976. *J. Bacteriol.* 128:928–34
28. Fiss, E. H., Hollifield, W. C., Neilands, J. B. 1979. *Biochem. Biophys. Res. Commun.* 91:29–34
29. Fredericq, P. 1951. *Antonie van Leeuwenhoek J. Microbiol. Serol.* 17:103–6

30. Garibaldi, J. A., Neilands, J. B. 1956 *Nature* 177:526
31. Gibson, F., McGrath, D. I. 1969. *Biochim. Biophys. Acta* 192:175–84
32. Greenwood, K. T., Luke, R. K. J. 1980. *Biochim. Biophys. Acta* 614:185–95
33. Greenwood, K. T., Luke, R. K. J. 1981. *Biochim. Biophys. Acta* 660:371–74
34. Griffiths, E., Humphreys, J. 1980. *Infect. Immun.* 28:286–89
35. Guterman, S. K. 1973. *J. Bacteriol.* 114:1217–24
35a. Hamed, M. Y., Hider, R. C., Silver, J. 1982. *Inorgan. Chim. Acta* 66:13–18
36. Hantke, K. 1981. *Mol. Gen. Genet.* 182:288–92
37. Hartmann, A., Braun, V. 1980. *J. Bacteriol.* 143:246–55
38. Henning, U. 1975. *Ann. Rev. Microbiol.* 29:45–60
39. Hider, R. C., Drake, A. F., Kuroda, R., Neilands, J. B. 1980. *Naturwissenshaft* 67:136–39
40. Hollifield, W. C., Fiss, E. H., Neilands, J. B. 1978. *Biochem. Biophys. Res. Commun.* 83:739–46
41. Hollifield, W. C., Neilands, J. B. 1978. *Biochemistry* 17:1922–28
42. Hussein, S., Hantke, K., Braun, V. 1981. *Eur. J. Biochem.* 117:431–37
43. Inouye, M., ed. 1979. *Bacterial Outer Membranes.* New York: Wiley. 534 pp.
44. Irvin, R. T., McAlister, T. J., Chan, R., Costerton, J. W. 1981. *J. Bacteriol.* 145:1386–95
45. Irvin, R. T., McAlister, T. J., Costerton, J. W. 1981. *J. Bacteriol.* 145:1397–403
46. Janzen, J. J., Stan-Lotter, H., Sanderson, K. E. 1981. *Can. J. Microbiol.* 27:226–35
47. Kadner, R. J., Bassford, P. J. 1978. In *Bacterial Transport,* ed. B. P. Rosen, pp. 413–62. New York: Marcel Dekker
48. Kadner, R. J., Heller, K., Coulton, J. W., Braun, V. 1980, *J. Bacteriol.* 143:256–64
49. Klebba, P. E., McIntosh, M. A., Neilands, J. B. 1982. *J. Bacteriol.* 149:880–88
50. Kloepper, J. W., Leong, J., Teintze, M., Scroth, M. N. 1980. *Curr. Microbiol.* 4:317–20
51. Laird, A. J., Ribbons, D. W., Woodrow, G. C., Young, I. G. 1980. *Gene* 11:347–57
52. Laird, A. J., Young, I. G. 1980. *Gene* 11:359–66
53. Lammers, P. J., Sanders-Loehr, J. 1982. *J. Bacteriol.,* in press
54. Lankford, C. E. 1973. *Crit. Rev. Microbiol.* 2:273–331
55. Leong, J., Neilands, J. B. 1976. *J. Bacteriol.* 126:823–30
56. Leong, S. A., Neilands, J. B. 1981. *J. Bacteriol.* 147:482–91
57. Llinas, M. 1973. *Struct. Bonding* 17:135–220
58. Lodge, J. S., Gaines, C. G., Arceneaux, J. E. L., Byers, B. R. 1980. *Biochem. Biophys. Res. Commun.* 97:1291–95
59. Luckey, M., Neilands, J. B. 1976. *J. Bacteriol.* 127:1036–37
60. Luckey, M., Nikaido, H. 1980. *Proc. Natl. Acad. Sci. USA* 77:167–76
61. Luckey, M., Pollack, J. R., Wayne, R., Ames, B. N., Neilands, J. B. 1972. *J. Bacteriol.* 111:731–38
62. Luckey, M., Wayne, R., Neilands, J. B. 1975. *Biochem. Biophys. Res. Commun.* 64:687–93
62a. Maskell, J. P. 1980. *J. Microbiol. Serol. Antonie van Leeuwenhoek* 44:257–67
63. Mayer, H., Schmidt, G. 1979. *Curr. Topics Microbiol. Immunol.* 85:99–153
64. McIntosh, M. A., Chenault, S. S., Earhart, C. F. 1979. *J. Bacteriol.* 137:653–57
65. McIntosh, M. A., Earhart, C. F. 1976. *Biochem. Biophys. Res. Commun.* 70:315–22
66. McIntosh, M. A., Earhart, C. F. 1977. *J. Bacteriol.* 131:331–39
67. McLennan, B. D., Buck, M., Humphreys, J., Griffiths, E. 1981. *Nucleic Acids Res.* 9:2629–40
68. Moore, D. G., Earhart, C. F. 1981. *Infect. Immun.* 31:631–35
69. Moore, D. G., Yancey, R. J., Lankford, C. E., Earhart, C. F. 1980. *Infect. Immun.* 27:418–23
70. Muller, G., Winkelmann, G. 1981. *FEMS Microbiol. Lett.* 10:327–31
71. Murphy, T. P., Lean, D. R. S., Nalewajko, C. 1976. *Science* 192:900–2
72. Narrod, P., Williams, R. P. 1978. *Curr. Microbiol.* 1:281–84
73. Negrin, R. S., Neilands, J. B. 1978. *J. Biol. Chem.* 253:2339–42
74. Neilands, J. B. 1972. *Struct. Bonding* 11:145–70
75. Neilands, J. B., ed. 1974. *Microbial Iron Metabolism.* New York: Academic. 597 pp.
76. Neilands, J. B. 1977. In *Bioinorganic Chemistry II,* ed. K. N. Raymond, pp. 3–32. Washington: Am. Chem. Soc.
77. Neilands, J. B. 1981. *Ann. Rev. Biochem.* 50:715–31
78. Neilands, J. B. 1981 *Ann. Rev. Nutr.* 1:27–46
79. Neilands, J. B., Erickson, T. J., Rastetter, W. H. 1981. *J. Biol. Chem.* 256:3831–32

80. Neilands, J. B., Peterson, T., Leong, S. A. 1980. In *Inorganic Chemistry in Biology and Medicine,* ed. A. E. Martell, pp. 263–78. Washington: Am. Chem. Soc.
81. Nikaido, H. 1979. In *Bacterial Outer Membranes,* ed. M. Inouye, pp. 361–407. New York: Wiley
82. O'Brien, E. G., Gibson, F. 1970. *Biochim. Biophys. Acta* 215:393–402
83. Ohkawa, I., Shiga, S., Kageyama, M. 1980. *J. Biochem.* 87:323–31
84. Ong, S. A., Peterson, T., Neilands, J. B. 1979. *J. Biol. Chem.* 254:1860–65
85. Payne, S. M. 1980. *J. Bacteriol.* 143:1420–24
86. Payne, S. M., Finkelstein, R. A. 1978. *J. Clin. Invest.* 61:1428–46
87. Perry, R. D., San Clemente, C. L. 1979. *J. Bacteriol.* 140:1129–32
88. Peterson, T., Neilands, J. B. 1979. *Tetrahedron Lett.* 50:4805–4807
89. Pickett, C. L., Earhart, C. F. 1981. *Arch. Microbiol.* 128:360–64
90. Plastow, G. S., Holland, I. B. 1979. *Biochem. Biophys. Res. Commun.* 90:1007–14
91. Plastow, G. S., Pratt, J. M., Holland, I. B. 1981. *FEBS Lett.* 31:262–64
92. Pollack, J. R., Ames, B. N., Neilands, J. B. 1970. *J. Bacteriol.* 104:645–39
93. Pollack, J. R., Neilands, J. B. 1970. *Biochem. Biophys. Res. Commun.* 38:989–92
94. Postle, K., Reznikoff, W. 1979. *J. Mol. Biol.* 131:619–36
95. Quackenbush, R. L., Falkow, S. 1979. *Infect. Immun.* 24:562–64
96. Raymond, K. N., Carrano, C. J. 1979. *Acc. Chem. Res.* 12:183–90
97. Reeves, P. 1979. In *Bacterial Outer Membranes,* ed. M. Inouye, pp. 255–91. New York: Wiley
98. Regnier, P. 1981. *Biochem. Biophys. Res. Commun.* 99:844–54
99. Reynolds, P. R., Mottur, G. P., Bradbeer, C. 1980. *J. Biol. Chem.* 255:4313–19
100. Rogers, H. J., Synge, C., Woods, V. E. 1980. *Antimic Agents Chemother.* 18:63–68
101. Rosen, B. P., ed. 1978. *Bacterial Transport.* New York: Marcel Dekker. 684 pp.
102. Rosenberg, H., Young, I. G. 1974. In *Microbial Iron Metabolism,* ed. J. B. Neilands, pp. 67–82. New York: Academic
103. Schneider, R., Hartmann, A., Braun, V. 1981. *FEMS Microbiol. Lett.* 11:115–19
104. Sigel, S. P., Payne, S. M. 1982. *J. Bacteriol.* 150:148–55
105. Simpson, F. B., Neilands, J. B. 1976. *J. Phycol.* 12:44–48
106. Spiro, T. G., Saltman, P. 1969. *Struct. Bonding* 6:116–56
107. Stuart, S. J., Greenwood, K. T., Luke, R. K. J. 1980. *J. Bacteriol.* 143:35–42
108. Tait, G. H. 1975. *Biochem. J.* 146:191–204
109. Teintze, M., Hossain, M. B., Barnes, C. L., Leong, J., vander Helm, D. 1982. *Biochemistry* 20:6446–57
110. Uemura, J., Mizushima, S. 1975. *Biochim. Biophys. Acta* 413:163–76
111. van Tiel-Menkveld, G. J., deGraaf, F. K. 1980. *Abstr. Outer Membrane Proteins of Gram Negative Bacteria,* Lunteren Lectures on Molecular Genetics, Sept. 30–Oct. 3
111a. van Tiel-Menkveld, G. J., Oudega, B., Kempers, O., deGraaf, F. K. 1981. *FEMS Microbiol. Lett.* 12:373–80
112. Venuti, M. C., Rastetter, W. H., Neilands, J. B. 1979. *J. Med. Chem.* 22:123–24
113. Wagegg, W., Braun, V. 1981. *J. Bacteriol.* 145:156–63
114. Waring, W. S., Werkman, C. H. 1944. *Arch. Biochem.* 4:75–87
115. Warner, P. J., Williams, P. H., Bindereif, A., Neilands, J. B. 1981. *Infect. Immun.* 33:540–45
116. Wayne, R., Neilands, J. B. 1974. *Abstr. Papers, 168th Natl. Meet. Am. Chem. Soc., Atlantic City,* Sept., MICR 3
117. Wayne, R., Neilands, J. B. 1975. *J. Bacteriol.* 121:497–503
118. Weaver, C. A., Konisky, J. 1980. *J. Bacteriol.* 143:1513–18
119. Wettstein, F. O., Stent, G. S. 1968. *J. Mol. Biol.* 38:25–40
120. Williams, P. H. 1979. *Infect. Immun.* 26:925–32
121. Williams, P. H., Warner, P. J. 1980. *Infect. Immun.* 29:411–16
122. Winkelmann, G. 1974. *Arch. Microbiol.* 98:39–50
123. Winkelmann, G. 1979. *FEBS Lett.* 97:43–46
124. Winkelmann, G., Braun, V. 1981. *FEMS Microbiol. Lett.* 11:237–41
125. Woodrow, G. C., Langman, L., Young, I. G., Gibson, F. 1978. *J. Bacteriol.* 133:1524–26
126. Wookey, P. J., Hussein, S., Braun, V. 1981. *J. Bacteriol.* 146:1158–61
127. Wookey, P. J., Rosenberg, H. 1978. *J. Bacteriol.* 133:661–60
128. Yancey, R. J., Finkelstein, R. A. 1981. *Infect. Immun.* 32:592–99
129. Yancey, R. J., Finkelstein, R. A. 1981. *Infect. Immun.* 32:600–8
130. Yancey, R. J., Finkelstein, R. A. 1981. *Infect. Immun.* 32:609–13

Ann. Rev. Microbiol. 1982. 36:311–21

COPING WITH COMPUTERS AND COMPUTER EVANGELISTS

Micah I. Krichevsky

National Institute of Dental Research, National Institutes of Health, Bethesda, Maryland 20205

Most microbiologists make little or no use of computers. Those who do fall largely into three classes: clinical microbiologists, users of automated instruments, and generators of complex data bases.

The clinical microbiologist often was introduced to record keeping by the proximity of the hospital billing and patient record system. Recently, the clinical microbiologist has begun to use probability techniques to aid in the identification of possible pathogens. The calculations for the probabilities either are done uniquely in a computer or indirectly by consulting computer-generated lists of phenotypes by taxa.

Computers are finding use for process control in the laboratory and the fermentation plant. Instrumentation is appearing on the market that has built in computational capability. The controlled instruments include automated devices for identifying bacteria, determination of antibiotic resistance patterns, laboratory and pilot scale fermentors, and numerous instruments common to other disciplines such as scintillation counters, spectrophotometers, and gas-liquid chromatographs (often coupled to mass spectrometers). In all of these computer-instrument systems, the basic computations and the programming to perform them are supplied with the system.

Some of the heaviest use of computers in microbiology is for complex computations in such areas as numerical taxonomy, ecology, modeling of physiological pathways, and cellular interactions. This use of computers was anticipated by R. E. Buchanan. While Dean of the Agricultural Experiment Station at Iowa State University, he arranged contributory funding for the construction of the first computer. He foresaw the use of such a device

0066–4227/82/1001–0311$02.00

in solving differential equations used in modeling the growth of microorganisms (personal communication).

Those microbiologists requiring complex computations in their work were forced into the ranks of early computer users. The analogous situation is common to many areas of endeavor. Economists, mathematicians, actuaries, population geneticists, physicists, bankers, astronomers, all reached severely constricting limits without computers.

In microbiology, one of the earliest large scale uses of computers was for cluster analysis as part of the process termed numerical taxonomy. Although some early cluster analyses were performed by hand, the problems quickly demanded computers for their solution. More recently, other numerical and statistical techniques are being used in taxonomy and ecology. These include principle components analysis, discriminant analysis, probabilistic identification, and modeling, all of which make heavy demands on computer resources.

In addition to direct computer support of the microbiologist's laboratory efforts, computer resources find application in ancillary tasks. Literature searches through shared data bases (e.g. Medline operated by the U.S. National Library of Medicine) is a practical method of keeping up with one's specialty. Stand-alone word processors or time-shared text editors in large computers help in preparation of correspondence, reports, records, manuscripts, and the other paper work associated with the practice of microbiology. Some large culture collections are keeping records on computers and utilizing word processors to transform those records into their published catalogue of holdings.

Digital computers of a wide variety of sizes and capabilities have become widely available, through a combination of perceived needs and lower costs. This increased availability has resulted in a growth of users of computers. We now see whole sessions devoted to various applications of computer technology to microbiology at national and international meetings and workshops. No longer are the pioneers finding it necessary to prove the utility of computers in microbiology. The next phase has been entered wherein it is becoming increasingly fashionable to make use of computers. The computer has come of age as a useful tool as well as a status symbol in microbiology.

The status attribute of computer usage has resulted in examples (usually in larger institutions) of microbiologists working for the computers rather than the converse. Such situations can only provoke hostility and reduce productivity. In clinical microbiology, reports are generated by computers that swamp the reader in information. Periodic summary tables are issued whether or not they are desired. In reference and regulatory laboratories, phenotypic data are entered into computers in mindless profusion with no regard to the end use of the data. Conversely, clinical laboratories assess phenotypic characteristics of isolates, then store only the name of the organism and sometimes the antibiotic resistance pattern. This level of computer usage results in the classic "garbage in, garbage out" syndrome since the microbiologist generating the data received no reward for accurate and timely entry of data into the computer.

Since I make my livelihood by combining the disciplines of microbiology and computer technology, it is reasonable here to illustrate appropriate computer usage wherein the computer is not required but still serves the microbiologist well. In the area of computer-aided identification of bacteria, we find the modern-day equivalent of the John Henry legend. Microbiologists can, and most often still do, identify bacteria without resorting to computers or computer-generated lists. Computers can be effective in improving the identification qualitatively and quantitatively.

The U.S. Food and Drug Administration has a quality assurance program to monitor the ability of its field microbiologists to identify potentially harmful bacteria. Further, it has available 14 probability matrices for computer-assisted identification of commonly encountered heterotrophic bacteria (1). An informal comparison was made between an experienced laboratory staff and a group of neophyte microbiologists. The quality assurance scores of the experienced microbiologists had been much better than those achieved by the beginners. After the probability matrices were made available, the experienced personnel's scores did not change appreciably. However, the scores of the neophyte personnel improved to the approximate level of the experienced laboratory (F. A. Benedict, personal communication).

It is important to note that the computer was not replacing the microbiologist in the identification process. Rather, it was used as an informational aid. Proper use of the computer demands critical appraisal and interpretation of results by the microbiologist.

The aforementioned matrices were developed and are still used by the Bacteriology Department of the American Type Culture Collection (ATCC). The bacteriologists of the ATCC each have their areas of taxonomic specialty. They are often called upon to confirm the identity of strains in the Collection. It is estimated that use of the computer saves 2 to 6 hr of literature searching per culture. The differential is greatest with the pseudomonads. These savings accrue even when no identification is made by the computer, since the most common taxa are immediately eliminated as candidates (R. H. Gherna, personal communication).

In view of the usefulness of computers, as well as the increasingly common introduction of students to the mysteries of digital computation, more and more microbiologists will become convinced that computers are a must in their work. After a long lag phase, ever increasing numbers of microbiologists are anticipating using computers. Before a microbiologist can become computerized, or even make a rational decision as to whether or not to bother, many diverse parameters must be analyzed. Where computer information management is desired, choosing among the myriad of possible systems can be overwhelming for the newly converted. I discuss whether or not, and how, to get involved with computers and computer evangelists while minimizing the chances of getting burned.

First, some soul-searching is needed. A series of questions must be asked in turn. What is the pathway of information flow in place in the laboratory? What is its status? Is it working? Can information handling be improved without resorting to computers? What can not be done now? Is computer technology really needed or just fashionable?

In microbiology laboratory situations, computers will not save money, space, or personnel. Eventually, they will save time. They will increase information management and analysis capacity.

Some basic rules useful in contemplating use of computers are given by Shires (2).

Law I. If the task can be performed in any other reasonable way, avoid using a computer.

Law II. If thc computer can possibly be blamed for human error, it will be.

Law III. The information content of the output is inversely proportional to its weight.

An early decision that the microbioloigst must make is this: Should the microbiologist learn about computers or should the computer specialist learn about microbiology? Some of each is inevitable. However, it is far faster for the microbiologist to learn the rudiments of computers than the converse. Also, the more learned about computers, the better the microbiologist will be able to control the ultimate results.

The idea that a systems analyst possesses a universal framework, within which any information flow problem can be systematized, is misleading. At the very least, turning even the best computer scientists loose in the laboratory environment will yield delay and misunderstanding. The resulting system will often be inadequate and even counterproductive.

For systems analysis to do an efficient, useful job, the communication must be bidirectional. The microbiologist should become exposed to computer technology by such methods as short courses, trying a personal computer, reading source books, taking a beginning programming course (such as the BASIC language), or simply talking to people at length who have been through it before. It is not necessary to become a facile programmer.

In fact, there is the risk of becoming overconfident and wasting time writing programs best left to professionals. The aim is to learn the general organization, strengths, and weaknesses of computers as they apply to the tasks being considered.

Before investigating computer systems, an analysis of current and projected information flow should be done. Facets to be studied are data sources or generation, volume, data capture, information management, analysis, reports, and retrieval options. The results of this investigation should be used, first, to decide the desirability of computer facilities and, second, as the basis for specifications of a computer system.

The most desirable computer system would appear to be one that is ready made and tested for the particular tasks to be performed. Unfortunately, existing systems are obsolescent almost as soon as they are finished. Often, they are not completely compatible with the exact needs. In most cases, modifications will be required or serious compromise will result. Adaptation of an existing system may be possible and desirable. However, evaluating the resources required should be as carefully performed as if one is starting afresh.

In many cases, programmers find it more efficient to completely rewrite someone else's programs rather than to attempt modification. Since efficient programs are seldom compatible among dissimilar computers (even from the same manufacturers), it should be assumed that programming will be required. This function should be left to professional programmers, either hired staff or under contract. The cost of programming should be carefully assessed since such costs now are higher than the cost of the computer to run the programs. This cost ratio obtains for all sizes of computers.

The types of computers with which microbiologists are likely to be involved may be classified according to size and functional capability. The classification that follows is not for the computer purist. Further, there is great overlap in the border areas between classes.

The smallest computers are termed microcomputers. Usually, they can perform only one task at a time. That is, only a single program can be operational.

Microcomputers are often used as part of instrument packages. In these cases, the programs for instrument control and data reduction to manageable amounts are often supplied by the vendor. One of the problems that surfaces with such systems is that the end product of the data reduction process may be paper printout of laboratory results, which must then be laboriously managed in the traditional manner. An automated laboratory may be swamped with unmanageable data records.

Another incarnation of the microcomputer is the so-called personal computer. These computers can provide a great deal of computing power to the microbiologist. More and more programs are becoming available. For small data bases, they are a most useful tool. The biggest risk to the microbiologist are overstepping the limits of the machine and becoming so enamored of the computer that large amounts of time are spent in programming and discovering the wonderful world of computers. The smaller the computer, the more difficult it is to program efficiently. Also, a small computer can be tied up for hours performing calculations (if they fit at all) that are done in seconds in a large computer. One of the most startling examples of this problem is cluster analysis for numerical taxonomy. The number of comparisons required to construct a similarity triangle is $[n(n\text{-}1)]/2$, where n is the number of strains in the study. Adding strains to a study quickly overwhelms the capacity of microcomputers.

Another form of microcomputer is the word processor. Initially, such machines were limited to exactly the task implied by the name. Special purpose programs were installed to manipulate characters, words, lines of text, paragraphs, and pages. Word processors were the idiot savants of the computer world. Recently, one sees programs for personal computers that will do word processing (also termed text editing) and a word processor that can calculate. A dramatic increase in flexibility of a word processor can often be gained by incorporating the electronics to use them as computer terminals. Some configurations of word processors or personal computers are thus termed smart terminals. However, a dumb owner of a smart terminal is one who does not make the vendor demonstrate, prior to purchase, the compatibility of the microcomputer/terminal combination with the specific large computer to be used.

Another consideration for evaluating the applicability of a word processor to the projected needs is storage space and how it is used. The usual size manuscript will fit easily on a single floppy disk. (Floppy disks are cheap, reliable, magnetic storage media used in smaller computers.) However, it is a cumbersome exercise to keep track of data or text when it becomes distributed on multiple disks. In such cases, the operator becomes part of the program. It is not practical to cross disk boundaries to combine or search data sets.

The minicomputer differs functionally from the microcomputer in its ability to operate more than one program at a time. Thus, multiple instruments or users may interact with a single computer. This multiple task operation is possible by virtue of the memory size of the computer itself and the increased speed and complexity of the peripheral devices connected to the computer. The peripheral devices may include fixed disks, reel-to-reel magnetic tape drives, line printers, multiple terminals, etc. The advantages of the minicomputer are with the ability to store and process markedly more data than does the microcomputer and still be within the budget reach of a single laboratory.

Modifiability is an important requirement for computer systems. Systems must be designed for current needs, but users often fail to realize these needs will change. Often the mere presence of the computer will cause the change. For example, the availability of monthly reports leads to the idea that a weekly report would be very useful. The direction of change is almost always towards expansion.

The initial cost of the computer system may escalate rapidly and surprisingly. Careful needs assessment must be performed. Possible hidden expenses can come from the vendor-supplied programs not doing all the tasks required. The physical space and environmental conditions may be found inadequate, because of the size and the power requirements of such computers. Special air conditioning can be very costly. Power and electrical grounding requirements may exceed the available facilities, especially in older buildings. The costs of incrementally increasing system capability should be a factor in decision making unless it is certain the requirements will not exceed the initial system for the foreseeable future.

The largest computer system microbiologists are likely to utilize are those operated by a computer center. There are two modes of use to be considered, one markedly cheaper than the other. The cheaper mode is termed batch operation. This is the traditional way computer centers started. That is, the user brings the task to the computer center in machine-readable form (often punched cards), the job is placed in a queue, and the results are returned to the user after the job is executed. There is no mechanism for the user to interact with the job during execution. All options available within a program must be selected at the outset.

The second mode (which can coexist with the first) is called on-line or time-sharing operation. (These are not synonomous terms, but will be treated as such for this discussion.) The user of a time-sharing system interacts directly with the computer through a typewriter terminal (or other less common device such as a joy stick or light pen). Interactive time-sharing is done to make large computer facilities available to many users simultaneously at a cost that can fit into a laboratory budget.

Large computers have a useful attribute in most modern computer centers, i.e. their large libraries of available programs. Commonly, one finds data retrieval, statistical, modeling, curve fitting, and file management programs, or packages of programs.

Although large computers relieve the user of much data management responsibility, they have their disadvantages. The foremost among these is that users are at the administrative mercy of the computer center managers.

The center's problems become everybody's problems. If the center decides to change computers, everyone awaits with trepidation the conversion problems. Acquiring needed increased capacity becomes a community process. Often the users conform to the center's requirements rather than the more desirable converse.

With time-sharing systems, the user must have access to or invest in a terminal and telephone line. If the call is not local, the telephone charges must be considered, as well as those for computation. Care should be taken to monitor the accretion of data on disk storage. Storage costs mount up quickly in time-sharing systems. Also, on-line storage may be limited.

As a time-sharing computer becomes more popular, the response degrades. What starts as a highly responsive system becomes annoyingly slow. The systems tend to be complex to operate, which is the price of versatility.

There are two great economic advantages to time-sharing for the microbiologist who is a neophyte computer user. They are low initial costs and the requirement to pay only for the resources used in the learning process. I return to this point later.

Each type of computer system has its own cadre of evangelical advocates. The most common evangelists are the users of any type of system who are convinced that they have found the true way. Use their method, program, algorithm, computer, etc, and it will be easy! Their need for self-justification should not be fulfilled at another's expense.

Vendors of text-editing systems, micro- and minicomputers, and time-sharing and other service bureaus will dazzle with how easy their system is to operate. Not so: There is no free lunch. The price of versatility has to be richness (i.e. complexity) of instruction set. To use the full capabilities of any of these systems, time must be invested. If the computer moves mountains at the touch of a button, it is necessary to know how the machine is doing it and where the pieces will end up.

There is the class of evangelists comprised of new amateur users of computers. The new graduate from a BASIC programming course, or the owner of a personal computer, or the parent of a budding computer science genius may be honestly convinced computers can do anything with a few simple instructions. Underestimating the complexity of the problem is probably the single most common cause of disappointment with computers. This type of failure is avoidable by careful analysis of the needs.

Most microbiological data-processing requirements involve large amounts of data. The manipulations are not often complex. Rather, the number of operations is great. Because of the large number of observations that a microbiology laboratory will want to enter into the computer (the input) and the length of reports and summaries to be obtained from the computer (the output), a system should be chosen that is not input/output limited (I/O bound in the jargon of the trade). Consider a simple example. The data for 100 isolates described by 50 test results of each, isolate designations of 5 characters (ch), and 20-character description of source of isolation occupy 7500 characters of storage. Typical output printers would take 250 sec (at 30 ch/sec), 62.5 sec (at 120 ch/sec), or 15 sec (using a 400 line/min printer). At this level, the differences may not seem important. Typically, a number of microbiologists each must enter this amount of data on a daily basis (at a rate of less than 5 ch/sec). Then, each should print it out both for a written record and for the most important step of verification of accuracy. With slower printers the computer will be occupied for considerable amounts of time with these basic tasks. Further, the users of the system will have to be scheduled to accomodate the speed of the printer.

The strategies to overcome these difficulties in multiple user situations are to obtain a faster printer or multiple slow printers. The individual situation will dictate preference.

A second consequence of large amounts of data is the need for file management. Large multi-user computers must have available file-management facilities (although they vary widely in ease of use). Minicomputers with nonremovable disks often mimic the facilities of the large computers. This is reasonable since today's minicomputers are comparable in power to the large machines of a decade ago.

File-management problems appear with smaller minicomputers and microcomputers, largely because of cost reduction devices such as removable disks. The very attribute that makes such devices attractive causes the problem. Storing data on differing removable disks, with individually limited capacity, makes the date in files cumbersome to compare, combine, or even search. If archival files of data must be accessed and combined for specific studies, their structure must be carefully thought out to ascertain the degree of effort required if removable disks are contemplated. The disks may be cheap, but the labor to use them may be very expensive. Further, persons other than the operator often will not be able to access the data in the operator's absence.

The most frequent operations a microbiologist will require of a computer are the ones that should be foremost in the evaluation criteria of a computer system. As stated before, complex computation is not high on the list. Where a small, hence inexpensive, computer will meet the frequency test, it may be the best choice if combined with the facility to communicate with a large computer for the occasional number-crunching task.

Entry, storage, editing, and retrieval of specific records (e.g. data on single strain) or items (e.g. all mannitol-utilizing strains) is probably the most important computer-aided task that concerns the microbiologist. All of these involve accessing and reading specific files of data and then modifying files. If these tasks are easy to do, the computer system will be successful and accepted. If they are cumbersome and complicated, the computer often will be treated as an enemy.

Simple search logic will suffice for most of the microbiologist's needs. Any computer should be able to search on the basis of positive or negative results for yes/no data. Numerical data can be searched by greater than, less than, or equal to.

By combining these elements with a logical "and" (i.e. the elements on both sides of the "and" must be true), most search constructions will be satisfied. Those that are not can be satisfied with a combination of "and" and "or" (i.e. the elements on either side of the "or" must be true). There is little advantage to more complicated logic constructions if there is much cost associated with the enhanced logic.

A decision parameter may be based on how the computer handles missing data. Missing data conventions are important in both searching and statistics. Most microbiological data sets will contain missing data if only because some natural isolates will fail to grow on transfer.

Heirarchical searches can be performed by sequential search and elimination of unwanted elements if the computer allows retention of the results of a previous search in either main memory or a secondary holding file. Simple sequential searching is considerably easier for the microbiologist to learn than the more complicated strategies that add to the system costs.

As noted earlier, many complex analytic calculations are performed by microbiologists. However, simple calculations should be easy to perform as they will be used most of the time. Simple tabulations (i.e. numbers of positive, negative, and missing tests results for yes/no data and maximum, minimum, average, and standard deviation for numerical data) will answer the vast majority of questions. Blind application of complex statistical or numerical procedures to data wastes resources and can be misleading unless expert consultation is utilized.

In the constant struggle between good and evil, the computer's ability to aid in generation of reports and tables seems to fall on both ends of the spectrum, but seldom in the middle. Some of the worst computer-aided sins have been committed in the name of comprehensive (i.e. voluminous, unwanted, and unusable) reports. Universal good is obtained when the report contains only the information needed to answer the question asked. The facility with which a computer system and its associated staff can respond to individual report requests is an extremely important evaluation criterion. Inflexibility may well be the original sin of computer technology.

Any computer system being considered should be capable of computer-to-computer communication either at the time of installation or by later addition. For small computers, this capacity represents insurance against size obsolescence. For all computers, it allows taking advantage of programs resident in other systems in the most economical and expeditious manner. If the other computers are known, this communication should be a planning priority. The communication medium may be direct through wires, indirect through punched cards or paper tape, or magnetic tape or disk.

Be not led into temptation by the promise of natural language commands or data structures. In the former case, remember that English (by far the most common base language in use in computers) contains a great deal of ambiguity in construction as well as redundancy in words. The ambiguity leads to unanticipated and often undetected misunderstandings unless unnatural formalism is introduced into the command structure. The formalism leads back to programming languages.

The redundancy renders searching of natural language records difficult and inefficient. For example, the source of an isolate could be a a ROCK or STONE, a FINGER or DIGIT, a STREAM or RUN, SOIL or DIRT, PIG or SWINE, etc. The answer to this problem is the controlled vocabulary. The use of free text is limited as much as possible. Lists of approved (or key) words are maintained. Where feasible, data are entered through use of check lists or multiple-choice selections (using numbers or letters to denote choices). The simplicity with which data can be entered and searched should be heavily weighted in choosing among alternative computer systems.

Programming strategies also affect the choice of computer systems and how to handle such systems after the choice is made. Wherever possible, the programming should be done by professional programmers under detailed instruction by the microbiologist. Asking the programmer to present the microbiologist with a menu from which to select the programming options will usually result in the laboratory working for the computer. Programmers are human beings and will design programs based on their own tastes and experiences. If presented with good user-requirement specifications by the microbiologist, the end product will more likely live up to the computer's potential as a helpful tool in the laboratory.

A possible strategy for getting started with minimum capital investment and minimum risk is to use a time-sharing computer and contract for the programming. This can be done even for the instrument-generated data with the proper interfacing. Any data entered in such a system can be translated for a permanent system. Such a sequence will allow useful work to be performed while the microbiologist gains the experience required to make educated decisions.

Literature Cited

1. Johnson, R. 1979. Computer-Aided Identification. *FDA By-Lines* 9:235–50
2. Shires, D. B. 1974. *Computer Technology in the Health Sciences,* p. x. Springfield, IL: Thomas. 140 pp.

Ann. Rev. Microbiol. 1982. 36:323–43

INTESTINAL MICROBIOTA OF TERMITES AND OTHER XYLOPHAGOUS INSECTS[1]

John A. Breznak

Department of Microbiology and Public Health, Michigan State University, East Lansing, Michigan 48824

CONTENTS

INTRODUCTION

One of the most fascinating examples of nutritional symbiosis is that displayed between termites and their intestinal microbiota: a symbiosis that permits termites to live by xylophagy. Yet despite a long recognition of this symbiosis, we are only beginning to understand the microbiological and biochemical details of its basis. A major purpose of this review is to update

[1]This review is dedicated to Professor R. E. Hungate, for his pioneering work on rumen and gastrointestinal fermentations, and whose early, incisive studies on termites have been an inspiration to my students and me.

0066–4227/82/1001–0323$02.00

our information on the intestinal microbiota of termites, and that of some less widely known xylophagous species, with particular emphasis on the role of gut microbes in the insects' nutrition. In this review the term xylophagous is used broadly to include not only consumption of woody materials, but cellulose-rich detritus as well. Similarly, the terms symbiosis and symbiont(s) are used in the broad sense of de Bary (4), i.e. the permanent or semipermanent association of two dissimilar organisms. More specific terms are used where appropriate and where possible.

Several relatively recent reviews have appeared on the general biology and behavior of termites (71), their ecology (27, 75), nutrient dynamics (72), and microbial symbionts (23), and these are cited fairly frequently herein as general references. However, critical original research papers, old or recent, are cited specifically. The focus of the present review is on research published in the last 10 years.

BIOLOGY OF TERMITES

Termites are insects belonging to the order Isoptera and are characterized by their colonial behavior, the development of morphologically distinguishable caste members in a colony, their incomplete metamorphosis, and the formation, by alates, of fore- and hindwings that are similar in size and veination (71). However, the most-oft-noted attribute of termites is their xylophagous habit. Depending on the species, food preferences range from wood (either sound or extensively decayed) to leaves, grasses, humus, detritus, and herbivore dung. Some even feed on soil, whereas others have evolved the intriguing habit of cultivating fungus gardens as a nutrient resource (110). There are nearly 2000 recognized species that can be quite diverse in terms of their biology and behavior (70). Overall, however, the diet of termites can be characterized as rich in cellulose, hemicelluloses, and lignin, but poor in nitrogen, i.e. termites are oligonitrotrophs.

"Worker" larvae dominate the nutrient dynamics of a colony by virtue of sheer number. It is the workers that forage for, and process, most of the colony's food. In addition, workers feed the nutrient-dependent members of a colony such as newly emerged and early-instar larvae, soldiers (whose mandibles are differentiated for fighting, not feeding), and reproductives. This is done by regurgitating masticated (stomodeal) food, or by expelling hindgut contents (proctodeal food), for recipients. These processes, termed trophallaxis, are quite common.

Although the destructive potential of termites is often cited (54), their ecological significance is frequently overlooked. Termites, with the aid of their intestinal microbiota, constitute important decomposers of plant litter in our biosphere. In temperate climates, at sites where termites are known to be abundant, their impact on litter decomposition rivals that of "large decomposer" groups of invertebrates such as earthworms (75). In tropical regions however, decomposition of plant material by termites is similar to that effected by grazing mammals, simply because their biomass density is so large (10–20 g/m^2) (128).

DIGESTION IN TERMITES

The alimentary tract of termites consists of three main divisions: the foregut (stomodeum), the midgut (mesenteron), and the hindgut (proctodeum). Its histology, cytology, and phylogenetic variations have been described in detail (96). However, pertinent to this discussion is the location of the intestinal microbiota and sites of nutrient absorption. The clear bulk of the microbiota is housed in the hindgut (usually in a dilated portion of the proctodeum known as the paunch), which is also a major site of nutrient absorption. Since an enteric valve prevents refluxing of hindgut contents to the midgut, absorption by the latter is limited to soluble nutrients present in the passing food bolus, substances liberated from the food by salivary or midgut enzymes, or hindgut nutrients by trophallaxis.

Cellulose and hemicelluloses (xylans and mannans) undergo appreciable degradation on passage through the termite gut (65–99%) (42, 127), and the assimilation efficiency of wood feeders is quite good (54–93%) (127). Values for lignin digestion vary widely and are discussed later. The bulk of polysaccharide dissimilation occurs in the hindgut, and all available evidence indicates that the hindgut microbiota is the driving force of such activity (except perhaps for fungus-growing termites, which are discussed later). In phylogenetically lower termites (families Mastotermitidae, Kalotermitidae, Hodotermitidae, Rhinotermitidae, and Serritermitidae), which constitute about 25% of all species, the hindgut microbiota includes bacteria as well as unique genera and species of oxymonad, trichomonad, and hypermastigote protozoa found almost nowhere else in nature (55). Most of the protozoa are cellulolytic anaerobes and appear to be the primary agents of cellulose decomposition. The phylogenetically higher termites lack protozoa and must depend largely on bacterial fermentative activities, although surprisingly little work has been done on this group.

The general physico-chemical characteristics of the hindgut ecosystem in both groups of termites is similar. The microbe-packed regions of the hindgut have a pH usually of 6.0 to 7.5 (14, 43, 72) and appear anaerobic (14, 123) (compare 14 and 123) with an E_0' of –230 to –270 mV (123). Anaerobicity of the hindgut is consistent with the oxygen sensitivity of the hindgut protozoa, bacterial uricolysis, and bacterial N_2 fixation that occur in this region, as well as the demonstration of strict anaerobic bacteria, and methanogenesis (23), in hindguts (discussed below).

The importance of protozoa to cellulose utilization by lower termites was established in the early 1900s (reviewed in 23) and has been confirmed by some recent studies (81, 84). From early work by R. E. Hungate in particular (57, 59), the following scheme for symbiotic cellulose utilization emerged. Wood particles passed to the hindgut are endocytosed by protozoa, and the cellulose portion is fermented anaerobically within the protozoa to CO_2, H_2, and acetate. The latter compound, after secretion from the protozoa, is absorbed from the hindgut and oxidized aerobically by termites for energy. Consistent with this scheme is the dominance of acetate in hindgut fluid of lower termites. Odelson & Breznak (98) found that 94 mol% of all volatile fatty acids in hindgut fluid of *Reticulitermes flavipes* was acetate, which was present at a concentration of 74.7 mM. By using ^{14}C-labeled substrates, they were able to show that about 80% of the acetate derives from cellulose carbon, whereas about 20% derives from hemicelluloses (D. A. Odelson, J. A. Breznak, manuscript in preparation). Interestingly, acetate is also a dominant volatile fatty acid in hindgut fluid of the higher termites *Nasutitermes nigriceps* (76) and *Microcerotermes edentatus* (67). Acetate is not only an oxidizable energy source for termites, it is an important precursor for synthesis of amino acids (83), cuticular hydrocarbons (19), and terpenes (108).

Termites possess a variety of carbohydrases in their intestinal tract including, of course, cellulases and beta-glucosidases (72, 85, 87, 97, 122, 129), although there is evidence that some enzymes of the cellulase repetoire may be synthesized by the termites themselves. However, in almost every case such claims have been made by using carboxymethylcellulose or reprecipitated cellulose as the assay substrate (68, 85, 97, 104, 121, 129). These substrates will permit demonstration of C_x-type cellulases (active against noncrystalline cellulose and soluble derivatives or degradation products of cellulose), but not C_1-type cellulases (active against crystalline cellulose) (109). Only two reports give us reason to suspect that termites secrete a C_1-type cellulase: one regarding a higher termite (*Trinervitermes trinervoides*) (105), the other a primitive lower termite (*Mastotermes darwiniensis*) (122). It is not impossible for an insect to secrete its own C_1-cellulase. Lasker & Giese (73) showed quite convincingly that certain silverfish (*Ctenolepisma lineata*) could digest cellulose without the aid of their gut microbiota. However, even if termites do secrete certain cellulases, the importance of the hindgut microbiota to cellulose digestion (particularly the hindgut protozoa of lower termites) is undeniable (23).

TERMITE HINDGUT PROTOZOA

Without a doubt, one of the most significant breakthroughs in the study of termite hindgut protozoa has been made by Yamin, who recently achieved axenic cultivation of major cellulolytic species in vitro (131, 133). By using a cellulose-based medium of an ionic composition consistent with that of termite hindgut fluid, he successfully isolated *Trichomitopsis termopsidis* (131) and *Trichonympha sphaerica* (133) from hindguts of *Zootermopsis* sp. Several factors appear to have been critical to Yamin's success, which included the following: the use of strict anaerobic techniques; the inclusion of antibiotics in primary cultures to eliminate bacterial contaminants; and the incorporation of serum, yeast extract, glutathione, and dead (heat-killed) rumen bacteria in the culture medium along with cellulose. Resulting cultures were not only free of live extracellular bacteria, but of endosymbiotic bacteria as well. Both protozoan species required cellulose for growth: For *T. termopsidis,* cellulose could not be replaced by glucose, cellobiose, carboxymethylcellulose, glycogen, inulin, or rice starch (131). Intriguing was the absolute requirement by cells for killed, mixed rumen bacteria, which were endocytosed and digested within food vacuoles. Cells of *Escherichia coli, Bacillus subtilis, Pseudomonas maltophilia,* or a *Clostridium* sp. alone would not suffice. Undoubtedly, the rumen bacteria were the in vitro correlate of termite hindgut bacteria, on which the protozoa graze in vivo. It is not yet known which species of rumen or termite hindgut bacteria are critical to the protozoa's nutrition.

The requirement for cellulose by *T. termopsidis* and *T. sphaerica* implied that cells possess cellulases and dissimilate the polysaccharide for energy. This was subsequently verified by using [^{14}C]cellulose as substrate. Both species fermented cellulose to CO_2, H_2, and acetate (132, 133), the same major products identified by Hungate (57, 59), who used suspensions of protozoa harvested from termite hindguts. Moreover, crude extracts of *T. termopsidis* possessed carboxymethylcellulase (134), as well as cellobiase, coenzyme A-dependent pyruvate : ferredoxin oxidoreductase, and hydrogenase (132). The latter two activities were probably associated with hydrogenosomes, organelles that have been described as the anaerobic analogue of mitochondria (90) and that appear to be present in the protozoa (131, 133). Importantly, Yamin critically examined the importance of *T. termopsidis* to termite nutrition (134). When defaunated *Zootermopsis* (i.e. with protozoa removed) were reinfected with *T. termopsidis,* they survived appreciably longer than did non-reinfected controls and as well as those reinfected with a mixed population derived from hindgut contents. This experiment further established the mutualistic role of cellulolytic protozoa in termite nutrition. A noncellulolytic, apparently commensal flagellate (*Tricercomitis divergens*) has also been isolated from termites (*Cryptotermes cavifrons*) (130).

Investigations so far have employed pure cellulose as a growth substrate, so we still know little about protozoan metabolism of hemicelluloses or of the native substrate itself—wood particles. However, considering the extensive dissimilation of wood polysaccharides on passage through the gut, and the ability of protozoa to endocytose wood particles into food vacuoles, the protozoa probably hydrolyze hemicelluloses as well. Presumably, hydrolytic enzymes of these cells are quite effective at stripping wood polysaccharides from their complex with lignin. Whether or not protozoan activity accounts for all observed dissimilation of polysaccharide in situ is unknown, yet it seems unlikely. Yamin (132) and Hungate (59) found that 10–30% of the cellulose carbon fermented by protozoa in vitro could not be recovered in acetate and CO_2. Yamin suggested that some of the missing carbon was present as soluble intermediates (oligomers) of cellulose hydrolysis released from the protozoa. It may be advantageous for protozoa to release hydrolysis intermediates in situ, since, if usable by bacteria, the protozoa could effectively culture a population of bacteria on which they might subsequently feed. The requirement for killed rumen bacteria by axenic cultures, the mixed culture studies of Trager (120), and the common occurrence of bacteria attached to (or within) hindgut protozoa (22, 23) all lend credence to such speculation.

The ability to successfully culture *T. termopsidis* and *T. sphaerica* make them important objects for future research. Detailed studies of their biochemistry, physiology, and molecular biology are now possible and will add greatly to our understanding of their evolution and symbiotic interactions. Such pursuits may have practical consequences as well. It is not unreasonable to assume that studies of termite hindgut protozoa could suggest novel strategies for anaerobic bioconversion of wood (our most abundant renewable resource) to feedstocks, fuels, or vendable chemicals.

Response to JH Analogues

Growth, moulting, and morphogenesis in insects are processes regulated by complex interactions between neurosecretory secretions, hormones, and target tissues. One such hormone, juvenile hormone (JH), appears to be a farnesol derivative secreted by cells of the corpora allata (35). Moulting and caste differentiation in termites are influenced by the JH titer of the hemolymph (86). JH tends to suppress reproductive development in insects, and in termites it can induce morphogenesis of larvae and nymphs into soldiers (53, and references therein). Superfluous soldier production in a colony would be detrimental, because they are a nutrient-dependent caste. Consequently, JH analogues have been tried for use in termite control, and with unexpected and interesting effects on hindgut protozoa.

The JH analogue Methoprene triggered various concentration-dependent lethal effects in *Reticulitermes flavipes* (56). At low concentrations an equal proportion of presoldiers was produced. However, at high concentrations 80% of the mortality occurred in larvae and was not due to ecdysis failure, suggesting a possible effect on hindgut protozoa. Cell counts subsequently revealed that as little as 0.032 to 0.064 mg of Methoprene/food wafer caused virtually complete elimination of protozoa by the time presoldiers began to appear. The decrease did not appear to be due to expulsion of protozoa from hindguts, as dead protozoans were readily seen in gut contents of treated individuals. This study indicated that one mode of Methoprene action involved toxicity to protozoa, causing larvae to starve to death. Separate experiments (53) showed that Methoprene was not overtly toxic to termites at concentrations low enough to still induce defaunation. The biochemical basis for Methoprene toxicity to hindgut protozoa is not yet known, however.

Response to Wood Extractives

Termite researchers have long recognized that certain woods are resistant to termite attack, and resistance appears due in part to trace compounds in wood that are either repellant, distasteful, or toxic to termites (9). Some of these compounds have been identified as quinone derivatives (32). In quests for effective and environmentally safe chemicals for termite control, efforts have been made to screen various woods for antitermitic activity and extract the compounds responsible. In theory, application of such extractives to susceptible woods might repel or kill foraging termites and constitute a safe means of wood preservation. Recently, it has become apparent that toxicity of some woods, and their extractives, to lower termites is related to their deleterious effect on hindgut protozoa. Mauldin et al (82) found that when *Reticulitermes flavipes* was fed certain American woods, protozoan populations were completely eliminated or drastically reduced. Magnolia, holly, and white mulberry were effective in eliminating protozoa while allowing greater than 50% termite survival during the 3-week test period. These three species might be particularly useful under natural conditions, since enough time might be available for foraging termites to spread the toxic factor(s) through the colony by trophallaxis. Similar results were seen with *Coptotermes formosanus* (34), but in this case baldcyprus, yellow poplar, and northern white cedar were most effective. Among the protozoa of *C. formosanus, Pseudotrichonympha grassii* appeared most sensitive.

Further studies (33) revealed that for *R. flavipes,* solvent extracts of holly, yellow poplar, sassafras, and catalpa reduced protozoan populations from 40,000/termite to less than 500/termite in 9 days. The specific toxic moieties in such extracts remain to be identified, and clearly toxicity could be either direct or indirect. Nevertheless, continued studies along these lines may prove to be of great practical value.

TERMITE GUT BACTERIA

Information on termite gut bacteria up to 1974 was reviewed by Breznak (23), who underscored the paucity of knowledge about them. Fortunately, some significant advances have been made since then.

Ultrastructural studies have revealed that a dense and morphologically diverse bacterial flora colonizes the hindgut of lower and higher termites (15, 16, 18, 26). Less colonization is apparent in the midgut (16, 17, 26). Bacterial colonization of termite guts was found to be dominated by, but not restricted to, adhesive interactions—with gut tissue, with protozoa, and with each other. Recently, bacteria have been isolated from guts of lower and higher termites (44, 111, 117). Most isolates proved to be species of strict or facultative anaerobes including *Streptococcus, Bacteroides,* various Enterobacteriaceae, *Staphylococcus,* and *Bacillus.* In an effort to gain insight into the role of bacteria in the hindgut fermentation, Schultz & Breznak (112) studied mono- and co-cultures of two major isolates from *Reticulitermes flavipes:* a *Streptococcus* and a *Bacteroides.* It was found that in co-culture, glucose was fermented primarily by *Streptococcus,* yielding lactate. Lactate produced by the streptococci then supported the growth of *Bacteroides,* which fermented this compound to propionate, acetate, and CO_2. It was suggested that cross-feeding of lactate between such species might constitute one aspect of the overall fermentation in vivo.

A number of bacterial isolates have been found to possess C_x-type cellulase activity (69, 117, 118); however, to date there is no convincing evidence that bacteria are quantitatively important to cellulose hydrolysis in vivo—at least not in lower termites that possess cellulolytic protozoa.

Some of the most characteristic inhabitants of termite guts are spirochetes that range in size from 0.2×3.0 to 1.0×100 μm. Information on termite gut spirochetes has been reviewed by this author (22, 23), and a discussion of their taxonomy and attachment to protozoa is forthcoming (24). Unfortunately, we still know almost nothing about their role in the hindgut ecosystem, as none have been isolated in pure culture. However, they may be important to termite vitality. Eutick et al (45) observed decreased survival of *Nuastitermes exitiosus* when spirochetes were eliminated from the gut with antibiotics.

Role in N_2 Fixation

Oligonitrotrophy in termites is particularly apparent for species that can thrive on sound, decay-free wood containing as little as 0.03–0.1% N and exhibiting a C/N ratio of 1000/1 (38). Since termite tissues contain N in amounts similar to that of other animals (72, 102), one might intuitively suspect that termites possess efficient means to acquire and/or conserve N. This seems to be the case, and it appears that gut bacteria can aid termites in both these processes.

The use of N_2 by termites as an N source had been speculated ever since Cleveland's claim that termites could live "perhaps indefinitely on a diet of pure cellulose" (36). However, not until recently has this speculation actually received good experimental support. With the acetylene reduction assay (52), N_2 fixation has been inferred for a wide variety of termite species (Table 1). Since N_2 fixation is a strictly prokaryotic phenomenon, it has been generally assumed that the fixation is mediated by gut bacteria.

In Table 1, the inferred rates of N_2 fixation have been converted to a TDN equivalent, which is the time it would take termites to double their N content if N_2 fixation rates remained constant, and if N_2 was the sole source of N. A striking feature of the tabulated data is the wide variation in N_2 fixation ratcs, intcr- as wcll as intraspccifically. Considering rates in, say, *Reticulitermes flavipes* workers, it would be difficult to argue that N_2 fixation is important to this termite's N economy, since the TDN value is nearly 1000 years! A similar conclusion might be drawn for *Zootermopsis* sp., *Cryptotermes brevis,* and others. By contrast, N_2 fixation would seem more important to *Nasutitermes* species and others whose TDN values are relatively low. Whole colonies of *Nasutitermes corniger* revealed strikingly high rates of N_2 fixation, rates that in theory could provide all the N needed for one to two doublings of the colony population per year. This is undoubtedly significant, yet it would help greatly to know exactly how fast termite populations increase in the field. *Macrotermes* sp. in East Africa turnover four populations annually (40); however, good data for biomass production and colony growth of most other termites are scarce and sorely needed (95).

Several factors might account for the large inter- and intraspecific variations seen in Table 1. One of these is the N content of food being eaten by termites prior to assay. For example, Breznak et al (25) found that fixation rates of *Coptotermes formosanus* varied inversely with the amount of NO_3^- or NH_4^+ added to a filter paper food disc. Variations up to 200—fold were observed. Moreover, a significant change in fixation rate could be detected within 5 hr of a dietary shift. Since nitrogenase synthesis by bacteria is repressed by readily utilizable sources of combined N (28), the modulation of N_2 fixation activity in *C. formosanus* presumably reflected

Table 1 Nitrogen fixation (acetylene reduction) rates of live termites

Termites	Caste[a]	μg of N fixed/ [g (fresh wt) × day][b]	TDN (yr)[c]	Reference
Amitermes sp.	W	0.36	126	116
	S	0.45	100	116
Armitermes sp.	W + S	1.44	31	116
Coptotermes formosanus	W	0.16–49.39	283–1	23, 25
	S	0.03	1,507	25
Coptotermes lacteus	W	0.37–1.87	122–24	49
Cornitermes sp.	W + S	1.06	43	116
Cryptotermes brevis	BL	0.38	119	25
Cubitermes sp.	W	0.17	265	109a
	R	0.00	∞	109a
Heterotermes sp.	W	0.94	48	116
	S	0.00	∞	116
Incisitermes minor	W	1.00–18.87	45–2	10
	S	0.33	137	10
	R	0.66	68	10
Labiotermes sp.	W	0.16	282	116
Macrotermes ukuzii	W, S, R	0.00	∞	109a
Mastotermes darwiniensis	W	0.00–23.47	∞–2	49
Nasutitermes corniger[d]	W	6.00–8.00	8–6	106, 107
	S	0.90–28.40	50–2	106, 107
	Whole colony	27.40–81.68	2–0.5	106
Nasutitermes exitiosus	W	0.00–5.60	∞–8	49
Nasutitermes sp.	W	0.20–13.28	226–3	116
	S	0.87–7.44	52–6	116
	W + S	1.11–18.31	41–2	116
Neocapritermes sp.	W	0.00	∞	116
Reticulitermes flavipes	W	0.05	904	25
	S	0.02	2,260	25
Rhynchotermes speratus	W	3.5	13	107
	S	< 0.5	> 90	107
Trinervitermes trinervoides	W	6.86	7	109a
	S	4.48–4.73	10–9	109a
	R	0.00	∞	109a
Zootermopsis sp.	W + BL	0.06	753	25

[a]W, Workers; S, soldiers; BL, brachypterous larvae; R, reproductives.

[b]Values calculated assuming that N_2 fixation rates are one third that of acetylene reduction (52), and that fresh weight of termites = 6.7 × dry weight.

[c]TDN, Time required for termites to double their N content (see text). It is assumed that the N content of all termites is 11% (dry wt basis) (102).

[d]Mistakenly identified as *Nasutitermes ephratae* (107) and corrected in a subsequent paper (106).

a response of N_2-fixing gut bacteria to dietary N. Similarly, the difference between fixation rates of *Rhynchotermes speratus* and *N. corniger* collected from the field (Table 1) was attributed to the fact that the former feeds on material (leaf litter) higher in N than food of the latter (woody litter) (107). It is interesting to note that N_2 fixation activity in sea urchins (*Strongylocentrotus droebachiensis*) was also found to vary inversely with the N content of macrophytes on which they were feeding (51). The feeding habits and foraging preferences of some termites may obviate the need for N_2 fixation, simply because their diet contains adequate, albeit low, combined N. N budget studies by Hungate (58, 60) revealed that long-term laboratory cultures of *Zootermopsis, Incisitermes* (*Kalotermes*), and *Reticulitermes* species showed no evidence for N_2 fixation. For these termites it appeared that combined N in wood and soil, when acted upon by fungi, was sufficient to support good growth and reproduction. In fact, many of the Hodo- and Rhino- and some of the Kalotermitidae actually prefer wood that has undergone fungal decay (75). In Hungate's experiments it appeared that fungi had the effect of concentrating N in the wood, by degrading some of the wood polysaccharides, and by drawing N into the wood from surrounding soil during mycelial growth (58, 60). Thus, perhaps it is not surprising that the species studied by Hungate show only low to moderate rates when assayed by the acetylene reduction method (Table 1).

Age or developmental stage of termites may account for intraspecific variations in N_2 fixation activity. For example, Breznak (23) found that fixation rates of small larvae of *C. formosanus* (1.9 mg/larva, fresh wt) were 300-fold greater than that of larger, more fully developed larvae (4.0 mg/larva, fresh wt). Presumably the N needs of the smaller larvae were greater, and this was mirrored in greater fixation rates. A similar inverse relationship between body weight and N_2 fixation rate was observed with marine shipworms (family Teredinidae) (31). Other factors that might contribute to intraspecific variations are cyclic demands for N placed on colony members during bursts of reproductive activity (106), as well as physical disturbance to the termites being assayed (99, 106, 107).

So far in this discussion it has been implied that N_2 fixation is mediated by termite gut bacteria, but this implication is valid. Breznak et al (25, 99) showed that N_2 fixation in *C. formosanus* was associated with guts, not degutted bodies, and that the activity could be abolished by feeding termites antibacterial drugs. Moreover, N_2-fixing bacteria have been isolated from termite guts. French et al (49) isolated N_2-fixing *Citrobacter freundii* from Australian termite species, whereas Potrikus & Breznak (99) attributed N_2 fixation in *C. formosanus* to *Enterobacter agglomerans.*

Clearly, more work is needed on this critical question of termite nutrition. We still know nothing about the path by which N_2 may ultimately be assimilated by termites. Kinetic experiments with ^{15}N- or ^{13}N-labeled N_2

would be helpful here and should be a top priority for future research. The most definitive information is likely to be obtained by focusing on the tropical nasute species: These show the highest rates of N_2 fixation (Table 1), and their nests, often arboreal, are easy to obtain in toto (106, 107).

Role in N Recycling

The complementary side to N acquisition is N recycling, and there are three main ways in which this could be achieved by termites: storage/recycling of nitrogenous metabolic wastes; recycling of termite tissues (e.g. exoskeletons); and digestion and assimilation of gut microbes or lytic or secretory products thereof. The contribution of these processes to N economy in natural termite colonies has been largely speculative. However, recent studies indicate that one of these strategies (recycling of nitrogenous wastes) is important and bacterial mediated.

Uric acid (UA) is a common, and well suited, nitrogenous excretory product of many terrestrial insects (37). Because it is poorly soluble in water (65 μg/ml) (113), UA can be voided with feces as a nontoxic solid, thereby minimizing water loss to the insect (35). However, in 1938 Leach & Granovsky (74) advanced the intriguing notion that UA in termites might be degraded by the hindgut microbiota to a form of N reusable by the insects. Their hypothesis was feasible, because UA and other urinary metabolites are normally transported to the gut by Malpighian tubules that empty at the midgut-hindgut juncture (96). Consequently, UA would have to transit the microbe-packed hindgut before being voided to the exterior. Potrikus & Breznak (100–103) recently examined the Leach & Granovsky hypothesis (74) critically and established its validity. Working mainly with *Reticulitermes flavipes,* these investigators showed that termite tissues contain UA, as well as key enzymes for UA biosynthesis. However, little or no UA was observed in termite feces despite the fact that termite tissues lacked uricase or any other enzyme to degrade the purine. This apparent paradox was reconciled by the demonstration that uricolysis does, in fact, occur in termites, but as an anaerobic process mediated by hindgut bacteria. Furthermore, in vitro and in vivo studies with ^{14}C- and ^{15}N-labeled UA revealed that UA was transported to the gut from its site of synthesis and storage (fat body tissue) by Malpighian tubules, and that microbial uricolysis in vivo liberates N that is reused by the termites for biosynthesis. Thus, it appeared that UA was not necessarily a waste product of termites per se, but could function as an important storage form of N. It was calculated that synthesis of UA as an N reserve was an energetically sound investment for *R. flavipes,* and that subsequent bacterial uricolysis, in theory, could liberate enough N annually to support biosynthesis of termites equivalent to 30% of the colony biomass (103). The actual dynamics of UA biosynthesis and mobilization in natural colonies is an important question that remains to be answered. However, it seems safe to assume that UA synthesis will be favored when intake of dietary N exceeds that needed for biosynthesis, and mobilization will be favored when colony demand for N is high (e.g. during peak reproductive and growth periods). Another important question is the precise form of UA-N made available to the insects. NH_3, the major nitrogenous product of uricolysis by gut bacteria in vitro (see below), is one possibility since *R. flavipes* tissues possess glutamine synthetase (103). Alternatively, UA-N might first be assimilated by gut microbes to become available to termites by secretion (see next section) or lysis of cells in the gut, or through coprophagy or trophallaxis.

Significant populations of uricolytic bacteria were demonstrated in *R. flavipes* hindguts, and major isolates were identified as *Bacteroides termitidis, Streptococcus* sp., and *Citrobacter* sp. (100). Growth studies revealed that all isolates used UA as an energy source, but only under anaerobic conditions, and major products of UA fermentation by *B. termitidis* and *Streptococcus* strains were CO_2, acetate, and NH_3 (101). Since both termites and bacteria benefit nutritionally from UA cycling, the symbiosis was considered one of true mutualism (103).

Utilization of UA as an N reserve is not restricted to termites, however. In an elegant study with cockroaches (*Periplaneta americana*), Mullins & Cochran (92, 93) showed that UA accumulated by insects on high N diets was mobilized when they were shifted to a low N diet. Part of the UA-N mobilized could be used for oothecal production by females. Interesting from a behavioral aspect is the fact that male cockroaches (*Blattella germanica*) appear to make a paternal investment of UA-N to reproduction (94). During copulation, males coat the spermatophore deposited in the female with UA derived from the male's uricose glands. The paternally derived UA is ultimately metabolized within oothecae and apparently serves as a C and N source during embryogenesis. Microbial involvement in cockroach uricolysis is not as clearly defined as in termites, but it might be accomplished by gut bacteria (91), by mycetocyte bacteria present in fat body tissue cells (41), or by both (see 37 for further review).

Crickets (*Acheta domestica*) have recently been shown to harbor uricolytic gut bacteria (121); however, UA recycling in these insects was not studied. Perhaps the best insects to examine for bacterial-mediated UA-N recycling would be those, like termites, that subsist in N poor diets for all or part of their life. With regard to higher animals, uricolysis by gut bacteria has been postulated for birds (3, 89), monkeys (62), and man (124). In ruminants, a cycle superficially similar to that in termites has been established. However, in this case UA is replaced by urea, which is recycled by the blood stream and saliva to the rumen where it serves as a major N source for rumen microbes (126). Microbial protein so synthesized constitutes the ruminant's main protein source.

A seemingly important strategy for N conservation in termite colonies would also be the consumption of exuviae (shed exoskeletons) as well as dead, dying, or supernumery colony members (via cannibalism). The incidence of such behavior has been reviewed (72). A rich source of N in such food is chitin, a major component of insect cuticle. Growing insects must possess chitinase in moulting fluid to escape the confines of their integument during periodic ecdysis. For termites, possession of a digestive as well as moulting chitinase would seem advantagous, since it might allow them to conserve chitin N when exuviae are eaten. Chitinase has been demonstrated in some termites (109a, 119, 125), but since whole insects were used to prepare the enzyme extracts, the anatomical origin of such activity is unknown. By contrast, Rohrmann & Rossman (109a) found that most of the chitinase of *Macrotermes ukuzii* workers and nymphs was associated with the gut. Since *M. ukuzii* cultivates and consumes the combs of its symbiotic *Termitomyces* fungus, an intestinal chitinase may make fungal cell wall chitin available to this termite as a nutrient. These investigators also isolated a chitinolytic bacterium from the gut of *M. ukuzii,* but they were not certain what proportion of intestinal chitinase was attributable to gut symbionts. It would be interesting to examine termite gut microbes further as a source of chitinase.

The ability of termites to conserve N by actively digesting some of their gut microbiota is unknown. Lysis of protozoa in the hindgut of lower termites appears quantitatively insignificant as a means of N conservation (61), although proctodeal trophallaxis could conserve microbial N if recipients could digest the cells in the fore- and hindgut. Rates of growth, lysis, and turnover of bacteria in situ remain to be determined, but such information would greatly aid our evaluation of the nutritive potential of the bacterial cell material. The importance of coprophagy to N conservation is also unknown. Although the N content of termite feces is generally low (0.2–0.8%) (102), it could be a resource under some conditions. La Fage & Nutting (72) do not consider recycling of raw feces to be a general or economical practice, at least with respect to carbon nutrition.

Role in Amino Acid Synthesis

The ability of termite hindgut bacteria to supply amino acids to their host has been suggested by two important studies. Mauldin et al (83) found that normal synthesis of free and protein bound amino acids by *Coptotermes formosanus* fed [^{14}C]acetate was dependent on the combined presence of normal gut bacteria and two protozoa (*Holomastigotoides hartmanni* and *Spirotrichonympha leidyi*). A similar role for bacteria was suggested by Speck et al (114), who used *Reticulitermes santonensis* fed [^{14}C]glucose.

It is pertinent to note that in many rumen bacteria, amino acid biosynthesis is not subject to the feedback inhibition pattern seen in most heterotrophs (2). Moreover, Stevenson (115) found that all of 126 strains of major rumen bacteria tested excreted free amino acids during in vitro cultivation. Such excretion occurred during active growth and was not due to cell lysis. From this information it is tempting to speculate that hindgut bacteria might synthesize amino acids exported for use by the termites.

MYCOPHAGY AND ACQUIRED ENZYMES

External fungi appear to play an important role in the nutrition of many termites. This is true not only for termites that prefer fungus-degraded wood, but also for certain higher termites (family Macrotermitinae) that actually cultivate fungi in elaborate gardens (72, 110). Although such fungi do not strictly constitute part of the termites' gut microbiota, recent work on the fungus growers has revealed the importance of acquired enzymes to resource utilization and must be included in this review.

Macrotermes natalensis maintains in its nest structures known as fungus combs. The combs are sponge-like structures consisting of chewed, but undigested, plant fragments, covered with mycelia and conidiophores (nodules) of the symbiotic fungus *Termitomyces.* It had long been recognized that comb material, including nodules, was essential to the diet of *M. natalensis,* but for reasons unclear. However, in what will certainly be recognized as a classic study, Martin & Martin (78, 79) showed that the major site of plant cellulose decomposition in *M. natalensis* was the midgut, and that the critical C_1-cellulase was derived exclusively from the fungus nodules. The C_x-enzymes and beta-glucosidase were derived in part from the fungus nodules and in part from the termite midgut and salivary glands. The reliance of *M. natalensis* on *Termitomyces* now seems clear, i.e. the acquisition, by mycophagy, of digestive enzymes. A similar strategy was observed for *Macrotermes subhyalinus* by Abo-Khatwa (1). The use of enzymes acquired by mycophagy has far reaching implications for invertebrates that subsist on litter, detritus, and dead wood, and the interested reader is referred to the excellent review by Martin (77) for a more detailed discussion of this topic.

LIGNIN DEGRADATION IN TERMITES

The question of lignin degradation by termites and their gut microbiota is intriguing, since much of the termite alimentary tract appears anaerobic and natural mechanisms for anaerobic lignolysis are not known. Conclusions based on analyses of termite feces are conflicting: Some authors report as much as 83% lignolysis (reviewed in 72, 75), others report virtually none

(42). The fecal analysis approach is open to criticism, unless precautions are taken to eliminate coprophagy, and to insure that observed lignolysis was not actually occurring in fecal pellets (aerobically) between the time pellets were voided and subsequently collected for assay.

Recently, other approaches have been used. Mishra & Sen-Sarma (88) inferred lignin degradation by the hindgut microbiota of *Neotermes bosei* based on the presence of vanillin and veratraldehyde in hindgut contents of sound wood-fed insects. However, no quantitative data were reported, and lignase activity could not be demonstrated in vitro. On the other hand, French & Bland (48) inferred lignolysis for *Coptotermes lacteus* and *Nasutitermes exitiosus* by measuring incorporation of ^{14}C-label into tissues of termites fed wood previously grown with [3-^{14}C]cinnamic acid. However, the most convincing evidence comes from the work of Butler & Buckerfield (30). By measuring $^{14}CO_2$ evolution, they found that the higher termite *N. exitiosus* respired 14–32% and 15–63% of synthetic and maize lignins, respectively, which were labeled in various positions in the polymer (methoxy; C_2; ring). An important control was their demonstration that maximal $^{14}CO_2$ release required the presence of termites: Little $^{14}CO_2$ issued from culture vessels reincubated after removal of termites. Lignin degradation was undoubtedly initiated in the gut, although the site(s) was not identified nor was the relevance of gut organisms. One still wonders, however, how lignin degradation can proceed in a putatively anaerobic environment. Perhaps enough O_2 diffuses into the gut, or is carried with the food bolus, to sustain lignolysis. It is interesting to note that in higher termites anterior sections of the hindgut are alkaline and can be as high as pH 11 in *Cubitermes severus* (14). It may be that alkaline pretreatment of lignin, coupled with O_2 intrusions (or peroxides released from gut tissue? from the gut microbiota?), effect lignolysis in termite guts as is observed in reaction vessels (80). Other scenarios are possible, and more work is needed on this important issue of termite digestion.

GUT MICROBIOTA OF OTHER XYLOPHAGOUS INSECTS

There exist a variety of insects other than termites that feed on, bore through, or live in cellulose-rich materials of one sort or another and therefore fall under the rubric of xylophagy. A fairly large number have established rather spectacular relationships with ectosymbiotic fungi, relationships whose intricacy is underscored by morphological and behavioral adaptations of the insects that allow them to store, transmit, and propagate the fungi as a nutrient resource. Among such insects are the ambrosia beetles (Platypodidae and Scolytidae) and the siricid wood wasps. Although beyond the scope of this review, such ectosymbioses (47, 50) and the biochemical implications of mycophagy (77) have been discussed elsewhere. Still other xylophagous insects, notably the Anobiidae and Cerambycidae, possess evaginations of the alimentary tract whose cells actually house intracellular microbial symbionts (29, 66). In the cigarette beetle, *Lasioderma serricorne,* the endosymbionts appear to secrete vitamins and sterols for their host (63, 64).

Comparatively little work has been done on the intestinal microbiota of xylophagous species other than termites, although several reports are worth noting. Bayon and co-workers (5–8) have shown that larvae of *Oryctes nasicornis,* a scarabaeid beetle that feeds on woody detritus and dung, has a prominent proctodeal dilation displaying a microbial fermentation of cellulose to volatile fatty acids and methane. As in xylophagous termites, acetate dominates the volatile fatty acid pool and is absorbed into the hemolymph, undoubtedly for subsequent oxidation by the larvae. Both the proctodeum and mesenteron were sites of cellulolysis, and rates of methane evolution by live beetles {34–48 nmol/[hr X g (fresh wt)]} (7) were similar to those reported for termites (23). No microbiological studies were done, so the importance of gut tissue versus gut microbiota to cellulase secretion is unknown. However, the extent of dissimilation of cellulose on passage through the larval gut (24–39%) was only about one third that of most termites. This undoubtedly reflects the shorter retention time of food in the hindgut of scarabaeid larvae (6 hr) (5), as opposed to that of termites (26 hr; *Reticulitermes flavipes*) (D. A. Odelson, J. A. Breznak, manuscript in preparation).

Cockroaches, although as a group generally considered omnivorous, also contain a prominent and diverse bacterial flora in their hindgut (12, 13, 46). Many of the organisms are anaerobes (20) and possess C_x-type cellulases active against carboxymethylcellulose (39). That such enzymatic activity is not trivial is suggested by the work of Bignell (11), who demonstrated that 73% and 48% of ^{14}C-labeled dietary cellulose and hemicellulose, respectively, were respired to $^{14}CO_2$ by *Periplaneta americana.* Such activity was inhibitable by antibacterial drugs. Undoubtedly, microbial participation in polysaccharide degradation resulted in liberation of volatile fatty acids taken up from the colon and oxidized by the insect. Both physiological (21) and ultrastructural (13) studies on the colon of *P. americana* support this contention.

Larvae of aquatic craneflies (*Tipula abdominalis*) feed on detrital leaf litter and also have an enlarged hindgut colonized by a diverse bacterial community (65). Most of the isolatable bacteria were shown to be facultative anaerobes, although nothing is yet known on their importance to *T. abdominalis* nutrition.

The seemingly common occurrence of a dense and diverse fermentative microbiota, within an enlarged hindgut of xylophagous/detritivorous insects (29), suggests resource utilization strategies akin to that seen with termites. Such strategies may be of critical survival value, not only to individuals housing the microbiota, but also to those in food webs based on refractile lignocellulosic substrates.

CONCLUDING REMARKS

Important strides have been made in our understanding of the termite gut microbiota in terms of its community structure and composition, and the roles of microorganisms in the carbon, nitrogen, and energy requirements of the insect. Particularly satisfying is the current availability, in pure culture, of a number of major bacteria and cellulolytic protozoa, whose biochemistry can now be studied in greater detail and whose mechanistic involvement in the hindgut symbiosis assessed with greater precision. However, compared to the voluminous information on the rumen ecosystem, our understanding of most xylophagous insect gut ecosystems is embryonic. This is unfortunate, because such ecosystems are anatomically more analogous to that of man than is the rumen. Moreover, they provide foci for numerous research problems important not only to a better understanding of host-microbe interactions per se, but to ecological strategies for resource utilization and bioconversion processes as well.

Entomology and microbiology are quite distinct disciplines; problems important to one are often not appreciated, and rarely entertained, by scientists of the other. It seems that evolution of scientists into these specialties entails a basic divergence fairly early in one's training. There simply are not that many microbiologists working with insect systems, and vice versa. For those microbiologists with a basic interest in intestinal microecology, and in light of the information presented in this review, it might be useful to consider the gut of a xylophagous insect more seriously as a potential model system. Indeed, perhaps the time is right for more convergent thought and interaction between the disciplines in general. Therein lies much fertile scientific ground worth cultivating, and prospects for a bountiful harvest would seem good.

Acknowledgments

I wish to thank my students, D. A. Odelson, C. J. Potrikus, and J. E. Schultz, whose work and comments helped generate many of the ideas presented in this review. I also thank J. K. Mauldin and R. W. O'Brien for sending me preprints of unpublished papers. Some of the research reported herein was funded by grants from the National Science Foundation and the Michigan Agricultural Experiment Station of Michigan State University, the support from which is gratefully acknowledged.

This journal article no. 10228 from the Michigan Agricultural Experiment Station.

Literature Cited

1. Abo-Khatwa, N. 1978. *Experientia* 34:559–60
2. Allison, M. J. 1969. *J. Anim. Sci.* 29: 797–807
3. Barnes, E. M., Impey, C. S. 1974. *J. Appl. Bacteriol.* 37:393–409
4. de Bary, A. 1879. *Die Erscheinung der Symbiose.* Strasburg: Trubner
5. Bayon, C. 1980. *CR Acad. Sci. Paris* 290:1145–48
6. Bayon, C. 1980. *J. Insect Physiol.* 26: 819–28
7. Bayon, C., Etiévant, P. 1980. *Experientia* 36:154–55
8. Bayon, C., Mathelin, J. 1980. *J. Insect Physiol.* 26:833–40
9. Becker, G. 1971. *Wood Sci. Technol.* 5: 236–46
10. Benemann, J. R. 1973. *Science* 181: 164–65
11. Bignell, D. E. 1977. *Can. J. Zool.* 55: 579–89
12. Bignell, D. E. 1977. *J. Invert. Pathol.* 29:338–43
13. Bignell, D. E. 1980. *Tissue Cell* 12: 153–64
14. Bignell, D. E., Anderson, J. M. 1980. *J. Insect Physiol.* 26:183–88
15. Bignell, D. E., Oskarsson, H., Anderson, J. M. 1979. *Appl. Environ. Microbiol.* 37:339–42
16. Bignell, D. E., Oskarsson, H., Anderson, J. M. 1980. *J. Gen. Microbiol.* 117:393–403
17. Bignell, D. E., Oskarsson, H., Anderson, J. M. 1980. *J. Invert. Pathol.* 36:426–28
18. Bignell, D. E., Oskarsson, H., Anderson, J. M. 1980. *Zoomorphology* 96: 103–12
19. Blomquist, G. J., Howard, R. W., McDaniel, C. A. 1979. *Insect Biochem.* 9:371–74
20. Bracke, J. W., Cruden, D. L., Markovetz, A. J. 1978. *Antimicrob. Agents Chemother.* 13:115–20
21. Bracke, J. W., Markovetz, A. J. 1980. *J. Insect Physiol.* 26:85–89
22. Breznak, J. A. 1973. *CRC Crit. Rev. Microbiol.* 2:457–89
23. Breznak, J. A. 1975. In *Symbiosis (Soc. Exp. Biol. Symp. Ser. 29),* ed. D. H. Jennings, D. L. Lee, pp. 559–80. Cambridge: Cambridge Univ.
24. Breznak, J. A. 1983. *Bergey's Manual of Systematic Bacteriology,* In press 9th ed.
25. Breznak, J. A., Brill, W. J., Mertins, J. W., Coppel, H. C. 1973. *Nature* 244: 577–80
26. Breznak, J. A., Pankratz, H. S. 1977. *Appl. Environ. Microbiol.* 33:406–26
27. Brian, M. V. 1978. *Production Ecology of Ants and Termites.* Cambridge: Cambridge Univ. 409 pp.
28. Brill, W. J. 1975. *Ann. Rev. Microbiol.* 29:109–29
29. Buchner, P. 1965. *Endosymbiosis of Animals with Plant Microorganisms.* New York: Interscience. 909 pp.
30. Butler, J. H. A., Buckerfield, J. C. 1979. *Soil Biol. Biochem.* 11:507–13
31. Carpenter, E. J., Culliney, J. L. 1975. *Science* 187:551–52
32. Carter, F. L., Garlo, A. M., Stanley, J. B. 1978. *J. Agric. Food Chem.* 26: 869–73
33. Carter, F. L., Mauldin, J. K. 1981. *Mater. Org.* 16:175–88
34. Carter, F. L., Mauldin, J. K., Rich, N. M. 1981. *Mat. Org.* 16:29–38
35. Chapman, R. F. 1975. *The Insects: Structure and Function.* New York: Elsevier. 819 pp.
36. Cleveland, L. R. 1925. *Biol. Bull.* 48: 289–93
37. Cochran, D. G. 1975. In *Insect Biochemistry and Function,* ed. D. J. Candy, B. A. Kilbey, pp. 178–281. London: Chapman & Hall. 314 pp.
38. Cowling, E. B., Merrill, W. 1966. *Can. J. Bot.* 44:1539–54
39. Cruden, D. L., Markovetz, A. J. 1979. *Appl. Environ. Microbiol.* 38:369–72
40. Darlington, J. P. E. C. Cited in Ref. 106, p. 251
41. Donnellan, J. F., Kilby, B. A. 1967. *Comp. Biochem. Physiol.* 22:235–52
42. Esenther, G. R., Kirk, T. K. 1974. *Ann. Entomol. Soc. Am.* 67:989–91
43. Eutick, M. L., O'Brien, R. W., Slaytor, M. 1976. *J. Insect Physiol.* 22:1377–80
44. Eutick, M. L., O'Brien, R. W., Slaytor,

M. 1978. *Appl. Environ. Microbiol.* 35: 823–28
45. Eutick, M. L., Veivers, P., O'Brien, R. W., Slaytor, M. 1978. *J. Insect Physiol.* 24:363–68
46. Foglesong, M. A., Walker, D. H. Jr., Puffer, J. S., Markovetz, A. J. 1975. *J. Bacteriol.* 123:336–45
47. Francke-Grosmann, H. 1967. In *Symbiosis,* ed. S. M. Henry, 2:141–205. New York: Academic. 443 pp.
48. French, J. R. J., Bland, D. E. 1975. *Mat. Org.* 10:281–88
49. French, J. R. J., Turner, G. L., Bradbury, J. F. 1976. *J. Gen. Microbiol.* 95:202–6
50. Graham, K. 1967. *Ann. Rev. Entomol.* 12:105–26
51. Guerinot, M. L., Fong, W., Patriquin, D. G. 1977. *J. Fish. Res. Board Can.* 34:416–20
52. Hardy, R. W. F., Burns, R. C., Holsten, R. D. 1973. *Soil Biol. Biochem.* 5:47–81
53. Haverty, M. I., Howard, R. W. 1979. *Ann. Entomol. Soc. Am.* 72:503–8
54. Hickin, N. E. 1971. *Termites, A World Problem.* London: Hutchinson
55. Honigberg, B. M. See Ref. 71, 2:1–36
56. Howard, R. W., Haverty, M. I. 1978. *Sociobiology* 3:73–77
57. Hungate, R. E. 1939. *Ecology* 20: 230–45
58. Hungate, R. E. 1941. *Ann. Entomol. Soc. Am.* 34:467–89
59. Hungate, R. E. 1943. *Ann. Entomol. Soc. Am.* 36:730–39
60. Hungate, R. E. 1944. *Texas Acad. Sci. Proc. Trans.* 27:1–7
61. Hungate, R. E. 1955. In *Biochemistry and Physiology of Protozoa,* ed. S. H. Hutner, A. Lwoff, 2:159–99. New York: Academic. 388 pp.
62. Hunter, A., Givens, M. H. 1914. *J. Biol. Chem.* 17:37–53
63. Jurzitza, G. 1969. *Oecologia* 3:70–83
64. Jurzitza, G. 1974. *Oecologia* 16:163–72
65. Klug, M. J., Kotarski, S. 1980. *Appl. Environ. Microbiol.* 40:408–16
66. Koch, A. 1967. See Ref. 47, pp. 1–106
67. Kovoor, J. 1967. *CR Acad. Sci. Paris* 264:486–88
68. Kovoor, J. 1970. *Ann. Sci. Nat. Zool. Paris* 12:65–71
69. Krelinova, O., Kirku, V., Skoda, J. 1977. *Int. Biodeter. Bull.* 13:81–87
70. Krishna, K. See Ref. 71, 1:1–17
71. Krishna, K., Weesner, F. M. 1969/70. *Biology of Termites,* Vol. 1 & 2. New York: Academic. 598 pp.; 643 pp.
72. LaFage, J. P., Nutting, W. L. See Ref. 27, pp. 165–232
73. Lasker, R., Giese, A. C. 1956. *J. Exp. Biol.* 33:542–53
74. Leach, J. G., Granovsky, A. A. 1938. *Science* 87:66–67
75. Lee, K. E., Wood, T. G. 1971. *Termites and Soils.* New York: Academic. 251 pp.
76. Mannesmann, R. 1977. *Z. Angew. Entomol.* 83:1–10
77. Martin, M. M. 1979. *Biol. Rev.* 54:1–21
78. Martin, M. M., Martin, J. S. 1978. *Science* 199:1453–55
79. Martin, M. M., Martin, J. S. 1979. *Physiol. Zool.* 52:11–21
80. Marton, J. 1971. In *Lignins: Occurrence, Formation, Structure and Reactions,* ed. K. V. Sarkanen, C. H. Ludwig, pp. 639–94. New York: Interscience. 916 pp.
81. Mauldin, J. K. 1977. *Insect Biochem.* 7:27–31
82. Mauldin, J. K., Carter, F. L., Rich, N. M. 1981. *Mater. Org.* 16:15–28
83. Mauldin, J. K., Rich, N. M., Cook, D. W. 1978. *Insect Biochem.* 8:105–9
84. Mauldin, J. K., Smythe, R. V., Baxter, C. C. 1972. *Insect Biochem.* 2:209–17
85. McEwen, S. E., Slaytor, M., O'Brien, R. W. 1980. *Insect Biochem.* 10:563–67
86. Miller, E. M. See Ref. 71, 1:283–310
87. Mishra, S. C. 1980. *Mater. Org.* 15: 253–61
88. Mishra, S. C., Sen-Sarma, P. K. 1980. *Mater. Org.* 15:119–24
89. Mortensen, A., Tindall, A. 1978. *J. Physiol.* 284:159–60P
90. Müller, M. 1980. In *The Eukaryotic Microbial Cell,* ed. G. W. Gooday, D. Lloyd, A. P. J. Trinci, pp. 127–42. Cambridge: Cambridge Univ. 439 pp.
91. Mullins, D. E. 1974. *J. Exp. Biol.* 61:541–56
92. Mullins, D. E., Cochran, D. G. 1975. *Comp. Biochem. Physiol.* 50A:489–500
93. Mullins, D. E., Cochran, D. G. 1975. *Comp. Biochem. Physiol.* 50A:501–10
94. Mullins, D. E., Keil, C. B. 1980. *Nature* 283:567–69
95. Nielsen, M. G., Josens, G. See Ref. 27, pp. 45–53
96. Noirot, C., Noirot-Timothée, C. See Ref. 71, 1:49–88
97. O'Brien, G. W., Veivers, P. C., McEwen, S. E., Slaytor, M., O'Brien, R. W. 1979. *Insect Biochem.* 9:619–25
98. Odelson, D. A., Breznak, J. A. 1981. *Abstr. Annu. Meet. Am. Soc. Microbiol.,* p. 100
99. Potrikus, C. J., Breznak, J. A. 1977. *Appl. Environ. Microbiol.* 33:392–99
100. Potrikus, C. J., Breznak, J. A. 1980. *Appl. Environ. Microbiol.* 40:117–24

101. Potrikus, C. J., Breznak, J. A. 1980. *Appl. Environ. Microbiol.* 40:125–32
102. Potrikus, C. J., Breznak, J. A. 1980. *Insect Biochem.* 10:19–27
103. Potrikus, C. J., Breznak, J. A. 1981. *Proc. Natl. Acad. Sci. USA* 78:4601–5
104. Potts, R. C., Hewitt, P. H. 1973. *Insect. Soc.* 20:215–20
105. Potts, R. C., Hewitt, P. H. 1974. *Comp. Biochem. Physiol.* 47B:327–37
106. Prestwich, G. D., Bentley, B. L. 1981. *Oecologia* 49:249–51
107. Prestwich, G. D., Bentley, B. L., Carpenter, E. J. 1980. *Oecologia* 46:397–401
108. Prestwich, G. D., Jones, R. W., Collins, M. S. 1981. *Insect Biochem.* 11:331–36
109. Reese, E. T., Mandels, M. 1971. In *High Polymers,* ed. N. M. Bikales, L. Segal, 5:1079–94. New York: Interscience
109a. Rohrmann, G. F., Rossman, A. Y. 1980. *Pedobiologia* 20:61–73
110. Sands, W. A. See Ref. 71, 2:495–524
111. Schultz, J. E., Breznak, J. A. 1978. *Appl. Environ. Microbiol.* 35:930–36
112. Schultz, J. E., Breznak, J. A. 1979. *Appl. Environ. Microbiol.* 37:1206–10
113. Seegmiller, J. E., Laster, L., Howell, R. R. 1963. *N. Engl. J. Med.* 268:712–16
114. Speck, U., Becker, G., Lenz, M. 1971. *Z. Angew. Zool.* 58:475–91
115. Stevenson, I. L. 1978. *Can. J. Microbiol.* 24:1236–41
116. Sylvester-Bradley, R., Bandeira, A. G., Oliveira, L. A. 1978. *Acta Amazon.* 8:621–27
117. Thayer, D. W. 1976. *J. Gen. Microbiol.* 95:287–96
118. Thayer, D. W. 1978. *J. Gen. Microbiol.* 106:13–18
119. Tracey, M. V., Youatt, G. 1958. *Enzymologia* 19:70–72
120. Trager, W. 1934. *Biol. Bull.* 66:182–90
121. Ulrich, R. G., Buthala, D. A., Klug, M. J. 1981. *Appl. Environ. Microbiol.* 41:246–54
122. Veivers, P. C., Musca, A. M., O'Brien, R. W., Slaytor, M. 1982. *Insect Biochem.* 12:35–40
123. Veivers, P. C., O'Brien, R. W., Slaytor, M. 1980. *J. Insect Physiol.* 26:75–77
124. Vogels, G. D., Van der Drift, C. 1976. *Bacteriol. Rev.* 40:404–68
125. Waterhouse, D. F., Hackman, R. H., McKellar, J. W. 1961. *J. Insect Physiol.* 6:96–112
126. Wolin, M. J. 1979. In *Advances in Microbial Ecology,* ed. M. Alexander, 3:49–77. New York: Plenum. 225 pp.
127. Wood, T. G. See Ref. 27, pp. 55–80
128. Wood, T. G., Sands, W. A. See Ref. 27, pp. 245–92
129. Yamaoka, I., Nagatani, Y. 1975. *Zool. Mag.* 84:23–29
130. Yamin, M. A. 1978. *J. Parasitol.* 64:1122–23
131. Yamin, M. A. 1978. *J. Protozool.* 25:535–38
132. Yamin, M. A. 1980. *Appl. Environ. Microbiol.* 39:859–63
133. Yamin, M. A. 1981. *Science* 211:58–59
134. Yamin, M. A., Trager, W. 1979. *J. Gen. Microbiol.* 113:417–20

Ann. Rev. Microbiol. 1982. 36:345–70

THE BIOLOGY OF RICKETTSIAE

Emilio Weiss[1]

Naval Medical Research Institute, Bethesda, Maryland 20814

CONTENTS

INTRODUCTION

Concern over rickettsiae as an actual or potential public health menace has fluctuated widely. Rickettsial research, however, has always been a challenge to investigators interested in complex host-parasite relationships and specialized technologies. Previous reviews in *The Annual Review of Microbiology* reflect both the concern and the challenge. In 1947 Topping & She-

[1]The opinions and assertions contained herein are the private ones of the writer and are not to be construed as official or reflecting the views of the Navy Department or the naval service at large.

0066-4227/82/1001-0345$02.00

pard (130) detailed the remarkable advances made during World War II in the cultivation of rickettsiae and in vaccine development. In 1952 Bell & Philip (15) discussed epidemiology and the exciting new knowledge on metabolism and antibiotic susceptibility. In 1969 Ormsbee (102) stressed the importance of treating rickettsiae as full-fledged prokaryotic organisms. This review continues to characterize rickettsiae as typical bacteria, but it takes into account the renewed interest in ecology. Methods of isolation, cultivation, and identification have been presented in detail elsewhere (146) and are not discussed here.

As in the previous review (102) rickettsiae are defined as small, Gram-negative, fastidious bacteria that are associated with arthropods, are pathogenic for man, or are related to human pathogens. Two species have been removed from the genus *Rickettsia* and one has been added. The sennetsu agent was transferred to the genus *Ehrlichia,* whose properties it more closely resembles (119). The trench fever agent was placed in a separate genus and is designated *Rochalimaea quintana* (152). Although genetically related to the typhus rickettsiae (95), unlike the other rickettsiae it can be cultivated in axenic media of moderate complexity (135). *Rickettsia rhipicephali* has been added to the subset of spotted fever group rickettsiae, which presumably are not pathogenic for man (30). Other spotted fever group rickettsiae have been isolated, but their degree of relatedness to established strains is not known. There is no evidence that *Coxiella burnetii* is related to the other rickettsiae, but it is retained in the tribe *Rickettsieae* because it is a human pathogen that is studied by comparable methods.

The order *Rickettsiales* includes several other taxa for which considerable information has accumulated that indicates that they are not related to rickettsiae. Among them are the hemotrophic bacterial pathogens of man and animals (78, 79) and genera in the tribe *Ehrlichieae,* which are primarily pathogenic for domestic animals (119, 148). Also unrelated are the "guinea pig agents" (19), which have been identified as members of the genus *Legionella* (22, 64, 88).

ECOLOGY: NEW PERCEPTIONS

Rickettsia rickettsii and Related Nonpathogenic Rickettsiae

Man is only an incidental host to *Rickettsia rickettsii,* but the disease produced, Rocky Mountain spotted fever, is severe enough to be recognized, and statistics on human infection, with all their imperfections (for example, some of the early reports may have included cases of Colorado tick fever virus infection), are of value in the study of ecology. They clearly indicate that the geographic distribution of this rickettsia has changed during this century.

In a recent analysis of old records, Ormsbee (103) found no evidence that the disease had occurred in Montana prior to 1873, although in subsequent decades it terrorized the inhabitants of the west side of the Bitter Root Valley. It attracted the attention of H. T. Ricketts and eventually of Congress, and it led to the establishment of the Rocky Mountain Laboratory. From approximately 1930 the disease declined in the Mountain and Pacific States and cases in this area now account for a very small percentage of the total in the USA (Figure 1). At the same time, the disease, which was relatively uncommon in the East before 1930, began to rise in the South Atlantic States. The greatest percentage of cases now occurs in the piedmont area of North Carolina, South Carolina, and Virginia. The incidence has also recently increased in the South Central and mid-Atlantic States.

Many of the trends illustrated in Figure 1 can be attributed to changes wrought by man. Following the advent of broad-range antibiotics in the 1950s, there was a sharp decline in the number of recognized or reported cases. In the 1960s, increased suburbanization, increased dog population in suburban areas, and the opening of new outdoor recreational facilities have contributed to the increase in the number of cases to a high of 1153 in 1977. The number of reported cases has remained approximately the same since then (1163 in 1980).

The decline of the disease in the West remains largely unexplained, but two factors may have contributed to it. Although *R. rickettsii* is generally regarded as well established in the tick, and transovarian passage in the laboratory can be 100% effective, Burgdorfer & Brinton (26) have shown that continuous passage is detrimental. Beginning with the fifth filial generation, increasing numbers of engorged females die and oviposited eggs fail to develop. If these laboratory observations apply to events in nature, there must be a steady decline in the number of ticks that transmit the rickettsiae transovarially. This decline may or may not be matched by new infections of the ticks via rickettsemic vertebrate hosts. It is possible, though it has not been proven, that wild rodents and lagomorphs are not maintaining the level of infection in *Dermacentor andersoni,* the tick vector in the West. On the other hand, the dog remains a highly efficient source of infection of *Dermacentor variabilis,* the tick vector in the East. In fact, there is good evidence that the dog is highly susceptible to infection and develops clinical signs (54, 83, 125). The second factor that may limit *R. rickettsii* infection is competition with *Rickettsia montana, Rickettsia rhipicephali,* and other nonpathogenic rickettsiae. *R. rickettsii* has been isolated from the west side of the Bitter Root Valley, but only nonpathogenic rickettsiae have been isolated from the east side. There is no record of Rocky Mountain spotted fever having been acquired on the east side of the river (113). There is good evidence that ticks do not become dually infected: Infection with one species of rickettsia prevents infection with another (27). *R. montana* was originally isolated from *Dermacentor variabilis* and *Dermacentor andersoni* in eastern Montana (14, 80), but it has been recovered from other parts of the US and more recently from *D. variabilis* in Cape Cod, Massachusetts (53), Connecticut (84), and Ohio (82). *R. rhipicephali,* originally isolated from the brown dog tick, *Rhipicephalus sanguineus* (30), and in some cases from *D. variabilis* in Mississippi, Texas, and North Carolina, was recovered recently with surprising frequency from *D. andersoni* ticks collected in western Montana (113). It will be interesting to determine whether or not in future years the nonpathogenic rickettsiae actually spread to new areas and whether or not their spread coincides with a decline in the number of cases of Rocky Mountain spotted fever.

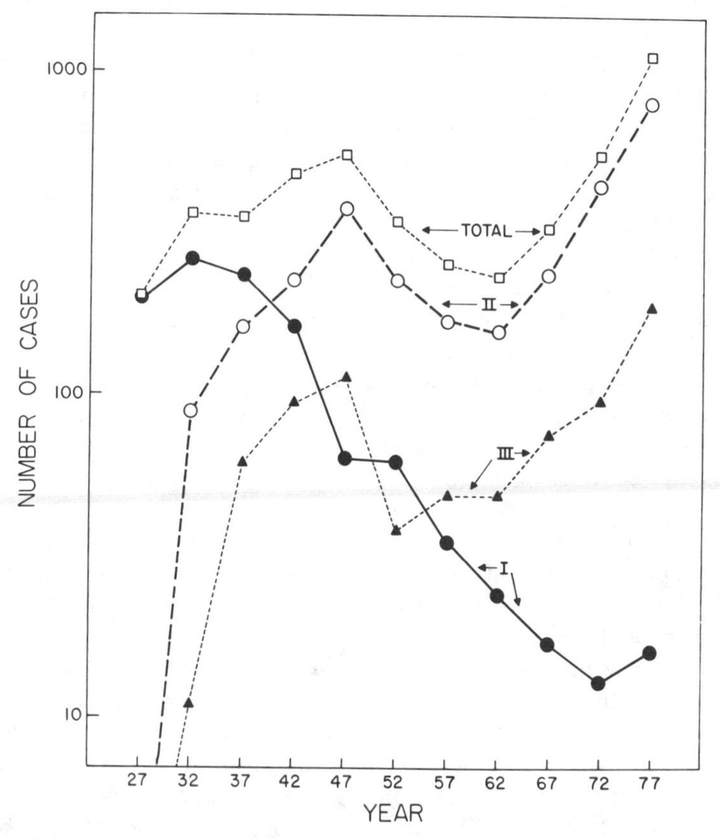

Figure 1 Incidence of Rocky Mountain spotted fever in the US from 1925 to 1979. Each symbol represents the mean of 5 years, the year indicated on the abscissa plus 2 years preceding and following it. I, Mountain and Pacific States; II, South Atlantic and South Central States; III, New England, mid-Atlantic, and North Central States. (From 118; Morbidity and Mortality Weekly Reports; J. E. McDade and C. Helmick, personal communication.)

Rickettsia prowazekii

Rickettsia prowazekii, the etiologic agent of epidemic typhus, is well established in man as its primary host, with the human body louse as its vector. In Ethiopia and other endemic areas of Africa where people share living quarters with their livestock, some evidence was developed that human infection may spill over into the domestic animals and that infection is maintained in the animal population by ticks (114–116). Subsequent surveys of blood specimens obtained from wild and domestic animals and from ticks collected in Egypt and Ethiopia, as well as ancillary tests of susceptibility of these animals to infection with *R. prowazekii,* have not provided supportive evidence (28, 29, 105, 106, 112). The view is now held that infection of domestic animals from a human source of *R. prowazekii* is not a common occurrence and does not contribute to the survival of *R. prowazekii* in nature.

In contrast, Bozeman et al (20) have recently discovered a widespread, self-sustaining natural infection with *R. prowazekii* in the southern flying squirrel, *Glaucomys volans,* in the eastern US. From December 1973 to the present a total of 15 strains of *R. prowazekii* have been isolated, 10 from flying squirrels and 5 from their ectoparasites, lice (three isolations) and fleas (two isolations) (21, 129). Extensive studies of four strains indicated that their biological and biochemical properties, including growth characteristics, metabolism, hemolytic activity, and protein migration patterns, are entirely comparable to those of human strains of *R. prowazekii.* Only minor differences were encountered in protein banding after isoelectric focusing (41, 169). No differences were encountered in the properties of the DNA (95). On the other hand, the flying squirrel strains could easily be distinguished from the related species, *Rickettsia typhi,* by protein migration patterns and DNA/DNA hybridization.

The mechanism of transmission of *R. prowazekii* between flying squirrels is remarkably similar to the mode of transmission in man. The flying squirrel louse is the most efficient vector. Since this ectoparasite cannot yet be reared in the laboratory, the course of rickettsial infection in the flying squirrel louse is not known, but it is as infectious and virulent for the human body louse as are the human strains. The flying squirrel flea is a less efficient vector. The rickettsiae do not consistently survive in ticks and mites (21, 129).

Since the discovery of the sylvatic cycle of *R. prowazekii,* indigenous cases of epidemic typhus fever have been reported in the US for the first time in over half a century. As of November 1981, 28 cases have been confirmed (21, 45, 89, 120; J. E. McDade, personal communication). Although the distribution of cases (Figure 2) reflects as much the lines of communication of interested investigators as the true incidence of the disease, it is obvious that the infection is widespread. All cases have occurred well within the area of distribution of the flying squirrel. In many cases it was clearly established that flying squirrels were on the premises or in the immediate environment. There is no real evidence of involvement of other animals, but such a possibility cannot be excluded.

There is a clear difference in seasonal incidence of infection between Rocky Mountain spotted fever and sylvatic typhus. The former is most prevalent during late spring and summer, at the time of greatest tick activity. Sylvatic typhus occurs most frequently during late fall and winter (Figure 2), when the flying squirrels seek the warmth of human habitation. In fact, several patients reported that their attics had been invaded by flying squirrels. The few cases that occurred during the warm months were probably acquired during camping trips or on the golf course. It is interesting to note that epidemic typhus of human origin is also a disease of the cold months, when heavy clothing aggravates louse infestation.

When did the flying squirrel become infected? Since the human and flying strains are so similar, it is a reasonable assumption that sylvatic typhus is of human origin and was acquired by the flying squirrel in an epidemic during the wars of the 18th century or during the massive immigrations of the 19th century. Why, of all the animals, the flying squirrel? A large number of other rodents were investigated for the presence of rickettsiae, but *R. prowazekii* was isolated only from the flying squirrel. It is true that this animal establishes an intimate relationship with man by invading attics, but it is difficult to believe it is the only species to establish such a close association. An answer to this question is not yet available.

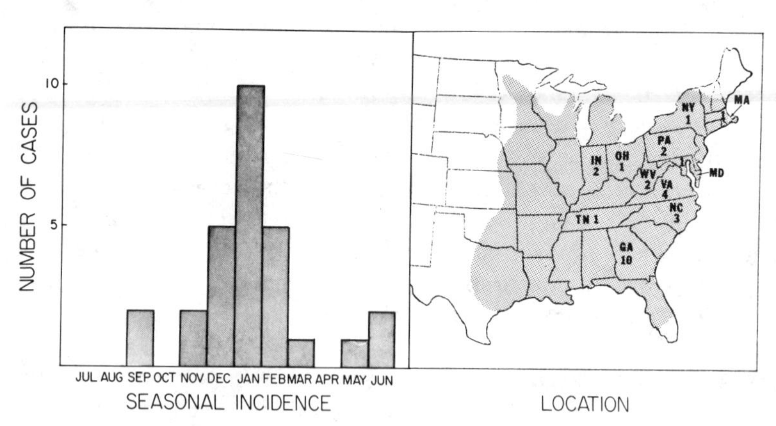

Figure 2 Incidence of reported sylvatic typhus (*Rickettsia prowazekii* infection) in the US. The shaded zone on the map is the area of distribution of the flying squirrel (*Glaucomys volans*). The number of cases in each state is indicated next to its abbreviation. (From 21, 45, 89, 120; J. E. McDade, personal communication.)

Rochalimaea vinsonii and Rochalimaea quintana

The small Grosse Isle in the St. Lawrence River in the Province of Quebec, Canada, served as a quarantine station and thousands of immigrants died there of typhus in the middle of the 19th century. In 1943 Baker (10) studied the local fauna in the belief that he might uncover a vestige of this epidemic. From the vole (*Microtus pennsylvanicus*) he repeatedly isolated a small Gram-negative rod that resembled a rickettsia, but—as he realized—it was not *Rickettsia prowazekii.* The Canadian vole agent was re-examined recently (91, 97, 150) and was shown to be genotypically and phenotypically similar, but not identical, to the agent of trench fever, *Rochalimaea quintana,* and the name *Rochalimaea vinsonii* was proposed (149).

Trench fever made its first explosive appearance during World War I; it reappeared on a more limited scale in Europe during World War II and has been reported from other parts of the world. *R. quintana* has an ecological niche identical to that of *R. prowazekii:* Man is its main reservoir and the human body louse is its vector. If *R. quintana* and *R. vinsonii* indeed have a common ancestor, Baker may have found a vestige of the epidemic of the previous century, since many of the typhus patients might have been infected concurrently with the trench fever agent. This hypothesis is supported by the fact that *R. vinsonii* or similar microorganisms have not been isolated from voles in other locations. There is, however, another explanation. When trench fever first appeared in 1915 the infection was believed to have been acquired from the European vole (121), *Arvicola terrestris,* related to *Microtus* (W. L. Jellison, personal communication), which was infesting the trenches in large numbers. If this is correct, a vole parasite might have been a common ancestor of both *R. vinsonii* and *R. quintana.* An investigation of other organisms of the general appearance of rickettsiae that are epicellular rather than intracellular in the arthropod host, such as *Wolbachia melophagi,* might be rewarding.

Other Rickettsiae

Reports of *Rickettsia typhi* and *Rickettsia akari* infections, murine typhus and rickettsialpox, respectively, have declined dramatically in the US and in other countries (118, 168). These declines may be due to reduced severity and recognition of the disease because of the availability of antibiotics, diminished interest in reporting these diseases, or better protections of dwellings against rodent infestations. Many aspects of the natural cycles of *R. typhi* and *R. akari* remain largely unknown and may not have changed (131). In contrast, reported *Rickettsia conorii* infection (boutonneuse fever) has increased sharply in certain regions of Italy. Whether these reports reflect increased awareness, a change in the ecology of the arthropod vector,

Rhipicephalus sanguineus, or the spread of the rickettsia to a new vector is not known (122). Some of the features of the complex ecology of *Rickettsia tsutsugamushi* have been reviewed recently (146). Bourgeois et al (18a) carefully analyzed the immunological responses of a group of scrub typhus patients in an endemic area and established that reinfection with a *R. tsutsugamushi* strain of different serotype, which may or may not lead to recognizable disease, is a relatively common occurrence.

There is no indication that the worldwide distribution of *Coxiella burnetii* has changed in any substantial manner. Its constant presence was recently demonstrated by outbreaks among individuals who used pregnant ewes in their research (47).

MORPHOLOGY AND STRUCTURE

Rickettsia and Rochalimaea

OUTER LAYERS The morphology of *Rickettsia* and *Rochalimaea* is typical of Gram-negative bacteria. These organisms do not have flagella, pili, or attachment proteins that can be recognized morphologically. Costerton et al (38) point out that virtually all bacterial cells are surrounded by a glycocalix (slime) in their natural environment, but the slime layer is easily lost during in vitro cultivation or other laboratory manipulations. Rickettsiae are no exception. The presence of a slime layer in rickettsiae has been suspected since the early days of cultivation in the yolk sac of chicken embryos: Crude yolk sac suspensions have a soluble antigen that is lost when the rickettsiae are purified. Furthermore, ultrathin sections of infected cells consistently reveal an electron-lucent zone surrounding each rickettsia. The morphological demonstration that *R. prowazekii, R. rickettsii,* and *R. tsutsugamushi* have substantial slime layers was achieved by the specific-antibody stabilization procedure on organisms gently liberated from their host cells. The slime layers stain with ruthenium red and silver methenamine, which is consistent with a polysaccharide nature, but the chemical nature of the slime layer still needs to be investigated (126, 127).

Rickettsiae of the typhus and spotted fever groups can be separated quite well from host cell components without appreciable loss of viability or metabolic activity by the Renografin density gradient procedure (147), even though they lose most of the slime. The loss of the microcapsular layer, however, might be quite detrimental, and this probably occurs in suspending fluids of low ionic strength. Dasch (39) showed that typhus rickettsiae suspended in water or in various buffers, pH 7.6, at 10 mM concentration, release into the suspending fluid a species-specific protein antigen (SPA). The amount of protein released is quite large, under optimal conditions as much as 10–15% of the total cell protein content, but it migrates as a single component in sodium dodecyl sulfate-polyacrylamide gel electrophoresis (SDS-PAGE). Although low-molecular-weight RNA fragments are also released, the cells are not disrupted, but, as shown in other experiments (143), when suspended in fluids of low ionic strength, rickettsiae lose most of their metabolic activity. Protein release can be abolished by $MgCl_2$, provided the cell envelope is not damaged by pretreatment with lysozyme or detergents. Although the anatomical origin of the SPA has not been identified, it is most likely the microcapsular layer.

R. tsutsugamushi, in contrast to other rickettsiae, adheres tenaciously to host cell components and is not purified without substantial loss of activity. This difference in adherence properties is reflected in the structure of the outer membranes (Figure 3). Although all rickettsiae have two unit membranes, the cytoplasmic membrane and the cell wall, in the typhus and spotted fever group the inner leaflet of the cell wall is the thicker of the two, whereas the reverse is true in the scrub typhus rickettsia (126). Apparently, physical-chemical means that remove host cell components from *R. tsutsugamushi* damage the cell wall.

DNA The base composition of the DNA of rickettsiae was studied by acid hydrolysis and chromatography at the time when the information on the constant ratios of adenine to thymine and guanine to cytosine was being developed. The molar percentages of guanine plus cytosine (G+C) for *R. prowazekii* (170) and *C. burnetii* (128) obtained then were only slightly

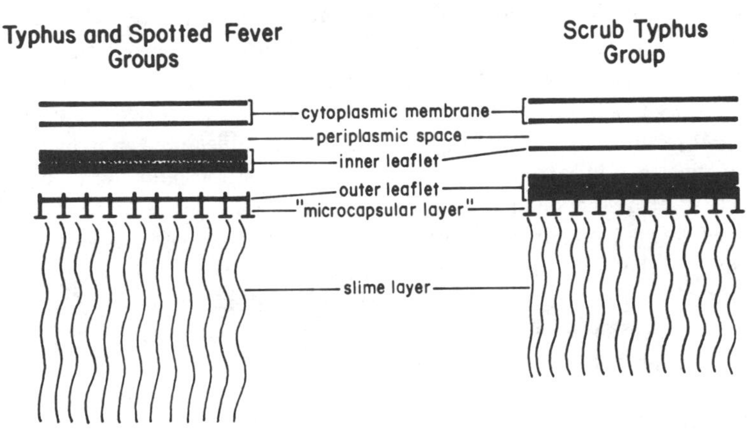

Figure 3 Diagrammatic representation of the outer layers of rickettsiae. (From 126.) Reproduced with the kind permission of the authors and publishers.

higher than those obtained later (95, 132) by thermal denaturation and buoyant density (29 to 30 for *R. prowazekii* and 43 for *C. burnetii*). Studies of the past decade (1, 96, 123, 132) have confirmed that there is a small but significant difference between the base ratios of typhus (including *Rickettsia canada*) and spotted fever group of rickettsiae (32 to 33 mol % G+C). The properties of the DNA of *R. tsutsugamushi* have not yet been investigated. The mol % G+C of the DNA of *Rochalimaea* is approximately 39 (132, 150). The genome size of two rickettsial species was studied by reassociation kinetics (73) and shown to be 40% of that of *Escherichia coli,* thus intermediate between *Neisseria* and *Chlamydia* (142).

The genetic relatedness of rickettsiae has been studied only recently (95–97) by reassociation kinetics and by the nick translation technique. No difference in DNA/DNA percentage of hybridization was encountered among strains of the same species of *Rickettsia prowazekii, Rickettsia typhi,* and *Rochalimaea quintana.* Interspecies percentage of hybridizations obtained thus far by reassociation kinetics are summarized in Table 1. As expected, *R. prowazekii* and *R. typhi* are closely related with 73% hybridization. Most unexpected was a small but highly significant degree of hybridization between the typhus group of rickettsiae and *R. quintana.* Since there is a 10% difference in base ratios, the degree of hybridization (31%) approaches the theoretical maximum.

The other results shown in Table 1 should be examined with caution since the number of replicate experiments has not yet been quite as extensive. However, it is apparent that the degrees of relatedness (about 40% hybridization) between the two typhus species (*R. prowazekii* and *R. typhi*) and *R. canada,* between *R. canada* and *R. rickettsii,* and between *R. rickettsii* and the typhus species are about the same. The degree of hybridization between the typhus group and *R. rickettsii* was not unexpected, but interrelationships within the spotted fever group rickettsiae and of this group to the typhus rickettsiae remain to be investigated. *R. canada* occupies an unusual position in the genus *Rickettsia.* It resembles the typhus group of rickettsiae in DNA base ratio, antigenic composition, and the manner in which it grows in chicken embryos and cell cultures. It resembles the spotted fever group in its ability to invade the host cell nucleus and to survive in the tick, but not in the louse (25, 153). It may have a wide distribution in nature, since a strain was recently isolated from California (81). As is also shown in Table 1, there is a moderate degree of DNA/DNA hybridization between *R. quintana* and *R. vinsonii.*

Coxiella burnetii

McCaul & Williams (87) recently reported that *Coxiella burnetii* undergoes a developmental cycle that involves vegetative and sporogenic differentiation. This conclusion was based primarily on observations on ultrathin

Table 1 Properties of the DNA of rickettsiae[a]

Rickettsia	Genome size[b]	Mol% G + C	DNA/DNA percent hybridization[c] P	T	C	R	Q	V
Rickettsia								
R. prowazekii	110	29	*97*	73	44	53	32	—
R. typhi	108	29	73	*101*	47	36	32	—
R canada	149	29	37	41	—	39	—	—
R. rickettsii	130	33	38	27	43	—	—	—
Rochalimaea								
R. quintana	103	39	—	—	—	—	*95*	37
R. vinsonii	134	39	—	—	—	—	—	*102*

[a] From Myers et al (95–97). Mean of values was obtained by reassociation kinetics. Mol% G + C is in agreement with previously reported values (132, 150).
[b] $\times 10^7$ daltons.
[c] Based on genome size shown in the same row. Homologous reactions are shown in italics. The species is indicated by its first letter.

sections of *C. burnetii* cells separated from yolk sac components by a procedure that avoided osmotic damage (156). It involved Renografin density gradient centrifugation, which was shown to preserve the integrity of other rickettsiae (147) and *Chlamydia* (56), and the inclusion of 0.25 M sucrose in the buffers used for the suspension of the rickettsiae and for the preparation of the gradients. No attempt was made to obtain synchronous growth. The cycle was traced from the examination of a population containing a variety of forms.

There are two main cell types, designated large and small cell variants (LCV and SCV), in such populations. Besides size, the two cell types differ in the width of the periplasmic space, wider in LCV, and in electron density of the cytoplasm, greater in SCV. Both types are capable of transverse binary fission. These observations are in agreement with those made previously by Wiebe et al (154), who demonstrated that the two cell types revert to a mixture when cultured separately. In addition, McCaul & Williams (87) described the formation of extra layers ("caps") in the periplasmic spaces of some LCV that developed as the result of unequal division of the cytoplasm into small, multilayered dense bodies. These sporelike bodies were eventually released by lysis from the LCV, enlarged to SCV, and finally divided and/or enlarged into LCV.

McCaul et al (86) showed that there is considerable difference in the physiologic properties of LCV and SCV, although both are infectious, as shown by Wiebe et al (154). The LCV are highly susceptible to osmotic shock, and a population consisting almost exclusively of SCV can be obtained by suspending and washing the mixture of cells in distilled water. As the number of LCV decreases, so does specific metabolic activity, as measured by CO_2 evolution from glutamate, but SCV clearly have some metabolic activity.

Potassium permanganate staining was particularly useful for the definition of ultrastructural detail, but this is probably not the entire explanation of why the cycle McCaul & Williams (87) describe was not seen by previous investigators. In the work of Wiebe et al (154), which most closely resembles that of McCaul & Williams, the use of hypertonic buffer and Celite may have damaged the cells just enough to discriminate against the LCV bearing the small dense bodies. Infection in mammalian cells, carefully described by Kordova et al (77) and Burton et al (31), differs from growth in the yolk sac, as pointed out by them and by Khavkin et al (72). Obviously, the work by McCall & Williams (87) needs to be repeated with more than one type of cell infected under defined conditions. Until this is done caution should be used in the selection of the epithet for the developmental cycle of *C. burnetii.* The small dense body resembles an endospore morphologically and good evidence indicates it is associated with the organisms' capability to survive in environments that destroy many nonsporulating bacteria (5). However, it does not have dipicolinic acid (87) and we know too little of the factors that regulate its formation and its germination to call it a spore. Any other designation might be equally misleading. Perhaps the most appropriate epithet is "sporelike."

GROWTH AND METABOLISM

Rickettsial Diversity

Although it is tempting to compare rickettsiae to each other and to use one as a model for another, this can be done profitably only when the great diversity of these organisms is taken into consideration. The most obvious one is the location of the intracellular rickettsiae: *Rickettsia* primarily in the cytoplasm, sometimes in the nucleus; *Coxiella* in the phagolysosome. The optimal pH for metabolic activity reflects their respective environments. *Rickettsia* appear to be free in the cytoplasm and not to be surrounded by a membrane of host origin, a property shared with some of the insect endosymbionts (24) and, possibly, with the protozoan parasite, *Trypanosoma cruzi* (101). The location of *Coxiella* is not unique: It is encountered among some of the mycobacteria and the protozoan parasite, *Leishmania donovani* (35). *Rochalimaea* grow quite well in cell culture, but adhere to the outer surface of the eukaryotic cells. The few rickettsiae seen in an intracellular location appear to be degenerating (91, 150). Another major difference is the absence of glycolytic enzymes in *Rickettsia* (37) and in *Rochalimaea* (150), but not in *Coxiella* (109).

Rochalimaea

Vinson & Fuller (135) clearly established in 1961 that *Rochalimaea quintana* can be grown on blood agar. During the preceding 45 years since its recognition it was grown only with enormous difficulty or, if grown axenically, it was not identified as the same organism (149). Subsequent studies (85, 94, 134) showed that aerobic conditions, increased CO_2 pressure, and a factor in red blood cells were essential, and a factor in serum was stimulatory. The erythrocyte requirement could be replaced by hemin or fetal bovine serum, but not calf serum. More recently, Weiss (145) cultivated both species of *Rochalimaea* in a liquid medium consisting of amino acids, fetal calf serum, yeast extract, and succinate. Sodium bicarbonate is required as a source of CO_2 by *R. quintana,* but not by *Rochalimaea vinsonii.* The generation time in this medium is approximately 6 h.

Rochalimaea and *Rickettsia* appear to derive energy from similar substrates, but they differ greatly in ability to transport them. Glutamate is the substrate utilized most rapidly by all rickettsiae studied thus far, whereas exogenous succinate is utilized at a low rate (143). The reverse is true in *Rochalimaea quintana:* Succinate is utilized rapidly and glutamate is utilized slowly (67). Curiously, glutamate is metabolized at a very high rate in the presence of another substrate, such as succinate, pyruvate, or glutamine, which, presumably, provide energy for the transport of glutamate. *R. vinsonii* differs in this respect, since it utilizes the four substrates, including glutamate, at approximately the same rate (149).

Both species of *Rochalimaea* have low levels of ornithine decarboxylase activity. In *R. quintana* the enzyme appears to be constitutive, but it is demonstrated in whole cells only when another substrate is furnished for the transport of ornithine. In *R. vinsonii,* the enzyme has the characteristics of an inducible activity. It can be demonstrated only after the cells have been incubated with medium plus ornithine. The optimal pH of activity is slightly lower in *R. vinsonii* than in *R. quintana* (151).

The main interest in the nutrition and metabolism of the *Rochalimaea* species lies in the possibility that they might serve as models for the genetically related but more fastidious species of *Rickettsia.* This can be done if some of the differences in transport properties, illustrated above, are taken into account.

Rickettsia

The interaction of a rickettsia with a host cell requires the active participation of both. Although the rickettsia may gain access to the cytoplasm directly from an extracellular position, most of the recent evidence suggests that the rickettsia is phagocytized first. Since the rickettsia stimulates its own phagocytosis, this is called induced phagocytosis (139). If the rickettsia is to survive, however, it must quickly escape from the phagosome into the cytoplasm. This phenomenon is best illustrated in those instances where escape is inefficient. Rikihisa & Ito (117) showed that shortly after infection of guinea pig polymorphonuclear leucocytes with *Rickettsia tsutsugamushi,* some rickettsiae are free in the cytoplasm, whereas others are still in the phagosome and some of these are degenerating. Gambrill & Wisseman (57) showed that escape from the phagosome is dependent on the virulence of the strain. Although virulent (Breinl) and avirulent (E) strains of *Rickettsia prowazekii* multiply equally well in irradiated chicken embryo cells (165), only the Breinl escapes the phagosome in human macrophages. A phospholipase appears to be involved in the mechanism of escape (161, 162).

Intracellular growth proceeds in different ways. As shown by Wisseman and his associates (164–166), the cycle in *R. prowazekii* conforms closely to the concept of a one-step growth curve. It starts immediately or after a lag, depending on the stage of growth of the seed. Generation time is approximately 9 h, and when the rickettsiae become very numerous, the host cells break up and release the rickettsiae. *Rickettsia rickettsii* escapes from the host cytoplasm into the surrounding medium early in the course of infection and infects other cells. This property may be responsible for the relatively large plaques *R. rickettsii* produces on cell culture monolayers. The occasional rickettsia that invades the nucleus multiplies there without escaping and produces heavy accumulations that contrast with the low density in the cytoplasm. *R. tsutsugamushi* achieves a very high density especially in the perinuclear region and often protrudes from the surface of the cell (117). A curious cycle was described in mouse peritoneal mesothelium cells (52), which is very common in *R. tsutsugamushi* infection (117). The organisms acquire a host membrane coat from the host cell plasma membrane and bud from the cell surface. Rickettsiae enveloped by this membrane enter other cells by phagocytosis and shed their coat as they escape from the phagosome.

The nutritional requirements of rickettsiae, as distinct from those of the host cell, have not been extensively studied. One surprising finding is that *R. tsutsugamushi,* when cultivated on irradiated chicken embryo cells with an organic buffer without bicarbonate, multiplies equally well in the presence or absence of atmospheric CO_2, whereas *R. prowazekii, R. typhi,* and *R. ricketsii* require CO_2 (76). Most of the other investigations have been concerned with radiolabeling rickettsial macromolecules. This is not difficult to do when a low level (usually 1 to 2 μ g/ml) of cycloheximide is added to the cultures, to reduce eukaryotic biosynthesis without being overly toxic (51, 143). A mixture of radiolabeled amino acids is used most commonly, but in some experiments [^{3}H]galactose effectively labeled surface components of *R. prowazekii* and *Rickettsia conorii* (108).

Recent studies of the metabolism of rickettsiae separated from host cells has been limited to the typhus rickettsiae. This work has been reviewed (144) and can be summarized as follows: Winkler (159) has shown that rickettsiae have a carrier-mediated transport system for ADP and ATP.

Rickettsiae achieve a considerable (5 mM) endogenous pool of adenine nucleotides, which is maintained against a low extracellular concentration. However, when the extracellular concentration of ATP is high, it is readily exchanged with the endogenous pool. This finding has both theoretical and practical implications. Although rickettsiae are not energy parasites and they derive their ATP from the metabolism of substrates such as glutamate, they can capture phosphorylated compounds that might be available in their intracellular environment. The practical application is a simple means of labeling rickettsiae by incubating them with ^{32}P-labeled ATP. Such rickettsiae have been used to study their interaction with host cells and particularly the process of internalization, since rickettsial ATP exchange with exogenous ATP is greatly reduced when the rickettsiae are inside a host cell (139, 140).

Possibly associated with their ability to transport phophorylated compounds is their relatively limited nucleotide metabolism. This was illustrated by Williams & Peterson (157) with the uridine and thymidine pathways, but it is obviously true of other nucleotides (155, 158). Nucleotide metabolism in rickettsiae appears to be confined to the mono-, di-, and triphosphates. There is no evidence that rickettsiae synthesize or degrade nucleoside monophosphates. Thus, their nutritional requirements must include at least the nucleoside monophosphates, which they can phosphorylate, but, probably, cannot manufacture de novo. A corollary of this property is that the concept of a constant energy charge (ATP + ½ ADP)/(ATP + ADP + AMP), does not apply. When the intracellular concentration of ATP declines there is a corresponding increase in ADP and AMP. Thus, there are much greater shifts in adenylate energy charge in rickettsiae compatible with survival than in other bacteria.

The mechanism of enzyme regulation in rickettsiae remains largely unknown. Phibbs & Winkler (111, 111a) recently isolated a key regulatory enzyme, citrate synthase, of *R. prowazekii* and showed that in its properties it more closely resembled an enzyme from a eukaryotic cell or a Gram-positive bacterium rather than another Gram-negative bacterium.

Coxiella

The interaction of *Coxiella burnetii* with host cells is highly complex. Although multiplication occurs in the presence of lysosomal hydrolases (3), excessive amounts of enzyme may be detrimental (32, 33) or may have a modulating effect on the developmental cycle. In addition, *C. burnetii* undergoes phase variation (23), and phase II rickettsiae are more infectious for L cells than are phase I (32).

Of particular interest are recent reports of persistent infection of L and Vero cells (33) and of several types of macrophagelike tumor cell lines (7).

The rickettsiae are always located in vacuoles; in some cases nearly all the cells are infected, some containing $>10^3$ rickettsiae. Despite heavy infection, the viability of the host cells, as determined by dye exclusion, remains high, and, obviously, the heavily infected population of cells are able to propagate. Persistent cultures were produced with both phase I and II rickettsiae, although some of the macrophage cell lines did not remain infected with phase I. A surprising finding was that phase I rickettsiae in L cells and in one of the macrophage cell lines eventually changed to phase II, a phenomenon previously encountered only in yolk sac. To what extent persistent infection plays a role in the survival of *C. burnetii* in nature is not known.

Until recently, investigations of metabolic activities of *C. burnetti* have been limited to experiments with cell extracts (109, 143). Such experiments continue to uncover key enzymatic functions. Christian & Paretsky (36) demonstrated the synthesis of RNA by a DNA-dependent RNA polymerase, as well as some nucleotide kinase activity with ATP as the activated phosphate donor. Baca (6) obtained polyuridylic acid-directed biosynthesis of polyphenylalanine, and Donahue & Thompson (43, 44) obtained the manufacture of a coliphage coat protein identical in electrophoretic and antigenic properties to the original protein. Thus, *C. burnetii* appears to have the essential components for the initiation, elongation, and termination of polypeptides. The rickettsial preparations in these experiments retained little if any mRNA, and some of the enzymatic activities were considerably lower than in *Escherichia coli,* but differences were minimized when optimal conditions were achieved in each case.

Although Paretsky and his associates (109) have had no difficulty demonstrating a large number of enzymatic activities with disrupted preparations, results with whole cells have been disappointing. At best, Ormsbee & Peacock (107) demonstrated low levels of O_2 uptake with pyruvate and some of the intermediates of the citric acid cycle. The reason for this apparent paradox has been resolved: The reactions were studied at an unfavorable pH. Hackstadt & Williams (61–63) showed that the optimal pH for transport and utilization of glutamate and glucose lies between 3 and 5. There is a sharp drop in activity above 5.5 and the cells are metabolically virtually inert with these substrates at the pH 7.0 of previous investigations. Of the substrates tested, pyruvate, succinate, and glutamate were metabolized most rapidly. Glucose was utilized at a low level, but its metabolism was stimulated by the addition of glutamate. At pH 4.5 the ATP pool was stabilized most effectively by glutamate. Interestingly, at pH 7.0 the ATP level was stable even without substrate, possibly because of the absence of endogenous metabolism at that pH.

To test the hypothesis that the low pH requirement reflects the natural environment of *C. burnetii,* the phagolysosome, chicken embryo fibroblasts were treated with lysosomal perturbants, i.e. chloroquine, methylamine, or ammonium chloride, which raise the pH of the phagolysosomes. The growth of *C. burnetii* in cells thus treated was greatly diminished (61).

The metabolism of *C. burnetii* needs to be reexamined in the light of the above-described findings and the evidence, summarized earlier, that it undergoes a developmental cycle.

PATHOGENESIS AND IMMUNOLOGY

Factors that Contribute to Pathogenesis

It is reasonably well established that one of the main factors responsible for rickettsial pathogenesis is the mechanism by which rickettsiae gain entry into a host cell. Rickettsial acute "toxicity" for mice or hemolysis of sheep or rabbit erythrocytes requires the physiologic integrity of the rickettsiae and can best be described as massive infection that results in injury to the endothelial cells of the mouse capillaries or abortive infection that results in the release of hemoglobin (141, 160, 163). Immediate cytotoxicity can also be produced by overlaying L cells with a high multiplicity of *Rickettsia prowazekii.* Winkler & Miller (162) showed that this phenomenon involves phospholipase activity, most likely a rickettsial enzyme, but, possibly, a eukaryotic enzyme activated by the rickettsiae. Walker & Cain (136) examined histologically the various zones of the plaques produced by *Rickettsia rickettsii* on chicken embryo cell monolayers and found the cytopathology in the intermediate zone, where the cells are exposed to the massive release of rickettsiae from the center, to have some of the features of the lesions observed in experimental animals and human autopsy sections.

Rickettsiae, as Gram-negative bacteria, are expected to have lipopolysaccharides (LPS). The LPS of *Coxiella burnetii* and *Rochalimaea quintana* have been described in detail (8, 9, 66), and the presence of LPS in the typhus rickettsiae has also been demonstrated (124). Many of the features of the pathology of rickettsial infection are consistent with LPS toxicity.

Walker & Henderson (137), in a study of immunosuppression in *R. rickettsii* infection of the guinea pig, found no evidence that immunopathological mechanisms contribute to pathogenesis, as appears to be the case in human Rocky Mountain spotted fever. On the other hand, *C. burnetii* endocarditis, which is a rare sequela of Q fever, involves prolonged antigen-antibody interaction and probably has a strong immunopathological component (2, 133).

Genetics of Susceptibility

It has been recognized early in the course of study of *Rickettsia tsutsugamushi* that this species consists of a highly heterogeneous group of strains, differing in antigenic composition and virulence for the guinea pig and mouse. The Karp strain is regarded as highly virulent, whereas the Gilliam strain is of variable virulence for random bred mice. It was recently shown by Groves et al (59, 60) that most inbred mouse strains are either highly susceptible or highly resistant to the Gilliam strain. Resistance was found to be due to a single autosomal, dominant gene, closely linked to the retinal degeneration locus on chromosome 5. Resistance to the Gilliam strain was unrelated to susceptibility to other strains. Nacy & Groves (98) compared the immunological responses of the resistant (BALB/c) and susceptible (C3H/He) mice to the Gilliam strain. In vitro there is no difference in resident macrophages of the two strains in their support of the growth of rickettsiae or in their activation with specific lymphokines. When the mice are inoculated intraperitoneally, however, there is a delay in the activation of the C3H macrophages, which is not due to deficient lymphokine production. Thus, susceptibility to infection appears to be due to defective macrophage regulation rather than function. A similar mechanism seems to be involved in the susceptibility to *Rickettsia akari* of certain inbred strains of mice (90).

It is not known whether genetic factors in the human population affect susceptibility to rickettsial disease. The possibility that Rocky Mountain spotted fever might be more severe in individuals with glucose-6-phosphate dehydrogenase deficiency is being investigated (138).

Humoral and Cell-Mediated Immunity

Antibodies have no direct killing effect on typhus rickettsiae and do not interfere with their infection of nonprofessional phagocytes, such as endodermal cells of the yolk sac or chicken embryo cell monolayers (58). Virulent rickettsiae grow quite well in cell cultures of human macrophages obtained from nonimmune and in one case from a convalescent individual. When the macrophages are first coated with cytophilic antirickettsial serum and then are exposed to rickettsiae, phagocytosis of rickettsiae is accelerated, but subsequent growth of rickettsiae is not impeded. Only when rickettsiae are first coated with antibodies and then are exposed to macrophage does the interaction lead to increased phagocytosis as well as to digestion of the rickettsiae. Even under these circumstances rickettsiae escape digestion in an occasional macrophage and multiply (11, 167).

The guinea pig has been widely used as an infection model for typhus rickettsiae. When they are injected intradermally with *Rickettsia typhi*, rickettsiae replicate locally and spread to the draining lymph nodes and subsequently cause systemic infection (92). If prior to injection the guinea pigs are inoculated with serum from a syngeneic immune guinea pig, they are protected against systemic infection, but not against the skin lesion. Complete protection, analogous to the response seen in animals recovering from infection, is obtained by the injection of spleen cells (93).

Experiments with *R. tsutsugamushi* were done with BALB/c mice, which can be immunized with the avirulent strain Gilliam and challenged with the virulent strain Karp. It was shown that mice can be passively protected by lymphocytes derived from immune donors. Lymphocytes obtained from peritoneal exudates become protective earlier in the course of the immunizing infection and are more effective than lymphocytes derived from the spleen (34). Peritoneal macrophages from BALB/c mice can be infected with the Gilliam strain in vitro. The number of infected cells can be reduced to 50% by pretreating the macrophages with antiserum and to 25% by pretreatment with lymphokines. An even greater curtailment is seen in macrophages derived from immune mice. Interestingly, under these conditions there is little difference in the progress of infection in the macrophages that do become infected (100). When lymphokines are added after infection and are maintained in the culture medium for 24 h, the number of intracellular rickettsiae declines sharply. Thus, lymphokines affect the macrophages in two different ways: They cause an immediate killing of rickettsiae and a reduction in the number of infected cells; when their action continues, most of the intracellular rickettsiae are killed also (99).

The cell-mediated immunity to *C. burnetii,* investigated in peritoneal guinea pig macrophages, is complicated by the difference in the properties of phase I and phase II rickettsiae. Phase I are more resistant than phase II to phagocytosis by macrophages from normal or immunized guinea pigs. After ingestion, the rickettsiae multiply and destroy macrophages from nonimmune guinea pigs or guinea pigs immunized with phase II organisms. When ingested by macrophages from guinea pigs immunized with phase I, phase I rickettsiae are destroyed, while phase II, after a short delay, proceed to multiply. Immune serum increases the rate of phagocytosis of the rickettsiae and intracellular destruction (74, 75), but at least one report (65) casts some doubt on the role of antiserum.

Most of the studies described above were directed toward the establishment of better methods of evaluation of experimental vaccines. It is obvious that a test of a single arm of the immune response might be misleading, even though in most cases such a test has provided the correct answer. A case in point is the mouse toxin-neutralization test, widely used for the determination of the potency of typhus vaccines. In this test the experimentally vaccinated animal, the guinea pig, provides the serum and the nonimmune mouse provides the necessary phagocytic cells. A more critical assessment might be provided by a vaccinated animal that must furnish both serum and cells, as is the case when a guinea pig resists intradermal challenge.

Experimental Vaccines

Vaccination against rickettsial diseases has a long history (130). Parker developed a vaccine against Rocky Mountain spotted fever in 1925 from phenolized infected tick tissue (see 130). Weigl and Castaneda prepared vaccines against epidemic typhus in the 1930s from infected lice and infected rodent lungs, respectively (see 130). Mass production of vaccines, however, began in the early 1940s, when Cox demonstrated that rickettsiae can be grown in the yolk sac of chicken embryos. Some of these vaccines were undoubtedly effective and have saved lives. They are not used today because rickettsial diseases, if promptly recognized and treated, are no longer lethal and because standards for safety and efficacy have become much more stringent. The relatively high levels of LPS and tissue cell components of previous vaccine preparations would not be acceptable in a vaccine submitted for approval now. However, if high safety and efficacy standards can be satisfied, vaccination would be recommended for defined populations at increased risk of infection. There are at present at least four experimental vaccines in various stages of development and evaluation. Each is prepared in a different manner, as dictated by the biological properties of the organisms.

The combined epidemic-murine typhus experimental vaccine of Dasch and his associates (17, 39, 40, 42) consists of the SPA (species-specific antigens), released from the rickettsiae in a hypotonic solution. Renografin purification (147) prior to SPA extraction eliminates most of the yolk sac components and the LPS is removed by high speed centrifugation. Old evidence (see 40) suggests that antigen preparations rich in SPA are the most effective immunogens. A recent study of sera and lymphocytes of individuals infected during previous decades suggests that SPA stimulation of lymphocyte transformation is a highly persistent immune response (18). Thus far the SPA antigens have been tested only in guinea pigs. A subcutaneous injection of 5 μg followed by a 2-μg booster 2 months later protects guinea pigs against intradermal challenge about as well as immunization by infection.

The chief difficulty encountered in the preparation of a vaccine against *Ricksettsia rickettsii* is that the small yield of rickettsiae from yolk sac provides only minimal protection (46). Kenyon and his associates (4, 68–71) obtained much more satisfactory yields by growing the rickettsiae in duck and chicken embryo cell monolayers in roller bottle cultures. The rickettsiae are inactivated with 0.1% Formalin and are purified by sucrose gradient. Experiments in guinea pigs and rhesus monkeys suggest that this experimental vaccine is considerably more immunogenic than antigens derived from yolk sac and confers protection to guinea pigs infected by aerosol and by subcutaneous injection. Of 16 human volunteers who received one to three injections of two dilutions of vaccine, the majority developed antibody and lymphocyte transformation responses.

Early attempts at vaccination of humans against *Rickettsia tsutsugamushi* were probably not successful in part because Formalin or other inactivating agents reduce immunogenicity and in part because strains of scrub typhus are antigenically quite heterogeneous. Eisenberg & Osterman (48–50) found that inactivation by gamma irradiation preserves the immunogenicity of the rickettsiae better than Formalin treatment. A judicious combination of strains confers reasonably good protection to mice challenged with a variety of strains.

Coxiella burnetii phase I, after a few passages in yolk sacs, changes to phase II and early vaccines consisted of phase II rickettsiae. They produced large local skin lesions. Recent vacine preparations are based on the observation of Ormsbee and his associates (55, 104, 110) that phase I organisms are 100–300 times more immunogenic than are phase II. Thus, the amount injected can be reduced to 30 μg per person. Sensitized persons, for whom the vaccine is toxic, can be identified by the intradermal injection of 0.2 μg. Even this small dose, which is without effect on unsensitized individuals, elicits a pronounced antibody response in those who were previously exposed. The results of field test in man are not yet available. Of interest, however, are the results of field tests in dairy cattle, one of the primary sources of human infection. Behymer and his associates (12, 13, 16) showed that although it did not prevent infection, the injection of 75 μg of phase I antigen mixed with Freund's incomplete adjuvant greatly reduced the shedding of rickettsiae by the cattle.

Considerable amounts of information has been obtained on rickettsial cell fractions and on humoral and cell-mediated immunologic responses in man and experimental animals. As suggested by the above examples, in most cases the information is not yet sufficiently well integrated for the production of vaccines that are entirely satisfactory.

CONCLUDING REMARKS

This review gave preference to topics that can be examined from two points of view: (*a*) What we have learned during the past decade and what we hope to learn in the next; (*b*) to what extent rickettsial research has benefited from advances in other fields and what has rickettsial research contributed to our patrimony of knowledge of biology.

In 1935 Zinsser (171) warned us that typhus is not dead and may break into the open again. We can echo his warning by saying that rickettsiae have deep ecological roots and, as we have learned in recent years, it is quite possible that not all have been recognized. What we hope to learn in the next decade is the identity of the next-of-kin of rickettsiae by analyses of ribosomal RNA nucleotide sequences or amino acid sequences in ribosomal proteins. This information will greatly enhance our understanding of the place of rickettsiae in the microbiological world.

Progress has also been made in the definition of the environment of rickettsiae within the eukaryotic cell. Thus, rickettsiae are excellent tools for the study of membrane function in microorganisms adapted to at least two "extreme" environments, the cytoplasm and the phagolysosome. A decade ago, the primary research priority might have been the growth of rickettsiae in an axenic medium. We still need to learn a great deal about rickettsial biosynthesis, but with the availability of recombinant DNA technology, perhaps the manufacture of large amounts of proteins needed for diagnostic and immunoprophylactic applications can be entrusted onto *Escherichia coli.*

Rickettsial infections, if promptly recognized and properly treated, are no longer dread diseases. One exception is Q fever endocarditis, which occurs in a few of the many who suffer clinical or subclinical *Coxiella burnetii* infection. Possibly, earlier treatment can be provided to these patients if a correlation is found between Q fever endocarditis and lymphocyte antigen type. For early recognition of all rickettsial infections, there is every indication that during the next decade hybridoma technology will generate powerful and specific diagnostic tools. Furthermore, monoclonal antibodies will be used for the identification and purification of antigens to be included in a vaccine.

I predict that the next decade will be a very busy one for rickettsiologists.

Acknowledgments

The recent work from my laboratory described in this review was supported by the Naval Medical Research and Development Command, Research Work Unit no. MR00001.001.1271. I am greatly indebted to several colleagues, and in particular to Gregory A. Dasch, who read a draft of this review and made valuable suggestions, to Joseph E. McDade and Chad Helmick, who furnished unpublished data, and to David J. Silverman and Charles L. Wisseman Jr., who provided an illustration from one of their publications for reproduction in this review.

Literature Cited

1. Anacker, R. L., McCaul, T. F., Burgdorfer, W., Gerloff, R. K. 1980. *Infect. Immun.* 27:468–74
2. Applefeld, M. M., Billingsley, L. M., Tucker, H. J., Fiset, P. 1977. *Am. Heart J.* 93:669–70
3. Ariel, B. M., Khavkin, T. N., Amosenkova, N. I. 1973. *Pathol. Microbiol.* 39:412–23
4. Ascher, M. S., Oster, C. N., Harber, P. I., Kenyon, R. H., Pedersen, C. E. Jr. 1978. *J. Infect. Dis.* 138:217–21
5. Babudieri, B. 1959. *Adv. Vet. Sci.* 5:81–82
6. Baca, O. G. 1978. *J. Bacteriol.* 136:429–32
7. Baca, O. G., Akporiaye, E. T., Aragon, A. S., Martinez, I. L., Robles, M. V., Warner, N. L. 1981. *Infect. Immun.* 33:258–66
8. Baca, O. G., Paretsky, D. 1974. *Infect. Immun.* 9:939–45
9. Baca, O. G., Paretsky, D. 1974. *Infect. Immun.* 9:959–61
10. Baker, J. A. 1946. *J. Exp. Med.* 84:37–51
11. Beaman, L., Wisseman, C. L. Jr. 1976. *Infect. Immun.* 14:1071–76

12. Behymer, D. E., Biberstein, E. L., Riemann, H. P., Franti, C. E., Sawyer, M. 1975. *Am. J. Vet. Res.* 36:781–84
13. Behymer, D. E., Biberstein, E. L., Riemann, H. P., Franti, C. E., Sawyer, M., Ruppanner, R., Crenshaw, G. L. 1976. *Am. J. Vet. Res.* 37:631–34
14. Bell, E. J., Kohls, G. M., Stoenner, H. G., Lackman, D. B. 1963. *J. Immunol.* 90:770–81
15. Bell, E. J., Philip, C. B. 1952. *Ann. Rev. Microbiol.* 6:91–118
16. Biberstein, E. L., Riemann, H. P., Franti, C. E., Behymer, D. E., Ruppanner, R., Bushnell, R., Crenshaw, G. 1977. *Am. J. Vet. Res.* 38:189–93
17. Bourgeois, A. L., Dasch, G. A. 1981. In *Rickettsiae and Rickettsial Diseases,* ed. W. Burgdorfer, R. L. Anacker, pp. 71–80. New York: Academic. 650 pp.
18. Bourgeois, A. L., Dasch, G. A., Strong, D. M. 1980. *Infect. Immun.* 27:483–91
18a. Bourgeois, A. L., Olson, J. G., Fang, R. C. Y., Huang, J., Wang, C. L., Chow, L., Bechthold, D., Dennis, D. T., Coolbaugh, J. C., Weiss, E. 1982. *Am. J. Trop. Med. Hyg.* 31:532–40
19. Bozeman, F. M., Humphries, J. W., Campbell, J. M. 1968. *Acta Virol.* 12:87–93
20. Bozeman, F. M., Masiello, S. A., Williams, M. S., Elisberg, B. L. 1975. *Nature* 255:545–47
21. Bozeman, F. M., Sonenshine, D. E., Williams, M. S., Chadwick, D. P., Lauer, D. M., Elisberg, B. L. 1981. *Am. J. Trop. Med. Hyg.* 30:253–63
22. Brenner, D. J., Steigerwalt, A. G., Gorman, G. W., Weaver, G. E., Feeley, J. C., Cordes, L. G., Wilkinson, H. W., Patton, C., Thomason, B. M., Sasseville, K. R. L. 1980. *Curr. Microbiol.* 4:111–16
23. Brezina, R. 1978. In *Rickettsiae and Rickettsial Diseases,* ed. J. Kazar, R. A. Ormsbee, I. N. Tarasevich, pp. 221–35. Bratislava: VEDA. 649 pp.
24. Buchner, P. 1965. *Endosymbiosis of Animals with Plant Microorganisms.* New York: Wiley. 909 pp.
25. Burgdorfer, W., Brinton, L. P. 1970. *Infect. Immun.* 30:112–14
26. Burgdorfer, W., Brinton, L. P. 1975. *Ann. NY Acad. Sci.* 266:61–72
27. Burgdorfer, W., Hayes, S. F., Mavros, A. J. See Ref. 17, pp. 588–94
28. Burgdorfer, W., Ormsbee, R. A., Hoogstraal, H. 1972. *Am. J. Trop. Med. Hyg.* 21:989–98
29. Burgdorfer, W., Ormsbee, R. A., Schmidt, M. L., Hoogstraal, H. 1973. *Bull. WHO* 48:563–69
30. Burgdorfer, W., Sexton, D. J., Gerloff, R. K., Anacker, R. L., Philip, R. N., Thomas, L. A. 1975. *Infect. Immun.* 12:205–10
31. Burton, P. R., Kordova, N., Paretsky, D. 1971. *Can. J. Microbiol.* 17:143–50
32. Burton, P. R., Stueckemann, J., Paretsky, D. 1975. *J. Bacteriol.* 122: 316–24
33. Burton, P. R., Stueckemann, J., Welsh, R. M., Paretsky, D. 1978. *Infect. Immun.* 21:556–66
34. Catanzaro, P. J., Shirai, A., Agniel, L. D. Jr., Osterman, J. V. 1977. *Infect. Immun.* 18:118–23
35. Chang, K.-P., Dwyer, D. M. 1978. *J. Exp. Med.* 147:515–30
36. Christian, R. G., Paretsky, D. 1977. *J. Bacteriol.* 132:841–46
37. Coolbaugh, J. C., Progar, J. J., Weiss, E. 1976. *Infect. Immun.* 14:298–305
38. Costerton, J. W., Irvin, R. T., Cheng, K.-J. 1981. *Ann. Rev. Microbiol.* 35: 299–324
39. Dasch, G. A. 1981. *J. Clin. Microbiol.* 14:333–41
40. Dasch, G. A., Bourgeois, A. L. See Ref. 17. pp. 61–70
41. Dasch, G. A., Samms, J. R., Weiss, E. 1978. *Infect. Immun.* 19:676–85
42. Dasch, G. A., Samms, J. R., Williams, J. C. 1981. *Infect. Immun.* 31:276–88
43. Donahue, J. P., Thompson, H. A. 1980. *J. Gen. Microbiol.* 121:293–302
44. Donahue, J. P., Thompson, H. A. 1981. *J. Bacteriol.* 146:808–12
45. Duma, R. J., Sonenshine, D. E., Bozeman, F. M., Veazy, J. M., Elisberg, B. L., Chadwick, D. P., Stocks, N. I., McGill, T. M., Miller, G. B., MacCormack, J. N. 1981. *J. Am. Med. Assoc.* 245:2318–23
46. DuPont, H. L., Hornick, R. B., Dawkins, A. T., Heiner, G. G., Fabrikant, I. B., Wisseman, C. L. Jr., Woodward, T. E. 1973. *J. Infect. Dis.* 128:340–44
47. Editorial. 1981. *Lab Animal* 10:24–38
48. Eisenberg, G. H. G., Osterman, J. V. 1977. *Infect. Immun.* 15:124–31
49. Eisenberg, G. H. G., Osterman, J. V. 1978. *Infect. Immun.* 22:80–86
50. Eisenberg, G. H. G., Osterman, J. V. 1979. *Infect. Immun.* 26:131–36
51. Eisenman, C. S., Osterman, J. V. 1976. *Infect. Immun.* 14:155–62
52. Ewing, E. P. Jr., Takeuchi, A., Shirai, A., Osterman, J. V. 1978. *Infect. Immun.* 19:1068–75
53. Feng, W. C., Murray, E. S., Burgdorfer, W., Spielman, J. M., Rosenberg, G., Dang, K., Smith, C., Spickert, C.,

Waner, J. L. 1980. *Am. J. Trop. Med. Hyg.* 29:691–94
54. Feng, W. C., Murray, E. S., Rosenberg, G. E., Spielman, J. M., Waner, J. L. 1979. *J. Clin. Microbiol.* 10:322–25
55. Fiset, P., Ormsbee, R. A. 1968. *Zentralbl. Bakteriol. Parasitenk. Infektionskr. Hyg. Abt. 1 Orig.* 206:321–29
56. Friis, R. R. 1972. *J. Bacteriol.* 110: 706–21
57. Gambrill, M. R., Wisseman, C. L. Jr. 1973. *Infect. Immun.* 8:519–27
58. Gambrill, M. R., Wisseman, C. L. Jr. 1973. *Infect. Immun.* 8:631–40
59. Groves, M. G., Osterman, J. V. 1978. *Infect. Immun.* 19:583–88
60. Groves, M. G., Rosenstreich, D. L., Taylor, B. A., Osterman, J. V. 1980. *J. Immunol.* 125:1395–99
61. Hackstadt, T., Williams, J. C. 1981. *Proc. Natl. Acad. Sci. USA* 78:3240–44
62. Hackstadt, T., Williams, J. C. See Ref. 17, pp. 431–40
63. Hackstadt, T., Williams, J. C. 1981. *J. Bacteriol.* 148:419–25
64. Hebert, G. A., Moss, C. W., McDougal, L. K., Bozeman, F. M., McKinney, R. M., Brenner, D. J. 1980. *Ann. Intern. Med.* 92:45–52
65. Hinrichs, D. J., Jerrells, T. R. 1976. *J. Immunol.* 117:996–1003
66. Hollingdale, M. R., Vinson, J. W., Herrmann, J. E. 1980. *J. Infect. Dis.* 141:672–79
67. Huang, K.-Y. 1967. *J. Bacteriol.* 93: 853–59
68. Kenyon, R. H., Acree, W. M., Wright, G. G., Melchior, F. W. Jr. 1972. *J. Infect. Dis.* 125:146–52
69. Kenyon, R. H., Kishimoto, R. A., Hall, W. C. 1979. *Infect. Immun.* 25:580–82
70. Kenyon, R. H., Pedersen, C. E. Jr. 1975. *J. Clin. Microbiol.* 1:500–3
71. Kenyon, R. H., Sammons, L. S. C., Pedersen, C. E. Jr. 1975. *J. Clin. Microbiol.* 2:300–4
72. Khavkin, T., Sukhinin, V., Amosenkova, N. 1981. *Infect. Immun.* 32: 1281–91
73. Kingsbury, D. T. 1969. *J. Bacteriol.* 98:1400–1
74. Kishimoto, R. A., Veltri, B. J., Shirey, F. G., Canonico, P. G., Walker, J. S. 1977. *Infect. Immun.* 15:601–7
75. Kishimoto, R. A., Walker, J. S. 1976. *Infect. Immun.* 14:416–21
76. Kopmans-Gargantiel, A. I., Wisseman, C. L. Jr. 1981. *Infect. Immun.* 31: 1277–80
77. Kordova, N., Burton, P. R., Downs, C. M., Paretsky, D., Kovacova, E. 1970. *Can. J. Microbiol.* 16:125–33
78. Kreier, J. P., Dominguez, N., Krampitz, H. E., Gothe, R., Ristic, M. 1981. In *The Prokaryotes,* ed. M. P. Starr, H. Stolp, H. G. Truper, A. Balows. H. G. Schlegel, pp. 2189–209. Berlin: Springer. 2284 pp.
79. Kreier, J. P., Ristic, M. 1981. *Ann. Rev. Microbiol.* 35:325–38
80. Lackman, D. B., Bell, E. J., Stoenner, H. G., Pickens, E. G. 1965. *Health Lab. Sci.* 2:135–41
81. Lane, R., Philip, R. N., Casper, E. A. See Ref. 17, pp. 575–84
82. Linneman, C. C. Jr., Schaeffer, A. E., Burgdorfer, W., Hutchinson, L., Philip, R. N. 1980. *Am. J. Epidemiol.* 111: 31–36
83. Lissman, B. A., Benach, J. L. 1980. *J. Am. Vet. Med. Assoc.* 176:994–95
84. Magnarelli, L. A., Anderson, J. F., Philip, R. N., Burgdorfer, W., Casper, E. A. 1981. *Am. J. Trop. Med.* 30:715–21
85. Mason, R. A. 1970. *J. Bacteriol.* 103: 184–90
86. McCaul, T. F., Hackstadt, T., Williams, J. C. See Ref. 17, pp. 267–80
87. McCaul, T. F., Williams, J. C. 1981. *J. Bacteriol.* 147:1063–76
88. McDade, J. E., Brenner, D. J., Bozeman, F. M. 1979. *Ann. Intern. Med.* 90:659–61
89. McDade, J. E., Shepard, C. C., Redus, M. A., Newhouse, V. F., Smith, J. D. 1980. *Am. J. Trop. Med. Hyg.* 29: 277–84
90. Meltzer, M. S., Nacy, C. A. 1980. *Cell. Immunol.* 54:487–90
91. Merrell, B. R., Weiss, E., Dasch, G. A. 1978. *J. Bacteriol.* 135:633–40
91a. Morbidity and Mortality Weekly Reports. 1981. *Annual Summary 1980,* Vol. 29, No. 54. Washington DC: US Dept. Health & Human Services
92. Murphy, J. R., Wisseman, C. L. Jr., Fiset, P. 1978. *Infect. Immun.* 21:417–24
93. Murphy, J. R., Wisseman, C. L. Jr., Fiset, P. 1980. *Infect. Immun.* 27:730–38
94. Myers, W. F., Osterman, J. V., Wisseman, C. L. Jr. 1972. *J. Bacteriol.* 109:89–95
95. Myers, W. F., Wisseman, C. L. Jr. 1980. *Int. J. Syst. Bacteriol.* 30:143–50
96. Myers, W. F., Wisseman, C. L. Jr. See Ref. 17, pp. 313–25
97. Myers, W. F., Wisseman, C. L. Jr., Fiset, P., Oaks, E. V., Smith, J. F. 1979. *Infect. Immun.* 26:976–83
98. Nacy, C. A., Groves, M. G. 1981. *Infect. Immun.* 31:1239–50
99. Nacy, C. A., Meltzer, M. S. 1979. *J. Immunol.* 123:2544–49

100. Nacy, C. A., Osterman, J. V. 1979. *Infect. Immun.* 26:744–50
101. Nogueira, N., Cohn, Z. 1976. *J. Exp. Med.* 143:1402–20
102. Ormsbee, R. A. 1969. *Ann. Rev. Microbiol.* 23:275–92
103. Ormsbee, R. A. 1979. *Rev. Infect. Dis.* 1:559–62
104. Ormsbee, R. A., Bell, E. J., Lackman, D. B., Tallent, G. 1964. *J. Immunol.* 92:404–12
105. Ormsbee, R., Burgdorfer, W., Peacock, M., Hildebrandt, P. 1971. *Am. J. Trop. Med. Hyg.* 20:117–24
106. Ormsbee, R. A., Hoogstraal, H., Yousser, L. B., Hildebrandt, P., Atalla, W. 1968. *J. Hyg. Epidemiol. Microbiol. Immunol.* 12:226–30
107. Ormsbee, R. A., Peacock, M. G. 1964. *J. Bacteriol.* 88:1205–10
108. Osterman, J. V., Eisemann, C. S. 1978. *Infect. Immun.* 21:866–73
109. Paretsky, D. 1968. *Zentralbl. Bakteriol. Parasitenk. Infektionskr. Hyg. Abt. 1 Orig.* 206:284–91
110. Peacock, M. G., Fiset, P., Ormsbee, R. A., Wisseman, C. L. Jr. See Ref. 23, pp. 593–601
111. Phibbs, P. V. Jr., Winkler, H. H. See Ref. 17, pp. 421–30
111a. Phibbs, P. V. Jr., Winkler, H. H. 1982. *J. Bacteriol,* 149:718–25
112. Philip, C. B., Hoogstraal, H., Reiss-Gutfreund, R., Clifford, C. M. 1966. *Bull. WHO* 35:127–31
113. Philip, R. N., Casper, E. A. 1981. *Am. J. Trop. Med. Hyg.* 30:230–38
114. Reiss-Gutfreund, R. J. 1956. *Bull. Soc. Pathol. Exot.* 49:946–1023
115. Reiss-Gutfreund, R. J. 1966. *J. Trop. Med. Hyg.* 15:943–49
116. Reiss-Gutfreund, R. J. 1968. *J. Hyg. Epidemiol. Microbiol. Immunol.* 12: 133–39
117. Rikihisa, Y., Ito, S. 1980. *Infect. Immun.* 30:231–43
118. Riley, H. D. Jr. 1981. *Curr. Probl. Pediatr.* 11:1–46
119. Ristic, M., Huxsoll, D. L. 1982. In *Bergey's Manual of Systematic Bacteriology,* ed. N. R. Krieg, pp. Baltimore: Williams & Wilkins. 1st ed. In press
120. Russo, P. K., Mendelson, D. C., Etkind, P. H., Garber, M., Berardi, V. P., Gilfillan, R. F. 1981. *N. Engl. J. Med.* 304:1166–68
121. Rutherford, W. J. 1916. *Brit. Med. J.* 2:386–87
122. Scaffidi, V. 1981. *Minerva Med.* 72: 2063–70
123. Schramek, S. 1974. *Acta Virol.* 18: 173–74
124. Schramek, S., Brezina, R., Kazar, J. 1977. *Acta Virol.* 21:439–41
125. Sexton, D. J., Burgdorfer, W., Thomas, L., Norment, B. R. 1976. *Am. J. Epidemiol.* 103:192–97
126. Silverman, D. J., Wisseman, C. L. Jr., 1978. *Infect. Immun.* 21:1020–23
127. Silverman, D. J., Wisseman, C. L. Jr., Waddell, A. D., Jones, M. 1978. *Infect. Immun.* 22:233–46
128. Smith, J. B., Stoker, M. G. P. 1951. *Brit. J. Exp. Pathol.* 32:433–41
129. Sonenshine, D. E., Bozeman, F. M., Williams, M. S., Masiello, S. A., Chadwick, D. P., Stocks, N. I., Lauer, D. M., Elisberg, B. L. 1978. *Am. J. Trop. Med. Hyg.* 27:339–49
130. Topping, N. H., Shepard, C. C. 1947. *Ann. Rev. Microbiol.* 1:333–50
131. Traub, R., Wisseman, C. L. Jr., Farhang-Azad, A. 1978. *Trop. Dis Bull.* 75:237–317
132. Tyeryar, F. J. Jr., Weiss, E., Millar, D. B., Bozeman, F. M., Ormsbee, R. A. 1973. *Science* 180:415–17
133. Uff, J. S., Evans, D. J. 1977. *Histopathology* 1:463–72
134. Vinson, J. W. 1966. *Bull. WHO* 35: 155–64
135. Vinson, J. W., Fuller, H. S. 1961. *Pathol. Microbiol.* 24(Suppl):152–66
136. Walker, D. H., Cain, B. G. 1980. *Lab. Invest.* 43:388–96
137. Walker, D. H., Henderson, F. W. 1978. *Infect. Immun.* 20:221–27
138. Walker, D. H., Kirkman, H. N. 1980. *J. Infect. Dis.* 142:771
139. Walker, T. S., Winkler, H. H. 1978. *Infect. Immun.* 22:200–8
140. Walker, T. S., Winkler, H. H. 1981. *Infect. Immun.* 31:289–96
141. Weiss, E. 1960. *Ann. NY Acad. Sci.* 88:1287–97
142. Weiss, E. 1971. In *Trachoma and Related Disorders Caused by Chlamydial Agents,* ed. R. L. Nichols, pp. 3–12. Amsterdam: Excerpta Medica.
143. Weiss, E. 1973. *Bacteriol. Rev.* 37: 259–83
144. Weiss, E. 1979. *Microbiol.* 1979:144–49
145. Weiss, E. See Ref. 17, pp. 384–400
146. Weiss, E. See Ref. 78, pp. 2137–60
147. Weiss, E., Coolbaugh, J. C., Williams, J. C. 1975. *Appl. Microbiol.* 30:456–63
148. Weiss, E., Dasch, G. A. See Ref. 78, pp. 2161–71
149. Weiss, E., Dasch, G. A. 1982. *Int. J. Syst. Bacteriol.* 32: In press
150. Weiss, E., Dasch, G. A., Woodman, D. R., Williams, J. C. 1978. *Infect. Immun.* 19:1013–20

151. Weiss, E., Mamay, H. K., Dasch, G. A. 1982. *J. Bacteriol.* 150:245–50
152. Weiss, E., Moulder, J. W. 1974. In *Bergey's Manual of Determinative Bacteriology,* ed. R. E. Buchanan, N. E. Gibbons, pp. 890–91. Baltimore: Williams & Wilkins. 8th ed.
153. Weyer, F., Reiss-Gutfreund, R. J. 1973. *Acta Trop.* 30:177–92
154. Wiebe, M. E., Burton, P. R., Shankel, D. M. 1972. *J. Bacteriol.* 110:368–77
155. Williams, J. C. 1980. *Infect. Immun.* 28:74–81
156. Williams, J. C., Peacock, M., McCaul, T. F. 1982. *Infect. Immun.* 32:840–51
157. Williams, J. C., Peterson, J. C. 1976. *Infect. Immun.* 14:439–48
158. Williams, J. C., Weiss, E. 1978. *J. Bacteriol.* 134:884–92
159. Winkler, H. H. 1976. *J. Biol. Chem.* 251:389–96
160. Winkler, H. H. 1977. *Infect. Immun.* 17:607–12
161. Winkler, H. H., Miller, E. T. 1980. *Infect. Immun.* 29:316–21
162. Winkler, H. H., Miller, E. T. See Ref. 17, pp. 327–33
163. Wisseman, C. L. Jr. 1968. *Zentralbl. Bakteriol. Parasitenk. Infektionskr. Hgy. Abt. 1 Orig.* 206:299–313
164. Wisseman, C. L. Jr., Edlinger, E. A., Waddell, A. D., Jones, M. R. 1976. *Infect. Immun* . 14:1052–64
165. Wisseman, C. L. Jr., Waddell, A. D. 1975. *Infect. Immun.* 11:1391–401
166. Wisseman, C. L. Jr., Waddell, A. D., Silverman, D. J. 1976. *Infect. Immun.* 13:1749–60
167. Wisseman, C. L. Jr., Waddell, A. D. , Walsh, W. T. 1974. *Infect. Immun.* 9:571–75
168. Wong, B., Singer, C., Armstrong, D., Millian, S. J. 1979. *J. Am. Med. Assoc.* 242:1998–99
169. Woodman, D. R., Weiss, E., Dasch, G. A., Bozeman, F. M. 1977. *Infect. Immun.* 16:853–60
170. Wyatt, G. R., Cohen, S. S. 1952. *Nature* 170:846–47
171. Zinsser, H. 1935. *Rats, Lice, and History.* Boston: Little, Brown. 301 pp.

Ann. Rev. Microbiol. 1982. 36:371–413

MEDIATORS OF ANAPHYLAXIS AND INFLAMMATION

Michael K. Bach

The Upjohn Company, Kalamazoo, Michigan 49001

CONTENTS

INTRODUCTION

The responses of a sensitized animal to the sensitizing antigen have been categorized into four major classes on the basis of the time that must elapse between the beginning of exposure and the manifestation of symptoms, and on the basis of the nature of the reaction (cellular or humoral) and the target cells or tissues involved. Of these four classes, the type-1 reactions are the most explosive in onset; they reach full expression in a matter of seconds or minutes from the time of exposure. Mechanistically, the initiating event in these anaphylactic reactions is the interaction of antibodies, usually of the immunoglobulin E (IgE) class, with the sensitizing antigen, whereas the antibody molecules are themselves bound to specific receptors on mast cells or basophils. A consequence of the antigen-antibody reaction is the activation of the cells, which results in the secretion of a series of preformed

0066-4227/82/1001-0371$02.00

mediators of anaphylaxis previously stored in the metachromatic granules of these cells. Secondarily, there ensues the production of another group of substances both by the mast cells and basophils and by other cells recruited into the reaction; these secondary mediators often exacerbate the severity of the overall reaction.

Release of Mediators of Anaphylaxis

Anaphylactic reactions have been described in many mammals (59, 87, 88, 129, 153, 160, 294) as well as in the fowl (60). The biochemical mechanism that couples the activation of mast cells for mediator release and the ultimate mediator release itself can only be considered very briefly (see 15, 149 for reviews of older data). The release reaction is initiated by the bridging of two IgE-specific receptors on mast cells or basophils (185). This can occur by the binding of a divalent antigen to IgE molecules that are themselves bound to two separate receptors, by the bridging of two IgE molecules with the divalent $F(ab')_2$ fragments of specific antibody to IgE, or more directly by the bridging of two receptors with the divalent $F(ab')_2$ fragment of specific antibody to the IgE receptor itself (135). Aside from possible changes in membrane polarization or partial aggregation of receptors [clustering (135)], the first recognizable biochemical change is the activation of a nascent serine esterase (133). This is followed by two other events that appear to occur in parallel: a transient activation of adenylate cyclase (133, 296); and a transient increase in methylation of membrane phospholipids (124, 133, 134), which leads to increased membrane fluidity (123). The newly formed phosphatidyl choline, the product of the methylation reaction, may serve as the phospholipid substrate from which arachidonic acid is released (125); the arachidonic acid, in turn, is the precursor of the leukotrienes (see below). In addition, its metabolism via a lipoxygenase pathway is necessary for mediator release to occur (177, 224). Phosphatidyl inositol (65) and diacylglycerol (151) have also been proposed as potential sources of arachidonate; different precursors and pathways of arachidonate release may be dominant in different cells or under different conditions of activation. The activation of adenylate cyclase and the resulting increase in the concentration of AMP results in changes in AMP-dependent protein kinase activity and in changes in the incorporation of ^{32}P into specific proteins (127, 274). Subsequent steps in the release reaction involve an increase in uptake of extracellular calcium ion (see 133, for example), changes in microfilaments and microtubules, and the fusion of perigranular membranes to the plasma membrane with the resulting exocytosis of the granule contents. Details of these latter reactions are now being elucidated with elegant electron microscope techniques (58, 84).

Substances Considered in this Review

"Mediators of anaphylaxis" used to be defined as the group of substances produced or released as a function of an anaphylactic challenge. It was implicit in this definition that these substances all played a role in the elicitation of the symptoms of anaphylaxis. Furthermore, since only mast cells and basophils were known to possess receptors for binding IgE antibody, it was assumed that all the mediators were derived from mast cells. This definition is no longer tenable. Not only is it now clear that mast cell granules contain a variety of substances released during an anaphylactic response that are not directly involved in the elicitation of the symptoms of anaphylaxis, but cells other than mast cells and basophils are capable of producing many of the secondary mediators of anaphylaxis; they may produce these substances upon recruitment by the mast cells or basophils and also under both anaphylactic (143) and non-anaphylactic conditions without the involvement of mast cells or basophils. Indeed, the demarcation between mediators of anaphylaxis and mediators of inflammation has become so diffuse that the choice of including many specific substances in this review is rather arbitrary. This review considers (Table 1) the preformed mediators of anaphylaxis, histamine, and serotonin, the well-established secondary mediators such as the leukotrienes (slow reacting substance of anaphylaxis), and various chemotactic factors (both the proteins and peptides from mast cells and those that are lipid in nature) and various hydrolytic enzymes found in mast cell granules. Space does not permit consideration of the extensive literature dealing with inhibitors and modulators of the production or actions of these substances.

MAST CELL-DERIVED PREFORMED MEDIATORS

The preformed mediators of anaphylaxis are the biologically active substances found in association with the granules of mast cells and basophils. The composition of this mixture differs both qualitatively and quantitatively as a function of the cell type being considered, and even more so as a function of the species of animal from which the cells are derived. Examples are the major neutral protease of mast cells [chymase in the rat, tryptase in man (247)], and the major proteoglycan, [heparin in mast cells; chondroitin sulfate in basophils (15, 84)]. The binding affinity of the mediators to the granule matrix differs from one mediator to the next. The biogenic amines, as well as eosinophil chemotactic factor of anaphylaxis (ECF-A), are readily dissociated from the membrane matrix. On the other hand, several hydrolytic enzymes and heparin require concentrated salt solutions before they can be solubilized (84, 299).

Table 1 Mediators of anaphylaxis and inflammation[a]

Mediator	Structural characteristics	Abbreviation	Assay(s)	Other functions	Inactivation	Ref.
Primary, Preformed, Granule-Associated						
Histamine	β-Imidazolylethylamine, $M_r = 111$	H	Contraction of guinea pig ileum; fluorometric; radio-enzymatic	Pro-inflammatory (H_1): Increased vascular permeability, chemokinesis, increased mucus production	Histaminase; histamine-methyltransferase	15
				Anti-inflammatory (H_2): Elevation of cyclic AMP; HSF production; feed-back regulation of delayed type hypersensitivity		8, 174
Seretonin	5-Hydroxytryptamine, $M_r = 176$	5-HT	Biologic, fluorometric and enzymatic	Increased vascular permeability	5-HT deaminase	15
Eosinophil chemotactic factors of anaphylaxis	Val-gly-ser-glu and ala-gly-ser-glu	ECF-A	Chemotactic attraction of eosinophils and neutrophils	Eosinophil and neutrophil deactivation; induced lysosomal enzyme release; increased expression of C_3b and C_4b receptors on eosinophils	—	146, 290
Intermediate mol. wt. ECF	Three peptides, $M_r = 2500$		Chemotactic attraction of eosinophils and neutrophils	Target cell deactivation	—	233
Neutrophil chemotactic factor	$M_r \geqq 750{,}000$	HMW-NCF	Chemotactic attraction of neutrophils	Neutrophil deactivation	—	12, 289, 291
Heparin (mast cell); Chrondroitin sulfate (basophil)	Macromolecular, acidic proteoglycans, $M_r = 750{,}000$		Antithrombin activation	Anticoagulant, anticomplement activity	—	15
Chymase (Rat mast cell)	$M_r = 25{,}000$		Hydrolysis of benzoyl tyrosine-ethyl ester	Proteolysis	—	298

Tryptase (Human mast cell)	M_r = 130,000		Hydrolysis of tosyl-arginine-methyl ester	Proteolysis	—	247
Kallikrein	M_r = 1,200,000	BK-A	Hydrolysis of Benz-Pro-Phe-Arg-p-Nitro-anilide	Bradykinin synthesis		194
Prekallikrein activator			Prekallikrein activation assayed as kallikrein (above)	Kinin, complement, and clotting system		196
Hageman factor activator			Hageman factor activation	Kinin, complement, and clotting system		196
N-Acetyl-β-D-glucoseminidase	M_r = 150,000		Hydrolysis of *p*-nitro phenyl-β-D-glucosaminide	—	—	15, 246
β-glucuronidase	M_r = 280,000		Hydrolysis of phenolphthalein-β-glucuronide			15, 246,
Arylsulfatase A	M_r = 115,000		Hydrolysis of *p*-nitro catechol sulfate	Degradation of collagen?	—	174
Vasoactive intestinal polypeptide	M_r = 1400	VIP	Radioimmunoassay	Relaxes smooth muscle		68
Probably Not Preformed Peptides, Cell Source Not Determined						
Histamine suppressor factor	M_r = 23,000 to 40,000	HSF	Inhibition of production of migration inhibitory factor (MIF)	Suppresses cellular immune responses		234, 235
Inflammation factor of anaphylaxis	M_r = 1400 based on aminoacid composition	IF-A		(From mast cell granules) neutrophil and monocyte infiltration		198, 199, 272
Prostaglandin-generating factors of anaphylaxis	M_r = 1450 based on aminoacid composition	PGF-A	Prostaglandin synthesis in lung segments	Recruitment of secondary mediators?		266
Secondary Mediators						
Pristaglandin D_2	OH, CO_2H, OH, OH	PGD_2	Gas chromatography-mass spectrometry (GC-MS)	Bronchoconstrictor potentiate leukotriene effects	Prostaglandin-15-dehydrogenase	231, 285, 287

Table 1 *(Continued)*

Mediator	Structural characteristics	Abbreviation	Assay(s)	Other functions	Inactivation	Ref.
Secondary Mediators (continued)						
Prostacyclin	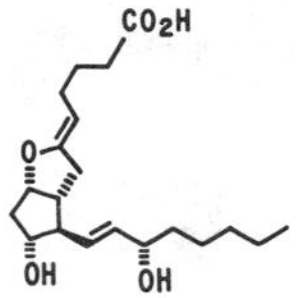	PGI_2	Radioimmunoassay of degradation product	Inhibit platelet aggregation	Spontaneously to 6-keto-$PGF_{1\alpha}$	91, 201
Thromboxane A_2	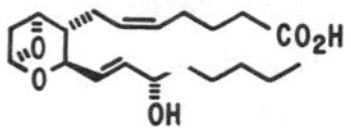	TxA_2	Radioimmunoassay of TxB_2	Bronchoconstriction, platelet aggregation	Spontaneously to TxB_2	91, 243
Prostaglandin E_2	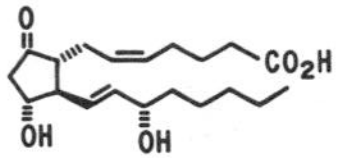	PGE_2	Radioimmunoassay	Inhibit histamine release; potentiate leukotriene effects	Prostaglandin-15-dehydrogenase	229, 243
Prostaglandin $F_{2\alpha}$	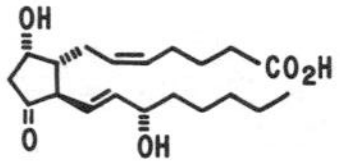	$PGF_{2\alpha}$	Radioimmunoassay		Prostaglandin-15-dehydrogenase	229, 243
5-Hydroperoxy-eicosatetraenoic acid	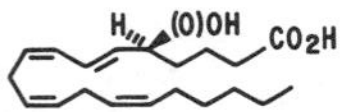	5-HPETE	HPLC or GC-MS	Potentiate histamine release chemotaxis and chemokinesis	Further lipoxygenolytic oxidation?	224
5-Hydroxyeicosatetraenoic acid		5-HETE	HPLC or GC-MS	As for 5-HPETE	As for 5-HPETE	224
12-Hydroperoxy-eicosatetraenoic acid	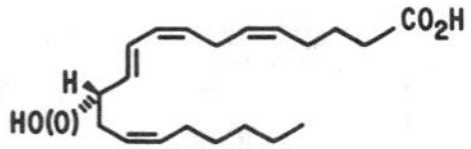	12-HPETE	HPLC or GC-MS	Inhibit platelet cyclooxygenase (and aggregation)		254

15-Hydroperoxy-eicosatetraenoic acid	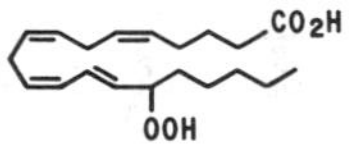 CO_2H; OOH	15-HPETE	HPLC or GC-MS	Inhibit PGI_2 synthesis		186
15-Hydroxyeicosa-tetraenoic acid	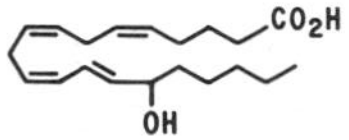 CO_2H; OH	15-HETE	HPLC or GC-MS	Inhibit 5-lipoxygenase, inhibit T cell mitogenesis		282 26, 96, 222
Leukotriene B_4	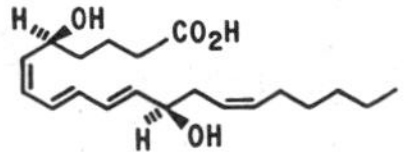 H, OH; CO_2H; H, OH	LTB_4	HPLC	Potent chemoattractant smooth muscle action on guinea pig parenchyma		44, 258
Leukotriene C_4	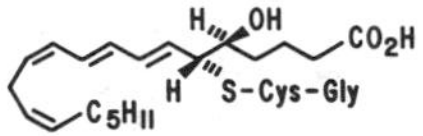 H, OH; CO_2H; H; S-Cys-Gly; C_5H_{11}	LTC_4	Bioassay, radioimmunoassay	Potent brochoconstrictor or vasoconstrictor? increase vascular permeability	Lipoxygenolysis via 15-lipxoygenase	213, 239, 257
Leukotriene D_4	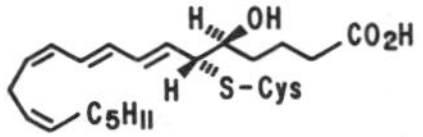 H, OH; CO_2H; H; S-Cys; C_5H_{11}	LTD_4	Bioassay, radioimmunoassay	Most potent constrictor of peripheral airways, vasodilator, increase vascular permeability	Lipoxygenolysis	213, 239, 257
Leukotriene E_4	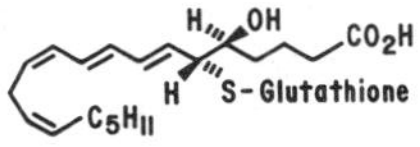 H, OH; CO_2H; H; S-Glutathione; C_5H_{11}	LTE_4	Bioassay, radioimmunoassay	Bronchoconstrictor, increase vascular permeability	Lipoxygenolysis	239, 257
Platelet-activating factor	$H_2COC_{16}H_{33}$ CH_3COOCH $CH_2OPO_3CH_2CH_2N^{\oplus}(CH_3)_3$	PAF; AGEPC PAF-acet-ether	Factor IV release from rabbit platelets	Platelet aggregation, neutrophil aggregation, TxA_2 generation, bronchoconstriction, acute apnea	Formation of 2-lyso AGEPC?	226

[a] Adapted and updated from (15).

Biogenic Amines

ASSAYS FOR BIOGENIC AMINES AND FOR THE RELEASE REACTION FROM MAST CELLS Histamine was classically bioassayed using the guinea pig ileum. This assay has been largely supplanted by modifications of the fluorometric assay for histamine using o-phthalaldehyde (105, 253). Automated, continuous flow and highly sensitive micromethods employing this procedure have been described (255) and adapted for the measurement of histamine in secretions (192). A fluorometric assay for the simultaneous estimation of histamine and serotonin has also been described (220). Even greater sensitivity can be achieved by using the more laborious radioenzymatic procedure (163).

The mediator release reaction can be followed by measuring histamine release. However, measurement of the release of preformed chemotactic factors (291) (see below) and hydrolytic enzymes (195, 197, 246, 247) offers greater selectivity or other advantages especially when clinical specimens are used. Finally, release of tritiated serotonin from prelabeled rat mast cells appears to parallel histamine release (180; M.K. Bach, unpublished results). Recently, however, it has become clear that this generalization is not necessarily correct; Askenase et al (11, 273) have reported that Amitriptyline and a variety of antidepressants could inhibit both IgE-dependent and compound 48/80-induced histamine release without affecting the release of tritiated serotonin from the cells. They have explained this dissociation by identifying two serotonin binding proteins in the cytoplasm of mast cells (10). Apparently the dissociation of bound serotonin from these proteins can occur under conditions where the exocytosis of mast cell granules is inhibited.

PHYSIOLOGIC, PATHOLOGIC, AND PHARMACOLOGIC EFFECTS OF BIOGENIC AMINES The concept that agonists and hormones act by way of binding to specific receptors on target cells has become so commonly accepted that it may be in danger of being taken for given without proof. In the case of histamine, there are at least two distinctly different receptors, H_1 and H_2. The H_1 receptors are considered to be responsible for the pro-inflammatory activities of histamine. They are selectively antagonized by typical antihistamines such as Diphenhydramine, Pyrilamine, etc. 2-Methyl-histamine is a relatively selective agonist for this receptor. The H_2 receptors tend to be associated with the anti-inflammatory actions of histamine (15). They are activated selectively by 4-methyl-histamine and especially by the analogue Dimaprit and selectively are antagonized by Burimamide, Metiamide, and Cimetidine.

Pro-inflammatory effects of the biogenic amines A variety of leukocytic cells possess receptors for histamine. Thus, mixed human blood leukocytes formed rosettes with ox erythrocytes to which histamine had been coupled via a serum albumin linker. The reaction was inhibited by relatively high concentrations of free histamine and by H_1 antihistamines, but not by H_2 antihistamines, histidine, or other histamine metabolites (260). Guinea pig alveolar macrophages also formed rosettes with histamine bound to erythrocytes in an H_1-specific reaction (79). Occupation of the histamine receptor on these cells by histamine-activated zymosan induced the formation of superoxide anion and chemiluminescence in these cells and was as effective in this regard as serum-activated zymosan (78).

The receptors for histamine on eosinophils present a more complex situation: Both histamine and other chemotactic materials (see below) induced an increase in the expression of receptors for the C3b component of complement (4, 146) on human eosinophils. The effect was specific for eosinophils as compared to neutrophils and was antagonized by H_1 anti-histamines. Although there was a concomitant increase in receptors for the C4b component of complement on the cells, there was no change in the abundance of receptors for the F_c portion of γ-globulin (146). A consequence of this increased expression of complement receptors was the finding that histamine augmented the complement-dependent, eosinophil-mediated killing of Schistosomules without affecting antibody-dependent, eosinophil-mediated or neutrophil-mediated killing of the same target cells (5). By contrast to these results, the chemotactic activity of histamine for eosinophils (64) did not involve a typical H_1 receptor in that imidazole acetic acid, a metabolite of histamine, was also quite active (142, 278). On the other hand, neither 2- nor 4-methyl-histamine were active. In addition to causing deactivation relative to a subsequent stimulation by itself, histamine also caused deactivation toward the chemotactic peptides of ECF-A (277). The chemotactic activity of histamine could not be demonstrated in marmoset skin nor in abraded human skin under conditions where ECF-A was active (278, 279).

When fragments of human lung were cultured for prolonged periods in vitro, and the incorporation of radioactive precursors into mucus was measured, it was found that relatively high concentrations of histamine (0.1 mM) caused a profound increase in mucus production, as did the H_2-specific agonists Dimaprit and Impromidine. The effect was blocked by Cimetidine. On the other hand, H_1 antihistamines themselves caused increased mucus secretion (251). Thus, this pro-inflammatory reaction appears to be H_2 dependent. Relatively high concentrations of histamine have been shown to potentiate the uptake of preformed, radiolabeled histamine in guinea pig basophils (271) and this reaction is partly blocked by H_1 antihistamines.

The effects of the intravenous infusion of histamine into human volunteers have been recently reported (145). Two symptoms, increased pulse rate and changes in lung mechanics, could be prevented by pretreatment with H_1 antihistamines. On the other hand, flushing, headache, and changes in the pulse pressure could only be prevented by combined premedication with an H_1 and an H_2 antihistamine.

Anti-inflammatory effects of biogenic amines Suppressor T cells possess receptors for histamine (250). Thus, the inclusion of histamine, or H_2-selective histamine analogues in the induction phase but not the assay phase of in vitro cytotoxic T cell reactions, markedly inhibited the development of the cytotoxic response (244). Similarly, the addition of histamine to mouse spleen cells in culture inhibited the development of T suppressor cells in the presence of Concanavalin A. Dimaprit, an H_2-selective agonist, was active in this system as well (245). Concanavalin A-activated lymphocytes have also been shown to produce an inhibitor of eosinophil chemotaxis under the influence of 10 nM histamine (159), though the nature of the inhibition was not characterized. A histamine-induced suppressor factor, designated HSF, has been described (233–235) in both guinea pig and human delayed-type hypersensitivity reactions. The factor inhibited the formation of migration inhibitory factor in in vitro assays and blocked delayed type skin reactions in vivo. Production of the factor was blocked by H_2 antihistamines.

The mouse presents an interesting model in which the selective activities of the two biogenic amines present in the mast cells of that species can be studied. Although both histamine and serotonin are released by mouse mast cells, murine endothelial cells are virtually unresponsive to histamine whereas they are quite sensitive to serotonin. On the other hand, the receptors on mouse T cells are specific for histamine and do not recognize serotonin. Nevertheless, the expression of a delayed-type hypersensitivity reaction in the mouse depends on the presence of mast cells at the site of the reaction and on the release of serotonin (7, 8, 95). These considerations have led Askenase et al to propose the existence of a biogenic amine cycle (8) in which, upon antigenic stimulation, T cells cause the release of biogenic amines from mast cells in their milieu (9). The secreted histamine then feeds back on histamine receptor-posessing T cells, which causes suppression of the response. At the same time, the secreted serotonin acts on the endothelial cells, which causes diapedesis, thus promoting the arrival of circulating cells to the reaction site.

Chemotactic Proteins and Peptides

Mast cell granules contain a number of chemotactic principles that differ considerably in molecular weight (264–266, 279). All the principles described thus far are relatively selective for cells from the granulocytic series as contrasted to mononuclear cells. The level of selectivity among granulocytes of different classes differs from one principle to the next. Chemotactic substances have a number of properties in common, the most important in terms of their function is their ability to inactivate cells from responding to them (284). After a cell has been exposed to a given chemotactic stimulus once, it does not respond to the same stimulus a second time. This deactivation may extend to other stimuli as well. The dose of chemotactic factor required for deactivation is generally smaller than that required for eliciting a chemotactic response. Although this has practical implications in terms of permitting economy in the assay of scarce chemotactic principles, the explanation is that the high local concentration in a chemotactic assay is needed to establish the concentration gradient on which the chemotactic response depends. The concentration of the chemoattractant actually encountered by the responding cells is probably similar to that which causes deactivation.

ASSAYS FOR CHEMOTAXIS The most commonly used methods for the measurement of chemotactic activity are based on the chambers described by Boyden (42), in which the presumed chemoattractant is separated from the responding cell population by a semipermeable membrane and the rate of migration of the responding cells into or through this membrane is determined after appropriate fixation and staining. Variations on this procedure that employ membranes with more uniform holes for the cells to crawl through, different chamber configurations that may be more economical in the amounts of chemoattractant required, or different methods for the enumeration of the cells that have migrated have been described. A radiochemical modification (98) that measures the number of ^{51}Cr-labeled responder cells that penetrate the barrier membrane and actually fall onto a second semipermeable membrane with much smaller pores requires a source of relatively pure populations of specific responder cells. Other methods for measuring chemotaxis have involved measurement of cell migration from a well cut in a thin agar block or from an agar droplet. The different methods appear to measure somewhat different properties of the cells.

In measurements of chemotaxis, it is necessary to discriminate between chemotactic activity of a substance and chemokinetic activity (152). The latter property, which merely reflects the tendency of a substance to increase the random migration of cells, is best measured by using a checkerboard design for chemotaxis measurements. The substance to be evaluated is incorporated in multiple doses on either or both sides of the membrane. In this manner it is possible to distinguish pure chemotactic activity, pure (positive or negative) chemokinetic activity, and mixed activity in the same preparation (152).

ECF-A ECF-A was first found in the effluent solutions obtained from antigen-challenged, actively sensitized guinea pig lung (148). Subsequently, it was shown (168) that the activity could be detected after challenge of human basophilic leukemia cells with the calcium ionophore A 23187 and that it was present in preformed state in extracts of rat mast cells and guinea pig and human lung (290). ECF-A consists of two closely related tetrapeptides, val-gly-ser-glu and ala-gly-ser-glu; these structures were confirmed by synthesis (99).

Actions of ECF-A In vivo correlates in primates of the chemoattractant properties of ECF-A for guinea pig and human eosinophils in vitro have been reported (278) both in marmosets and in humans. However, there is at least one report (73) of complete failure to elicit chemotactic activity with both synthetic tetrapeptides on human or guinea pig eosinophils or on human neutrophils, in the face of activity with lipid chemotactic factors on the same cells. ECF-A can induce a marked increase in the expression of receptors for the C3b and C4b components of complement (146), as can histamine (see above). The effect is selective for eosinophils and does not affect expression of similar receptors on neutrophils. A corollary to this activity is the enhancement of complement-dependent, eosinophil-mediated killing of Schistosomules (5). Capron et al (56) noted that eosinophil-dependent, antibody-mediated cytotoxicity of unopsonized Schistosomules was dependent on the presence of mast cells in the cell mixture. They subsequently reported (54) that ECF-A restored the ability of mast cell-depleted cell suspensions to effect this killing. These workers also confirmed certain of the conclusions from the structure-activity studies with ECF-A analogues (55) (see below).

Addition of ECF-A to zymosan caused a dose-dependent increase in superoxide anion production in human eosinophils. By contrast to results with the synthetic peptide, formyl-met-leu-phe, which caused similar increases in neutrophils along with induction of chemiluminescence, the stimulation of eosinophils with ECF-A resulted in no chemiluminescence. ECF-A appeared to inhibit the stimulatory activity of formyl-met-leu-phe on neutrophils (35).

Structure-activity relationships Evaluation of the ECF-A peptides with human eosinophils has shown that response occurs over a narrow dose range, which differs from one individual to the next (277). A detailed analysis of the structural requirements for chemotactic activity has been performed (40, 261, 277). The N-terminal tripeptides, val-gly-ser and ala-gly-ser, are competitive inhibitors of the binding of ECF-A and have little if any chemotactic activity of their own. On the other hand, the C-terminal tripeptide, gly-ser-glu, caused a time- and concentration-dependent inhibition of the chemotactic response to subsequently applied ECF-A tetrapeptides (40). Modifications of the hydrophobic portion of the N-terminal amino acid suggested that there was need for the proper spatial fitting; replacement with leu caused enhanced activity whereas replacement with phe resulted in apparently lower affinity. The N-terminal amino group itself was not required for activity (261).

The central two amino acids appear to serve largely as spacers (40, 277). Thus, omission of gly resulted in a 50% reduction in potency whereas inversion of the gly and ser residues, or replacement of ser by asp, resulted in no change in activity. Building a permanent kink into the spacer amino acids by replacing gly with pro resulted in increased affinity. Blocking of the carboxylic acid function at the C-terminal position, replacing gly by D-glu or L-asp or by an amide-blocked glutamine, resulted in reduced activity of the peptides.

All these considerations have led to the proposal of a receptor that involves two points of attachment of the peptides: one involves the N-terminal amino acid and is a sterically permissive, hydrophobic region; the other involves the C-terminal amino acid, is stereospecific, and relatively tight-fitting, and requires an ionic interaction with the free carboxyl group. Based on the activities of the tripeptides, the functions of the two attachment sites are different: Occupation of the site for the C-terminal portion leads to a tighter (irreversible) interaction, which in the absence of activation of the N-terminal site results in inhibition of response.

OTHER MAST CELL-DERIVED CHEMOTACTIC FACTORS The list of other chemotactic factors is growing rapidly. It is not always established that they are truly derived from mast cells, but this seems to be their most likely source.

Intermediate molecular weight eosinophil chemotactic factors The chemotactic activity for eosinophils in an intermediate molecular weight region on chromatograms of mast cell extracts was purified by gel filtration followed by ion exchange chromatography and reverse-phase high-pressure liquid chromatography. This led to the identification of three peptides (41). All were found to be associated with mast cell granules, to cause deactivation against rechallenge by themselves or by the C5-derived anaphylatoxin fragment, and to have the same molecular weight, approximately 2500. All had chemotactic activity and none had chemokinetic activity. Their maximum stimulation of eosinophil migration was approximately eightfold whereas their maxium stimulation for neutrophils was only approximately twofold.

High-molecular-weight neutrophil chemotactic factor High-molecular-weight netrophil chemotactic factor is released into the circulation when limbs of patients suffering from cold urticaria are immersed into an ice-cold solution (291). It is absent from the venous blood of the contralateral control limb. Peak production is reached by 2 min. and activity returns to baseline by 15 min. This time course is identical to the time course of release of ECF-A and histamine in the same situation. The factor is specific for granulocytes as compared to mononuclear cells; it can cause two and one half times as much stimulation of neutrophil migration compared to eosinophil migration, and this maximum stimulation is achieved using only one quarter the dose of chemoattractant. The persistence of the high-molecular-weight neutrophil chemotactic factor in the circulation has led to the suggestion that this material may play a role in the induction of the persistent or late phase of IgE-mediated reactions. Molecular weight, based on gel filtration chromatography, is estimated as 750,000 (289). It is not clear if the factor is identical to a ragweed antigen-induced neutrophil-specific chemotactic factor, which appears to have very similar properties (12).

Inflammation factor of anaphylaxis Injection of purified granules from rat mast cells into rat skin resulted in a slowly developing inflammatory reaction, which showed a neutrophil infiltrate at 1-2 hr and peaked at 8-24 hr with a predominantly mononuclear infiltrate (272). A soluble factor, obtained from the isolated granules by extraction with salt, mimicked the properties of the total granules. A low-molecular-weight and a high-molecular-weight active component were identified in the granule extract (198) based on ultrafiltration through a membrane with a 10,000-dalton cut-off. The low-molecular-weight substance is a peptide, M_r 1400, containing 12 amino acids with the empirical composition asp_2, glu_2, thr, ser, gly, ala, val, ile, leu, phe (199).

VIP Vasoactive intestinal polypeptide (VIP), which has received a great deal of attention in the last few years, has recently been reported to be released in parallel with histamine from cells challenged either with compound 48/80 or with ionophore A 23187. Mast cells were reported to contain 1.4 ng of VIP per million cells. Immunoflourescence studies showed an association of flourescence due to binding of antibody to VIP to mast cell granules (68). This study, which remains unconfirmed, poses the interesting possibility that in addition to secreting the pro-inflammatory substances, mast cells may secrete a substance that could tone down the response at the same time.

Hydrolases and Other Granule-Associated Enzymes

Rat serosal mast cells release approximately 60 μg of protein/10^6 cells upon challenge with the ionophore A 23187 (246). Only approximately half of this total protein can be accounted for when all the known proteins secreted are summed. Of the secreted proteins a neutral protease is by far the most abundant. In the rat this is a chymase, M_r 25,000, whereas in man it is a tryptase, M_r 130,000 (247). A different chymase (298), the so-called protease II, is found, distinctly, in atypical mast cells of the gut and lung. Other enzymes released from the granules are a β-D-hexosaminidase, α-D-glucuronidase, α-D-galactosidase, and an aryl-sulfatase. In addition, a small amount of superoxide-generating capacity appears to be associated with the granules and a peroxidase capable of inactivating leukotrienes has been shown to be released from compound 48/80-challenged mast cells (121). The release of several of the hydrolases does not account for the total content of these enzymes in the mast cells. Furthermore, in the case of arylsulfatase, both the A and B forms of the enzyme are found in the cells, but only the A form is released during anaphylactic challenge (174). This has led to the suggestion that in addition to the secretory granules, mast cells also contain nonsecretory lysosomes. The acid phosphatase of the cells may be associated with these organelles (246).

In addition to the tryptase (247), human basophils and mast cells release a series of specific proteases. One of these, basophil kallikrein-like activity (194), M_r 1,200,000, cochromatographed with an activity that can generate kinin from either heated or fresh plasma. It is a serine esterase, inhibitable by trasylol and diisopropyl phospho-fluoridate; its specificity for synthethic substrates (p-toluenesulfonylarginine methyl ester, bensoylarginine methylester, and acetyl lysine methyl ester, but not acetyl tryosine methyl ester or alanyl alanine methyl ester) also resemble those of plasma kallikrein. A series of other arginine esterases of lower molecular weight were also found in the reaction mixtures. Release of these enzymes occurred in parallel with histamine release upon stimulation of the cells with IgE and antigen (195, 197). One of these enzymes resembled plasma kallikrein in its ability to activate Hageman Factor. In addition, a prekallikrein activator and a Hageman Factor Activator were also identified (196).

SECONDARY MEDIATORS OF ANAPHYLAXIS AND INFLAMMATION

This section considers some of the substances that are not present preformed, and that depend in their release on a primary stimulus. The rapidly growing family of metabolites of arachidonic acid, which are exquisitely potent and specific in their activities, comprise a major portion of the list of secondary mediators.

Products of the Arachidonic Acid Cascade

Our present knowledge of the multiple pathways by which arachidonic acid can be metabolized is summarized in Figure 1. Arachidonic acid is primarily present in the body esterified to the second carbon of glycerol in membrane phospholipids. Activation of the cell leads to the activation of a calcium-requiring phospholipase A_2 that liberates arachidonic acid. Recently, a modulator protein, named lipomodulin, which regulates the activity of the lipid mobilizing enzymes (126), has been identified.

The primary attack on arachidonic acid is by one of the specific lipoxygenases (101). The product of 5-lipoxygenase action is 5-hydroperoxy-6,8,11,14-eicosatetraenoic acid (5-HPETE). This molecule can be converted to 5-hydroxy-eicosatetraenoic acid (5-HETE). In addition, it can be converted, with the abstraction of water, to 5,6-oxido-7,9,11,14-eicosate-traenoic acid, which has been named leukotriene A_4 (LTA_4) (240). LTA_4 can be attacked in turn by water spontaneously giving rise to two isomers of 5(S),6-dihydroxy-eicosatetraenoic acid and to one isomer of 5(S)-12-dihydroxyeicosatetraenoic acid. In addition, the epoxide ring can be opened enzymatically to form 5(S),12(R)-dihydroxy-(6Z),8(E),10(E),14(Z)-eicosatetraenoic acid or leukotriene B_4 (LTB_4) (239). Finally, LTA_4 can react with glutathione, under the influence of glutathione S-transferase, to form 5(S)-hydroxy,6(R)-S-glutathionyl, 7(E),9(E), 11(Z), 14(Z)-eicosatetraenoic acid or leukotriene C_4 (LTC_4). LTC_4 can be metabolized by the usual glutathione metabolizing enzymes, γ-glutamyl transpeptidase and then amino peptidase yielding leukotriene D_4 (LTD_4) and leukotriene E_4 (LTE_4) respectively. LTC_4, LTD_4, and LTE_4 have been known collectively as slow reacting substance of anaphylaxis (SRS-A), which had been recognized as a mediator of anaphylaxis for over 40 years (46, 150).

The ultimate result of a primary attack on the 11 position of arachidonate by cyclooxygenase is the formation of the prostaglandin endoperoxide, PGH_2. This molecule is the intermediate for the formation of the prostaglandins, prostacyclin (PGI_2), and thromboxane A_2 (TxA_2). Both PGI_2 and TxA_2 are rapidly converted to biologically inactive metabolites (187, 286).

The 12-lipoxygenase and the 15-lipoxygenase cause the formation of 12- and 15-HPETE and ultimately 12- and 15-HETE when they act on arachidonic acid. In addition, these enzymes can react with products from other pathways of arachidonate metabolism. Introduction of the 15-hydroxy function is essential for the activity of the cyclooxygenase products, although this does not involve the 15-lipoxygenase. However, formation of

5(S),12(S)-dihydroxy-eicosatetraenoic acid (DHETE) by reaction of 12-lipoxygenase with 5-HETE yields a compound of low biologic activity (170, 258). This is also the case for the 5,15-dihydroxy acid (175).

Both LTB_4 and 5,12-DHETE can be further metabolized by omega oxidation. A 5,12,19- and a 5,12,20-trihydroxy-eicosatetraenoic acid have been identified (170, 175); furthermore, 5(S),12(R)-dihydroxy-20-carboxy-eicosatetraenoic acid has been reported (238). The products of omega oxidation of LTB_4 have been reported to have 10 times the biologic activity of LTB_4 on guinea pig lung (238).

Finally, subsequent to formation of 15-HPETE, in a reaction that is quite analogous to the formation of LTA_4 from 5-HPETE, both 14,15- and 8,15-dihydroxy-eicosatetraenoic acids have been identified (144, 173, 175). As yet no biologic activity for these molecules has been reported.

The capabilities of different cells to produce the many metabolites of arachidonic acid cannot be considered here in detail. A few examples of the complexities involved will have to suffice.

1. The main metabolite of arachidonic acid in mast cells is PGD_2 (231), a product not formed in most other cells.
2. Both mast cells and basophils can produce leukotrienes (136, 171, 172, 297); cell-free enzyme preparations from rat basophilic leukemia cells have been the main source of enzymes for studying the biosynthesis of these substances (25, 141).
3. The stimulus used to activate cells markedly influences the nature of the arachidonate metabolites formed. Thus, although rat peritoneal mononuclear cells are capable of forming TxA_2 and leukotrienes upon stimulation with the calcium ionophore A 23187 (17, 207), other stimuli that activated these cells in other respects (25), or that successfully induced leukotriene formation in murine peritoneal macrophages (45, 236) or in rat alveolar macrophages, failed to induce leukotriene synthesis in these cells.
4. In the presence of the ionophore A 23187, metabolites of LTA_4 appear to predominate in neutrophils. However, with the synthetic chemotactic peptide formyl-met-leu-phe, the major products were the polyhydroxy-products that had been oxidized in the omega position (238).

PRODUCTS DERIVED FROM CYCLOOXYGENASE ACTION ON ARACHIDONATE Anaphylactic challenge of perfused lungs from sensitized guinea pigs results in the production of a variety of products of the cyclooxygenase pathway of arachidonate metabolism. Initial characterization of the products was by differential bioassay with a cascade arrangement (228) and different smooth muscle preparations that have selective responses to

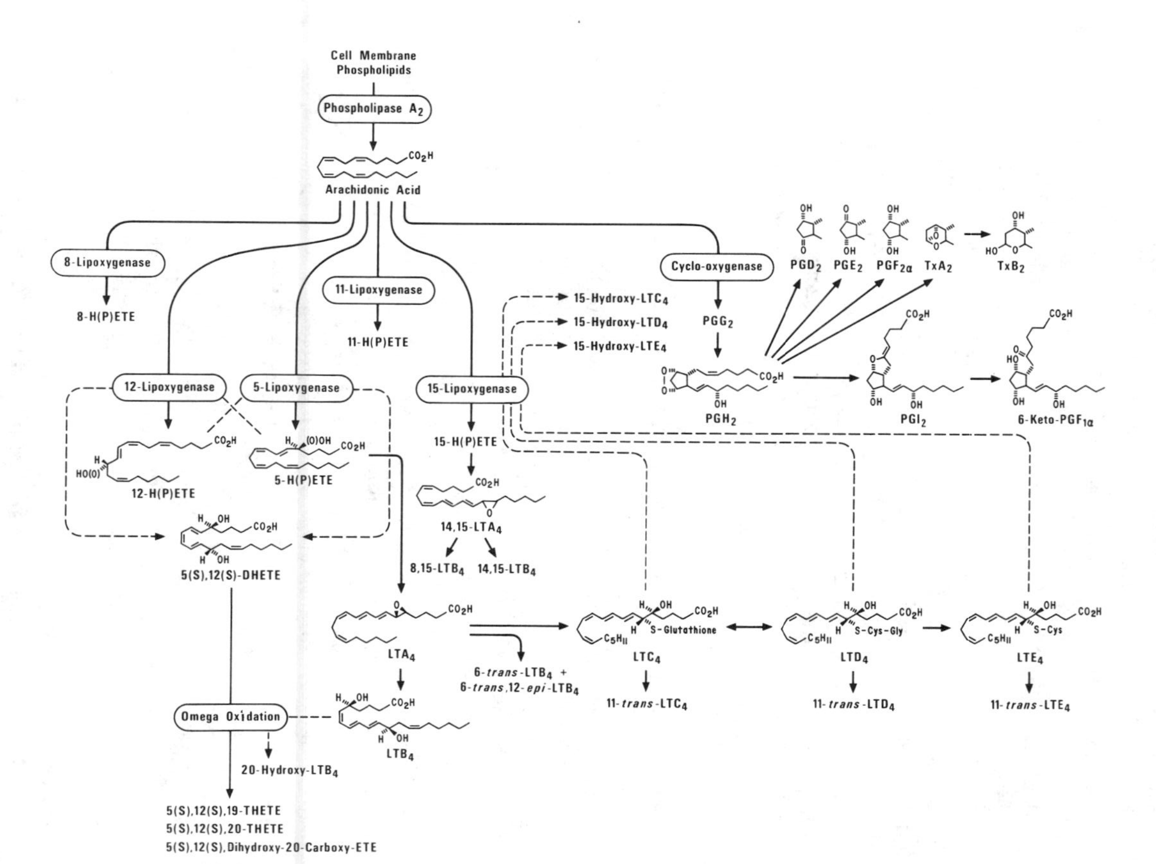

Figure 1 The arachidonic acid cascade.

various products (277); subsequent studies further characterized PGE_2-like material by radioimmunoassay (85). A more detailed analysis of the origin of the various products has revealed (1, 243) that the ratio of the amounts of PGE_2 and $PGF_{2\alpha}$ produced is markedly different in lung tissue and in bronchioles. In the parenchymal tissue of the lungs there were approximately equal amounts of PGI_2 and $PGF_{2\alpha}$ formed with approximately one quarter as much TxB_2 and only on tenth as much PGE_2. In the bronchioles, on the other hand, PGI_2 production was half as great as in the parenchyma. $PGF_{2\alpha}$, PGE_2, and TxB_2 production were even lower relative to PGI_2 production on this tissue.

Mechanisms of induction of formation of cyclooxygenase products Production of the cyclooxygenase products can be dissociated from the release of the other mediators; thus their formation must be a secondary event (85, 86). Pharmacologic studies with selective antagonists to histamine and SRS-A suggested that production of PGE_2 could be inhibited in antigen-challenged guinea pig tracheas both by H_1 antihistamines and by the selective end organ antagonist of SRS-A, FPL-55712 (14, 49). An inhibitor of the muscarinic cholinergic type, atropine, was also able to inhibit the formation of PGE_2. Studies with partly purified SRS-A preparations and with synthetic LTC_4 demonstrated that infusion of this mediator into guinea pigs resulted in the production of PGE_2, $PGF_{2\alpha}$, PGI_2, 12-HETE, and TxA_2 (34, 91, 178, 201, 202). Histamine also caused the formation of TxA_2 and this occurred more extensively when SRS-A, synthetic LTC_4, or histamine were infused into lungs from immunized guinea pigs than it did in lungs of unimmunized guinea pigs (34, 91). Production of PGI_2 occurred predominantly in the lung vasculature and the tracheas, whereas the parenchymal tissue produced primarily TxA_2 (91). Furthermore, low concentrations of LTC_4, which caused little TxA_2 formation when given alone, markedly potentiated the production of this substance when synthesis was induced by histamine or bradykinin (91). A similarly increased formation of cyclooxygenase products could also be demonstrated when purified rat peritoneal macrophages were incubated with LTC_4 or LTD_4 (90).

Another means of inducing the formation of cyclooxygenase products has recently come to light: The anaphylactic challenge of human lung results in the liberation of a factor that in turn can cause the formation of prostaglandins and TxA_2 in unchallenged guinea pig or human lung fragments (267). This factor was not found in water-lysed lung supernatants and therefore was presumably formed during anaphylaxis. The factor, which has been named prostaglandin-generating factor of anaphylaxis (PGF-A), has M_r 1400 and an amino acid composition glu_6, asp_3, gly_3, ser, thr (266). The mode of action of this factor remains unknown.

Actions of cyclooxygenase products in asthma and inflammation The extensive literature on the actions of the prostaglandins in inflammation and asthma are not reviewed here. In general, PGE_2 is considered to have anti-inflammatory activity whereas $PGF_{2\alpha}$ has pro-inflammatory activity. PGD_2, which was long neglected, was recently shown to be a potent bronchoconstrictor (221, 285, 287), more active than $PGF_{2\alpha}$. PGI_2, which by itself caused only changes in respiration rate, potentiated the response of monkeys to very low doses of PGD_2 (221). The pathologic manifestations of systemic mastocytosis could be related to overproduction of PGD_2 by the abnormally large numbers of mast cells in these patients. Thus, treatment with aspirin, to inhibit cyclooxygenase and therefore PGD_2 formation, was found to be highly beneficial in the management of this disease (232).

The apparently pro-inflammatory activity of PGI_2 in augmenting the response of rhesus monkeys to PGD_2 contrasts with such anti-inflammatory actions of PGI_2 as the inhibition of platelet aggregation caused by TxA_2. Very low concentrations of PGI_2 were also found to protect rats from antigen-induced anaphylaxis (48). TxA_2 plays a critical role in guinea pig anaphylaxis (103, 108); however, its role in rat anaphylaxis is not known. Thus, a definitive explanation for the effects of PGI_2 in the rat is not presently possible.

PRODUCTS DERIVED FROM ARACHIDONATE VIA LIPOXYGENASE ACTION: HYDROXY ACIDS Both the hydroperoxyacids and the hydroxyacids have many potent activities on leukocytes (97). Chemotactic factors that had the properties of lipids were identified in the exudates of immunologically challenged rat peritoneal cells (280). The nonselective inhibitor of arachidonate metabolism, 5,8,11,14-eicosatetraynoic acid (ETYA), inhibited formation of these principles, which suggests that arachidonate was at least involved in their formation. Subsequently it was shown that incubation of cell-free enzyme preparations from human polymorphonuclear leukocytes with arachidonate gave rise to 11-HETE, 5-HETE, 8-HETE, 12-HETE, and 9-HETE in order of decreasing abundance. Some dihydroxyacids, which were not further characterized, were also found (101). All these compounds were chemotactic for polymorphonuclear leukocytes, but the concentrations required to demonstrate activity differed markedly, ranging from 8 ng/ml for 5-HETE to 2000–5000 ng/ml for 11- and 12-HETE. The compounds also markedly increased random migration and the concentrations required to elicit this response were lower than those that caused chemotaxis (97). Even 12-HETE was reportedly as active as the chemotactic peptides from C5 in this regard (102).

In addition to affecting the motility of leukocytes, the hydroxyacids caused the increased expression of receptors for the C3b portion of complement on the susceptible cells (97) and also increased the release of lysosomal enzymes (270). Here, 5-HETE and 12-HETE were active in the same dose range, although the magnitude of the effect of 5-HETE was approximately twofold that of 12-HETE. Both hydroxyacids were clearly less active than was the more traditionally used serum-activated zymosan, however.

The lack of specificity among the hydroxyacids argues against the role of a receptor and therefore raises problems relative to their mode of action. A possible explanation was proposed by the finding (268, 269) that the hydroxyacids were rapidly taken up by human neutrophils. When radiolabeled 12-HETE was added to such cells, 26% of the radioactivity was taken up in a 30-min period. Some 90% of this amount was recovered undegraded from the triglycerides and the remainder was recovered from phospholipids, particularly phosphatidyl choline. It was suggested that the incorporation of the hydroxyacids into the lipids might increase membrane fluidity, thus accounting for some of the activities of these molecules.

Rabbit peritoneal macrophages appear to produce 15-HETE as the predominant metabolite of arachidonic acid (62). A lipoxygenase from these cells has been purified 250-fold and has been shown to produce exclusively 15-HPETE (193). Interestingly, rabbit reticulocytes also contain a lipoxygenase that produced 15-HPETE selectively (47). This enzyme appears to be the inhibitor of respiration present in reticulocytes and absent from mature erythrocytes (230). A radioimmunoassay that can detect 1.4 pg of 15-HPETE has been described (132). 15-HETE was by far the predominant metabolite of arachidonic acid in incubations of lung specimens of one asthmatic patient and no 5-HETE or 5-HPETE were found (107). Both 5-HETE and 15-HETE caused slow, histamine-like contractions of human bronchi. However, although 5-HETE was approximately equipotent to histamine and, furthermore, potentiated the response of the bronchi to histamine, 15-HETE was less potent (66). In parenchymal strips from guinea pig lung, both 15-HETE and 15-HPETE caused dose-dependent contractions, the hydroperoxide was equipotent to histamine, and 15-HETE was only 20% as active. On the other hand, 15-HPETE caused a relaxation of helically cut guinea pig tracheas and was somewhat less potent than was PGE_2 in this regard (75).

Regulatory functions of hydroperoxy- and hydroxyacids In addition to the presumably direct effects on various leukocytes of the products of lipoxygenase action, these molecules also have potent indirect effects. Thus, 15-HPETE selectively inhibited the conversion of PGH_2 to PGI_2 with a 50% inhibitory concentration of approximately 1 μM (186). Similarly, 12-HPETE, but not 12-HETE, inhibited platelet cyclooxygenase (ID_{50}, 3 μM), whereas one tenth of that concentration actually acted as a potent positive feedback regulator, causing an increase in the activity of the 12-lipoxygenase. These latter activities were specific for the 12-hydroperoxide; 11-hydroperoxide was only one quarter as active on the 12-lipoxygenase and half as active on the cyclooxygenase and the 8- and 9-hydroperoxides were inactive even at 150 times the ID_{50} for 12-HPETE (254). 5-HPETE, which was not evaluated in this study, has been recently reported to potentiate histamine release from human basophils. Its action resembles that of other agents that cause a reduction in the cAMP levels in these cells, such as cyclooxygenase inhibitors (224). 5-HETE was somewhat less active than was 5-HPETE in this system by contrast to the platelet system (above) where the hydroxyacid was actually inactive. This action of 5-HPETE and 5-HETE may explain the dependence of the mediator release reaction on an intact lipoxygenase pathway in the cells.

15-HETE has been reported to be a selective inhibitor of the 5-lipoxygenase and of the 12-lipoxygenase of platelets (281, 282). Structure-activity studies have shown that the 15-hydroxy-8,11,13-eicosatrienoic acid is more active than is 15-HETE and that the 11,13-dienoic acid and the fully saturated 15-hydroxy arachidic acid are less active. Inhibition of the 5-lipoxygenase in lymphocytes by 15-HETE resulted in the concomitant inhibition of LTB_4 formation and inhibition of Concanavalin A-induced mitogenesis (26, 96). Indeed, although 5-HETE was found to cause increased chemokinesis in human T cells, this activity could also be blocked by 15-HETE and by 15-HPETE at approximately the same concentrations (222).

LTB_4 and ECF Leukotriene B_4 (LTB_4), a dihydroxyacid, is derived from LTA_4 and is the product of a single lipoxygenolytic attack on arachidonate. It is the only enzymatically produced dihydroxyacid metabolite of LTA_4 and displays potent biologic activity (44, 104). The stereochemistry of the double bonds at positions 6, 8, and 10 was established by total synthesis and by the demonstration that only one of the four isomers synthesized possessed the high degree of biologic activity identical to that of biologically prepared LTB_4 (259).

Rabbit, rat (44), and human neutrophils produced LTB_4 upon stimulation with the ionophore A 23187. In zymosan-activated human neutrophils, LTB_4 production could be inhibited by maneuvers that increased the intracellular levels of cAMP (61). This was observed both by using PGI_2, a known inducer of adenylate cyclase, as well as by using an inhibitor of the cAMP-degrading phosphodiesterase. On the other hand, LTB_4, particularly in combination with an inhibitor of phosphodiesterase, caused a marked increase in the cellular concentration of cAMP in these cells. Thus, this mediator can participate in the modulation of cellular function by affecting cAMP levels in a manner reminiscent of that of the prostaglandins. There have also been preliminary reports that LTB_4 can increase cellular cyclic guanosine-3',5'-monophosphate levels in a manner reminiscent of activities reported for some of the hydroperoxyacids (44).

LTB_4 caused profound increases in chemokinesis and chemotaxis in human neutrophils, guinea pig eosinophils, human mononuclear cells, and rat macrophages (44, 92, 176, 262) and was some 100 to 1000 times more potent in eliciting these responses than were the monohydroxyacids. It was equipotent with formyl-met-leu-phe and some 10 times more potent than were the chemotactic peptides derived from C5. A number of in vivo demonstrations of these in vitro responses have been described. Thus, injection of LTB_4 intraperitoneally into guinea pigs stimulated the accumulation of neutrophils and mononuclear cells without causing the accumulation of eosinophils (262). When injected into rabbit skin (57), 25–100 ng of LTB_4 caused a dose-dependent accumulation of leukocytes. 12-HETE or 12-HPETE (1 μg) only caused 5% as much cell accumulation. A particularly promising new method involves perfusion of LTB_4 into the eyes of rabbits, with the second eye of the same animal serving as a control (38). In this system, LTB_4 activity could be detected after infusion of 25 ng whereas 100 ng of formyl-met-leu-phe were required. 5-HETE, 5-HPETE, 12-HETE, PGE_1, and PGE_2 were inactive even when 500 ng were infused. Along with effects on chemotaxis, LTB_4 has also been shown to enhance the expression of receptors to C3b (44) and to induce the release of lysosomal enzymes from human neutrophils (104).

LTB_4 causes the aggregation of neutrophils (44). A possible in vivo correlate of this is the observation that infusion of exceedingly low concentrations of LTB_4 (1 nM) intravenously into the hamster cheek pouch resulted in slowing of the motion of leukocytes in the vasculature and, eventually, their increased adhesion to the vessel walls (74, 261a). LTB_4 had no effect on the vascular caliber or vascular permeability of the hamster cheek pouch (74). On the other hand, intradermal injection of LTB_4 (1 to 100 ng) into guinea pig or rabbit skin or (100 ng) into monkey or human skin (43, 169) caused increased plasma exudation, but it had no effect on blood flow. The effects in guinea pig, rabbit, or rat skin were markedly potentiated by the simultaneous injection of 10 ng of PGE_2. In monkey and human skin, the effects of LTB_4 were markedly augmented by the simultaneous injection of 1 μg of PGD_2. In addition, LTB_4 caused development of tenderness and infiltration of neutrophils at the injection site in human volunteers 4–6 hr after the injection. These reactions were also potentiated by PGD_2.

A very sensitive and quantitative assay for LTB_4 based on the contraction of parenchymal strips derived from guinea pig lung has been proposed

(258). LTB_4 was approximately 90 times more potent than was histamine in causing contractions, being active at doses of 10 to 500 ng. The contractions could be differentiated pharmacologically from contractions caused by histamine, serotonin, acetylcholine, or LTC_4 and LTD_4. However, the contractions were completely inhibited by moderate doses of a cyclooxygenase inhibitor (258). The proposition that LTB_4-induced contractions were due to the secondary production of a spasmogen derived from arachidonate via the cyclooxygenase pathway was strengthened by the observation that infusion of LTB_4 into the pulmonary artery of isolated guinea pig lung preparations resulted in the formation of rabbit aorta contracting substance activity (a mixture of TxA_2 and PGH_2), and that this could be prevented by treatment of the tissue with Indomethacin (258).

Taken together, the observations that known inflammatory stimuli can induce LTB_4 formation in cells known to participate in the inflammatory response, and that exogenously applied LTB_4 has strongly pro-inflammatory activity in animal models and in man, lead to the conclusion that LTB_4 may play a major role in the inflammatory response. Further support for this conclusion comes from clinical studies. It was found (154) that the concentration of LTB_4 was elevated in the synovial fluid of patients suffering from rheumatoid arthritis and spondilitis, but not in those patients suffering from non-inflammatory arthropathies. Anti-inflammatory therapy, namely intraarticular injection of corticosteroids, markedly lowered the concentration of LTB_4 subsequently found in the synovial fluid. In another study (67) it was found that sputum specimens of 16 of 25 patients suffering from cystic fibrosis contained LTB_4, whereas normal sputum did not.

In 1975, Czarnetzki et al (72) described an ECF produced when human neutrophils were challenged with the ionophore A 23187. Generation of this factor (M_r 500), which was produced de novo, could also be induced immunologically, required calcium ions, was potentiated by arachidonate, and was insensitive to inhibitors of DNA or protein synthesis (156). Human serum was reported to contain both a factor that inactivated ECF (155) and a factor that enhanced its activity (69). The reported selectivity for eosinophils over neutrophils (158), the lack of activity on mononuclear cells, and the claimed failure of ECF to cause neutrophils to aggregate have led one investigator to conclude that this substance was different from LTB_4 (157). However, this has been disputed (70).

A variety of other chemotactic factors, which act on granulocytes and are produced by various tissues, have been described (44). One example is a chemotactic factor for neutrophils produced by alveolar macrophages (94, 131). This factor is reported to have a molecular weight of 400 to 600, to be heat stable, and to be stable to acid and alkali. It is also resistant to proteolytic enzymes and is partly extracted into organic solvents. Although gas chromatography has ruled out that it is a monohydroxyacid, the possibility that it is a dihydroxy acid, and in particular LTB_4, has not been addressed.

PRODUCTS DERIVED FROM ARACHIDONATE VIA LIPOXYGENASE ACTION: LEUKOTRIENES Because of the differences that exist both in the structure and in the biologic properties between LTB_4 and the leukotrienes derived from the reaction of glutathione with LTA_4, these latter molecules are considered separately in this section. Several excellent reviews are available on many aspects of this work (213, 239, 257).

The traditional assay for SRS-A involves the quantitative measurement of the contractions elicited on a segment of longitudinal muscle from the terminal portion of the guinea pig ileum. Selectivity for SRS-A is achieved by blocking responses to histamine with an H_1 antihistamine and responses to cholinergic stimuli with an antimuscarinic agent (204). A selective end organ antagonist for the leukotrienes, FPL 55712 (14), has been described. With the establishment of the structures of the leukotrienes and their total synthesis, efforts have intensified to develop workable and selective radioimmunoassays (119, 162). Such assays now promise to effect a profound change in the future course of work in this area.

Structure, biosynthesis, and metabolism of the leukotrienes From the early studies (46, 203), it was clear that SRS-A was a lipid and was highly polar, that it was extremely potent and was apparently present in minute concentrations, that it was destroyed by acid and by a variety of chemical reagents, and that it bound quite firmly to serum albumin. The turning points in studies with SRS-A were the findings that various cells could be induced to produce a substance resembling SRS-A with the calcium ionophore A 23187 (16, 137, 168), and that production of SRS-A could be markedly enhanced by the incorporation of relatively high concentrations of cysteine into the incubations (205, 206). The latter observation, along with the finding of sulfur in impure preparations by spark source mass spectroscopy (209) and the fact that SRS-A could be inactivated upon incubation with arylsulfatase (208), led to speculation that there may be a sulfate ester, or at least a sulfur atom, in the molecule. Development of high-resolution thin-layer chromatography (216, 297) and reverse-phase high-pressure liquid chromatography (19) led to the discovery that the spasmogen was composed of at least two biologically active substances and that, kinetically, one of these appeared to be derived from the other (23). The conclusion that SRS-A was derived from arachidonic acid was based on inhibitor studies (20), radiolabel incorporation (20, 136), and the observation that soybean lipoxygenase could selectively and effectively inactivate SRS-A (256). The absorption of SRS-A in the ultraviolet region (189, 208) led Samuelsson's group to propose a linear triene structure, similar to that of the 5,12-dihydroxy-eicosatetraenoic acids (239). Finally, the demonstration that radioactivity from [^{35}S]cysteine was incorporated into the molecule (191) led to the proposal that the 6-S-cysteinyl derivative of 5-hydroxy-eicosatetraenoic acid was a slow reacting substance with the proposed name of leukotriene C. Comparison to synthetic materials (113, 114, 165, 188, 211) led to a revision in nomenclature and to a definitive assignment of structures for LTC_4, LTD_4, and LTE_4. Proof of the identity of the biologically produced slow reacting substances to the leukotrienes, regardless of their cell or tissue source or the initiating stimuli used in their production, was accomplished by use of cochromatography, amino acid analysis after degradation, mass spectroscopy, and the demonstration of comparable rates of inactivation and the formation of spectroscopically comparable products upon reaction with lipoxygenase (21, 22, 190).

Both the conversion of LTA_4 and of 5-HPETE into spasmogenically active leukotrienes have been reported in rat basophil leukemia (RBL-1) cells (218, 229). The conversion of LTC_4 to LTD_4 has been accomplished with these cells as well as with commercially available γ-glutamyl transpeptidase (211). Finally, the formation of LTE_4 was demonstrated in human plasma (214, 219). The biosynthetically prepared analogue of LTC_4, which lacks the double bond at position 14, LTC_3, was converted to LTD_3 and LTE_3 in various tissues (112). High concentrations of glutathione found in liver and kidney actually caused conversion of LTD_3 to LTC_3 by shifting the equilibrium of the γ-glutamyl transpeptidase reaction. The absolute requirement for glutathione for the biosynthesis of LTC_4 has also been shown by depletion studies (215, 237) and in cell-free preparations from RBL-1 cells (138) and from rat liver (25).

The apparent inactivation of leukotrienes by arylsulfatase is not limited to any one form of the enzyme (174). Inactivating capacity of various tissues was reported to correlate with their content of total arylsulfatase, and the arylsulfatase activity actually cochromatographed on molecular sieves with the leukotriene-inactivating activity (147). Nevertheless, in the absence of a sulfate ester in the molecule it was difficult to explain how this enzyme might effect the inactivation. The explanation ultimately depended on the observation that LTE_4 was considerably less active when assayed on the guinea pig ileum than was either LTC_4 or LTD_4 and that leucine amino peptidase could cleave LTD_4 to yield LTE_4 (263). LTE_4 turned out to be resistant to inactivation by arylsulfatase, and commercial arylsulfatase was shown to be contaminated with a leucine amino peptidase (217), which could be effectively inhibited by high concentrations of cysteine and other β-mercapto carboxylic acids (18, 263).

Sources of the leukotrienes Traditionally, the induction of leukotriene production by IgE made the belief that these mediators are produced by mast cells almost idiomatic. However, studies with cells from the peritoneal cavity of rats revealed that different cells were involved when different inducers of leukotriene production were employed. Although mast cells were required for leukotriene generation in an IgE-dependent reaction, these cells were not needed when the inducing stimulus was the heat-stable IgG_a antibody. Instead, neutrophils and complement then appeared to be required (210). Mononuclear cells in the peritioneal cavity were at least as active as mast cells when the ionophore was used to induce synthesis. Given the much greater abundance of monocytes, they are likely to account for the major portion of the leukotrienes produced (17, 207). Further support for a role for mononuclear cells in leukotriene formation came from the finding that only a mixture of isolated mononuclear cells and mast cells could approximate the capacity of unfractionated, disaggregated human lung cells to produce SRS-A after stimulation in an IgE-dependent reaction (167). Recent data have suggested, however, that highly purified mast cells isolated from enzymatically disaggregated human lung were capable of producing very large amounts of leukotrienes (172). Rabbit platelets have been reported to produce leukotrienes upon stimulation with platelet-activating factor (30), although platelets are not known to contain a 5-lipoxygenase. This enzyme is essential for leukotriene formation.

Actions of the leukotrienes Intravenous administration of partly purified SRS-A to conscious guinea pigs caused a selective decrease in lung compliance without any marked effects on lung resistance (80). No such selective activity on the peripheral airways was seen when histamine, bradykinin, or $PGF_{2\alpha}$ were infused into the same animals. The in vitro correlate to these observations is the finding (83) that SRS-A had a dose ratio for equivalent activity on parenchymal strips and tracheal spirals of 100 whereas the ratio for histamine was 1.0. Further studies revealed that only LTD_4 had this selective activity for the parenchymal strips (23), where it was approximately 100 times more potent than was LTC_4 (81, 165). LTC_4, in turn, was 100 times more potent than was histamine. By contrast, LTC_4 and LTD_4 were equipotent on tracheal spirals and were 1000 times more potent than was histamine, although the maximum contraction that could be caused by LTD_4 was only 35% as great as the maximum contraction caused by histamine or by LTC_4. LTE_4 was approximately equipotent to LTC_4. Repeat administration of LTC_4 caused tachyphylaxis (75, 166). Dose response curves for LTD_4 on parenchymal strips were biphasic. Only the relatively shallow, low-dose portion of the curve, below 10^{-11} M agonist, could be antagonized by the selective inhibitor, FPL 55712.

Further pharmacologic evaluations of the smooth muscle contractile activities of the leukotrienes have been reported (120, 128). In addition to the activities on lung parenchyma and tracheal tissues, the leukotrienes contracted the guinea pig uterine horn and enhanced dermal vascular permeability. They were inactive on the rat uterine horn, the rabbit duodenum, or rabbit bronchi at concentrations up to 100 nM. An in vivo model of lung function in which the activities of the leukotrienes have been evaluated is the measurement of the change in insufflation pressure in anesthetized guinea pigs (Konzett Rössler assay). In this model, the increased insufflation pressure, which reflects constriction of the airways, is reflected in an increase in the overflow volume of liquid when a constant volume of air is artificially pumped into the airways. Intravenous administration of the leukotrienes caused profound increase in insufflation pressure (120, 295).

The effects of the leukotrienes on human lung tissue have also been investigated with bronchiolar parenchymal and venous and arterial muscle preparations (76, 117). LTC_4 was approximately equipotent to LTD_4 and both were several hundred fold more active than was LTA_4. Contractions were reversed by isoproterenol (76) and there appeared to be significant tachyphylaxis when the same tissue was exposed to repeated doses of either agonist. The pulmonary veins were more responsive than were the pulmonary arteries.

Intradermal injection of 0.1 pmol of LTC_4 or LTD_4 caused increases in the vascular permeability of guinea pig skin as visualized by the extravasation of intravenously administered colloidal dye. This is some 1000 times less compound than the amount of histamine that was required to achieve the same increases in permeability (120). LTE_4 caused increases in vascular permeability in rat skin under similar conditions (295). By contrast, LTC_4 has been reported to cause a vasoconstriction rather than vasodilatation (81). This was visualized by pre-injecting the guinea pigs with dye prior to the intradermal injection of the leukotriene. In the hamster cheek pouch, the local administration of 0.3 to 20 nmol of LTC_4 or LTD_4 caused a dose-dependent and rapid vasoconstriction of the arterioles. The potency of the leukotrienes is very similar to that of angiotensin in eliciting the same response. In contrast to the effects of angiotensin, however, the vasoconstrictor activity of the leukotrienes lasted a short time and was followed by a marked increase in extravasation of colloidal material (74). Intra-arterial instillation of partly purified SRS-A in the cat caused a dose-dependent, but modest, increase in blood flow in the nasal mucosa (3). By contrast, and in agreement with the observations in the hamster cheek pouch, the leukotrienes caused a dose-dependent decrease in blood flow in guinea pig skin as measured by ^{133}Xe clearance (223). It is particularly interesting that low doses of PGE_1, which by themselves caused no colloid exudation, markedly potentiated the effects of the leukotrienes in guinea pig skin (254). PGD_2

augmented the development of the dose-dependent wheal and flare response to injected LTD_4 in human volunteers (169).

The leukotrienes play a highly selective role in cardiac function. Intracardiac administration of synthetic LTC_4 or LTD_4 caused a negative inotropic effect and reduced coronary flow, and at higher doses it caused an arteri-oventricular conduction block. In addition, LTD_4 but not LTC_4 potentiated the positive chronotropic effect of histamine on spontaneously beating guinea pig atria (50, 161).

Mucus is crucial to the proper elimination of foreign matter from the airways, and improper production of mucus (plugging) is life threatening in the severe asthmatic. The leukotrienes have profound effects on both mucus production and mucus transport. The hydroxy-eicosatetraenoic acids are potent stimulators of mucus production in cultured explanted segments of human lung (179). Leukotrienes in the picomolar concentration range were the most potent stimulators of the incorporation of radiolabeled precursors into mucus in such cultures (178a, 179). Infusion of LTC_4 into the exposed cranial thyroid artery of the dog has resulted in a dose-dependent and profound increase in the production of mucus as visualized by enumeration of "hillocks" on tantalum powder-coated tracheas of living dogs (H. G. Johnson, R. A. Chinn, A. W. Chow, M. K. Bach, and J. A. Nadel, submitted for publication). The role of the leukotrienes in mucus transport can also be deduced from the effects of FPL 55712 (14) on mucus flow in dogs and humans (2, 283).

In addition to their direct effects on smooth muscle and exocrine gland function, the leukotrienes also appear to affect nerve cells. Thus, pressure injection of minute amounts of LTC_4 into Purkinje cells in urethane-anesthetized rats caused large and long-lasting increases in the excitation of these cells (212). The enhancement of mucus secretion in the dog (H. G. Johnson, R. A. Chinn, A. W. Chow, M. K. Bach, and J. A. Nadel submitted for publication) was inhibited by atropine and hexamethonium but not by vagotomy, and the induction of TxA_2 formation in isolated guinea pig lungs by superfusion with leukotrienes was inhibited by atropine in an apparently nonmuscarinic action of this anticholinergic agent (33).

Initial studies have been reported on the effects of leukotrienes on pulmonary function in primates and in man. In the monkey, instillation of leukotrienes caused profound increases in resistance and decreases in dynamic compliance that persisted for longer than 1 hr (24; M. K. Bach, J. R. Brashler, M. L. McNee, H. G. Johnson, M. A. Johnson, and D. R. Morton Jr., manuscript in preparation). Aerosol administration to monkeys and humans, on the other hand, caused much more modest changes that lasted a relatively short time (130, 164).

Preliminary results suggest that the leukotrienes also play a role in the more cellular aspects of inflammation and the immune response. Thus, the leukotrienes caused a dose-dependent inhibition of the ability of human buffy coat cells to adhere to plastic and, concomitantly, inhibited the leukocyte adherence inhibition, which is ordinarily seen as a correlate of a tumor-specific immune response (275). Such activities of the leukotrienes are of particular interest in view of the ability of both neutrophils and macrophages to produce these mediators upon stimulation with pro-inflammatory stimuli.

The role of the leukotrienes in the etiology of pulmonary disease remains conjectural. In addition to the demonstration that they can mimic certain symptoms of disease, it is important to demonstrate that these substances are present in the target tissue under conditions where disease is expressed and not in healthy tissue. This requirement has been satisfied in part. A slow-contracting substance was identified in the sputum of asthmatics over 50 years ago (118). This observation was rediscovered many years later (276) when a slow-contracting substance was also found in the sputum of patients suffering from nonatopic chronic bronchitis. More recently, LTD_4 was also found in the sputum of patients with cystic fibrosis (67), a disease in which the mucus is abnormal.

Structure activity relations Preparations of all three naturally occurring leukotrienes invariably contain small amounts of substances that chromotograph very near to the major product. These contaminants are the 11-*trans* isomers of the major products (22, 63, 190). The 11-*trans* isomers were generally somewhat less active than the 11-*cis* isomers (165). It is still not clear whether or not the 11-*trans* isomers are formed biosynthetically. However, their abundance tends to increase in purified preparations upon standing even at –80°C. The conversion of the 11-*cis* to the 11-*trans* isomer may be due to the reversible addition of a thiyl radical to the C-12 position of the triene system, resulting in epimerization (13).

The enzymes involved in the biosynthesis of the leukotrienes can convert the 5,8,11-eicosatrienoic and the 5,8,11,14,17-eicosapentaenoic acid to the corresponding LTC_3 and LTC_5, and these, in turn, can be metabolized to give LTD_3, LTE_3, LTD_5, and LTE_5 (109–111). Furthermore, the cell-free leukotriene-generating system was capable of converting 18- and 19-carbon triethylenic fatty acids to hydroxyacids and to spasmogenically active products provided they met the minimum requirement of having three unsaturations at positions 5, 8, and 11 (139). Certain combinations of chain length and positional isomers of polyenoic acids are selective inhibitors for different steps in the biosynthesis of the leukotrienes in this system (140).

All the possible stereoisomers of LTC_4, LTD_4, and LTE_4 based on the double bond configuration at positions 9 and 11 and the chirality of the substituents at positions 5 and 6 have been prepared (27). In general, the order of biologic activity was 9-*trans,* 11-*cis* > 9-*cis,* 11-*trans* >> 9-*cis,* 11-*cis* >> 9-*trans,* 11-*trans* = O. Furthermore, the natural 5(S),6(R) configuration was generally more active than the 5(R),6(S) configuration.

From comparisons of the activities of analogues of LTC_4 and LTD_4 on the guinea pig ileum and guinea pig lung parenchymal preparations, several points have emerged (82, 166).

1. The hydrophobic region of the molecule, i.e. carbons 9 and beyond, is important to activity. However, the details of the stereochemistry of the double bonds in this region and even the absolute number of the double bonds present are not critical. On the other hand, the double bond at carbon 7 is quite critical.
2. Displacement of the hydroxyl and thiol ether linkages from positions 5 and 6 (eg 11-S-R, 12-hydroxy 14-S-R,15-hydroxy, or 5-hydroxy, 12-S-R) where R is glutathione or cys-gly resulted in compounds with very low activity.
3. In terms of the nature of the R group on the thiol ether, both a free amino group and a free glycine carboxyl group were found to be necessary for the full activity of the molecule. The same was also true for the hydroxyl group in position 5. Furthermore, epimers at positions 5 and 6 were markedly less active than the natural forms of the molecule (see also 27). One isomer of the sulfoxide of LTD_4 had approximately one tenth the activity of native LTD_4. This casts doubt on the proposal that SRS-A is a molecule in which the sulfur atom is oxidized (200).

Aside from their activities on smooth muscles, structural analogues of the leukotrienes are of interest in another context. Although the leukotrienes are not inactivated by arylsulfatase, these molecules are potent competitive inhibitors of this enzyme when chromogenic substrates such as *p*-nitrocatechol sulfate are used to estimate the activity of the enzyme (288). Structure-activity studies for this inhibitory activity indicated that here, too, the presence of a substituent at position 6 and its chirality were critical (293). However, by contrast to the structural requirements for spasmogenic activity, the presence of the triene series of conjugated double bonds was critical for the inhibitory activity on the enzyme whereas it was not critical for the spasmogenic activity of the leukotrienes.

PAF

Progress with Platelet-Activating Factor (PAF) from its first identification to proof of its structure by synthesis was considerably faster than was progress with the leukotrienes.

The release of a mediator from leukocytes during an anaphylactic challenge that causes histamine release from platelets was recognized in 1972 (31). The developments that led to a proposed structure 8 years later (32, 77), as well as many of the biologic activities of PAF, have been reviewed (226). Thus far there has been no unified nomenclature established for PAF; the substance is referred to as PAF in this review unless the discussion pertains to the chemically defined AGEPC, which refers only to the chemical entity that has been proven to be identical to the PAF from rabbit leukocytes. Although activities of AGEPC appear to mimic those of PAFs that can be obtained from other cell sources, there has been no formal proof of the chemical identity of all these factors.

STRUCTURE, BIOSYNTHESIS, AND METABOLISM OF PAF In 1977, Benveniste suggested that PAF may be 1-lyso-phosphatidyl choline (29), but progress in the purification of the actual mediator was the limiting factor in the definite elucidation of its structure. The final proof that rabbit leukocyte PAF was 1-O-alkyl,2-acetyl-sn-glyceryl,3-phosphoryl choline (32, 77) was based on partial synthesis of molecules of this structure beginning with plasmalogens and the demonstration of their identity with respect to physical, chemical, and biologic properties to biosynthetic PAF (115, 225). Serendipidously, the very same molecule was also identified by Blank et al as the antihypertensive factor in rats they had been studying for over 20 years (39).

PAF is not present preformed in cells. Its production occurs in response to a secretory stimulus that can be either immunologic in nature (e.g. antigen, anti-IgE, or the anaphylatoxins derived from C3 or C5), inflammatory in nature (e.g. cationic proteins from neutrophil granules), or nonspecific (e.g. the calcium ionophore) (29). There is modulation of the production of PAF by cellular levels of cAMP (51). The lipid precursor for AGEPC appears to be phospholipid. In rat peritoneal exudate cells production was inhibited by *p*-bromophenacylbromide, a relatively selective inhibitor of phospholipase A_2 (184), and this inhibition was reversed by the addition of the biologically inactive 2-lyso analogue of AGEPC, which presumably is the precursor molecule. Incorporation of radioactivity from tritiated acetyl CoA into PAF has also been reported in such cell suspensions as well as in cell-free preparations (184). A β-lipoprotein fraction from human and rabbit serum rapidly inactivates AGEPC (89). The mode of action of this factor is unknown but may represent the cleavage of the ester linkage at the second carbon of glycerol to form 2-lyso-AGEPC.

SOURCES OF PAF Production of PAF has been shown in cells and tissues from humans, hogs, mice, rats, guinea pigs, rabbits, and dogs (28). Peripheral leukocytes, peritoneal cells, lung, kidney, skin, spleen, liver, and ileum slices were examined; however, cells and tissues from rabbits and man have been studied most. Basophils were first implicated in PAF production by virtue of its formation during an anaphylactic challenge of rabbit leukocyte suspensions. Indeed, the production of AGEPC in such suspensions was correlated with the basophil content of the preparation (37). By contrast, however, human basophils failed to produce AGEPC under the same conditions and evidence that these human cells, or mast cells from any species, can produce PAF is indirect at best (28). However, the demonstration that cells from a patient suffering from basophil leukemia were able to secrete a PAF upon stimulation with the calcium ionophore supports the conclusion that basophils have the capacity to produce such a mediator (168). The only evidence suggesting a possible capability of mast cells to produce PAF comes from observations in man, where PAF activity in cutaneous inflammatory disorders was associated with the production of all the other mast cell-derived mediators of anaphylaxis (264). Inflammatory cells (both peripheral mononuclear cells, tissue macrophages, and neutrophils) are major producers of PAF (6, 36, 183, 241). Production of PAF in neutrophils has been dissociated from degranulation and superoxide production by demonstrating that cells from patients suffering from chronic granulomatous disease, in which release of lysosomal enzymes is abnormal, nevertheless produced PAF normally (36).

ACTIONS OF AGEPC Activation of rabbit platelets by AGEPC resulting in the secretion of platelet factor 4 (182) has been reported. AGEPC can be assayed very sensitively by monitoring platelet factor 4 release by radioimmunoassay (226). In addition to release of factor 4, production of TxA_2 in platelets is also induced by AGEPC and that is presumably the manner in which AGEPC causes platelet aggregation (249). There is, as yet, no direct evidence for the existence of a true receptor for AGEPC on platelets or on any other cells that are targets for its action. However, the structure activity relations (see below) and the observation that AGEPC was bound to both platelets (248) and neutrophils (52) are evidence in support of the existence of such receptors.

Activation of neutrophils by AGEPC resulted in the aggregation of the cells that, in vivo, caused an acute neutropenia resembling that observed when immune complexes were injected (53, 122). This reaction was insensitive to inhibitors of cyclooxygenase, but it was inhibited by steroids, serine esterase inhibitors, and ETYA, which suggests that neutrophil activation involved a surface protease and a lipoxygenase in a manner reminiscent of the activation of mast cells for mediator release. Lysosomal enzyme release and ECF formation occurred when neutrophils were treated with AGEPC

(28). Injection of minute amounts of AGEPC into rabbits or baboons, intravenously, caused an acute, severe, and transient thrombocytopenia, and neutropenia (106, 181). Among the physiologic changes were rapid onset of shallow breathing, and transient apnea lasting for 30 to 60 sec, which was followed by increased lung resistance and decreased-dynamic lung compliance. In terms of cardiovascular action, a brief period of bradycardia, increased right ventricular pressure, and decreased aortic pressure was followed by systemic hypotension. Doses of AGEPC in excess of a few micrograms were uniformly lethal to rabbits in less than 2 min. All these changes are characteristic of systemic anaphylaxis, thus supporting the conclusion that this mediator may play a key role in the elicitation of the symptoms of anaphylaxis. The effects of AGEPC are not entirely due to direct actions of the mediator on the affected organs. The pulmonary changes, except for the apnea, but not the cardiovascular changes, could be prevented if the rabbits were depleted of platelets prior to challenge with AGEPC (28). Pretreatment with cyclooxygenase inhibitors increased the lethality of injected AGEPC in rabbits, and PGI_2 protected human platelets from the effects of AGEPC. Thus it appears that PGI_2 may have a protective function in the homeostatic control of pulmonary function.

AGEPC also has activity on smooth muscle per se. When added to a guinea pig ileum preparation, AGEPC caused a profound and irreversible contraction that was not sensitive to classical antihistamines, anticholinergics, or inhibitors of the leukotrienes or of cyclooxygenase (28). When injected into rabbit, rat, guinea pig, or baboon skin, AGEPC caused a dose-dependent increase in vascular permeability that was visualized as extravasation of a blue colloidal dye. Similar reactions (i.e. wheal formation) were also seen in human skin. Concomitantly with the increased extravasation, but apparently not merely because of it (28), AGEPC also caused the accumulation of neutrophils at the injected skin sites. It is tempting to speculate that this chemotactic activity may be due to formation of LTB_4 at these sites (71).

STRUCTURE ACTIVITY RELATIONSHIPS Although 2-lyso AGEPC is inactive, the 2-proprionyl derivative was as active as the 2-acetyl derivative (77). The 2-butyryl derivative was considerably less active, and the 2-maleiyl and 2-succinyl derivatives were approximately 5% as active as AGEPC (100). The ester linkage itself was not essential. Thus the 2-acetyl group could be replaced by a 2-O-methyl ether group with retention of much of the activity. Indeed, even exchange of the substituents on the one and two positions of the glycerol (i.e. 1-O-methyl-2-O-alykyl-glyceryl-phosphoryl-choline) retained much of the activity of the molecule. Regarding the length of the alkyl ether chain at carbon one, natural AGEPC consists of a mixture that is better than 90% 18-carbon alkane and less than 10% 16-carbon alkane (115). On the other hand, the 16-carbon alkane-containing molecule is some three times more active than the 18-carbon alkane-containing molecule (116). Shorter chain alkanes at position (1) such as a 10- and 12-carbon substituent were markedly less active (116). There is some question regarding the requirement for the chirality at the 2 position of glycerol. The sn-1 analogue has been prepared (116) and has been reported to be active. This may have trivial explanations, however, because of the low level of activity and uncertainty regarding the ability of phospholipase A_2 to act on sn-1 molecules that have a short chain at this position.

There is considerable specificity in the requirement for a phosphorylcholine polar group in the molecule. Neither the phosphatidic acid, phosphoryl ethanol, nor phosphorylethanolamine analogues of AGEPC possessed biologic activity. On the other hand, the mono-, di-, and tri-methylethanolamine (i.e. phosphorylcholine) compounds were approximately equiactive (242).

SECOND-ORDER REACTIONS; PROSPECTS FOR THE FUTURE

Throughout this review instances have been documented where one mediator either stimulates or inhibits the formation or action of another. These second-order events attest to the complexity of the anaphylactic and the inflammatory response in the whole animal. There are many examples of cell recruitment in these reactions. In addition to the chemotactic and chemokinetic substances that activate, attract, or ultimately desensitize and immobilize various cell classes at the sites of the reaction, there must be as yet unknown principles elaborated that specifically activate certain pathways in other cells. For example, the signal that activates leukotriene formation in rat mononuclear cells is not yet identified, and the role of leukotrienes in non-anaphylactic inflammatory reactions is barely recognized. Another example is the biogenic amine cycle, which has been proposed to explain the multiple effects of histamine and serotonin in the regulation of the cellular immune response and the effects of the lymphokine products of the cellular immune response on the release of the amines (8).

The products of the arachidonate cascade appear to be particularly adept at either augmenting or inhibiting one another's effects. The opposite effects of PGI_2 on the one hand and TxA_2, or PGD_2, on the other have been noted. These actions even extend to effects of an otherwise unrelated secondary mediator such as AGEPC (226). Similarly, augmentation of LTC_4 or

LTD_4 activity by induced generation of TxA_2 has been mentioned (90, 91, 201). In addition to having a modulating influence on the formation of products of the cyclooxygenase pathway, the leukotrienes actually inhibited their own synthesis in fragments of human lung (292). In this respect, the leukotrienes are reminiscent of the hydroxy- and hydroperoxy-eicosatetraenoic acids, which are effective modulators of a number of different steps in the arachidonate cascade. Finally, there is the role of the metabolites of arachidonic acid in modulating the release of the preformed mediators. 5-Lipoxygenase activity is required for histamine release (224), and histamine release is enhanced when inhibitors of cyclooxygenase are added to the incubations (85).

Our understanding of mediator interactions at the target organ is at its infancy. The present euphoria of having pure leukotrienes available, at long last, has led to a veritable deluge of studies seeking to characterize the pharmacologic properties of these potent substances. But the actions of the pure single agonists cannot account for the total pathophysiologic picture of the disease in which these substances are believed to play a role. As a minimum assumption, it must be considered that the disease syndromes must be a reflection of the interactions of all the mediators elaborated. Thus, much further work remains to be done, studying mediator interactions in detail and considering all the mediators that are now recognized, as well as the many new ones that will be identified in the future, before an understanding of the mechanisms of anaphylaxis, asthma and inflammation can be expected.

ACKNOWLEDGMENTS

I am indebted to P. W. Askenase, F. Berti, A. Capron, B. Czarnetzki, E. J. Goetzl, J. A. Grant, M. A. Kaliner, L. M. Lichtenstein, H. H. Newball, and B. Samuelsson for making prepublication copies of their most recent work available for inclusion in this review.

Literature Cited

1. Adkinson, N. F. Jr., Newball, H., Findlay, S., Adams, G. K. III, Lichtenstein, L. M. 1979. *Monogr. Allergy* 14:122–25
2. Ahmed, T., Greenblatt, D. W., Birch, S., Marchette, B., Wanner, A. 1981. *Am. Rev. Resp. Dis.* 124:110–14
3. Änggard, A., Strandberg, K. 1981. *Acta Physiol. Scand.* 111:329–33
4. Anwar, A. R. E., Kay, A. B. 1980. *Clin. Exp. Immunol.* 42:196–99
5. Anwar, A. R. E., McKean, J. R., Smithers, S. R., Kay, A. B. 1980. *J. Immunol.* 124:1122–29
6. Arnoux, B., Duval, D., Benveniste, J. 1980. *Eur. J. Clin. Invest.* 10:437–41
7. Askenase, P. W. 1977. *Progr. Allergy* 23:199–320
8. Askenase, P. W. 1979. *J. Allergy Clin. Immunol.* 64:79–89
9. Askenase, P. W., Bursztajn, S., Gershon, M. D., Gershon, R. K. 1980. *J. Exp. Med.* 152:1358–74
10. Askenase, P. W., Tamir, H., Theoharides, T. C., Gershon, M. B. 1981. *Fed. Proc.* 40:1022 (Abstr.)
11. Askenase, P. W., Theoharides, T. C. 1980. *Fed. Proc.* 39:905 (Abstr.)

12. Atkins, P. C., Norman, M., Zweitman, B., Rosenblum, F. 1979. *J. Allergy Clin. Immunol.* 64:251–58
13. Atrache, V., Sok, D.-E., Pai, J.-K., Sih, C. J. 1981. *Proc. Natl. Acad. Sci. USA* 78:1523–26
14. Augstein, J., Farmer, J. B., Lee, T. B., Sheard, P., Tattersall, M. L. 1973. *Nature New Biol.* 245:215–17
15. Austen, K. F. 1979. *Harvey Lect.* 73:93–161
16. Bach, M. K., Brashler, J. R. 1974. *J. Immunol.* 113:2040–44
17. Bach, M. K., Brashler, J. R. 1978. *J. Immunol.* 120:998–1005
18. Bach, M. K., Brashler, J. R. 1978. *Life Sci.* 23:2119–26
19. Bach, M. K., Brashler, J. R., Brooks, C. D., Neerken, A. J. 1979. *J. Immunol.* 122:160–65
20. Bach, M. K., Brashler, J. R., Gorman, R. R. 1977. *Prostaglandins* 14:21–38
21. Bach, M. K., Brashler, J. R., Hammarström, S., Samuelsson, B. 1980. *Biochem. Biophys. Res. Commun.* 93: 1121–26
22. Bach, M. K., Brahsler, J. R., Hammarström, S., Samuelsson, B. 1980. *J. Immunol.* 125:115–17
23. Bach, M. K., Brashler, J. R., Johnson, M. A., Drazen, J. M. 1980. *Immunopharmacology* 2:361–73
24. Bach, M. K., Brashler, J. R., Johnson, H. G., McNee, M. L. 1981. *SRS-A and Leukotrienes,* ed. P. J. Piper, pp. 161–80. London: Wiley. 282 pp.
25. Bach, M. K., Brashler, J. R., Morton, D. R., Steel, L. K., Kaliner, M. A., Hugli, T. E. 1982. *Adv. Prostaglandin Thromboxane Res.* 9:103–14
26. Bailey, J. M., Low, C. E., Pupillo, M., Bryant, R. W., Vanderhoek, J. Y. See Ref. 25, pp. 341–53
27. Baker, S. R., Boot, J. R., Dawson, W., Jamieson, W. B., Osborne, D. J., Sweatman, W. J. F. See Ref. 25, pp. 223–27
28. Benveniste, J. 1978. In *Immediate Hypersensitivity, Modern Concepts and Developments,* ed. M. K. Bach, pp. 625–34. New York: Marcel Dekker, 848 pp.
29. Benveniste, J., Camussi, G., Mencia-Huerta, J. M., Polonsky, J. 1977. *Ann. Immunol.* 128C:259–60
30. Benveniste, J., Hadji, L., Jouvin, E., Mencia-Huerta, J. M., Pirotzky, E., Roubin, R. 1981. *Fed. Proc.* 40:1022 (Abstr.)
31. Benveniste, J., Henson, P. M., Cochrane, C. G. 1972. *J. Exp. Med.* 136:1356–77
32. Benveniste, J., Tencé, M., Varenne, P., Bidault, J., Boullet, C., Polonsky, J. 1979. *CR Acad. Sci. Ser D* 289:1037–40
33. Berti, F., Folco, G. C., Giachetti, A., Malandrino, S., Omini, C., Viganò, T. 1980. *Br. J. Pharmacol.* 68:467–72
34. Berti, F., Folco, G. C., Omini, C. 1981. *Atherosclerosis Res.* 8:109–23
35. Beswick, P. H., Kay, A. B. 1981. *Clin. Exp. Immunol.* 43:399–407
36. Betz, S. J., Henderson, P. M. 1980. *J. Immunol.* 125:2756–63
37. Betz, S. J., Lotner, G. Z., Henson, P. M. 1980. *J. Immunol.* 125:2749–55
38. Bhattacherjee, P., Eakins, K. E., Hammond, B. 1981. *Br. J. Pharmacol.* 73:254P–55P
39. Blank, M. L., Snyder, F., Byers, L. W., Brooks, B., Muirhead, E. E. 1979. *Biochem. Biophys. Res. Commun.* 90: 1194–1200
40. Boswell, R. N., Austen, K. F., Goetzl, E. J. 1976. *Immunol. Commun.* 5: 469–79
41. Boswell, R. N., Austen, K. F., Goetzl, E. J. 1978. *J. Immunol.* 120:15–20
42. Boyden, S. 1962. *J. Exp. Med.* 115:453–66
43. Bray, M. A., Cunningham, F. M., Ford-Hutchinson, A. W., Smith, M. J. H. 1981. *Br. J. Pharmacol.* 72:483–86
44. Bray, M. A., Ford-Hutchinson, A. W., Smith, M. J. H. See Ref. 24, pp. 253–70
45. Bretz, U., DeWald, B., Payne, T., Schnyder, J. 1980. *Br. J. Pharmacol.* 71:631–34
46. Brocklehurst, W. E. 1962. *Progr. Allergy* 6:539–58
47. Bryant, R. W., Bailey, J. M., Schewe, T., Rapoport, S. M. 1982. *Abstr. Int. Symp. Leukotrienes and other Lipoxygenase Products, Florence, Italy, June 10–12, 1981,* p. 41
48. Burka, J. F., Garland, L. G. 1977. *Br. J. Pharmacol.* 61:697–99
49. Burka, J. F., Paterson, N. A. M. 1980. *J. Pharm. Pharmacol.* 32:869–70
50. Burke, J. A., Levi, R., Guo, Z. G., Corey, E. J. 1982. *J. Pharmacol. Exp. Ther.* 221:235–41
51. Bussolino, F., Benveniste, J. 1980. *Immunology* 40:367–76
52. Camussi, G., Bussolino, F., Tetta, C., Brusca, R., Ragni, R. 1980. *Panminerva Med.* 22:1–5
53. Camussi, G., Tetta, C., Bussolino, F., Caligaris Cappio, F., Coda, R., Masera, C., Segoloni, G. 1981. *Int. Arch. Allergy Appl. Immunol.* 64:25–41
54. Capron, M., Bazin, H., Joseph, M., Capron, A. 1981. *J. Immunol.* 126: 1764–68
55. Capron, M., Capron, A., Goetzl, E. J., Austen, K. F. 1981. *Nature* 289:71–73

56. Capron, M., Torpier, G., Capron, A. 1979. *J. Immunol.* 123:2220–30
57. Carr, S. C., Higgs, G. A., Salmon, J. A., Spayne, J. A. 1981. *Br. J. Pharmacol.* 73:253P–54P
58. Caulfield, J. P., Lewis, R. A., Hein, A., Austen, K. F. 1980. *J. Cell Biol.* 85:299–311
59. Chand, M., DeRoth, L., Eyre, P. 1979. *Br. J. Pharmacol.* 66:511–16
60. Chand, N., DeRoth, L. 1980. *Am. J. Vet. Res.* 41:101–5
61. Claesson, H.-E., Lundberg, U. See Ref. 47, p. 45
62. Claeys, M., Coene, M.-C., Herman, A. G., Van der Planken, M., Jouvenaz, G. H., Nugteren, D. H. See Ref. 47, p. 46
63. Clark, D. A., Goto, G., Marfat, A., Corey, E. J., Hammarstrom, S., Samuelsson, B. 1980. *Biochem. Biophys. Res. Commun.* 94:1133–39
64. Clark, R. A. F., Gallin, J. I., Kaplan, A. P. 1975. *J. Exp. Med.* 142:1462–76
65. Cockroft, S., Gomperts, B. D. 1979. *Biochem. J.* 178:681–87
66. Copas, J. L., Gardiner, P. J., Borgeat, P. See Ref. 47, p. 52
67. Cromwell, O., Walport, M. J., Morris, H. R., Taylor, G. W., Hodson, M., Batten, J., Kay, A. B. 1981. *Lancet* 2:164–65
68. Cutz, E., Chan, W., Track, N. S., Goth, A., Said, S. I. 1978. *Nature* 275:661–62
69. Czarnetzki, B. M. 1979. *Int. Arch. Allergy Appl. Immunol.* 60:315–24
70. Czarnetzki, B. M. 1981. *Int. Arch. Allergy Appl. Immunol.* 66S:172–73
71. Czarnetzki, B. M., Benveniste, J. 1981. *Agents and Actions.* 11:549–50
72. Czarnetzki, B. M., König, W., Lichtenstein, L. M. 1975. *Nature* 258:725–26
73. Czarnetzki, B. M., Zimmermann, R. E. 1981. *Int. Arch. Allergy Appl. Immunol.* 65:23–26
74. Dahlén, S.-E., Björk, J., Hedqvist, P., Arfors, K.-E., Hammarström, S., Lindgren, J. -Å., Samuelsson, B. 1981. *Proc. Natl. Acad. Sci. USA* 78:3887–91
75. Dahlén, S. E., Hedqvist, P., Hamberg, M. See Ref. 47, p. 48
76. Dahlén, S.-E., Hedqvist, P., Hammarström, S., Samuelsson, B. 1980. *Nature* 288:484–86
77. Demopoulos, C. A., Pinckard, R. N., Hanahan, D. J. 1979. *J. Biol. Chem.* 254:9355–58
78. Diaz, P., Jones, D. G., Kay, A. B. 1979. *Nature* 278:454–56
79. Diaz, P., Jones, D. G., Kay, A. B. 1979. *Clin. Exp. Immunol.* 35:462–69
80. Drazen, J. M., Austen, K. F. 1974. *J. Clin. Invest.* 53:1679–85
81. Drazen, J. M., Austen, K. F., Lewis, R. A., Clark, D. A., Goto, G., Marfat, A., Corey, E. J. 1980. *Proc. Natl. Acad. Sci. USA* 77:4354–58
82. Drazen, J. M., Lewis, R. A., Austen, K. F., Toda, M., Brion, F., Marfat, A., Corey, E. J. 1981. *Proc. Natl. Acad. Sci. USA* 78:3195–98
83. Drazen, J. M., Lewis, R. A., Wasserman, S. I., Orange, R. P., Austen, K. F. 1979. *J. Clin. Invest.* 63:1–5
84. Dvorak, A. M. See Ref. 30, pp. 369–405
85. Engineer, D. M., Niederhauser, U., Piper, P. J., Sirois, P. 1978. *Br. J. Pharmacol.* 62:61–66
86. Engineer, D. M., Piper, P. J., Sirois, P. 1976. *Br. J. Pharmacol.* 57:460P–61P
87. Eyre, P., Deline, T. R. 1976. *Arch. Int. Pharmacodyn.* 222:141–48
88. Eyre, P., Lewis, A. J. 1973. *Br. J. Pharmacol.* 48:426–37
89. Farr, R. S., Cox, C. P., Wardlow, M. L., Jorgensen, R. 1980. *Clin. Immunol. Immunopathol.* 15:318–30
90. Feuerstein, N., Foegh, M., Ramwell, P. W. 1981. *Br. J. Pharmacol.* 72:389–91
91. Folco, G. C., Omini, C., Vigano, T., Brunelli, G., Rossoni, G., Berti, F. See Ref. 25, pp. 153–67
92. Ford-Hutchinson, A. W., Bray, M. A., Doig, M. V., Shipley, M. E., Smith, M. J. H. 1980. *Nature* 286:264–65
93. Deleted in proof
94. Gadek, J. E., Hunninghake, G. W., Zimmerman, R. L., Crystal, R. G. 1980. *Am. Rev. Respir. Dis.* 121:723–33
95. Gershon, R. K., Askenase, P. W., Gershon, M. D. 1975. *J. Exp. Med.* 142:732–47
96. Goetzl, E. J. 1981. *Biochem. Biophys. Res. Commun.* 101:344–50
97. Goetzl, E. J. 1981. *Med. Clin. N. America* 65:809–28
98. Goetzl, E. J., Austen, K. F. 1972. *Immunol. Commun.* 1:421–30
99. Goetzl, E. J., Austen, K. F. 1975. *Proc. Natl. Acad. Sci. USA* 72:4123–27
100. Goetzl, E. J., Derian, C. K., Tauber, A. I., Valone, F. H. 1980. *Biochem. Biophys. Res. Commun.* 94:881–88
101. Goetzl, E. J., Sun, F. F. 1979. *J. Exp. Med.* 150:406–11
102. Goetzl, E. J., Tashjian, A. H. Jr., Rubin, R. H., Austen, K. F. 1978. *J. Clin. Invest.* 61:770–80
103. Greenberg, R., Steinbacher, T., Antonaccio, M. J. 1980. *Fed. Proc.* 39:1103 (Abstr.)
104. Hafstrom, I., Palmblad, J., Malmsten, C. L., Radmark, O., Samuelsson, B. 1981. *FEBS Lett.* 130:146–48

105. Hakanson, R., Rönnberg, A.-L. 1974. *Anal. Biochem.* 60:560–67
106. Halonen, M., Palmer, J. D., Lohman, I. C., McManus, L. M., Pinckard, R. N. 1980. *Am. Rev. Respir. Dis.* 122:915–24
107. Hamberg, M., Hedqvist, P., Radegran, K. 1980. *Acta Physiol. Scand.* 110: 219–21
108. Hamberg, M., Svensson, J., Hedqvist, P., Strandberg, K., Samuelsson, B. 1976. *Adv. Prostaglandin Thromboxane Res.* 1:495–501
109. Hammarström, S. 1980. *J. Biol. Chem.* 255:7093–94
110. Hammarström, S. 1981. *J. Biol. Chem.* 256:2275–79
111. Hammarström, S. 1981. *Biochim. Biophys. Acta* 663:575–77
112. Hammarström, S. See Ref. 25, pp. 83–101
113. Hammarström, S., Murphy, R. C., Samuelsson, B., Clark, D. A., Mioskowski, C., Corey, E. J. 1979. *Biochem. Biophys. Res. Commun.* 91:1266–72
114. Hammarström, S., Samuelsson, B., Clark, D. A., Goto, G., Marfat, A. 1980. *Biochem. Biophys. Res. Commun.* 92:946–53
115. Hanahan, D. J., Demopoulos, C. A., Lieher, J., Pinckard, R. N. 1980. *J. Biol. Chem.* 255:5514–16
116. Hanahan, D. J., Munder, P. G., Satouchi, K., McManus, L. M., Pinckard, R. N. 1981. *Biochem. Biophys. Res. Commun.* 99:183–88
117. Hanna, C. J., Bach, M. K., Pare, P. D., Schellenberg, R. R. 1981. *Nature* 290:343–44
118. Harkavy, J. 1930. *Arch. Intern. Med.* 45:641–46
119. Hayes, E. C., Young, R. N., Maycock, A. L., Rokach, J., Rosenthal, A. S. See Ref. 47, p. 49
120. Hedqvist, P., Dahlén, S.-E., Gustafsson, L., Hammarström, S., Samuelsson, B. 1980. *Acta Physiol. Scand.* 110:331–33
121. Henderson, W. R., Kaliner, M. 1979. *J. Immunol.* 122:1322–28
122. Henson, P. M., Pinckard, R. N. 1977. *Monogr. Allergy* 12:13–26
123. Hirata, F., Axelrod, J. 1978. *Nature* 275:219–20
124. Hirata, F., Axelrod, J. 1980. *Science* 209:1082–90
125. Hirata, F., Corcoran, B. A., Venkatasubramanian, K., Schiffmann, E., Axelrod, J. 1979. *Proc. Natl. Acad. Sci. USA* 76:2640–43
126. Hirata, F., Schiffmann, E., Venkatasubramanian, K., Salomon, D., Axelrod, J. 1980. *Proc. Natl. Acad. Sci. USA* 77: 2533–36
127. Holgate, S. T., Lewis, R. A., Austen, K. F. 1980. *J. Immunol.* 124:2093–99
128. Holme, G., Brunet, G., Piechuta, H., Masson, P., Girard, Y., Rokach, J. 1980. *Prostaglandins* 20:717–28
129. Holme, G., Piechuta, H. 1981. *Immunology* 42:19–24
130. Holroyde, M. C., Altounyan, R. E. C., Cole, M., Dixon, M., Elliott, E. V. See Ref. 25, pp. 237–42
131. Hunninghake, G. W., Gadek, J. E., Fales, H. M., Crystal, R. G. 1980. *J. Clin. Invest.* 66:473–83
132. Hwang, D. H., Bryant, R. W. See Ref. 47, p. 59
133. Ishizaka, T. 1981. *J. Allergy Clin. Immunol.* 67:90–96
134. Ishizaka, T., Hirata, F., Ishizaka, K., Axelrod, J. 1980. *Proc. Natl. Acad. Sci. USA* 77:1903–6
135. Ishizaka, T., Ishizaka, K., Conrad, D. H., Froese, A. 1978. *J. Allergy Clin. Immunol.* 61:320–30
136. Jakschik, B. A., Falkenhein, S., Parker, C. W. 1977. *Proc. Natl. Acad. Sci. USA* 74:4577–81
137. Jakschik, B. A., Kulczycki, A. Jr., MacDonald, H. H., Parker, C. W. 1977. *J. Immunol.* 119:618–22
138. Jakschik, B. A., Lee, L. H. 1980. *Nature* 287:51–52
139. Jakschik, B. A., Sams, A. R., Sprecher, H., Needleman, P. 1980. *Prostaglandins* 20:401–10
140. Jakschik, B. A., DiSantis, D. M., Sankarappa, S. K., Sprecher, H. See Ref. 25, pp. 127–35
141. Jakschik, B. A., Sun, F. F., Lee, L.-H., Steinhoff, M. M. 1980. *Biochem. Biophys. Res. Commun.* 95:103–10
142. Jones, D. G., Kay, A. B. 1977. *Int. Arch. Allergy Appl. Immunol.* 55: 277–82
143. Joseph, M., Tonnel, A. B., Capron, A., Voisin, C. 1980. *Clin. Exp. Immunol.* 40:416–22
144. Jubiz, W., Radmark, O., Lindgren, J. A., Malmsten, C., Samuelsson, B. 1981. *Biochem. Biophys. Res. Commun.* 99:976–86
145. Kaliner, M., Sigler, R., Summers, R., Shelhamer, J. 1981. *J. Allergy Clin. Immunol.* 68:365–71
146. Kay, A. B., Anwar, A. R. E. 1980. In *The Eosinophil in Health and Disease,* ed. A. A. F. Mahmoud, K. F. Austen, pp. 207–28. New York: Grune & Stratton. 364 pp.
147. Kay, A. B., Roberts, E. M., Jones, D. G. 1976. *Immunology* 30:83–87

148. Kay, A. B., Stechschulte, D. J., Austen, K. F. 1971. *J. Exp. Med.* 133:602–19
149. Kazimierczak, W., Diamant, B. 1978. *Prog. Allergy* 24:295–365
150. Kellaway, C. H., Trethewie, E. R. 1940. *Q. J. Exp. Physiol.* 30:121–45
151. Kennerly, D. A., Sullivan, T. J., Sylwester, P., Parker, C. W. 1979. *J. Exp. Med.* 150:1039–44
152. Keller, H. U., Wilkinson, R. C., Abercrombie, M., Becker, E. L., Hirsch, J. G., Miller, M. E., Ramsey, W. S., Zigmond, S. H. 1977. *J. Immunol.* 118: 1912–14
153. Kelly, J. F., Cugell, D. W., Patterson, R., Harris, K. E. 1974. *J. Lab. Clin. Med.* 83:738–49
154. Klickstein, L. B., Shapleigh, C., Goetzl, E. J. 1980. *J. Clin. Invest.* 66:1166–70
155. König, W., Czarnetzki, B. M. 1978. *Int. Arch. Allergy Appl. Immunol.* 57:399–410
156. König, W., Czarnetzki, B. M., Lichtenstein, L. M. 1978. *Int. Arch. Allergy Appl. Immunol.* 56:364–75
157. König, W., Kunau, H. W., Borgeat, P. See Ref. 25, pp. 301–14
158. König, W., Tesch, H., Frickhofen, N. 1978. *Eur. J. Immunol.* 8:434–37
159. Kownatzki, E., Till, G., Gemsa, D. 1979. *Monogr. Allerg* 14:130–33
160. Lett-Brown, M. A., Thueson, D. O., Grant, J. A. 1981. *Int. Arch. Allergy Appl. Immunol.* 64:241–48
161. Levi, R., Burke, J. A. 1980. *Eur. J. Pharmacol.* 62:41–49
162. Levine, L., Morgan, R., Lewis, R. A., Austen, K. F., Clark, D. A., Marfat, A., Corey, E. J. 1981. *Proc. Natl. Acad. Sci. USA* 78:7692–96
163. Levy, D. A., Widra, M. 1973. *J. Lab. Clin. Med.* 81:291–97
164. Lewis, R. A., Austen, K. F., Drazen, J. M., Soter, N. A., Figueriedo, J. C., Corey, E. J. See Ref. 25, pp. 137–51
165. Lewis, R. A., Drazen, J. M., Austen, K. F., Clark, D. A., Corey, E. J. 1980. *Biochem. Biophys. Res. Commun.* 96: 271–77
166. Lewis, R. A., Drazen, J. M., Austen, K. F., Toda, M., Brion, F., Marfat, A., Corey, E. J. 1981. *Proc. Natl. Acad. Sci. USA* 78:4579–83
167. Lewis, R. A., Drazen, J. M., Corey, E. J., Austen, K. F. See Ref. 24, pp. 101–17
168. Lewis, R. A., Goetzl, E. J., Wasserman, S. I., Valone, F. H., Rubin, R. H., Austen, K. F. 1975. *J. Immunol.* 114:87–92
169. Lewis, R. A., Soter, N. A., Corey, E. J., Austen, K. F. 1981. *Clin. Res.* 29:492A (Abstr.)
170. Lindgren, J. Å., Hansson, G., Samuelsson, B. 1981. *FEBS Lett.* 128:329–35
171. Lichtenstein, L. M. See Ref. 47, p. 31
172. Lichtenstein, L. M. 1981. *Int. Cong. Recent Adv. Clin. Immunol., Naples, June,* ed. G. Marone. Academic Press. In press
173. Lundberg, U., Radmark, O., Malmsten, C., Samuelsson, B. 1981. *FEBS Lett.* 126:127–32
174. Lynch, S. M., Austen, K. F., Wasserman, S. I. 1978. *J. Immunol.* 121: 1394–99
175. Maas, R. L., Brash, A. R., Oates, J. A. See Ref. 25, pp. 29–44
176. Malmsten, C. L., Palmblad, J., Udén, A.-M., Radmark, O., Engstedt, L., Samuelsson, B. 1980. *Acta Physiol. Scand.* 10:449–51
177. Marone, G., Kagey-Sobotka, A., Lichtenstein, L. M. 1979. *J. Immunol.* 123:1669–77
178. Mathe, A. A., Strandberg, K., Yen, S. S. 1977. *Prostaglandins* 14:1105–15
178a. Marom, Z., Shelhamer, J. H., Bach, M. K., Morton, D. R. Jr., Kaliner, M. 1982. *Am. Rev. Resp. Dis.* In press
179. Marom, Z., Shelhamer, J. H., Kaliner, M. 1981. *J. Clin. Invest.* 67:1695–702
180. Mazingue, C., Dessaint, J. P., Capron, A. 1978. *J. Immunol. Methods* 21: 65–77
181. MacManus, L. M., Hanahan, D. J., Demopoulos, C. A., Pinckard, R. N. 1980. *J. Immunol.* 124:2919–24
182. McManus, L. M., Morley, C. A., Levine, S. P., Pinckard, R. N. 1979. *J. Immunol.* 123:2835–41
183. Mencia-Huerta, J. M., Benveniste, J. 1979. *Eur. J. Immunol.* 9:409–15
184. Mencia-Huerta, J. M., Ninio, E., Landes, A., Godfroid, J. J., Benveniste, J. 1981. *Fed. Proc.* 40:1015 (Abstr.)
185. Metzger, H., Bach, M. K. See Ref. 30, pp. 561–88
186. Moncada, S., Gryglewski, R. J., Bunting, S., Vane, J. R. 1976. *Prostaglandins* 12:715–37
187. Moncada, S., Vane, J. R. 1979. In *Prostacyclin,* ed. J. R. Vane, S. Bergström, pp. 5–16. New York: Raven. 453 pp.
188. Morris, H. R., Taylor, G. W., Piper, P. J., Samhoun, M. N., Tippins, J. R. 1980. *Prostaglandins* 19:185–201
189. Morris, H. R., Taylor, G. W., Piper, P. J., Tippins, J. R. 1978. *FEBS Lett.* 87:203–6
190. Morris, H. R., Taylor, G. W., Piper, P. J., Tippins, J. R. 1980. *Nature* 285: 104–6
191. Murphy, R. C., Hammarstrom, S.,

Samuelsson, B. 1979. *Proc. Natl. Acad. Sci. USA* 76:4275–79
192. Myers, G., Danlon, M., Kaliner, M. 1982. *J. Allergy Appl. Immunol.* 67:305–11
193. Narumiya, S., Salmon, J. A., Flower, R. J., Vane, J. R. See Ref. 25, pp. 77–82
194. Newball, H. H., Berninger, R., Talamo, R. C., Lichtenstein, L. M. 1979. *J. Clin. Invest.* 64:457–65
195. Newball, H. H., Meier, H. L., Lichtenstein, L. M. 1980. *J. Invest. Dermatol.* 74:344–48
196. Newball, H. H., Meier, H. L., Kaplan, A. P., Sevak, S. D., Cochrane, C. G., Lichtenstein L. M. 1981. *Trans Assoc. Am. Physicians.* 94:126–34
197. Newball, H. H., Talamo, R. C., Lichtenstein, L. M. 1979. *J. Clin. Invest.* 64:466–75
198. Oertel, H., Kaliner, M. 1981. *J. Allergy Clin. Immunol.* 68:238–45
199. Oertel, H. L., Kaliner, M. 1981. *J. Immunol.* 127:1398–403
200. Ohnishi, H., Kusuzume, H., Kitamura, Y., Yamaguchi, K., Nobuhara, M., Suzuki, Y., Yoshida, S., Tomioka, H., Kumagai, A. 1980. *Prostaglandins* 20: 655–66
201. Omini, C., Folco, G. C., Viganó, T., Rossoni, G., Brunelli, G., Berti, F. 1981. *Pharmacol. Res. Commun.* 13: 633–40
202. Omini, C., Sautebin, L., Galli, G., Folco, G. C., Berti, F. 1980. *Adv. Prostaglandin Thromboxane Res.* 7:927–31
203. Orange, R. P., Austen, K. F. 1969. *Adv. Immunol.* 10:105–44
204. Orange, R. P., Austen, K. F. 1976. *Methods Immunol. Immunochem.* 5: 145–49
205. Orange, R. P., Chang, P.-L. 1975. *J. Immunol.* 115:1072–77
206. Orange, R. P., Moore, E. G. 1976. *J. Immunol.* 116:392–97
207. Orange, R. P., Moore, E. G., Gelfand, E. W. 1980. *J. Immunol.* 124:2264–67
208. Orange, R. P., Murphy, R. C., Austen, K. F. 1974. *J. Immunol.* 113:316–22
209. Orange, R. P., Murphy, R. C., Karnovsky, M. L., Austen, K. F. 1973. *J. Immunol.* 110:760–70
210. Orange, R. P., Valentine, M. D., Austen, K. F. 1967. *Science* 157:318–19
211. Örning, L., Hammarström, S., Samuelsson, B. 1980. *Proc. Natl. Acad. Sci. USA* 77:2014–17
212. Palmer, M. R., Mathews, R., Murphy, R. C., Hoffer, B. J. 1980. *Neurosci. Lett.* 18:173–80
213. Parker, C. W. 1979. *J. Allergy Clin. Immunol.* 63:1–14
214. Parker, C. W., Falkenhein, S. F., Huber, M. M. 1980. *Prostaglandins* 20:863–86
215. Parker, C. W., Fischman, C. M., Wedner, H. J. 1980. *Proc. Natl. Acad. Sci. USA* 77:6870–73
216. Parker, C. W., Huber, M. M., Hoffman, M. K., Falkenhein, S. F. 1979. *Prostaglandins* 18:673–86
217. Parker, C. W., Koch, D. A., Huber, M. M., Falkenhein, S. F. 1980. *Prostaglandins* 20:887–908
218. Parker, C. W., Koch, D., Huber, M. M., Falkenhein, S. F. 1980. *Biochem. Biophys. Res. Commun.* 94:1037–43
219. Parker, C. W., Koch, D., Huber, M. M., Falkenhein, S. F. 1980. *Biochem. Biophys. Res. Commun.* 97:1038–46
220. Parsons, J. F., Safford, R. J., Philp, J. 1980. *J. Immunol. Methods* 33:183–93
221. Patterson, R., Harris, K. E., Greenberger, P. A. 1980. *J. Allergy Clin. Immunol.* 65:269–73
222. Payan, D. G., Goetzl, E. J. 1981. *J. Clin. Immunol.* 1:266–70
223. Peck, M. J., Piper, P. J., Williams, T. J. 1981. *Prostaglandins* 21:315–21
224. Peters, S. P., Siegel, M. I., Kagey-Sobotka, A., Lichtenstein, L. M. 1981. *Nature* 292:455–57
225. Pinckard, R. N., McManus, L. M., Demopoulos, C. A., Halonen, M., Clark, P. O., Shaw, J. O., Kniker, W. T., Hanahan, D. J. 1980. *J. Reticuloendothel. Soc.* 28:95S–103S
226. Pinckard, R. N., McManus, L. M., Hanahan, D. J. 1982. *Adv. Inflammation Res.* In press
227. Piper, P. J. 1977. *Pharmacol. Ther. B* 31:75–98
228. Piper, P., Vane, J. 1971. *Ann. NY Acad. Sci.* 180:363–83
229. Rådmark, O., Malmsten, C., Samuelsson, B. 1980. *Biochem. Biophys. Res. Commun.* 96:1679–87
230. Rapoport, S. M., Schewe, T., Wiesner, R., Halangk, W., Ludwig, P., Janicke-Höhne, M., Tannert, C., Hiebsch, C., Klatt, D. 1979. *Eur. J. Biochem.* 96:545–61
231. Roberts, L. J. II, Lewis, R. A., Oates, J. A., Austen, K. F. 1979. *Biochim. Biophys. Acta* 575:185–92
232. Roberts, L. J. II, Sweetman, B. J., Lewis, R. A., Austen, K. F., Oates, J. A. 1980. *N. Engl. J. Med.* 303:1400–404
233. Rocklin, R. E. 1976. *J. Clin. Invest.* 57:1051–58
234. Rocklin, R. E. 1977. *J. Immunol.* 118:1734–38
235. Rocklin, R. E. 1979. *Monogr. Allergy* 14:134–37

236. Rouzer, C. A., Scott, W. A., Cohn, Z. A., Blackburn, P., Manning, J. M. 1980. *Proc. Natl. Acad. Sci. USA* 77:4928–32
237. Rouzer, C. A., Scott, W. A., Griffith, O. W., Hamill, A. L., Cohn, Z. A. 1981. *Proc. Natl. Acad. Sci. USA* 78:2532–36
238. Samuelsson, B. See Ref. 25, pp. 1–17
239. Samuelsson, B., Borgeat, P., Hammarström, S., Murphy, R. C. 1980. *Adv. Prostaglandin Thromboxane Res.* 6: 1–18
240. Samuelsson, B., Hammarström, S. 1980. *Prostaglandins* 19:645–49
241. Sánchez-Crespo, M., Alonso, F., Egido, J. 1980. *Immunology* 40:645–55
242. Satouchi, K., Pinckard, R. N., McManus, L. M., Hanahan, D. J. 1981. *J. Biol. Chem.* 256:4425–32
243. Schulman, E. S., Adkinson, N. F., Demers, L. M., Fitzpatrick, F. A. 1980. *Fed. Proc.* 39:444 (Abstr.)
244. Schwartz, A., Askenase, P. W., Gershon, R. K. 1980. *Immunopharmacology* 2:179–90
245. Schwartz, A., Sutton, S. L., Askenase, P. W., Gershon, R. K. 1981. *Cell. Immunol.* 60:426–39
246. Schwartz, L. B., Austen, K. F. 1980. *J. Invest. Dermatol.* 74:349–53
247. Schwartz, L. B., Lewis, R. A., Seldin, D., Austen, K. F. 1981. *J. Immunol.* 126:1290–94
248. Shaw, J. O., Henson, P. M. 1980. *Am. J. Pathol.* 98:791–810
249. Shaw, J. O., Klusick, S. J., Hanahan, D. J. 1981. *Biochim. Biophys. Acta* 663:222–29
250. Shearer, G. M., Melmon, K. L., Weinstein, Y., Sela, M. 1972. *J. Exp. Med.* 136:1302–7
251. Shelhamer, J. H., Marom, Z., Kaliner, M. 1980. *J. Clin. Invest.* 66:1400–8
252. Shelhamer, J. H., Marom, Z., Sun, F., Bach, M. K., Kaliner, M. 1982. *Chest.* 81:36S
253. Shore, P. A., Burkhalter, A., Cohn, V. H. Jr. 1959. *J. Pharmacol. Exp. Therap.* 127:182–86
254. Siegel, M. I., McConnell, R. T., Abrahams, S. L., Porter, N. A., Cuatrecasas, P. 1979. *Biochem. Biophys. Res. Commun.* 89:1273–80
255. Siraganian, R. P. 1974. *Anal. Biochem.* 57:383–94
256. Sirois, P. 1979. *Prostaglandins* 17:395–404
257. Sirois, P., Borgeat, P. 1980. *Int. J. Immunopharmacol.* 2:281–93
258. Sirois, P., Roy, S., Borgeat, P. 1981. *Prostaglandins Med.* 6:153–59
259. Sirois, P., Roy, S., Borgeat, P., Picard, S., Corey, E. J. 1981. *Biochem. Biophys. Res. Commun.* 99:385–90
260. Smart, L. M., Kay, A. B. 1981. *Clin. Exp. Immunol.* 44:581–86
261. Smith, J. A., Goetzl, E., Austen, K. F. 1980. *Peptides: Structure and Biological Function,* ed. E. Gross, J. Meienhofer, pp. 753–56. Rockford, IL: Pierce Chem. Co. 1079 pp.
261a. Smith, M. J. H. See Ref. 25, pp. 283–92
262. Smith, M. J., Ford-Hutchinson, A. W., Bray, M. A. 1980. *J. Pharm. Pharmacol.* 32:517–18
263. Sok, D.-E., Pai, J.-K., Atrache, V., Sih, C. J. 1980. *Proc. Natl. Acad. Sci. USA* 77:6481–85
264. Soter, N. A., Austen, K. F. 1976. *J. Invest. Dermatol.* 67:313–19
265. Soter, N. A., Austen, K. F. 1977. *Fed. Proc.* 36:1736–41
266. Steel, L., Kaliner, M. 1981. *J. Biol. Chem.* 256:12692–98
267. Steel, L. K., Waxdal, M., Cohen, S., Kaliner, M. 1980. *Fed. Proc.* 39:444 (Abstr.)
268. Stenson, W. F., Parker, C. W. 1979. *Prostaglandins* 18:285–92
269. Stenson, W. F., Parker, C. W. 1979. *J. Clin. Invest.* 64:1457–65
270. Stenson, W. F., Parker, C. W. 1980. *J. Immunol.* 124:2100–4
271. Stewart, J., Kay, A. B. 1980. *Clin. Exp. Immunol.* 40:423–26
272. Tannenbaum, S., Oertel, H., Henderson, W., Kaliner, M. 1980. *J. Immunol.* 125:325–35
273. Theoharides, T. C., Askenase, P. W. 1980. *Eur. J. Cell Biol.* 22:181 (Abstr.)
274. Theoharides, T. C., Sieghart, W., Greengard, P., Douglas, W. W. 1980. *Science* 207:80–81
275. Thomson, D. M. P., Phelan, K., Bach, M. K. 1982. In *1st Int. Conf. Prostaglandins and Cancer,* pp. 679–84. New York: Liss
276. Turnbull, L. S., Turnbull, L. W., Leitch, A. G., Crofton, J. W., Kay, A. B. 1977. *Lancet* 2:526–29
277. Turnbull, L. W., Evans, D. P., Kay, A. B. 1977. *Immunology* 32:57–63
278. Turnbull, L. W., Kay, A. B. 1977. *Immunology* 31:797–802
279. Valone, F. H. 1980. *Clin. Immunol. Immunopathol.* 15:52–65
280. Valone, F. H., Goetzl, E. J. 1978. *J. Immunol.* 120:102–8
281. Vanderhoek, J. Y., Bryant, R. W., Bailey, J. M. 1980. *J. Biol. Chem.* 255: 10064–65
282. Vanderhoek, J. Y., Bryant, R. W., Bailey, J. M. See Ref. 47, p. 94

283. Wanner, A., Zarzecki, S., Hirsch, J., Epstein, S. 1975. *J. Appl. Physiol.* 39:950–57
284. Ward, P. A., Becker, E. L. 1968. *J. Exp. Med.* 127:693–709
285. Wasserman, M. A., DuCharme, D. W., Griffin, R. L., DeGraaf, G. L., Robinson, F. G. 1977. *Prostaglandins* 13: 255–69
286. Wasserman, M. A., Griffin, R. L. 1977. *Eur. J. Pharmacol.* 46:303–13
287. Wasserman, M. A., Griffin, R. L., Marsilisi, F. B. 1980. *Prostaglandins* 20: 703–15
288. Wasserman, S. I., Austen, K. F. 1977. *J. Biol. Chem.* 252:7074–80
289. Wasserman, S. I., Center, D. M. 1979. *J. Allergy Clin. Immunol.* 64:231–34
290. Wasserman, S. I., Goetzl, E. J., Austen, K. F. 1974. *J. Immunol.* 112:351–58
291. Wasserman, S. I., Soter, N. A., Center, D. M., Austen, K. F. 1977. *J. Clin. Invest.* 60:189–96
292. Weichman, B. M., Hosteley, L. S., Bostick, S. P., Mucitelli, R. M., Krell, R. D., Gleason, S. G. 1982. *J. Pharmacol. Exp. Ther.* 221:295–302
293. Weller, P. F., Lewis, R. A., Corey, E. J., Austen, K. F. 1981. *Fed. Proc.* 40:1023 (Abstr.)
294. Wells, P. W., Eyre, P. 1972. *Immunol. Commun.* 1:105–11
295. Welton, A. F., Crowley, H. J., Miller, D. A., Yaremko, B. 1981. *Prostaglandins* 21:287–96
296. Winslow, C. M., Austen, K. F. 1982. *Fed. Proc.* 40:22–29
297. Yecies, L. D., Wedner, H. J., Johnsons, S. M., Jakschik, B. A., Parker, C. W. 1979. *J. Immunol.* 122:2083–89
298. Yoshida, N., Everitt, M. T., Neurath, H., Woodbury, R. G., Powers, J. C. 1980. *Biochemistry* 19:5799–804
299. Yurt, R. W., Leid, R. W. Jr., Spragg, J., Austen, K. F. 1977. *J. Immunol.* 118:1201–7

Ann. Rev. Microbiol. 1982. 36:415–33

MICROBIOLOGICAL MODELS AS SCREENING TOOLS FOR ANTICANCER AGENTS: POTENTIALS AND LIMITATIONS

Richard J. White

Lederle Laboratories, Medical Research Division, American Cyanamid Company, Pearl River, New York 10965

CONTENTS

INTRODUCTION

Cancer is a collection of over 100 diseases, and in spite of our ability to successfully treat certain types of malignant tumors, others remain somewhat refractory. This is especially true for some of the more commonly occurring solid tumors, e.g. breast, lung, and colorectal cancer. Chemotherapy plays an important role in cancer treatment both alone and in combination with surgery and radiation (8). There is an urgent need for novel chemotherapeutic agents with a broadened, or different, spectrum of antitu-

0066-4227/82/1001-0415$02.00

mor activity and/or diminished toxicity to the host. This review examines the role in vitro microbiological test systems can play in the quest for new antitumor agents, emphasizing their potentials and limitations. The screening process consists of the evaluation of a large number of compounds and subsequent selection for further testing of a small number that have greater clinical potential, the majority being rejected as inactive. An ideal screen would select only those compounds with clinical activity in man (true positives) and would reject everything else (true negatives). In reality, of course, every screen will miss some actives (false negatives) and some of the compounds selected will turn out to be inactive (false positives). A measure of the efficiency of a screening system is the frequency with which it yields false negatives and false positives.

NEED FOR IN VITRO MODELS

At present, activity in an animal tumor model is a prerequisite for progression of any new drug to the clinic. This requirement would seem reasonable since all drugs with established clinical efficacy possess activity in several animal tumor models. Furthermore, a single animal model, such as murine P388 leukemia, is capable of detecting activity for the majority (96%) of clinically active drugs (28). However, this is not surprising since these drugs were selected for clinical trials based on activity in such models. It is not known how many false negatives there have been, i.e. drugs inactive against the typical murine leukemias used as in vivo screens, but active against one or more of the solid tumors in mice or humans (27, 28, 62).

In view of the importance attached to animal models, why is it necessary to consider use of in vitro screening systems at all? To answer this question, consideration must be given to the types of compounds to be screened. Compounds under investigation can be of synthetic or natural origin and the problems faced in these two cases are rather different. Synthetic compounds are usually available as defined chemical entities and can be tested directly in vivo. Natural products, on the other hand, are initially tested as complex mixtures of which the active component may only represent 0.1% of total solids (33). Therefore, to be useful in screening natural products, a test system must be both sensitive and selective, i.e. capable of detecting low concentrations but not subject to interference by high levels of inactive or even antagonistic compounds. Available in vivo animal models are not usually as sensitive as their in vitro counterparts; furthermore, they are costly and time-consuming (4, 5, 34). All three of these factors severely limit their applicability to direct initial screening of natural products found in microbial fermentation broths and plant extracts. Once the active component has been isolated from such crude mixtures, it is on a par with synthetic materials and can be more meaningfully tested in vivo. In vitro test systems used for the initial detection of activity are commonly referred to as prescreens. The National Cancer Institute (USA) has played an important role in helping develop and evaluate new prescreening techniques, and the results of this effort for the natural products area have recently been reviewed (10, 11).

SPECIFIC MICROBIOLOGICAL PRESCREENS

A considerable number of microbiological test systems have been used, or have potential to be used, as prescreens for anticancer agents. These different approaches are described separately, along with their specific potentials and limitations. This is followed by some more general comments on microbiological models.

Virus-Based Prescreens

PROPHAGE INDUCTION In his classical review of lysogeny in 1953, Lwoff (44) pointed out the potential of prophage induction as a screen for mutagenic, carcinogenic, and carcinostatic agents. Investigators at Bristol Laboratories were the first to report use of prophage induction to prescreen microbial fermentation broths for anticancer antibiotics (42). The original findings have been confirmed and extended by the Bristol group (36, 37, 54) and by workers at other laboratories (13, 14, 25, 65). Over a period of years, the prophage induction test has been improved with respect to sensitivity and practicability. In the original test, lysogenic bacteria were first incubated in liquid medium with the test sample to allow induction of the prophage, and in a second stage, the bacteriophage produced were enumerated by a plaque assay procedure (54). A simplified single stage test was later described by Japanese workers (40). In this system, samples were placed directly on the surface of agar containing a defined ratio of lysogenic and indicator bacteria. Areas of increased plaque formation were evident around inducing samples; with certain antibiotics an antibacterial effect produced a clear zone inside an area of increased plaque formation. As a means of increasing the sensitivity of the prophage induction assay, Moreau and co-workers (48) introduced mutations affecting permeability (envA) and DNA repair (uvrB). The combined effect of these two mutations was to increase the sensitivity toward aflatoxin B metabolites by 4000-fold. Moreau et al (48) also introduced the use of microsomal activation in conjunction with prophage induction. Although the objective of this work was to improve the applicability of prophage induction as a screen for mutagens and potential carcinogens, the modifications introduced are of interest in the detection of anticancer agents as well.

Normally, quantitation of prophage induction involves counting plaques. The recent advent of biochemical versions of the prophage induction assay now permits a more rapid quantitation by direct measurement of enzyme activity. In the three versions of this test so far described, the structural gene for a specific bacterial enzyme was placed under the negative control of the prophage repressor, and transcription was initiated at the efficient λ prophage P_L promoter. In the first of these (64), the bacterial gene was carried by a temperate phage (λ or φ80), which was used to superinfect an *Escherichia coli* strain lysogenic for the homologous prophage. The enzyme involved in this case was anthranilate synthetase, coded for by the trp ED genes of the tryptophan operon, which was carried by the superinfecting phage. Normal transcription from the tryptophan promoter was prevented by adding tryptophan to the growth medium. Transcription of the tryptophan operon occurred only as a result of read-through from the prophage P_L promoter when induction occurred. Thus, the process of prophage induction could be studied by following the synthesis of anthranilate synthetase, directed by genes belonging to the tryptophan operon.

In the second biochemical prophage induction assay (43), the host's galactose operon was put under negative control of the prophage λ repressor by a deletion of intervening genes. The normal bacterial promoter for this operon was inactivated by a mutation, and thus transcription could only be initiated at the prophage P_L promoter. As a measure of prophage induction, galactokinase activity was assayed.

In a third version of this test (12), a λ prophage with an integrated lacZ gene was used. The normal host lacZ gene was deleted. Synthesis of β-galactosidase (the lacZ gene product) was under the negative control of the prophage λ repressor. In the latter tests involving galactokinase and B-galactosidase, the prophages were cryptic because of deletions of genes under the control of the right-hand promoter P_R. Although transcription of the genes under the control of promoter P_L occurred on induction (release from repression), no replication, lysis, or phage production could occur. In the first biochemical test described, which involved anthranilate synthetase (64), production of intact phage and enzyme occurred as a result of induction. The major difference between the three biochemical induction assays is the ease with which the respective enzymes can be assayed. On this basis, the test monitoring induction of β-galactosidase is the most convenient. The sensitivity of any version of the prophage induction assay can be improved by incorporation of appropriate permeability and DNA repair mutations such as those described by Moreau et al (48).

Irrespective of the version of the prophage induction test used, the same types of compounds can be detected. All the compounds that caused induction interfere with DNA synthesis. The correlation between prophage induction and anticancer activity was impressive both for a series of related compounds (1, 14, 35) and for compounds of diverse structural classes (35, 36, 54). However, similar correlations have also been reported for prophage induction and carcinogenicity for a series of related nitroquinolines (15). The number of false positives generated by such a screen was rather small and not a practical problem (20, 54). For instance, nalidixic acid is a prophage inducer at high concentrations but is (presumably) devoid of anticancer activity (35). Other examples of false positives are the sulfhydryl-containing compounds such as cysteine (36). In most cases, however, the concentration required for induction was high (>50 μg/ml) and was unlikely to be encountered during a natural products screening program. A more serious problem is the number of false negatives, i.e. compounds that do not cause induction but do have anticancer activity. Although this is to be expected for those compounds whose mechanism of action does not involve any direct effect on DNA, e.g. vincristine, a microtubule inhibitor, this is somewhat surprising for others that are known to interact with DNA, e.g. actinomycin D. One possible explanation for the failure of such compounds to cause induction is that they are potent transcription inhibitors, thus precluding production of virus or synthesis of enzyme.

Antibiotics known to cause induction of prophage in lysogenic bacteria are listed in Table 1, whereas Table 2 lists those antibiotics that failed to cause induction at concentrations that did not inhibit bacterial growth. The fact that an antibiotic is qualitatively classified as a prophage inducer does not necessarily mean this will be an effective means of detecting it in microbial fermentations. Certain of the antibiotics listed as inducers in Table 1 are unlikely to occur in fermentation broths at sufficiently high

Table 1 Induction of lysogenic bacteria: Anticancer fermentation products with prophage inducing activity[a]

Product	Minimum inducing concentration (μg/ml)	Product	Minimum inducing concentration (μg/ml)
Mitomycin C	0.004	Baumycin	0.2
Neocarzinostatin	0.006	Sibiromycin	0.2
Phleomycin	0.006	Pluramycin	0.5
Streptonigrin	0.008	Adriamycin	0.8
Bleomycin	0.0125	Carminomycin	0.8
Hedamycin	0.0125	Daunomycin	0.8
Porfiromycin	0.032	Streptozocin	1.0
Azaserine	0.032	Iyomycin	4.0
Figaroic acid	0.05	Valinomycin	12.5
Macromomycin	0.05		

[a] From Bradner (4).

Table 2 Induction of lysogenic bacteria: Antibiotics that fail to induce prophage at tolerated levels[a]

Antibiotic	Antibiotic
Actinomycin D	Pyrromycin type anthracyclines
Anthramycin	Pyrromycin
Aureolic acids	Mussettamycin
Borrelidin	Marcellomycin
Kundraymycin	Cinerubin A
Anguidine	Trypanomycin
Muconomycin A	Violamycin
Sterigmatocystin	
Aclacinomycin	

[a] From Bradner (4).

concentrations to cause induction. One group that used a biochemical prophage induction assay as a prescreen for anticancer agents suggested an arbitrary threshold concentration of 25 μg/ml to indicate which compounds were likely to be detected under their assay conditions (20). Given these detection limits, antibiotics such as daunorubicin, adriamycin, and streptozocin might be missed. Because of this, any change that would lead to an increased sensitivity would be advantageous.

ANTIPHAGE ACTIVITY The application of prophage induction to screening for anticancer agents revealed that inducing compounds were also capable of inhibiting phage multiplication. This in turn led to the evaluation of antiphage activity as a prescreen. Aoki & Sakai (2, 61) have described a test system using *Bacillus subtilis* and a clear plaque mutant of the temperate bacteriophage SP10. Conditions were arranged so that assay plates gave confluent lysis. Using this assay, various antibiotics and synthetic compounds were tested for their ability to inhibit lysis. Of the 33 agents tested, 28 showed antiphage activity at concentrations that still permitted growth of *B. subtilis.* Although the test proved effective in detecting all the antitumor antibiotics tested (no false negatives reported), it also gave positive results for other antibiotics that are known not to have antitumor activity, such as erythromycin, kanamycin, tetracycline, and chloramphenicol. The authors maintained that antiphage activity was a better indication of antitumor potential than prophage induction because of its ability to pick up antitumor antibiotics that were not prophage inducers, such as actinomycin D and chromomycin A_3.

BIP TEST Fleck (16) has described a prescreen in which several different effects can be simultaneously monitored, namly antibacterial activity, antiphage activity, and inhibition of establishment of lysogeny (referred to as the BIP test, from the German words Bakterienwachstum, Induktionsgeschehen, Phagenvermehrung). In this test, a nonlysogenic *Escherichia coli* was infected with a high multiplicity of phage λ and the resultant mixture was considered by Fleck (16) to contain lysogenic, λ-sensitive and λ-resistant bacteria. All compounds tested that are known to cause the induction of prophage also caused increased plaque formation in the BIP test (17). This test was used to screen for natural products that increased or decreased plaque formation and consequently it detected a greater number of positives than did prophage induction alone. From a total of 15,000 crude fermentation broths tested for BIP and prophage induction activity, the former gave 4.8% positives and the latter gave 2.0%. The frequency of prophage-induction positives was similar to that recently quoted for application of a biochemical version of this assay (17). However, comparisons of this sort can be misleading because of the different producer organisms, fermentation media, and assay conditions used.

ACTIVITY AGAINST ANIMAL VIRUSES The correlation between antiviral and cytotoxic activity of microbial fermentation broths has been studied by screening for activity against Newcastle disease virus growing in primary cultures of chick embryo fibroblasts (67). By examining only those broths devoid of antibacterial activity (previously ascertained) but that inhibited Newcastle disease virus, most of the known antiviral antibiotics could be avoided. A total of 850 actinomycete strains screened for activity yielded 12% actives against Newcastle disease virus and only 25% of these had antibacterial activity. These same cultures screened for cytotoxicity in the absence of antiviral activity gave 15% actives, of which 56% had antibacterial activity. Among the different bacteria tested it was concluded that *Sarcina lutea* gave the best correlation with antiviral and cytotoxic effects. A correlation between antibacterial activity against *S. lutea* and anticancer activity was emphasized (67).

Alanosine, an anticancer antibiotic produced by *Streptomyces alanosinicus,* was fortuitously discovered during a screen for activity against animal viruses (49). It is currently undergoing clinical trials.

PHAGE MUTAGENESIS Ikeda & Iijima have described a phage mutagenesis test for the detection of anticancer antibiotics (40). This test involved the use of phage T2 and assayed its conversion to a host range mutant (T2h). *Escherichia coli* B is susceptible to T2 whereas strain B/2 is resistant. Mutation of T2 to T2h enables the phage to produce plaques on *E. coli* B/2. In the test, a mixture of T2 phage with *E. coli* B and B/2 (the latter predominating) was seeded in an agar layer and samples were applied directly to the surface. Mutation (T2 → T2h) was detected as a zone of lysis surrounding samples. The anticancer antibiotics alazopeptin, azaserine, mitomycin C, porfiromycin, and roseolic acid were all shown to be mutagenic. Nakamura et al (50) used this test to screen 1500 strains of soil actinomycetes for their ability to produce mutagenic antibiotics. A total of 16 strains was found to produce antibiotics with mutagenic activity. Mitomycins were the antibiotics most frequently detected, and other antibiotics such as fervenulin, azaserine, and D-4-amino-3-isoxazolidone were also discovered. This last antibiotic gave turbid rather than clear lytic zones and was not thought to be mutagenic (50). This compound is an inhibitor of cell wall synthesis and the turbid lytic zone must have been produced by an effect other than mutation.

Prokaryotic Prescreens

ANAEROBIC BACTERIA It has been known for many years that cancer cells generally have a much increased rate of glycolysis and a decreased rate of respiration in comparison with their normal counterparts. This was the basis for a theory of carcinogenesis developed by Warburg over 50 years ago (72). The respiration deficiency of cancer cells has provided a rationale for the development of several screening approaches to detect compounds with selective anticancer activity. It was on this basis that a variety of anaerobic bacteria were tested for their ability to act as models of cancer cells in detecting antitumor agents (6, 9). Bradner & Clarke (6) concluded that activity against one of the anaerobes tested, *Clostridium feseri,* correlated sufficiently well with the anticancer activity to merit further investigation as a screening tool. However, other workers (9) concluded that clostridia were no more predictive in detecting anticancer agents than was *E. coli.*

RESPIRATION-DEFICIENT MUTANTS Another, less-explored approach stemming from the anaerobic nature of the cancer cell's metabolism has been the use of respiration-deficient mutants of bacteria. Gause (21, 22) has described mutants of *Staphylococcus aureus* that have impaired respiration. The exact nature of the mutations involved was not clear as there were defects other than impaired respiration. These mutants were described as having a selective vulnerability to inhibitors of nucleic acid synthesis. Unfortunately, very little information is available concerning the actual application of such mutants to screening for anticancer agents.

ANTIMETABOLITE TEST SYSTEMS The early success in cancer chemotherapy achieved with folic acid analogues, e.g. methotrexate, encouraged the application of microbial screening systems to detect other antimetabolites. *Lactobacillus casei* has complex nutritional requirements and by changing the concentrations of appropriate growth factors, it becomes a sensitive detection system for specific antimetabolites. Hitchings & Elion (39) have described a system for the detection of inhibitors of nucleic acid biosynthesis by limiting the supply of (or omitting altogether) folic acid, purines, and pyrimidines from the growth medium of *L. casei.* Under these conditions, antimetabolites of vitamins and amino acids were without effect as the corresponding normal metabolites were all present in excess. Hanka and co-workers at the Upjohn Company have employed a less complex bacterial test system for antimetabolites (31, 34). In this version the bacterial assay organism (*Escherichia coli* or *Bacillus subtilis*) was grown in two different agar media: one a complex nutrient medium, the other a chemically defined medium with glucose as the sole carbon source. Samples were tested against the bacteria in the two different media. Characteristically, antimetabolite activity was indicated by a larger zone of inhibition on the defined versus the complex medium; one of the many components present in the complex medium was responsible for reversing the activity. In a series of additional tests, it was possible to define exactly which metabolite was responsible for the reversal. Pools of related compounds, e.g. amino acids, purines, and vitamins, were first tested for their ability to reverse activity. Finally, the individual members of a pool were checked for their abilities to reverse the antibacterial effect. This system of antimetabolite screening has proved effective in discovering several anticancer compounds (29, 30). It is of interest that one of the compounds, AT-125 (acivicin), was identified as an antimetabolite of histidine in the bacterial system but was subsequently found to be antagonized by glutamine in cell culture (32). Acivicin is currently undergoing clinical trials. A comparison of the prescreen for antimetabolites with an L-1210 cell culture (derived from a murine leukemia) prescreen (33) revealed a smaller percentage of positives for antimetabolites (5% versus 19% for L-1210). Subsequent testing of the positives from both in vitro prescreens for in vivo activity against the murine leukemia P388 also showed a lower percentage of actives for antimetabolites (2% versus 28% for L-1210). However, the number of significant new anticancer antibiotics found by these prescreens still gave support to the antimetabolite approach. The poor overall correlation of antimetabolite activity against bacteria with in vivo activity against an animal tumor system is hardly surprising. For specific types of antimetabolites, e.g. purines and pyrimidines as detected by the *L. casei* system (39), the correlation could be better in view of the known sensitivity of cancer cells to agents interfering with nucleic acid synthesis. More recently, *L. casei* has been used in a study on a series of novel β-hydroxyamino acids (51). A number of N-acyl derivatives were synthesized and tested for their abilities to inhibit the growth of *L. casei* in a riboflavin assay medium. Although these amino acid analogues were ranked with respect to antibacterial activity, no information was published on their activity against animal tumors. The inhibitory activity of the N-trifluoracetyl derivatives of amino acids against *L. casei* was not reversed by the corresponding natural metabolite (51). Thus, although they were structural analogues, they did not behave as antimetabolites in this system.

BACTERIAL MUTAGENESIS AND ANTIMUTAGENESIS The observation that many of the clinically active anticancer drugs are mutagenic in bacteria (3) can provide the rationale for detection of other drugs with similar properties. Microbial mutagenesis assays such as the Ames test (3), which utilizes the back mutation from histidine auxotrophy to prototrophy in *Salmonella typhimurium,* could be employed as a prescreen for antitumor agents. No one has yet described the results of applying such a test, although the reverse principle has been used to find an antimutagen. Takeuchi et al have described the isolation of a methionine analogue from a streptomycete fermentation broth (68) that apparently reversed the mutagenic activity of N-methyl-N'-nitro-N-nitrosoguanidine. This compound, L-2-amino-5-methyl-5-hexenoic acid, primarily inhibited protein synthesis and its effects could be reversed by methionine. Its apparent antimutagenic activity could have been a reflection of the partial inhibition of cell growth that it caused (68). In an attempt to demonstrate the potential anticarcinogenic activity of this compound, its effect on the transformation of chick embryo fibroblasts by Rous sarcoma virus was tested. Equivocal results were obtained that could have been caused by inhibition of growth.

DNA REPAIR MUTANTS A similar rationale to that for using bacterial mutagenesis as a prescreen can be applied to the use of mutants that are defective in their ability to repair DNA damage. Such mutants become hypersensitive to mutagenic antibiotics. A strain of *Escherichia coli* has been constructed that contains mutations in three genes concerned with DNA repair, i.e. uvrA, ruv, and tolC (52, 53). This triple mutant was very susceptible to substances that produce lesions in DNA, such as bleomycin, mitomycin C, and adriamycin; increases in susceptibility were 100-, 50-, and 70-fold, respectively. The value of such a strain in screening for carinostatic and carcinogenic compounds, as well as being a sensitive bioassay system, was discussed (53). However, no results on its application as an anticancer prescreen have been published. Another *E. coli* repair deficient mutant, polA, could also be utilized (5). This test organism was originally proposed for the rapid detection of mutagens and carcinogens (63). The parental strain and polA mutant showed no differential sensitivity to common antibiotics such as cycloserine and chloramphenicol; however, these strains showed marked differences in sensitivity to anticancer antibiotics. Other groups used rec⁻ mutants of *Bacillus subtilis* as test organisms (41, 46, 60, 73). In one of these studies, the authors reported the detection and isolation of the known anticancer antibiotic chartreusin using the *B. subtilis* rec⁻ mutant as a prescreen (46). In another study (73), the behavior of different anthracyclines was compared with the rec⁻ assay. The anthracyclines tested could be divided into two groups depending on whether or not they exhibited increased antibacterial activity against rec⁻ mutants as compared with wild type. For example, aclacinomycin A showed similar activity against both strains, whereas adriamycin was approximately four times more active against the rec⁻ mutant.

F⁺ TEST Fleck (17) described a simple screening test that detected agents that inactivated or eliminated plasmids from bacteria. The test utilized an F⁺ strain of *Escherichia coli* (1885 F⁺ or W1665 F⁺) and the male-specific RNA phage MS2. The F⁺ fertility factor carries genes for F pili synthesis, and if the F⁺ plasmid is inactivated or eliminated, no pili are produced. Therefore, the *E. coli* cells lacking pili become resistant to MS2 phage since it normally uses pili as adsorption sites for initiation of infection. The F⁺ test was run as a double-layer agar diffusion assay in which samples were placed directly on an agar layer containing the F⁺ host bacteria. Following incubation, an MS2-containing agar layer was applied, and after an additional 5–6 hr of incubation, the F⁺ containing bacteria were all lysed by the phage. If a compound inactivated or eliminated the F⁺ plasmid, a zone of growth surrounded the sample. As a control, an HFr strain was run in parallel; compounds that allowed growth on HFr and F⁺ had direct antiphage activity, whereas those active on F+ plates must have inactivated or eliminated the F⁺ plasmid. None of the leads obtained with this system had progressed to in vivo testing when the assay was described. As a positive control, N-benzyl acridine orange was used in this test (17). The F⁺ test was considered capable of detecting potential carcinostatic- and phage-inhibiting compounds that work by inactivating or eliminating plasmids from their host strains.

CROWN GALL TUMOR FORMATION The Gram-negative microorganism *Agrobacterium tumefaciens* is known to cause a neoplastic lesion in plants called the Crown Gall tumor. In a recent publication (19), a variety of natural products were tested for their abilities to inhibit initiation of Crown Gall tumor formation on potato disks. A definite correlation between inhibition of Crown Gall tumor initiation and activity against murine P388 leukemia was claimed. The relevance of this tumor initiation system to cancer in animals is not clear (7), and one might predict that it will detect prophylactic rather than therapeutic activity (anticarcinogenic rather than carcinostatic).

Eukaryotic Prescreens

YEAST PERMEABILITY MUTANTS In 1978, Gause & Laiko (24) reported on the potential use of a *Saccharomyces cerevisiae* mutant with a distorted membrane as a prescreen for anticancer agents. This mutant, FL599-lB, was originally isolated by LaCroute (see 24) and was shown to be more susceptible to antifungal antibiotics such as cycloheximide than was the parent. The susceptibility of this strain to a variety of anticancer antibiotics has been evaluated (see Table 3). Not surprisingly, antibacterial antibiotics, such as penicillin, streptomycin, and chloramphenicol, were negative in this test (24), which supported the notion that it showed some selectivity for anticancer antibiotics. No information has been published concerning the application of this mutant to screening for novel anticancer agents; it was used, however, in the development of improved semisynthetic derivatives of the antibiotic carminomycin (24). In an attempt to increase the sensitivity of this screening system, a radiometric version has been devised (66). This test used the catabolism of [^{14}C]glucose to $^{14}CO_2$ as an index of the inhibitory effect of a compound on the yeast mutant. Although only a small number of antibiotics were evaluated (66) with this more sensitive radiometric test, the original observations of Gause & Laiko (24) were confirmed.

Table 3 Effect of anticancer antibiotics on the parent culture of *Saccharomyces cerevisiae* FL 200 and membrane mutant FL 599–1B[a]

	MIC (μg/ml)[b]	
Antibiotic	Parent culture[c]	Mutant[c]
Actinomycin D	>100	5.0
Mitomycin C	>100	7.0
Daunorubicin	>100	6.0
Adriamycin	>100	5.0
Carminomycin	60	1.0
Streptonigrin	>100	0.3
Bleomycin	5	0.5
Sibiromycin	>100	2.0
Echinomycin	>100	0.7
Olivomycin A	>100	>100

[a] From Gause & Laiko (24).
[b] MIC, Minimal inhibitory concentration.
[c] Parent, FL 200; mutant, FL 599–1B.

YEAST RESPIRATION-DEFICIENT MUTANTS For the same basic reasons as using anaerobic- and respiration-deficient bacteria, respiration-deficient mutants of yeast have also been evaluated as prescreens for antitumor agents. Respiration-deficient mutants of yeast are isolated as petite colonies and have defective mitochondria. Petite mutants do not respire and have enhanced glycolysis. DiPaolo & Rosenfield (9) found that many of the compounds with established anticancer activity in animal models were inactive against such respiratory-deficient yeast mutants. They concluded that further studies were required to evaluate this approach. Gause & Laiko (23) have described the application of a respiration-deficient mutant of *Torulopsis globosa.* This mutant was used to screen 7820 freshly isolated cultures of actinomycetes and it yielded 7.6% positives. Although the identity of these compounds was not reported, it is interesting to note that this mutant detected activities not observed in many other prescreens.

CILIA REGENERATION IN TETRAHYMENA Rosenbaum & Carlson (56) have shown that the microtubule inhibitor colchicine blocked the regeneration of cilia by mechanically deciliated *Tetrahymena pyriformis.* Further investigations by a group at Takeda Chemical Industries (69, 70) showed that maytansine, vinblastine, and the ansamitocins also caused an inhibition of cilia regeneration. Although protein synthesis inhibitors such as cycloheximide showed some inhibition of cilia regeneration, it was not complete. Potent inhibitors of nucleic acid synthesis, such as actinomycin D, chromomycin A_3, daunomycin, ethidium bromide, and 5-fluorouracil, had no effect on regeneration but they did inhibit growth. Such a test system could provide a selective prescreen for novel anticancer agents that are microtubule inhibitors.

COMPARATIVE STUDIES

There have been a limited number of studies in which a variety of microbiological test systems have been compared for their abilities to detect anticancer agents. In one such experiment 14 different organisms (*Escherichia coli, Serratia marcescens, Bacillus subtilis, Staphylococcus aureus, Mycobacterium smegmatis, Mycobacterium phlei, Lactobacillus casei, Torulopsis utilis, Saccharomyces cerevesiae, Kloeckera brevis, Penicillium notatum, Aspergillus fumigatus, Streptomyces griseus,* and *Streptomyces antibioticus*) were tested for susceptibility to 28 different compounds (26, 55). The correlation between the microbiological and anticancer activities of the compounds studied was poor. In a later study, Foley and co-workers (18) investigated the inhibitory effects of a series of 200 compounds from the Cancer Chemotherapy National Service Center in 16 microbiological systems. The microorganisms used were *Streptococcus faecalis* (2 media), *Lac-*

tobacillus casei (2 media), *Escherichia coli* (2 media), *Lactobacillus arabinosus* (2 media), *Leuconostoc citrovorum, Lactobacillus fermentis, Candida albicans, Saccharomyces carlsbergensis, Tetrahymena pyriformis, Glaucoma scintillans, Colpidium camylum,* and *Neurospora crassa.* The series of compounds tested was not random but was carefully selected to include agents with known anticancer activity and agents devoid of it but possessing other biological activities. In this case the results indicated a more positive correlation between antimicrobial and anticancer activities. In fact, the majority (95) of compounds judged active by the animal tumor screens could be detected based on their antimicrobial activity by as few as four of the assay organisms used. For the same series of compounds, it was also calculated that there was a considerable number of false positives (60–65%) and only a few (5%) false negatives.

The experiences of the Upjohn Company in screening for anticancer antibiotics over a period of more than 20 years support the utility of microbial prescreens. A summary of this period is given in Table 4 (33). In the last few years the Upjohn group has relied upon the use of multiple screening tests combining antimicrobial and cell culture assays. A significant number of agents found later to have anticancer activity was initially detected and isolated on the basis of antimicrobial activity alone. Such was the case for the actinomycins (29, 71) and streptozocin (38), both of which are clinically interesting antibiotics. The biological activity of *cis*-platinum was first detected with *E. coli* and only at a later date did its anticancer potential become apparent (57, 58).

In an analysis of fermentation broths that were screened over 6-years using four different in vitro prescreens, Fleck (17) reported that 8% of the strains of microorganisms investigated produced compounds that were active in one or more of the tests. A total of more than 15,000 strains was screened comprising representatives of the different genera of the *Actinomycetales.* The prescreens compared were prophage induction, BIP test, a DNA repair-deficient mutant of *E. coli,* and the F^+ test (all described in earlier sections). The contribution of positives by the different prescreens was 2.0, 4.8, 0.9, and 0.2%, respectively. Only a limited number of the leads generated by these prescreens had progressed to animal tumor testing at the time that analysis was published. However, for the crude or crystalline products tested, 100% of the prophage induction and DNA repair test positives, as well as 80% of the BIP test positives, were active in vivo. None of the F^+ test positives had been further evaluated. Fleck (17) concluded that the BIP test was preferable to simple prophage induction because it detected all the prophage inducers and additional compounds with anticancer activity.

In a recent review of the different prescreens employed over a period of 16 years at Bristol Laboratories, Bradner (4) reported that induction of lysogenic bacteria was the most successful prescreen and assay tool used. It was exquisitely sensitive to certain agents and was remarkably tolerant of the complex fermentation media so frequently employed in such screening programs.

LIMITATIONS OF MICROBIAL PRESCREENS

Direct Versus Indirect Effects

If the anticancer activity of a compound depends on some host-mediated effect, it will not normally be detected by a microbiological test system. Examples of host-mediated effects are metabolism of the drug to an active form and modulation of the immune response. Some success has been achieved in attempting to model mammalian metabolism by subjecting test compounds to incubation with microsomal enzyme preparations from rat liver. Metabolic transformation of the drug by such a mixed bag of microsomal enzymes is an attempt to model in vivo transformation of drug in the host animal. The most publicized use of such enzyme preparations has been in mutagen detection systems. They have proved capable of revealing the

Table 4 Detection and assay of fermentation-derived anticancer agents by the Upjohn Company[a]

Approximate dates	Detection systems	Assay method	Examples of antitumor agents isolated
Pre–1958	Antimicrobial	Antimicrobial	Cycloheximide Chartreusin (Streptovaricins) Streptovitacin A Streptozocin Porfiromycin (Steffimycin) Zorbamycin
1958–1964	Cell culture (KB)	Antimicrobial KB	Pactamycin Sparsomycin Tubercidin Nogalamycin
1965–1970	Antimicrobial (antimetabolite)/ in vivo		5–Azacytidine AT–125 Hydroxy AT–125 AT–111
1971–1975	Cell culture (L1210)/ in vivo	Cell culture/antimicrobial in vivo	CC–1014
1976	Cell culture/spec. antimicrobial/in vivo	Cell culture/antimicrobial	

[a]From Hanka et al (33).

mutagenicity and suspected carcinogenicity of a variety of chemicals, including chemotherapeutic agents (3).

A major limitation of microbial prescreens is their inability to predict any kind of therapeutic index; this information can only be derived at a later stage with in vivo experiments. This particular limitation is perhaps a consequence of our inability to define an appropriate target that would provide the basis for selective toxicity.

Carcinostatic and Carcinogenic?

It is rather disturbing to realize that most of the tests described here as prescreens for anticancer agents are also those recommended for use as screens for mutagenic/carcinogenic compounds (59). Most of the clinically effective drugs used today have been shown to be carcinogenic in animals. Only a limited amount of information is available so far on the long-term effects of these compounds in humans. Should we be using the tests described to reject rather than select compounds? At present adriamycin is one of the most widely prescribed chemotherapeutic agents for cancer: It is carcinogenic in animals (45). No one is going to suggest withdrawing adriamycin from the clinic unless we have suitable alternatives. Anthracycline analogue programs are directed toward diminishing cardiotoxicity, considered to be the more important limitation of such drugs in the short term.

POTENTIALS OF MICROBIAL PRESCREENS

The important role microbial prescreens have played in the discovery of potentially active anticancer agents is perhaps more a reflection of the restricted way in which the compounds act, rather than being a measure of the effectiveness of microbes as models of cancer cells. The majority of anticancer agents interact directly with nucleic acid or inhibit its biosynthesis, and it is perhaps the overall similarity of these processes in microbial and cancer cells that has been responsible for the utility of microbial test systems (59). In addition to their direct role in the detection of anticancer agents, microbes have frequently provided a convenient means for assay during isolation and purification (30). The ease with which microbial mutants specifically resistant to selected anticancer agents can be isolated has provided an aid towards understanding how these drugs work, and how resistance might arise in humans (59). Resistant mutants have also proved very useful in helping classify unknown compounds found during screening programs (62).

Optimistically, we might hope that current cancer biology research will lead to the identification of some unique feature that characterizes cancer cells and will become the basis for their selective destruction. Current techniques permitting the cloning and expression of selected mammalian genes in microorganisms could lead to the construction of specific strains for screening purposes, once the crucial function and corresponding gene have been identified. Perhaps, a somewhat more realistic viewpoint is that the continued empirical approach to screening will lead to the discovery of agents that exhibit some degree of selectivity. The mechanism of such selectivity may in turn lead to an increased understanding of cancer biology. The lesson to be learned from the continuing success of antibacterial screening programs is the need to become more and more selective in the approaches used. The trend is certainly not to devise screening systems capable of detecting all known clinically active antibacterial agents, but rather to devise highly selective tests that ignore the majority of activities and only detect those compounds with specific mechanisms of action. However, in the case of bacteria there is a sound biochemical rationale for the selective toxicity of compounds that act on specific targets such as the bacterial cell wall. For the cancer cell, no known analogous target exists. For the present, we have to be satisfied with developing screening strategies suggested by the most effective existing anticancer agents, i.e. interference with nucleic acid sysnthesis.

CONCLUDING REMARKS

The conclusions reached over 20 years ago concerning the applicability of non-tumor systems to screening for anticancer agents are still valid today (18, 26, 62). There is no single non-tumor system that is capable of detecting all known agents with established clinical activity. However, the animal tumor models themselves are also subject to similar limitations (62). We remain ignorant of the number of false negatives generated by such in vivo screening systems. If one wishes to detect all known anticancer agents, then the only approach is to use a combination of prescreens; however, the practicability and utility of such an approach is questionable.

The principal advantage of microbial systems in the search for novel anticancer agents is their rapidity, low cost, and in many cases extreme sensitivity. This is particularly important when dealing with the complex mixtures of natural products encountered during microbial and plant screening programs. In reality, the microbe is being used more as a model of rapidly proliferating cells, both normal and cancerous. However imperfect a model the microbial cell is, the fact remains that clinically useful antibiotics have been, and continue to be, discovered as a result of its use in prescreens. The most important criterion for judging any screening system is its capacity to detect novel compounds with potential clinical utility, and not its ability to detect known active agents.

Literature Cited

1. Anderson, W. A., Moreau, P. L., Devoret, R., Maral, R. 1980. *Mutation Res.* 77:197–208
2. Aoki, H., Sakai, H. 1967. *J. Antibiot.* 20:87–92
3. Benedict, W. F., Baker, M. S., Haroun, L., Choi, E., Ames, B. N. 1977. *Cancer Res.* 37:2209–13
4. Bradner, W. T. 1978. *Antibiot. Chemother.* 23:4–11
5. Bradner, W. T. 1980. In *Cancer and Chemotherapy,* ed. S. T. Crooke, A. W. Preskayko, 1:313–24. New York: Academic. 373 pp.
6. Bradner, W. T., Clarke, D. A. 1958. *Cancer Res.* 18:299–304
7. Braun, A. C. 1972. *Prog. Exp. Tumor Res.* 15:165–87
8. DeVita, V. T. Jr., Young, R. C., Canellos, G. P. 1975. *Cancer* 35:98–110
9. DiPaolo, J. A., Rosenfield, R. 1958. *Cancer Res.* 18:1214–20
10. Douros, J., Suffness, M. 1980. *Recent Results Cancer Res.* 70:21–44
11. Douros, J., Suffness, M. 1981. *Cancer Treat. Rev.* 8:63–87
12. Elespuru, R. K., Yarmolinsky, M. B. 1979. *Environ. Mutagen.* 1:65–78
13. Endo, H., Ishizawa, M., Kamiya, T., Kuwano, M. 1963. *Biochim. Biophys. Acta* 68:502–5
14. Endo, H., Ishizawa, M., Kamiya, T., Sonoda, S. 1963. *Nature* 198:258–60
15. Epstein, S. S., Saporoschetz, I. B. 1968. *Experientia* 24:1245–48
16. Fleck, W. 1968. *Z. Allg. Mikrobiol.* 8:139–44
17. Fleck, W. 1974. *Post. Hig. I Med. Dosw.* 28:479–98
18. Foley, G. E., McCarthy, R. E., Binns, V. M., Snell, E. E., Guirard, B. M., Kidder, G. W., Dewey, V. C., Thayer, P. S. 1958. *Ann. NY Acad. Sci.* 76:413–38
19. Galsky, A. G., Wilsey, J. P. 1980. *Plant Physiol.* 65:184–85
20. Garretson, A. L., Elespuru, R. K., Lefriu, I., Warnick, D., Wei, T., White, R. J. 1981. *Dev. Ind. Microbiol.* 22:211–18
21. Gause, G. F. 1958. *Science* 127:506–8
22. Gause, G. F. 1966. *Front. Biol.* 1:12–85
23. Gause, G. F., Laiko, A. V. 1974. *Prog. Chemother. Antibacterial Antiviral Antineoplast. Proc. Int. Cong. Chemother. 8th,* III:116–17
24. Gause, G. F., Laiko, A. V. 1978. *Antibiot. Chemother.* 23:21–25
25. Gelderman, A. H., Lincoln, T. L., Cowie, D. B., Roberts, R. B. 1966. *Proc. Nat. Acad. Sci. USA* 55:289–97
26. Gellhorn, A., Hirschberg, E. 1955. *Cancer Res. Suppl.* 3:1–13
27. Goldin, A., Johnson, R. K., Venditti, J. M. 1980. *Antibiot. Chemother.* 28:1–7
28. Goldin, A., Venditti, J. M., Macdonald, J. S., Muggia, F. M., Henney, J. E., DeVita, V. T. Jr. 1981. *Eur. J. Cancer* 17:129–42
29. Hackmann, C. 1952. *Z. Krebsforsch.* 58:607–13
30. Hanka, L. J. 1972. *Adv. Appl. Microb.* 15:147–56
31. Hanka, L. J. See Ref. 23, pp. 118–22
32. Hanka, L. J. 1979. *Cancer Treat. Rep.* 63:1133–36
33. Hanka, L. J., Kuentzel, S. L., Martin, D. G., Wiley, P. F., Neil, G. L. 1978. *Recent Results Cancer Res.* 63:69–76
34. Hanka, L. J., Martin, D. G., Neil, G. L. 1978. *Lloydia* 41:85–97
35. Heinemann, B. 1971. *Chem. Mutagens* 1:234–66
36. Heinemann, B., Howard, A. J. 1964. *Appl. Microbiol.* 12:234–39
37. Heinemann, B., Howard, A. J. 1965. *Antimicrob. Agents Chemother.*, pp. 488–92
38. Herr, R. R., Eble, T. E., Bergy, M. E., Jahnke, H. K. 1960. *Antibiot. Ann.* 7:236–40
39. Hitchings, G. H., Elion, G. B. 1955. *Cancer Res. Suppl.* 3:66–68
40. Ikeda, Y., Iijima, T. 1965. *J. Gen. Appl. Microbiol.* 11:129–35
41. Kada, T., Tutikawa, K., Sadaie, Y. 1972. *Mutat. Res.* 16:165–74
42. Lein, J., Heinemann, B., Gourevich, A. 1962. *Nature* 196:783–84
43. Levine, A., Moreau, P. L., Sedgwick, S. G., Devoret, R. 1978. *Mutat. Res.* 50:29–35
44. Lwoff, A. 1953. *Bacteriol. Rev.* 17: 269–337
45. Marquardt, H., Philips, F. S., Sternberg, S. S. 1976. *Cancer Res.* 36: 2065–69
46. Matsui, K., Murakawa, S., Takahashi, T. 1980. *Agric. Biol. Chem.* 44:919–20
47. Meek, E. S. 1970. *Recent Results Cancer Res.* 28:1–78
48. Moreau, P., Bailone, A., Devoret, R. 1976. *Proc. Natl. Acad. Sci. USA* 73:3700–4
49. Murthy, Y. K. S., Thiemann, J. E., Coronelli, C., Sensi, P. 1966. *Nature* 211:1198–99
50. Nakamura, S., Omura, S., Hamada, M., Nishimura, T., Yamaki, H., Tanaka, N., Okami, Y., Umezawa, H. 1967. *J. Antibiot.* 20:217–22
51. Otani, T. T., Briley, M. R. 1979. *J. Pharmacol. Sci.* 68:1366–69

52. Otsuji, N., Higashi, T., Kawamata, J. 1972. *Biken J.* 15:49–59
53. Otsuji, N., Horiuchi, T., Nakata, A., Kawamata, J. 1978. *J. Antibiot.* 31:794–96
54. Price, K. E., Buck, R. E., Lein, J. 1964. *Appl. Microbiol.* 12:428–35
55. Reilly, H. C. 1955. *Cancer Res. Suppl.* 3:63–65
56. Rosenbaum, J. L., Carlson, K. 1969. *J. Cell. Biol.* 40:415–25
57. Rosenburg, B., VanCamp, L., Krigas, T. 1965. *Nature* 205:698–99
58. Rosenburg, B., VanCamp, L., Trosko, J. E., Mansour, V. H. 1969. *Nature* 222:385–86
59. Rosenkranz, H. S. 1973. *Ann. Rev. Microbiol.* 27:383–401
60. Sadaie, Y., Kada, T. 1976. *J. Bacteriol.* 125:489–500
61. Sakai, H., Aoki, H. 1969. *Prog. Antimicrob. Anticancer Chemother. Proc. Int. Cong. Chemother. 6th,* 1:470–73
62. Schabel, F. M. Jr., Pittillo, R. F. 1961. *Adv. Appl. Microbiol.* 3:223–56
63. Slater, E. E., Anderson, M. D., Rosenkranz, H. S. 1971. *Cancer Res.* 31:970–73
64. Smith, C. L., Oishi, M. 1976. *Mol. Gen. Genet.* 148:131–38
65. Specht, I. 1965. *Arch. Mikrobiol.* 51:9–17
66. Speedie, M. K., Fique, D. V., Blomster, R. N. 1980. *Antimicrob. Agents Chemother.* 18:171–75
67. Takatsuki, A., Hosoda, J., Tamura, G., Arima, K. 1969. *J. Antibiot.* 22:171–73
68. Takeuchi, M., Iinuma, H., Takeuchi, T., Umezawa, H. 1979. *J. Antibiot.* 32:1118–24
69. Tanida, S., Hasegawa, T., Higashide, E. 1980. *Agric. Biol. Chem.* 44:1847–53
70. Tanida, S., Higashide, E., Yoneda, M. 1979. *Antimicrob. Agents Chemother.* 16:101–3
71. Waksman, S. A., Tishler, M. 1942. *J. Biol. Chem.* 142:519–28
72. Warburg, O. 1956. *Science* 123:309–14
73. Yoshimoto, A., Oki, T., Inui, T. 1978. *J. Antibiot.* 31:92–94

Ann. Rev. Microbiol. 1982. 36:435–65

MECHANISM OF INCORPORATION OF CELL ENVELOPE PROTEINS IN ESCHERICHIA COLI

Susan Michaelis and Jon Beckwith

Department of Microbiology and Molecular Genetics, Harvard Medical School, Boston, Massachusetts 02115

CONTENTS

INTRODUCTION

Bacterial protein synthesis takes place in the cytoplasm, yet many proteins are located outside the cytoplasm in the cell envelope or in the cell's external environment. These noncytoplasmic proteins are referred to as exported

0066-4227/82/1001-0435$02.00

proteins. This review discusses what is currently known in bacteria concerning the molecular mechanisms that facilitate the process of protein passage into and through membranes. Emphasis is placed on the recent genetic studies in the Gram-negative bacterium *Escherichia coli* that are relevant to export.

Many exported proteins are synthesized via a precursor form containing an amino-terminal extension called the signal sequence (or signal peptide). This signal sequence is removed by proteolytic cleavage during or after export to yield the mature product. Geneticists have isolated mutants whose mutations alter the structure of the signal sequence of certain proteins. These mutations produce profound effects on the export of these proteins, thus proving the importance of the signal sequence for mediating transmembrane passage of proteins. Genetic selection techniques have also been used to obtain mutants simultaneously affected in the localization of many proteins. The mutations in these strains are unlinked to the structural genes encoding the proteins whose export they influence and may identify genes for cellular machinery involved in protein export. For detailed discussions of particular genetic strategies employed in obtaining mutants defective in protein export, the reader is referred to several recent reviews (6, 40, 46, 108, 109). Other facets of protein export not covered in depth here are reviewed elsewhere (28, 54, 91).

A wealth of information has been amassed concerning the export process in higher eukaryotic cells (12, 104). Eukaryotes and prokaryotes appear to share common mechanisms for certain aspects of protein export. The most striking evidence for this similarity comes from studies in which the gene for a eukaryotic secreted protein, preproinsulin, has been introduced into *E. coli.* The eukaryotic signal sequence of preproinsulin serves to promote transfer of the polypeptide across the prokaryotic cytoplasmic membrane (122). The signal peptide is cleaved at the correct position to yield proinsulin (121). Ovalbumin, another eukaryotic-secreted protein, can also be secreted by *E. coli* (39). Less extensive experiments on the reverse situation suggest that when the gene for a prokaryotic β-lactamase is introduced into yeast cells, the β-lactamase is correctly processed (99).

Much of our understanding of protein export has come from eukaryotic studies where the initial events of export can be reproduced in vitro (10–12). The powerful genetic tools available in *E. coli* have facilitated the selection and analysis of mutants that allow us to further dissect the process of protein export. The fruitful exchange of ideas between eukaryotic cell biologists and prokaryotic molecular geneticists has led to rapid advances in this field.

STRUCTURE OF THE GRAM-NEGATIVE BACTERIAL CELL ENVELOPE

The Gram-negative cell envelope is a complex structure composed of several layers. The cytoplasm is bounded by a cytoplasmic membrane, also called the inner membrane. This membrane contains proteins involved in energy metabolism, nutrient transport, and synthesis of peptidoglycan, lipids, and lipopolysaccharide (73, 100). A second distinct membrane, the outer membrane, surrounds Gram-negative bacteria. The proteins here facilitate entry of nutrients and ions; many of these proteins also serve as bacteriophage and colicin receptors (46, 53, 91). The periplasm is an aqueous compartment between the inner and outer membranes. It contains binding proteins required for nutrient transport as well as hydrolytic enzymes such as phosphatases, nucleases, proteases, and penicillinases (4). The peptidoglycan is a rigid layer of highly cross-linked polysaccharide chains lying between the inner and outer membranes (81). The peptidoglycan layer contains no proteins (although a fraction of the outer membrane lipoprotein is covalently linked to it). Together, the peptidoglycan, inner membrane, outer membrane, and periplasm comprise the cell envelope. Proteins must be exported and properly localized to each of these latter three compartments.

Gram-negative bacteria synthesize proteins that can leave the cell and are lytic or toxic to certain prokaryotic or eukaryotic target cells. These include (*a*) toxins, such as *Escherichia coli* heat-labile toxin (21, 95, 117), *E. coli* heat-stable toxins (116), and *Vibrio cholerae* enterotoxin (86); (*b*) haemolysins, for example, α-haemolysin (88, 118); and (*c*) bacteriocins such as colicins (60, 82, 97, 131) and cloacins (125). The location of these proteins in cells that synthesize them has not been unambiguously determined in most cases. Likewise, whether or not they are localized by a true export process is not always clear. Certain colicins, such as colicins E1 and E3, may be cytoplasmic constituents that are liberated only when the cell producing them lyses (60). These colicins do not appear to have a signal sequence (60, 131). On the other hand, several toxins can be considered to be exported proteins. They do contain signal sequences. They are located in the cell envelope or may actually be excreted through the cell envelope into the surrounding medium.

Assignment of proteins to a particular compartment is based on operational criteria. Periplasmic proteins are defined as those released from cells upon osmotic shock or spheroplasting. However, the operational criteria break down to some extent here, since these different procedures yield somewhat different patterns of proteins (4). When shocked cells are broken open the soluble cytoplasmic proteins can be separated from proteins in the particulate membrane fraction by centrifugation. The method employed for cell disruption, as well as the salt concentration, are important in these experiments. Inner and outer membranes are separable in sucrose gradients because they differ in density. In addition, inner and outer membrane proteins can be separated on the basis of their solubility properties in the presence of detergents (46). However, for membrane proteins, too, each separation procedure yields different protein profiles (24). It is important to emphasize that depending on the cell fractionation method employed, the cellular location to which a particular protein is assigned may vary. In our concluding remarks we discuss the relevance of these problems to studies on protein export.

MODELS FOR PROTEIN EXPORT

Various models for the mechanism of protein export have been presented. It is probable that no single model will be applicable to the export of all bacterial proteins. Two very different models are described below. They are intended to serve as useful frameworks within which available data can be viewed.

According to the now familiar signal hypothesis proposed by Blobel and co-workers (10, 12), proteins would be transferred across the membrane in the following way in Gram-negative bacteria: As a protein destined for export is being synthesized, the signal sequence emerges from the ribosome and binds to the membrane, thus causing attachment to the membrane of polysomes making this protein. According to the original model, this binding is suggested to recruit membrane proteins, which, together with ribosomal components, constitute a membrane pore through which the growing polypeptide chain is co-translationally extruded. After its extrusion to the noncytoplasmic face of the membrane, the signal peptide is cleaved off. For periplasmic proteins, the remainder of the protein continues its passage and is finally secreted. In the case of some cytoplasmic membrane proteins, the passage may not be completed and the protein remains embedded in the membrane.

The signal hypothesis envisions that the polypeptide chain crosses the plane of the membrane vectorially in an extended form. According to this model the role of the signal sequence is to mediate the initial events of export. It brings the synthesizing ribosomes to the membrane facilitating the formation of an export pore.

It has been demonstrated that alkaline phosphatase, a periplasmic enzyme, is secreted in a manner compatible with the signal hypothesis. It is made on membrane-bound ribosomes (115). Furthermore, at least a fraction of growing chains of nascent alkaline phosphatase can be labeled as they traverse the inner membrane by reagents that cannot penetrate this barrier (115). The Gram-positive bacterial proteins *Corynebacterium diphtheriae* diphtheria toxin and *Bacillus licheniformis* penicillinase have also been shown to be co-translationally secreted by this method (113, 114). Other *Escherichia coli* periplasmic proteins, maltose-binding protein and arabinose-binding protein, as well as the outer membrane LamB protein, are synthesized on membrane-bound ribosomes (98). These and other data (presented below) are consistent with a co-translational mode of export for these proteins. In no case for bacteria have the pore proteins invoked by the signal hypothesis been identified. However, the genetic data discussed in a subsequent section indicate that some kind of export machinery may exist.

A very different model that does not require an ordered vectorial export process nor any special cellular export machinery has been proposed by Wickner (135). This membrane trigger hypothesis emphasizes the role of self-assembly of proteins in mediating their own export. It allows for synthesis of a protein molecule to be completed before its export begins. The role of the signal sequence in this scenario is to promote the folding of the newly made precursor into a conformation that is soluble and competent for export. Upon reaching the membrane the protein is "triggered" into a new conformation that allows it to spontaneously insert into or through the phospholipid bilayer unaided by any sort of pore. Cleavage of the signal sequence would facilitate this conformational change and render it irreversible. The observation that precursor of bacteriophage M13 coat protein correctly inserts into liposomes containing nothing but phospholipids and purified processing enzyme supports this model (132).

STRUCTURE AND ROLE OF THE SIGNAL SEQUENCE

Demonstration of Precursor Forms of Prokaryotic Proteins

Since the precursor form of an exported protein contains extra amino acids at the amino terminus, its molecular weight is greater than that of the corresponding mature protein. Thus precursor species can be detected by sodium dodecyl sulfate-polyacrylamide gel electrophoresis, which separates proteins according to size. It is difficult to detect precursors in whole cells in vivo since they are short-lived. However, when cells are pulse labeled very briefly (10 sec or less), the precursors for a number of noncytoplasmic proteins have been seen (cf, for example, 63, 64). These precursors are probably true intermediates in the export pathway since they can be chased rapidly and quantitatively into the mature species. In cases in which export and processing are completely co-translational, full-sized precursor would never exist. This may be the case for *ampC*-encoded β-lactamase (63).

In vitro protein synthesizing systems have been used to identify the initial translation products of the genes for a number of noncytoplasmic proteins. Plasmid-containing minicells and phage infection of UV-irradiated cells have also been used to detect precursors. In some instances an actual precursor form of a protein has not been directly demonstrated but has been indirectly inferred when the amino acid sequence of the mature protein and the DNA sequence encoding this protein are known.

With these techniques many bacterial proteins appear to be made via a precursor form. These include the following: (*a*) inner membrane proteins —bacteriophage M13 (fd, fl) major coat proteins (17, 119), M13 minor coat protein (105), and penicillin-binding proteins, PB5 and PB6 (96); (*b*) periplasmic proteins—alkaline phosphatase (51, 52, 66), maltose-binding protein (7, 98), arabinose-binding protein (98, 136), ribose-binding protein (41), leucine-specific- binding protein (93), leucine-isoleucine-valine-binding protein (71, 92), histidine-binding protein from *Salmonella typhimurium* (50), lysine-arginine-ornithine-binding protein from *S. typhimurium* (50), Tn*1* or pBR322-encoded TEM β-lactamase (68, 120), and *ampC*-encoded β-lactamase (43, 61, 62); (*c*) outer membrane proteins—LamB (49, 80, 98), OmpA (5, 20, 37, 84), OmpF (20, 37; M. Hall and M. Berman, personal communication), lipoprotein (56, 85), and F factor-encoded TraT protein (38); and (*d*) toxins—heat-labile toxin, subunit A (117) and subunit B (21, 95), and heat-stable toxin ST 1 (116). Unless otherwise specified the proteins listed above are encoded by *Escherichia coli.* An amino-terminal signal sequence has been directly demonstrated or indirectly inferred from DNA or amino acid sequencing data for all the above proteins with the exception of penicillin-binding proteins PB5 and PB6, ribose-binding protein, and TraT protein. It seems likely that the extra amino acids present in the precursor forms of these proteins represent an amino-terminal extension.

An important lesson can be learned from the investigation of TraT protein. In initial studies precursor was not detected (1). TraT precursor was, however, visualized in a later study when a different gel electrophoresis system was employed (38). Similarly, we have found in our laboratory that precursor and mature alkaline phosphatase co-migrate in some electrophoresis systems but are separable in others. Therefore, failure to detect a precursor form for a particular protein does not constitute proof that it is not made in precursor form.

Structure of Signal Peptides

With amino acid or DNA sequencing techniques the primary structure of a large number of bacterial signal sequences has been determined. These are listed in Table 1. If these sequences are compared, little homology is found among them. An exception to this rule is seen for lipoprotein and arabinose-binding protein whose signal peptides are very similar (136). Since these two proteins are localized to different cell compartments it may well be that the signal sequence plays no role in determining a protein's ultimate destination.

When the degree of polarity is considered there is a striking similarity among all the signal peptides. They all have a positively charged basic amino-terminal region of two to eight amino acids, which is abruptly followed by a long stretch of uncharged, mainly hydrophobic amino acids. The amino acid before the cleavage site has a short side chain. Similarity in structure does not extend past the cleavage site. Signal peptides of eukaryotic exported proteins exhibit similar features (2, 12).

The predictive rules developed by Chou & Fasman (19), when applied to various signal sequences, indicate that these peptides can exist in highly ordered helical conformations containing very little random coiling (2). Conformational studies have been performed for the purified signal sequence from the eukaryotic parathyroid hormone (101). The secondary structure of this peptide was examined by circular dichroism. The two conformations experimentally observed were highly ordered and agreed well with the structures expected by using Chou & Fasman's prediction system.

Bedouelle & Hofnung (8, 9) have attempted to relate the structure of the hydrophobic segment of the signal peptide to its function. In this theoretical treatment, they have defined a parameter, hydrophobic axis length, which is a function both of the number of contiguous uncharged residues in this region and of its predicted secondary structure. They propose that a hydrophobic axis length of a particular minimum length is required to allow the signal peptide to insert into the membrane.

Overview of Exported Proteins

It is clear from examination of the catalogue above that many outer membrane and periplasmic proteins are initially synthesized in precursor form. In fact, in every case carefully examined, precursors have been demonstrated for proteins of these two compartments. As mentioned above, several toxins also have amino-terminal extensions. The presence of a signal sequence in all these noncytoplasmic proteins suggests that common molecular mechanisms may be involved in their export.

On the other hand, inner membrane proteins present a more complicated picture. Two inner membrane penicillin-binding proteins, PB5 and PB6, are apparently made in precursor form (96) (but see qualification above). The membrane disposition of analogous penicillin-binding proteins from *Bacillus* species has been determined. From these results it is clear that the latter are ectoproteins with the bulk of their polypeptides extending into the

Table 1 Prokaryotic signal sequences

Protein	Charged segment	Hydrophobic segment[a]	Reference
Inner membrane proteins			
Phage fd, major coat protein	MET LYS LYS SER LEU VAL LEU LYS	ALA SER VAL ALA VAL ALA THR LEU VAL PRO MET LEU SER PHE ALA ↓ ALA GLU GLY	119
Phage fd, minor coat protein	MET LYS LYS	LEU LEU PHE ALA ILE PRO LEU VAL VAL PRO PHE TYR SER HIS SER ↓ ALA GLU THR	105
Periplasmic proteins			
Alkaline phosphatase	MET LYS	GLN SER THR ILE ALA LEU ALA LEU LEU PRO LEU LEU PHE THR PRO VAL THR LYS ALA ↓ ARG THR PRO	51, 66
Maltose-binding protein	MET LYS ILE LYS THR GLY ALA ARG	ILE LEU ALA LEU SER ALA LEU THR THR MET MET PHE SER ALA SER ALA LEU ALA ↓ LYS ILE GLU	7
Leucine-specific-binding protein	MET LYS ALA ASN ALA LYS	THR ILE ILE ALA GLY MET ILE ALA LEU ALA ILE SER HIS THR ALA MET ALA ↓ ASP ASP ILE	93
Leucine-isoleucine-valine-binding protein	MET ASN ILE LYS GLY LYS	ALA LEU LEU ALA GLY CYS ILE ALA LEU ALA PHE SER ASN MET ALA LEU ALA ↓	71
Histidine-binding protein of *Salmonella typhimurium*	MET LYS LYS	LEU ALA LEU SER LEU SER LEU VAL LEU ALA PHE SER SER ALA THR ALA ALA PHE ALA ↓ ALA ILE PRO	50
Lysine-arginine-ornithine-binding protein of *S. typhimurium*	MET LYS LYS	THR VAL LEU ALA LEU SER LEU LEU ILE GLY LEU GLY ALA THR ALA ALA SER TYR ALA ↓ ALA LEU PRO	50
ampC β-lactamase	MET PHE LYS	THR THR LEU CYS ALA LEU LEU ILE THR ALA SER CYS SER THR PHE ALA ↓ ALA PRO GLN	60

TEM β-lactamase of PBR322	MET SER ILE GLN HIS PHE ARG	VAL ALA LEU ILE PRO PHE PHE ALA ALA PHE CYS LEU PRO VAL PHE ALA↓HIS PRO GLU	120
Arabinose-binding protein	MET LYS X THR LYS	LEU VAL LEU GLY ALA VAL ILE LEU THR ALA GLY LEU SER X GLY [ALA] X [ALA]↓GLU ASN LEU	136
Outer membrane proteins			
Lipoprotein	MET LYS ALA THR LYS	LEU VAL LEU GLY ALA VAL ILE LEU GLY SER THR LEU LEU ALA GLY↓CYS SER SER	56, 85
LamB	MET MET ILE THR LEU ARG LYS	LEU PRO LEU ALA VAL ALA VAL ALA ALA GLY VAL MET SER ALA GLN ALA MET ALA↓VAL ASP PHE	49
OmpA	MET LYS LYS	THR ALA ILE ALA ILE ALA VAL ALA LEU ALA GLY PHE ALA THR VAL ALA GLN ALA↓ALA PRO LYS	5, 84
OmpF	MET MET LYS ARG	ASN ILE LEU ALA VAL ILE VAL PRO ALA LEU LEU VAL ALA GLY THR ALA ASN ALA↓ALA GLU ILE	M. Hall, M. Berman, personal communication
Toxins			
Heat-labile toxin, B-subunit	MET ASN LYS VAL LYS	CYS TYR VAL LEU PHE THR ALA LEU LEU SER SER LEU TYR ALA HIS GLY↓ALA PRO GLN	21
Heat-labile toxin, A-subunit	MET LYS	ASN ILE THR PHE ILE PHE PHE ILE LEU LEU ALA SER PRO LEU TYR ALA↓ASN GLY ASP	117
Heat-stable toxin, ST1	MET LYS LYS	LEU MET LEU ALA ILE PHE ILE SER VAL LEU SER PHE PRO SER PHE SER GLN SER ?THR GLU SER	116

[a] ↓ Denotes site of proteolytic cleavage.

aqueous periplasmic environment. Only a short hydrophobic carboxy-terminal region anchors them to the membrane (133, 134). If PB5 and PB6 are similar to their Gram-positive counterparts, they face the same requirement as periplasmic proteins of transferring a major portion of their protein chains completely across the inner membrane. The presence of a signal sequence is therefore not surprising. These ectoproteins probably differ greatly in their export pathway from other inner membrane proteins.

Bacteriophage M13 coat protein is found in the bacterial inner membrane and is made in precursor form (17, 18). However, this protein is unusual and may not serve as a valid model system for understanding inner membrane proteins. M13 coat protein resides in the inner membrane only transiently. Its ultimate destination is the non-membranous viral coat. To accommodate its two different modes of residence this protein may be structurally different from typical inner membrane proteins. In addition, M13 coat protein is an unusually small protein of 5620 daltons.

In contrast to M13 coat protein, the integral membrane protein lactose permease, encoded by the *lacY* gene, is not made as a precursor. The amino-terminal amino acid sequence of lactose permease synthesized in vitro is identical to that of lactose permease isolated from the cytoplasmic membrane (30). DNA sequencing corroborates the absence of an amino-terminal cleaved peptide (15). Lactose permease must attain a highly ordered tertiary structure in the membrane to function in its role as a lactose/proton-symporter. Its entire amino acid sequence is known from DNA sequencing, and it is highly hydrophobic (15). The mechanism of localization of this protein into the inner membrane is not known, but the apparent absence of a signal sequence and presence of several long stretches of hydrophobic amino acids suggests that this protein may simply partition into the membrane. The integral membrane protein complex, F_o, of the *E. coli* ATP synthase contains three subunits. The DNA of the genes coding for these subunits has been sequenced. The amino-terminal amino acid sequence determined for two of the isolated subunits, b and c, has been compared to the amino acid sequence predicted from the DNA sequencing data. As in the case of lactose permease, this analysis indicates that these two inner membrane proteins contain no signal sequence (87). Subunit c has a highly hydrophobic amino acid composition, which may be responsible for its incorporation into the cytoplasmic membrane. Similar studies of the three inner membrane components of the histidine transport system in *Salmonella typhimurium* indicate that none of these proteins have signal peptides (G. Ames, personal communication). Taken together, it appears that there is a fundamental difference in the mode of export between many inner membrane proteins and other exported proteins. Mutational analysis (i.e. isolation of deletion and point mutants) may help to establish what regions of these proteins are required for proper insertion into the inner membrane.

Genetic Studies

Genetic studies have provided confirmation of the essentiality of the signal sequence for protein export. Two types of questions have been posed by bacterial geneticists to determine the role of the signal sequence. (*a*) If a cytoplasmic protein is provided with a signal sequence, can it be exported? (*b*) If the signal sequence of a noncytoplasmic protein is mutated, how is export affected?

GENE FUSIONS Gene fusions have been selected in vivo or constructed in vitro. These fusions are designed to encode hybrid proteins in which the amino-terminal portion of an exported protein, including its signal sequence, is substituted for a very short amino-terminal portion of the cytoplasmic enzyme β-galactosidase. Much of this work has been previously reviewed (6, 40, 46, 108, 109) and is only briefly described here.

The major finding from these studies is that the cytoplasmic protein β-galactosidase can be redirected to two different compartments of the *E. coli* cell envelope. This conclusion is drawn from the following evidence: In certain *lamB-lacZ* fusion-bearing strains containing a significant portion of the *lamB* gene, the hybrid protein is exported to the outer membrane (111), which is the normal location of the LamB protein. One class of *malF-lacZ* fusions produces a hybrid protein localized to the inner membrane, which is where the *malF* gene product normally resides (110). The *malE* gene codes for maltose-binding protein and *phoA* encodes alkaline phosphatase. Both products are found in the periplasm. However, in strains bearing *malE-lacZ* or *phoA-lacZ* fusions, β-galactosidase activity is found in the inner membrane and not in the periplasm (3; S. Michaelis, L. Guarente, H. Inouye, and J. Beckwith, manuscript in preparation). In fact, in no case has any hybrid protein containing a β-galactosidase domain been detected in the periplasm. One explanation for the properties of these fusions is that export is initiated but then aborted because the amino acid sequence of β-galactosidase is incompatible with passage through the membrane (123). These hybrid proteins remain embedded in the inner membrane. In fact, certain of the *lamB-lacZ*- and *malE-lacZ*-encoded hybrid proteins appear to block membrane export sites, causing accumulation of precursors of many exported proteins (34, 57). These strains can be used to detect the normally short-lived precursors of these proteins.

In the case of *lamB-lacZ, malE-lacZ,* and *phoA-lacZ* gene fusions, mutations have been isolated that prevent the hybrid protein from being localized to the outer or inner membranes. In all cases where the mutation resides in the gene fusions, it maps in the signal sequence portion of the hybrid genes (3, 34, 35).

In summary, these findings from gene fusion studies have established that a cytoplasmic protein can be guided to the cell envelope and further that a functional signal sequence is required [but is not necessarily sufficient (83)] to mediate this event.

ANALYSIS OF SIGNAL SEQUENCE MUTATIONS For bacterial proteins, the signal peptide is short in comparison with the rest of the protein and its DNA coding sequence therefore presents a small target for mutagenesis. Furthermore, even a protein with significantly altered export properties need not exhibit a strong phenotype. For these reasons signal sequence mutations have been isolated in indirect ways, in many cases taking advantage of special properties of fusion strains. Genetic engineering techniques in which in vitro mutagenesis can be directed to a specific segment of cloned DNA have also been used to obtain mutants.

A number of mutants now exist and are discussed below. The first series of mutations described below in the first four sections lie in the hydrophobic core of the signal peptide. The others, in the final two sections, are located in the basic amino-terminal region.

Maltose-binding protein and LamB protein By taking advantage of unusual properties of certain fusion-bearing strains, a selection has been devised for isolation of *lamB* and *malE* signal sequence mutations. It is described elsewhere (3, 34, 35). Both LamB protein and maltose-binding protein whose signal peptides have been mutationally altered cannot be exported and their precursor forms accumulate in the cytoplasm.

These mutations have been analyzed by DNA sequencing (7, 33). All four of the *lamB* point mutations and four of five *malE* point mutations produce changes from hydrophobic or weakly hydrophilic amino acids to charged amino acids. These mutations strongly support the contention that the hydrophobicity of this region is crucial for export. However, more recently, signal sequence mutants of LamB have been isolated that introduce charged amino acids into the hydrophobic core, but which do not have serious effects on LamB export (S. Emr and T. Silhavy, unpublished results). One of the *malE* point mutations causes substitution of proline for leucine. Since proline is a helix breaker, secondary structure of the signal peptide may be affected in this mutant. Two deletions were found in the *lamB* gene that shorten the hydrophobic region of the signal sequence by either 4 or 12 amino acids and which could affect secondary structure or the degree of overall hydrophobicity.

Alkaline phosphatase The hybrid protein encoded by a *phoA-lacZ* fusion is found in the cytoplasmic membrane (S. Michaelis, L. Guarente, H. Inouye, and J. Beckwith, manuscript in preparation). Surprisingly, this strain is phenotypically Lac$^-$. This may be due to aberrant folding or tetramerization of the β-galactosidase moiety when the hybrid protein is in the membrane. Selection for Lac$^+$ derivatives has yielded mutants in which the hybrid protein is now cytoplasmic. When these mutations are recombined onto an otherwise wild-type *phoA* gene, alkaline phosphatase can no longer be exported and precursor alkaline phosphatase accumulates in the cytoplasm. DNA sequencing for one mutation has shown that a leucine codon has been changed to an arginine codon in the hydrophobic region of the *phoA* signal sequence (S. Michaelis, D. Oliver, and J. Beckwith, manuscript in preparation).

β-Lactamase Two mutations that dramatically affect export of the TEM β-lactamase encoded by Tn*1* have been reported by Koshland (67). They were obtained by a combination of site-specific in vitro mutagenesis and classical genetic techniques. In one of these mutants a stretch of seven amino acids in the hydrophobic core of the signal sequence has been replaced by seven different amino acids. One of these new residues is arginine, and therefore a positively charged amino acid has been introduced into the otherwise hydrophobic segment. In this mutant export is completely blocked and precursor β-lactamase accumulates in the cytoplasm. In the second mutant three amino acids at the boundary between the charged amino-terminal region and the hydrophobic region are changed. This alteration again results in a shorter hydrophobic core, due to introduction of a charged residue. Export of β-lactamase occurs in this case, but at a rate dramatically slower than wild type. Three other mutants obtained by these investigators substitute neutral or hydrophilic amino acids in the hydrophobic region of the signal peptide. These changes affect processing but not traversal of β-lactamase through the lipid bilayer (67).

Lipoprotein Not all disruptions of the hydrophobic core have dramatic effects on export as is indicated by the *lamB* mutations mentioned above. A mutation was fortuitously obtained in the prolipoprotein gene that introduces a charged amino acid into this region (74). Only prolipoprotein is found in this strain. Therefore, the mutation affects processing. However, the major portion of the prolipoprotein in this strain is found in the outer membrane, indicating that export has occured (75, 76). This lipoprotein mutant shows that export is not dependent on processing.

LamB The first mutations in the basic positively charged amino-terminal region of the signal peptide have recently been isolated (106). Two of these mutations affect translation initiation, probably by introducing secondary structure into the messenger RNA, which interferes with ribosome binding. For one mutation that changes arginine to serine in the charged region, the translation initiation defect can be corrected by introduction of a second mutation near the ribosome-binding site. In this double mutant less LamB protein is made. Studies with gene fusions have led to the conclusion that the rate of completion of translation, rather than its initiation, is probably affected (45). Based on these results, Hall & Schwartz (45) have proposed a model that suggests that translation and export of LamB are coupled. This model is discussed below.

Lipoprotein Inouye and co-workers (55) have constructed mutations in the lipoprotein gene by in vitro site-specific mutagenesis. Synthetic oligonucleotides were employed to make these alterations. These manipulations generated a series of mutants with decreasing positive charge in the amino-terminal region of the lipoprotein signal sequence. In these mutants positively charged residues were deleted or replaced with neutral or negatively charged amino acids. As the charge decreased from (+2) in wild-type precursor to (+1), (0), and (–1), the total amount of production of prolipoprotein plus lipoprotein decreased. It may be that this is due to an effect on the rate of translation or rate of translation initiation. In most mutants, although less lipoprotein is made, its export is not significantly affected. However, in the mutant with a negatively charged amino terminus, export of lipoprotein is drastically altered. Precursor lipoprotein accumulates and is only very slowly processed to the mature form. This is the only mutant of this series that appears to affect export per se.

CONCLUSIONS FROM GENETIC STUDIES The genetic studies above lead to the conclusion that the signal sequence plays an essential role in protein export. Introduction of a charged amino acid into the hydrophobic core of LamB protein, maltose-binding protein, alkaline phosphatase, and β-lactamase signal sequences in some cases causes the precursor forms of these proteins to accumulate in the cytoplasm. This indicates that for these mutants, export cannot be initiated. However, the introduction of a charged amino acid in this region does not always lead to a serious defect in export (see cases of lipoprotein and LamB protein above). Therefore, the position at which the charge is introduced may be important (8). Furthermore, alkaline phosphatase in its wild-type signal sequence contains a lysine residue near the processing site.

On the other hand, mutants in which the positive charge of the basic amino-terminal region of the signal sequence has been altered exhibit a different phenotype. For the LamB mutant and all but one of the lipoprotein mutants of this class, precursors do not accumulate. Instead, translation of these precursors is affected, since in each case less total protein is made. It will be necessary to determine unambiguously whether the step affected is initiation of translation or continuation of translation. These mutations are all located in the vicinity of the ribosome-binding site and therefore a direct effect on translation initiation would not be unexpected. However, in the LamB double mutant it appears that a post-initiation step of translation may be affected. A larger collection of amino-terminal mutations should help clarify the effect of such alterations.

For the lipoprotein and β-lactamase mutants, in vitro mutagenesis was specifically directed to the DNA segment coding for the signal sequence. However, in the case of LamB protein, maltose-binding protein, and alkaline phosphatase, mutations that cause amino acid alterations at any position in the early portion of the proteins that could affect export might theoretically have been obtained.

It is striking that all the mutations detected affect only the structure of the signal sequence. A priori there is no reason to expect that the information directing a protein to traverse the membrane must be confined to that portion of the protein finally cleaved off, but so far this is the case. However, it must be kept in mind that the selection for export-deficient mutants employing fusion strains has certain biases. In particular, the special procedures previously used may select mutations affecting only certain very early steps in export. Also, these selections have required that there be no serious effects on translation of the protein, an approach now brought into question by the suggestion of some coupling between secretion and translation (45, 126–128) (see next section).

Different selection techniques and isolation of more mutants should help determine whether or not all the information required for initiation of export is located on the amino-terminal side of the cleavage site. In addition, the recent use of in vitro mutagenesis techniques should allow the accumulation of a less biased collection of mutants.

Possible Roles For the Signal Sequence in Initiating Export

Several models have been proposed to relate the structure of the signal peptide to its function. One of these, the loop model of Inouye & Halegoua (54), proposes that the basic, positively charged amino-terminal region of this peptide allows attachment of the nascent precursor (and thereby the polysomes synthesizing it) to the negatively charged inner surface of the cytoplasmic membrane by ionic interaction. Subsequently, the hydrophobic core region can insert into the hydrophobic interior of the bilayer, forming a loop structure that initates the export process.

According to this model, if the charge at the amino-terminal region were negative instead of positive, as in one of the lipoprotein mutants described above, the signal sequence could no longer bind to the membrane. Export could not begin and precursor should accumulate in the cytoplasm, which it does in this mutant. An additional role for the signal sequence has been proposed by Hall & Schwartz (45). It is based on their data for the particular LamB double mutant mentioned above. In their view, synthesis and export of LamB are obligatorily coupled. They propose the existence of a "stop translation" amino acid sequence early in the precursor that encompasses, but extends past, the signal sequence. Once synthesized, it binds to the ribosome and prevents translation from continuing unless export is initiated. The model is not described here in any detail since it becomes rather complex in explaining the properties of various signal sequence mutants. However, some such coupling of export and synthesis seems required by the properties of the LamB mutant. In addition, recent results that lend some credence to the model have been obtained in vitro in studies of a eukaryotic system. Walter et al have isolated a membrane protein complex they call signal recognition protein (126–128). This protein specifically binds to polysomes that synthesize a eukaryotic-secreted protein and not to ribosomes that make a cytoplasmic protein. An intact signal sequence is required to mediate this binding. Once bound (and in the absence of membranes) signal recognition protein causes arrest of translation of the secreted protein after 70 amino acids have been incorporated. There is no evidence, at this point, for the existence of such a signal receptor protein in prokaryotes.

ROLE OF THE SEQUENCE OF THE MATURE PORTION OF A PROTEIN IN EXPORT

According to the signal hypothesis, the mature portion of a protein may be thought of as a passenger, transported passively through the membrane after the signal peptide initiates a vectorial transfer process. It should be emphasized that vectorial transfer is consistent with both co- and post-transitional modes of export. In contrast, for the membrane trigger hypothesis, a particular tertiary conformation, determined by the entire protein sequence, is required before export can begin. In this case, the carboxy terminus of a protein would play an active role in its export. However, in all cases examined so far, the evidence shows that the carboxy-terminal portion of exported proteins is not necessary for their export.

The role of the carboxy-terminal portion of several periplasmic proteins has been studied genetically. Chain termination mutations, such as ambers and some frameshifts, cause production of polypeptides that are missing their normal carboxy termini. In many cases such nonsense fragments are unstable and are rapidly degraded. Two amber mutants of the maltose-binding protein were found to synthesize relatively stable polypeptides either one third or nine tenths the size of normal maltose-binding protein. Their export properties have been biochemically determined (58). Both precursor and mature species are detected, indicating that the carboxy end of maltose-binding protein is not required for processing. The mature species of the larger amber fragment is located in the periplasm, since it is released from cells by osmotic shock. Although the mature form of the smaller fragment is not shockable, its accessibility to externally added trypsin indicates that it, too, has traversed the inner membrane. It seems likely that export is normal and that localization is affected for other reasons. Abnormal proteins, including amber fragments, are often insoluble in water and form protein aggregates. Therefore, the smaller amber fragment could be associated with the external surface of the cytoplasmic membrane instead of existing free in the periplasm. The conclusion that can be drawn from this study is that the carboxy end of the mature protein does not play an essential role in cxport or processing of maltose-binding protein.

Similar results have been obtained by Koshland & Botstein (68) in an examination of a series of β-lactamase chain-termination mutants. Again, both precursor and mature forms of the nonsense fragments are made, indicating that processing shows no requirement for an intact carboxy region. Osmotic shock was performed and in no case are the mature peptides found in the periplasm; instead they are cell associated. Results from cell fractionation initially led to the conclusion that the β-lactamase fragments were in the cytoplasm and had not been secreted. However, as already mentioned, cell fractionation can be unreliable. Subsequent analysis of the location of these peptides indicated that they could be digested by externally added trypsin and therefore they must have traversed the cytoplasmic membrane (67). These results are comparable to the findings with the short maltose-binding protein amber fragment. Thus, it appears that the carboxy terminus of β-lactamase is necessary to promote solubility or release from the membrane, but it is not required for this protein to be exported or processed.

Finally, nonsense fragments of an arginine-binding protein have been found in the periplasm (16) and a premature termination product of an outer membrane constituent, OmpA protein, is still capable of export and localization into the outer membrane (14).

Although portions of a protein other than the signal sequence may not play an active role in export to the periplasm, their structure may facilitate or, at the very least, fail to interfere with the process. It may be, for instance, that only some primary sequences are compatible with passage through a lipid bilayer. From thermodynamic arguments, certain stretches of amino acids might be expected to interfere with transmembrane passage (36, 123, 124). β-Galactosidase, which is a very large cytoplasmic protein, is predicted to contain such nonpermissive amino acid stretches (123), and it cannot be dragged through the membrane even when it is provided with a signal sequence and its export is initiated. In this case, then, the carboxy-terminal portion of the hybrid protein actively interferes with its export. So although the carboxy terminus of a periplasmic protein may not actively promote export and its absence is not missed, the presence of a wrong sequence could prevent export.

Some kind of a stop transfer sequence may explain why certain proteins remain embedded in, or attached to, the membrane. Such a stop transfer sequence could reside in or near the segment of a protein that spans the phospholipid bilayer (12). For example, a eukaryotic viral protein, vesicular stomatitis virus glycoprotein (65, 102), is thought to be vectorially discharged through the membrane in a co-translational fashion until a carboxy-terminal stop transfer region reaches the membrane (77). At this point, further secretion is halted. The vesicular stomatitis virus glycoprotein remains permanently in a trans-membrane configuration. Such stop transfer sequences have not been demonstrated for prokaryotic membrane proteins, although a similar dissociation sequence has been hypothesized for the LamB protein (46). If they exist, it should be possible to alter them mutationally, perhaps causing membrane proteins to be secreted. It might also be possible to create or insert stop transfer sequences in proteins to evaluate their effect on export.

EXPORT MACHINERY

For proteins that are co-translationally exported, the signal hypothesis predicts that both membrane and ribosomal proteins are recruited to form a membrane channel through which these proteins are extruded. Early biochemical studies in a eukaryotic system identified transmembrane glycoproteins, called ribophorins, that physically associate with membrane-bound polysomes (69, 70, 104). More recently another protein complex, signal recognition protein, has been found to specifically bind to polysomes synthesizing a secreted protein (126–128). Ribophorins and signal recognition protein may be components of the hypothesized eukaryotic translocation apparatus.

On the other hand, it is theoretically possible to dispense with a need for part or even all of this machinery. Based on thermodynamic considerations, von Heijne & Blomberg (123, 124) calculated it is energetically feasible for proteins to go directly through the bilayer without a proteinaceous pore. For this direct transfer to be accomplished, it is hypothesized that energy for the process is supplied through an interaction between ribosomes and a membrane protein. Engelman & Steitz (36) proposed an even simpler theoretical scheme, which envisions proteins going directly through the membrane without the participation of ribosomal or membrane proteins.

In vitro experiments with bacteriophage M13 coat protein lend plausibility to this latter model. Vesicles made simply from *Escherichia coli* phospholipids and radiochemically pure leader peptidase (see below) correctly integrate and process purified M13 precursor coat protein (132). Although the purity of components in this system remains to be unambiguously demonstrated, these observations indicate that precursor coat protein can insert without the aid of a membrane or ribosome export apparatus, at least in vitro. However, due to the small size of this protein and its probable post-translational mode of export (42) (see also 17, 18), results from this system cannot be generalized to other *E. coli* proteins, particularly those co-translationally exported.

We are still very far from understanding the role and nature of export machinery, or whether it exists at all. However, biochemical and genetic data to be discussed below point to an important role both for ribosomes and for membrane components in mediating the export of many *E. coli* proteins.

Biochemical Studies of Export Machinery

It has recently been demonstrated by Smith (112) that alkaline phosphatase and diphtheria toxin can be co-translationally secreted in vitro into inverted inner membrane vesicles made from *E. coli.* These proteins are processed to their mature form. When the outer (cytoplasmic) surface of the vesicles is subjected to protease treatment prior to the initiation of translation, alkaline phosphatase and diphtheria toxin are made but are not secreted into the vesicles, nor are they processed. However, if protease is added after translation is in progress and ribosomes are already vesicle bound, sequestration into the vesicles and processing do occur. This experiment indicates that inner membrane proteins that reside on the cytoplasmic surface of the membrane may be required for polysome attachment and/or subsequent export of proteins.

The normally energized membrane state and an appropriate membrane fluidity are required for proper protein export and processing. Precursors can be accumulated in whole cells under a variety of conditions that affect the physiological state of the membrane. Phenylethyl alcohol perturbs membrane fluidity and results in accumulation of the precursors of matrix protein, OmpA protein, (44), and β-lactamase (67). Procaine, a drug that alters membrane fluidity, causes precursor alkaline phosphatase to accumulate (72). Growth of unsaturated fatty acid auxotrophs in the presence of elaidate causes accumulation of precursor alkaline phosphatase (94) and OmpA (29). In the presence of energy uncouplers such as carbonyl cyanide-*m*-chlorophenyl hydrazone and 2,4-dinitrophenol, or in oxygen-depleted *uncA* mutants, proton-motive force is dissipated. With these treatments precursors of a large number of proteins have been shown to accumulate. These include M13 coat protein (25, 27), B-subunit of heat-labile enterotoxin (95), leucine-binding protein, β-lactamase (22), OmpF protein, LamB protein, arabinose-binding protein, and maltose-binding protein (37).

Genetic Studies

prlA In mutants bearing alterations in the LamB signal sequence, this protein never reaches the outer membrane but instead is found in the cytoplasm in precursor form. Such strains are unable to grow on maltodextrin as sole carbon source and do not allow plating of bacteriophage lambda, since LamB is required to be in the outer membrane to mediate both of these functions. Suppressor mutations that restore export of mutant LamB have been obtained by selection for growth on maltodextrin (32). Such mutations map at three unlinked loci and are called *prlA, prlB,* and *prlC* (*prl* for protein localization). *prlA* mutations result in the strongest suppression and have been extensively investigated.

When cell fractionation and immuno-precipitation are performed on strains bearing signal sequence mutations, the precursor form of LamB is the major species found and it is in the cytoplasm. A striking difference is seen when a *prlA* mutation is also present in these strains. Now, much of the LamB protein has been processed to the mature form and is found in the outer membrane. Certain alleles of *prlA* suppress certain signal sequence mutations better than others do (31). In some of the mutants suppression is quite dramatic; nearly 100% of the LamB protein is exported despite the presence of a mutant signal sequence.

Suppression is not specific to *lamB* mutations. In *prlA malE* (signal sequence defective) double mutants, the maltose-binding protein is processed and is correctly localized, in this case to the periplasm (31). More recently, *prlA* has been shown to phenotypically suppress *phoA* signal sequence mutations (S. Michaelis, unpublished results). *prlA* maps within an operon coding for ribosomal proteins (32). It is very closely linked to the *rpsE* gene. Although an altered ribosomal protein has not yet been detected in *prlA* strains, it is likely that the gene for a ribosomal component or ribosome-associated protein has been mutated. Whether suppression results directly from this alteration, or whether the mutant protein in turn affects some other cell component involved in export, is difficult to determine. Several modes of action could be envisioned for the mutant *prlA* product. It might enhance the efficiency of some step in export and thereby reduce the normally stringent requirement for a perfect signal sequence. Alternatively, it might interfere with normal export at some step, perhaps translation. Slower translation could allow the extra time required by a defective signal sequence to initiate export. It should be noted in this regard that no gross alterations in translation (such as misreading) have been detected in *prlA* mutants. These strains are healthy and grow well. The finding that a protein encoded in a ribosomal gene cluster participates in suppression of a signal sequence defect is consistent with a co-translational mode of export for LamB protein, maltose-binding protein, and alkaline phosphatase.

Although many signal sequence alterations are sufficient to inhibit export, they do not block processing if the block in export is circumvented by the *prlA* mutation. The requirements for secondary structure or hydrophobicity must be different for initiating export than for serving as a substrate for processing. *prlB* and *prlC* mutants can export proteins with altered signal sequences, but in these cases the proteins are incorrectly processed (31). This may indicate that the proteins are exported via an unusual route.

secA and secB A second genetic approach for identifying components of the cell's export machinery is to isolate mutants impaired in the export of many normal cell envelope proteins. Since such mutations might be expected to render cells nonviable, conditional mutants have been sought. The basis for a selection technique is the finding that a hybrid protein inserted into the membrane can have different enzymatic properties from the same protein when it is located in the cytoplasm (89, 90). A hybrid protein comprised of the amino-terminal portion of maltose-binding protein joined to β-galactosidase is found in the inner membrane. In this location the protein has low enzymatic activity and the strain producing it is unable to grow on lactose as a carbon source. It may be that active β-galactosidase tetramers do not form efficiently or that the membrane environment induces partial denaturation of this protein. Selection for derivatives that can grow on lactose yields mutants in which the hybrid protein has become cytoplasmic. Some of these are due to signal sequence mutations in the *malE* portion of the fusion gene. Others are genetically unlinked to the fusion gene and among these are conditional-lethal mutations.

These mutations define two new genes called *secA* and *secB* (standing for secretion defective) (89, 90). The *secA* mutant strain has been extensively characterized following removal of the fusion gene by genetic means. At 30°C the *secA* mutant grows normally. At 42°C it is unable to grow on any medium, forms filaments, and accumulates precursors of many proteins, including maltose-binding protein, alkaline phosphatase, OmpF protein, and LamB protein. The mutation is not strictly conditional since a small amount of precursor is detected even at 30°C. This phenotype is required by the selection that is performed at 30°C. Not all exported proteins are affected in the *secA* mutant, since a number of periplasmic proteins are still properly localized at the nonpermissive temperature.

The *secA* gene lies at one end of a cluster of genes (at minute 2.5 on the *E. coli* map) involved in cell division and envelope biosynthesis. Other previously known mutations that map in this gene cluster have no effect on protein export. Since these other mutants do not accumulate precursors, it is unlikely that the export defect in the *secA* mutant is a secondary effect of blocking cell division.

Because the selection for *secA* was based on internalization of a hybrid protein, it is likely that export itself and not processing is blocked in the mutant. The simplest explanation for the properties of the *secA* mutation is that it causes alteration of a protein component of the cell's export machinery. However, the possibility that the physiological state of the membrane has been altered as a result of the mutation cannot be excluded. Suppressors of the lethality of *secA* have recently been obtained. Some of these map near *prlA* in the same cluster of genes encoding ribosomal proteins (E. Brickman and D. Oliver, unpublished results). Antibody directed against the *secA* gene product has been made by using the technique developed by Shuman et al (107). With this antibody preparation, it has been shown that *secA* codes for a 90,000-dalton protein located in the cytoplasmic membrane (D. Oliver, unpublished results).

The *secB* mutants are more difficult to characterize since no conditional lethal mutations exist at this locus. They also have pleiotropic effects, causing the accumulation of maltose-binding protein and *OmpF* precursors. They map at minute 80 on the *E. coli* chromosome (90).

TS 215 After localized mutagenesis of a ribosomal gene cluster, Ito and co-workers (59) have isolated a temperature-sensitive lethal mutation, *ts215,* that affects export of certain proteins. When the strain carrying the *ts215* mutation is subjected to high temperature, the rate of conversion of precursor to mature product for maltose-binding protein OmpA protein, and OmpF protein is drastically reduced. Ribosomal protein L15 is present in decreased amounts at the high temperature. However, the lesion does not map in the gene encoding L15 but instead is tightly linked to a different gene, rpsE. prlA maps in a similar position. Whether these two different mutations are allelic is not known. The observation that the *ts215* mutation lies in a ribosomal gene cluster and also causes an export or processing defect again implies a role for ribosomes in export.

perA and tpo Several investigators have isolated mutants in which a number of exported proteins are found in greatly reduced amounts in the cell envelope (23, 129, 130). In contrast to the results found with the *secA* and *ts215* mutants, the precursors of the proteins affected do not accumulate in these mutants.

The *perA* mutant was identified by virtue of its reduced amount of alkaline phosphatase activity (130). The protein profile of the *perA* mutant compared with that of a wild-type strain demonstrates that seven periplasmic and three outer membrane proteins are present in reduced amounts in the mutant. Other investigators have subsequently found that three iron-regulated proteins, which normally appear in the outer membrane upon iron starvation, do not appear in the *perA* strain (78). Transcriptional regulation of *phoA* is not defective in the *perA* mutant. The reduced amounts of alkaline phophatase seen are due to some post-transcriptional effect. *perA* maps at the previously known *ompB* locus at minute 74 on the *E. coli* map.

In an independent study, Wandersman and co-workers (129) obtained a mutant with reduced amounts of both LamB and OmpF proteins by selecting for resistance to a bacteriophage that can use either of these proteins as a receptor. This *tpo* mutation also maps at the *ompB* locus. Their studies show that *tpo* and *perA* are allelic. They found that transcription of the maltose operon (which encodes LamB) is aberrant in *tpo* or *perA* strains.

The *ompB* locus previously has been shown to encode products that affect the transcription of certain outer membrane proteins (47, 48). The above studies indicate they may exert post-transcriptional effects as well. Whether or not they play a role in export is not yet clear.

expA Another mutant has been isolated in which the appearance of several exported proteins is simultaneously affected. The mutational lesion in this strain maps at minute 22 in a gene called *expA* (23). This mutant was identified on the basis of decreased enzymatic activity of two periplasmic acid phosphatases. Ten periplasmic and three outer membrane proteins are found in decreased amounts in the cell envelope of the *expA* mutant, whereas its inner membrane protein profile is indistinguishable from that of a wild-type strain. Whether the lack of appearance of the affected proteins is the result of a transcriptional or post-transcriptional defect is not known.

All the mutations described in this section result in alterations in the expression or appearance of certain noncytoplasmic proteins. For *prlA, ts215* and *secA* and *secB* mutants, the defective gene products may well be directly involved in the export process and thus define components of an export apparatus. *perA, tpo,* and *expA* mutants are not as well understood. They may be affected in the appearance of envelope proteins because of some generalized transcriptional or post-transcriptional regulatory defect, specific to exported proteins. Alternatively, their defective gene products may prevent export by interference.

Although mechanisms can be postulated on the basis of mutational analysis, direct biochemical confirmation is required to establish this mechanism. Development of a prokaryotic in vitro system that can faithfully reproduce the export process will be required to understand the exact steps in export that are mediated by the gene products identified by the selection and analysis of mutants.

PROCESSING

Processing in Vivo

Recent advances have provided insights into the mechanism of cleavage of precursor molecules concomitant with their export. In an elegant study, Josefsson & Randall (63, 64) have shown that some proteins are processed both co-translationally and post-translationally in vivo. In these experiments, proteins are pulse labeled and immuno-precipitation is performed to obtain a particular protein species. In the case of maltose-binding protein, for example, when this precipitate is electrophoresed on a gel that separates proteins according to size, an array of molecules is detected including precursor and mature maltose-binding protein and incomplete peptide chains. The fact that precursor is observed at all indicates that some percentage of maltose-binding protein molecules must be processed post-translationally since this precursor can be chased into mature maltose-binding protein.

However, co-translational processing is also occurring. In situ limited proteolysis is performed on the above array of maltose-binding protein-related polypeptides and these digestion products are electrophoretically separated in a second dimension. The peptides characteristic of the amino terminus of both precursor and mature forms of maltose-binding protein can be identified. In this way it is possible to demonstrate that among the incomplete nascent chains some contain processed amino termini whereas others still have their signal peptide attached.

A similar result to that obtained for maltose-binding protein was observed for the periplasmic proteins alkaline phosphatase and arabinose-binding protein. This work demonstrates that these three proteins are processed both co-translationally and post-translationally in vivo. For all these proteins removal of the signal peptide is a relatively late event occurring after these polypeptides have been elongated to at least 80% of their final length. One protein investigated, TEM β-lactamase, was shown to be processed entirely post-translationally. Interestingly, another periplasmic penicillinase, the *Escherichia coli ampC*-encoded β-lactamase, appears to be processed entirely co-translationally.

The results from this study on the kinetics of processing can be looked on as evidence for the co-translational export of some proteins. The data shows that the signal sequence of several of the proteins examined can be removed before synthesis of these proteins is completed. Although it is formally possible that processing of nascent peptides could occur in the cytoplasm, several pieces of evidence make it seem likely that this cleavage occurs outside of the cytoplasmic compartment. For instance, at high temperature in the *secA* mutant, export is abolished. In this case precursor and not the mature forms of proteins appear in the cytoplasm. The observation that these non-exported but otherwise normal proteins are not processed suggests that the processing enzyme(s) may not be accessible to the inside of the cell. When *prlA* mutants overcome the export defect resulting from a signal sequence mutation, proteins are always both exported and cleaved to mature form. This strong correlation between export and processing further supports the contention that only after export has been achieved can processing occur. For these reasons it seems likely that since maltose-binding protein, arabinose-binding protein, and alkaline phosphatase are co-translationally processed, they must also be co-translationally exported. For alkaline phosphatase this has been directly demonstrated by other experiments discussed above (115).

On the other hand, the studies of Josefsson & Randall indicate that processing does not take place until relatively late in the synthesis of the polypeptide chain, revealing a greater complexity to the export process than might have been imagined (63, 64).

Mutations That Affect Processing

Mutants have been isolated for several proteins in which the precursor forms of these proteins can traverse the lipid bilayer, but they are processed at a slow rate or not all. The mutation in the hydrophobic portion of the signal sequence-coding region of the lipoprotein gene (mentioned above) results in complete inhibition of conversion of the precursor to mature lipoprotein (74). Nonetheless, most of the precursor lipoprotein made in this strain is found in the outer membrane, indicating that processing and not export is affected (75, 76). This mutation alters an amino acid seven residues away from the cleavage site.

Three mutations in the gene coding for β-lactamase that affect processing have been isolated by Koshland (67). Trypsin accessibility experiments indicate that the mutant precursors traverse the lipid bilayer in these mutant strains. However, conversion of the precursor β-lactamase to mature form is slow or is completely blocked depending on the particular mutation present. All of the mutations lie in the signal sequence-coding region, two to five amino acids before the cleavage site. In one mutant a proline residue is changed to serine. In a second mutant this same proline is changed to leucine. In the latter mutant processing is not seen to occur at all (in the course of the experiment), whereas in the former case, processing occurs but at a rate slower than for wild type. In the third mutant a stretch of four amino acids that includes the above proline is replaced by four new amino acids. In this case processing is also completely blocked. These mutants indicate that the domain recognized by the processing enzyme lies at least in part in the signal sequence.

In the case of bacteriophage M13 coat protein, a mutant in which the second amino acid of the mature coat protein is altered exhibits a slow rate of precursor cleavage (13, 103). This is the only example known in which an alteration downstream from the cleavage site affects processing. This finding suggests that for M13 coat protein, part of the domain recognized by the processing enzyme extends beyond the signal sequence.

Isolation of Leader (Signal) Peptidase

A processing function present in *E. coli* membranes was first identified in studies of M13 coat protein (17, 79). When bacteriophage M13 DNA is incubated in an in vitro coupled transcription-translation system in the absence of membranes, precursor M13 coat protein is a major product made. If *E. coli* membranes are added to the translation mixture this precursor is cleaved to mature form. Both inner and outer membranes can effect this cleavage (17, 79). Wickner's group has recently purified this processing activity, now called leader peptidase, from *E. coli* membranes. Purified leader peptidase has been shown to cleave purified precursor M13 coat protein in an endoproteolytic fashion, which yields mature coat protein and an intact signal peptide (138). Leader peptidase exhibits the striking property of being present in both inner and outer membrane fractions in roughly equal amounts (137). This is the first example of an *E. coli* membrane protein with this sort of dual distribution. Purified leader peptidase has also been shown to process the precursor forms of several other exported proteins in vitro (137).

By screening through a library of cloned *E. coli* DNA, a plasmid causing 30-fold overproduction of leader peptidase activity has been identified (26). This leader peptidase exhibits the same biochemical properties as the purified leader peptidase described above. Both inner and outer membrane preparations reflect this increased level of activity. Whether this plasmid encodes leader peptidase itself or some positive regulator that increases its level of production has not yet been shown. If this plasmid does encode leader peptidase it should be possible to map this gene on the *E. coli* chromosome, and this information should facilitate isolation of mutants. With such genetic studies it can be determined whether there are one or more processing activities in *E. coli.*

CONCLUDING REMAKRS

Considerable progress has been made in the study of protein export in bacteria. Mutant analysis has revealed the essential role of the signal sequence in the export process and has shown that the carboxy-terminal portion of a number of proteins is not essential for their passage through the cytoplasmic membrane. In vivo and in vitro genetic approaches should provide in the near future a detailed picture of the important aspects of signal sequence structure and should indicate whether or not any portions of the amino acid sequence of exported proteins other than the signal sequence play a significant role in the process. Such approaches should also contribute to understanding how proteins are directed to the different envelope compartments, e.g. what distinguishes a periplasmic protein from an outer membrane protein? Kinetic studies on the processing of envelope proteins have provided greater insight into the details of export. Genetic analysis is now being extended to the export machinery of the cell. New approaches to isolating pleiotropically defective secretion mutants and the availability of the cloned gene for a signal peptidase should allow a more detailed dissection of the steps involved in the transfer of proteins through membranes.

However, studies on phenomena occurring at the level of membranes are very complex—considerably more complex than, for instance, studies on gene expression and regulation. It is much more difficult to separate out direct and indirect effects in this highly ordered structure, which is comprised of so many different molecules, which can vary in such properties as fluidity, and which is involved in so many cellular processes, including the transport of small molecules and cell growth and division. Furthermore, fractionation procedures to determine the cellular location of proteins (e.g. intermediates in the export process) are fraught with difficulties and artifacts (57, 67, 68).

These problems apply particularly to the question of whether proteins are exported to the cell envelope co-translationally or post-translationally. These are important studies because the answer will have implications for such issues as the energetics of the transfer process. Although much of the evidence supports the co-translational model, no definitive experiment distinguishes the two. In the case of β-lactamase and M13 coat protein (42, 67), and under certain abnormal conditions for other proteins, (22, 55, 57, 59), some of the studies suggest that post-translational transfer of these proteins can occur. Such studies have relied to a large extent on existing cell fractionation procedures. One explanation for the current body of data is that co-translational export is the normal mode, but that under abnormal conditions the post-translational mode can be used.

In the case of β-lactamase, the apparent post-translational export might be ascribed to the inability of the bacterium to handle by the normal mechanism this protein, which evolved in some other organism. Alternatively, it may be that hitherto unrecognized properties of the Gram-negative bacteria and their membranes have generated artifacts that further complicate fractionation studies. One of the challenges in this field is the need to improve approaches and techniques for studying the prokaryotic cell membranes.

ACKNOWLEDGMENTS

We wish to thank Jeffrey Felton for critically reading and editing this review, and Ann McIntosh and Rosemary Bacco for their undaunted efforts in preparing the manuscript.

Literature Cited

1. Achtman, M., Manning, P. A., Edelbluth, C., Herrlich, P. 1979. *Proc. Natl. Acad. Sci. USA* 76:4837–41
2. Austen, B. M. 1979. *FEBS Lett.* 103:308–13
3. Bassford, P., Beckwith, J. 1979. *Nature* 277:538–41
4. Beacham, I. R. 1979. *Int. J. Biochem.* 10:877–83
5. Beck, E., Bremer, E. 1980. *Nucl. Acids Res.* 8:3011–27
6. Beckwith, J. R., Silhavy, T. J. 1982. *Methods Enzymol.* In press.
7. Bedouelle, H., Bassford, P. J. Jr., Fowler, A. V., Zabin, I., Beckwith, J., Hofnung, M. 1980. *Nature* 285:78–81
8. Bedouelle, H., Hofnung, M. 1981. In *Membrane Transport and Neuroreceptors,* pp. 399–403. New York: Liss
9. Bedouelle, H., Hofnung, M. 1981. In *Intermolecular Forces,* ed. B. Pullman, pp. 361–72. The Netherlands: Reidel Publ.
10. Blobel, G., Dobberstein, B. 1975. *J. Cell Biol.* 67:835–51
11. Blobel, G., Dobberstein, B. 1975. *J. Cell Biol.* 67:852–62
12. Blobel, G., Walter, P., Chang, C. N., Goldman, B. M., Erickson, A. H., Lingappa, V. R. 1979. *Symp. Soc. Exp. Biol.* 33:9–36
13. Boeke, J. D., Russel, M., Model, P. 1980. *J. Mol. Biol.* 144:103–16
14. Bremer, E., Beck, E., Hindennach, I., Sonntag, I., Henning, U. 1980. *Mol. Gen. Genet.* 179:13–20
15. Buchel, D. E., Gronenborn, B., Muller-Hill, B. 1980. *Nature* 283:541–45
16. Celis, R. T. F. 1981. *J. Biol. Chem.* 256:773–79
17. Chang, C. N., Blobel, G., Model, P. 1978. *Proc. Natl. Acad. Sci. USA* 75:361–65
18. Chang, C. N., Model, P., Blobel, G. 1979. *Proc. Natl. Acad. Sci. USA* 76:1251–55
19. Chou, P. Y., Fasman, G. D. 1978. *Adv. Enzymol.* 47:45–148
20. Crowlesmith, I., Gamon, K., Henning, U. 1981. *Eur. J. Biochem.* 113:375–80

21. Dallas, W. S., Falkow, S. 1980. *Nature* 288:499–501
22. Daniels, C. J., Bole, D. G., Quay, S. C., Oxender, D. L. 1981. *Proc. Natl. Acad. Sci. USA* 78:5396–400
23. Dassa, E., Boquet, P. -L. 1981. *Mol. Gen. Genet.* 181:192–200
24. Dassa, E., Ferlat, G., Boquet, P. -L. 1978. *Biochem. Biophys. Res. Commun.* 81:616–22
25. Date, T., Goodman, J. M., Wickner, W. T. 1980. *Proc. Natl. Acad. Sci. USA* 77:4669–73
26. Date, T., Wickner, W. 1981. *Proc. Natl. Acad. Sci. USA* 78:6106–10
27. Date, T., Zwizinski, C., Ludmerer, S., Wickner, W. 1980. *Proc. Natl. Acad. Sci. USA* 77:827–31
28. Davis, B. D., Tai, P. -C. 1980. *Nature* 283:433–38
29. DiRienzo, J. M., Inouye, M. 1979. *Cell* 17:155–61
30. Ehring, R., Beyreuther, K., Wright, J. K., Overath, P. 1980. *Nature* 283:537–40
31. Emr, S. D., Bassford, P. J. 1982. *J. Biol. Chem.* In press.
32. Emr, S. D., Hanley-Way, S., Silhavy, T. J. 1981. *Cell* 23:79–88
33. Emr, S. D., Hedgpeth, J., Clement, J. -M., Silhavy, T. J., Hofnung, M. 1980. *Nature* 285:82–85
34. Emr, S. D., Schwartz, M., Silhavy, T. J. 1978. *Proc. Natl. Acad. Sci. USA* 75:5802–6
35. Emr, S. D., Silhavy, T. J. 1980. *J. Mol. Biol.* 141:63–90
36. Engleman, D. M., Steitz, T. A. 1981. *Cell* 23:411–22
37. Enquist, H. G., Hirst, T. R., Harayama, S., Hardy, S. J. S., Randall, L. L. 1981. *Eur. J. Biochem.* 116:227–33
38. Ferrazza, D., Levy, S. B. 1980. *J. Biol. Chem.* 255:8955–58
39. Fraser, T. H., Bruce, B. J. 1978. *Proc. Natl. Acad. Sci. USA* 75:5936–40
40. Garwin, J. L., Beckwith, J. 1982. In *Membranes and Transport 1982,* ed. A. Martonosi, pp. 315–21. New York: Plenum
41. Garwin, J. L., Beckwith, J. 1982. *J. Bacteriol.* 149:789–92
42. Goodman, J. M., Watts, C., Wickner, W. 1981. *Cell* 24:437–41
43. Grundstrom, T., Jaurin, B., Edlund, T., Normark, S. 1980. *J. Bacteriol.* 143: 1127–34
44. Halegoua, S., Inouye, M. 1979. *J. Mol. Biol.* 130:39–61
45. Hall, M. N., Schwartz, M. 1982. *Ann. Microbiol. Inst. Pasteur* 133A:123–27
46. Hall, M. N., Silhavy, T. J. 1981. *Ann. Rev. Genet.* 15:91–142
47. Hall, M. N., Silhavy, T. J. 1981. *J. Mol. Biol.* 151:1–15
48. Hall, M. N., Silhavy, T. J. 1981. *J. Mol. Biol.* 146:23–43
49. Hedgpeth, J., Clement, J. -M., Marchal, C., Perrin, D., Hofnung, M. 1980. *Proc. Natl. Acad. Sci. USA* 77:2621–25
50. Higgens, C. F., Ames, G. F. -L. 1981. *Proc. Natl. Acad. Sci. USA* 78:6038–42
51. Inouye, H., Barnes, W., Beckwith, J. 1982. *J. Bacteriol.* 149:434–39
52. Inouye, H., Beckwith, J. 1977. *Proc. Natl. Acad. Sci. USA* 74:1440–44
53. Inouye, M., ed. 1979. *Bacterial Outer Membranes: Biogenesis and Functions* New York: Wiley. 534 pp.
54. Inouye, M., Halegoua, S. 1980. *CRC Crit. Rev. Biochem.* 7:339–71
55. Inouye, S., Franceschini, T., Nakamura, K., Soberon, X., Itakura, K., Inouye, M. 1982. *Proc. Natl. Acad. Sci. USA* In press.
56. Inouye, S., Wang, S., Sekizawa, J., Halegoua, S., Inouye, M. 1977. *Proc. Natl. Acad. Sci. USA* 74:1004–8
57. Ito, K., Bassford, P. J., Beckwith, J. 1981. *Cell* 24:707–17
58. Ito, K., Beckwith, J. R. 1981. *Cell* 25:143–50
59. Ito, K. 1982. *Ann. Microbiol. Inst. Pasteur* 133A:101–4
60. Jakes, K. S., Model, P. 1979. *J. Bacteriol.* 138:770–78
61. Jaurin, B., Grundstrom, T. 1981. *Proc. Natl. Acad. Sci. USA* 78:4897–901
62. Jaurin, B., Grundstrom, T., Edlund, T., Normark, S. 1981. *Nature* 290:221–25
63. Josefsson, L. G., Randall, L. L. 1981. *Cell* 25:151–57
64. Josefsson, L. G., Randall, L. L. 1981. *J. Biol. Chem.* 256:2504–7
65. Katz, F. N, Rothman, J. E., Lingappa, V. R., Blobel, G., Lodish, H. F. 1977. *Proc. Nal. Acad. Sci. USA* 74:3278–82
66. Kikuchi, Y., Yoda, K., Yamasaki, M., Tamura, G. 1981. *Nucl. Acids Res.* 9:5671–78.
67. Koshland, D. 1982. *A genetic analysis of β-lactamase.* PhD thesis. Mass. Inst. Technol., Cambridge.
68. Koshland, D., Botstein, D. 1980. *Cell* 20:749–60
69. Kreibich, G., Freienstein, C. M., Pereyra, B. N., Ulrich, B. L., Sabatini, D. D. 1978. *J. Cell Biol.* 77:488–506
70. Kreibich, G., Ulrich, B. L., Sabatini, D. D. 1978. *J. Cell Biol.* 77:464–87
71. Landick, R., Oxender, D. L. See Ref. 40, pp. 000–00

72. Lazdunski, C., Baty, D., Pages, J. M. 1979. *Eur. J. Biochem.* 96:49–57
73. Lieve, L., ed. 1973. *Bacterial Membranes and Walls.* New York: Marcel Dekker
74. Lin, J. J. C., Kanazawa, H., Ozols, J., Wu, H. C. 1978. *Proc. Natl. Acad. Sci. USA* 75:4891–95
75. Lin, J. J. C., Kanazawa, H., Wu, H. C. 1980. *J. Biol. Chem.* 255:1160–63
76. Lin, J. J. C., Kanazawa, H., Wu, H. C. 1980. *J. Bacteriol.* 141:550–57
77. Lingappa, V. R., Katz, F. N., Lodish, H. F., Blobel, G. 1978. *J. Biol. Chem.* 253:8667:70
78. Lundrigan, M., Earhart, C. F. 1981. *J. Bacteriol* 146:804–7
79. Mandel, G., Wickner, W. 1979. *Proc. Natl. Acad. Sci. USA* 76:236–40
80. Marchal, C., Perrin, D., Hedgpeth, J., Hofnung, M. 1980. *Proc. Natl. Acad. Sci. USA* 77:1491–95
81. Mirelman, D. See Ref. 53, pp. 115–66
82. Mock, M., Schwartz, M. 1978. *J. Bacteriol.* 136:700–7
83. Moreno, F., Fowler, A. V., Hall, M., Silhavy, T. J., Zabin, I., Schwartz, M. 1980. *Nature* 286:356–59
84. Movva, N. R., Nakamura, K., Inouye, M. 1980. *J. Biol. Chem.* 255:27–29
85. Nakamura, K., Inouye, M. 1979. *Cell* 18:1109–17
86. Nichols, J. C., Tai, P. -C., Murphy, J. R. 1980. *J. Bacteriol.* 144:518–23
87. Nielsen, J., Hansen, F. G., Hoppe, J., Friedl, P., von Meyenburg, K. 1981. *Mol. Gen. Genet.* 184:33–39
88. Noegel, A., Rdest, U., Goebel, W. 1981. *J. Bacteriol.* 145:233–47
89. Oliver, D. B., Beckwith, J. 1981. *Cell* 25:765–72
90. Oliver, D., Kumamoto, C., Quinlan, M., Beckwith, J. 1982. *Ann. Microbiol. Inst. Pasteur* 133A:105–10
91. Osborn, M. J., Wu, H. C. P. 1980. *Ann. Rev. Microbiol.* 34:369–422
92. Oxender, D. L., Anderson, J. J., Daniels, C. J., Landick, R., Gunsalus, R. P., Zurawski, G., Selker, E., Yanofsky, S. 1980. *Proc. Natl. Acad. Sci. USA* 77:1412–16
93. Oxender, D. L., Anderson, J. J., Daniels, C. J., Landick, R., Gunsalus, R. P., Zurawski, G., Yanofsky, C. 1980. *Proc. Natl. Acad. Sci. USA* 77:2005–9
94. Pages, J. M., Poivant, M., Varenne, S., Lazdunski, C. 1978. *Eur. J. Biochem.* 86:589–602
95. Palva, E. T., Hirst, T. R., Hardy, S. J. S., Holmgren, J., Randall, L. 1981. *J. Bacteriol.* 146:325–30
96. Pratt, J. M., Holland, B., Spratt, B. G. 1981. *Nature* 293:307–9
97. Pugsley, A. P., Rosenbusch, J. P. 1981. *J. Bacteriol.* 147:186–92
98. Randall, L. L., Hardy, S. J. S., Josefsson, L. -G. 1978. *Proc. Natl. Acad. Sci. USA* 75:1209–12
99. Roggenkamp, R., Kustermann-Kuhn, B., Hollenberg, C. P. 1981. *Proc. Natl. Acad. Sci. USA* 78:4466–70
100. Rosen, B. P., ed. 1978. *Bacterial Transport.* New York: Marcel Dekker. 684 pp.
101. Rosenblatt, M., Beaudette, N. V., Fasman, G. D. 1980. *Proc. Natl. Acad. Sci. USA* 77:3983–87
102. Rothman, J. E., Lodish, H. F. 1977. *Nature* 269:775–80
103. Russel, M., Model, P. 1981. *Proc. Natl. Acad. Sci. USA* 78:1717–21
104. Sabatini, D. D., Kreibich, G. Morimoto, T., Adesnik, M. 1982. *J. Cell Biol.* 92:1–22
105. Schaller, H., Beck, E., Takanami, M. 1978. In *The Single Stranded DNA Phages,* ed. D. Denhardt, D. Dressler, D. Ray, pp. 139–53. Cold Spring Harbor: Cold Spring Harbor Lab.
106. Schwartz, M., Roa, M., Debarbouille, M. 1981. *Proc. Natl. Acad. Sci. USA* 78:2937–41
107. Shuman, H. A., Silhavy, T. J., Beckwith, J. R. 1980. *J. Biol. Chem.* 255: 168–74
108. Silhavy, T. J., Bassford, P. J. Jr., Beckwith, J. R., See Ref. 53, pp. 203–54
109. Silhavy, T. J., Beckwith, J. R. 1982. *Methods Enzymol.* In press
110. Silhavy, T. J., Casadaban, M., Shuman, H. A., Beckwith, J. R. 1976. *Proc. Natl. Acad. Sci. USA* 73:3432–27
111. Silhavy, T.J., Shuman, H. A., Beckwith, J., Schwartz, M. 1977. *Proc. Natl. Acad. Sci. USA* 74:5411–15
112. Smith, W. P. 1980. *J. Bacteriol.* 141: 1142–47
113. Smith, W. P., Tai, P. -C., Davis, B. D. 1979. *Biochemistry* 18:198–202
114. Smith, W. P., Tai, P. C., Murphy, J. R., Davis, B. D. 1980. *J. Bacteriol.* 141:184–89
115. Smith, W. P., Tai, P. -C., Thompson, R. C., Davis, B. D. 1977. *Proc. Natl. Acad. Sci. USA* 74:2830–34
116. So, M., McCarthy, B. J. 1980. *Proc. Natl. Acad. Sci. USA* 77:4011–15
117. Spicer, E. K., Kavanaugh, W. M., Dallas, W. S., Falkow, S., Konigsberg, W. H., Schafer, D. E. 1981. *Proc. Natl. Acad. Sci. USA* 7:50–54.

118. Springer, W., Goebel, W. 1980. *J. Bacteriol.* 144:53–59
119. Sugimoto, K., Sugisaki, H., Okamoto, T., Takanami, M. 1977. *J. Mol. Biol.* 111:487–507
120. Sutcliffe, J. G. 1978. *Proc. Natl. Acad. Sci. USA* 75:3737–41
121. Talmadge, K., Kaufman, J., Gilbert, W. 1980. *Proc. Natl. Acad. Sci. USA* 77: 3988–92
122. Talmadge, K., Stahl, S., Gilbert, W. 1980. *Proc. Natl. Acad. Sci. USA* 77: 3369–73
123. von Heijne, G. 1980. *Eur. J. Biochem.* 103:431–38
124. von Heijne, G., Blomberg, C. 1979. *Eur. J. Biochem.* 97:175–81
125. van Tiel-Menkveld, G. J., Veltkamp, E., DeGraaf, F. K. 1981. *J. Bacteriol.* 146:41–48
126. Walter, P., Blobel, G. 1981. *J. Cell Biol.* 91:551–56
127. Walter, P., Blobel, G. 1981. *J. Cell Biol.* 91:557–61
128. Walter, P., Ibrahimi, I., Blobel, G. 1981. *J. Cell Biol.* 91:545–50
129. Wandersman, C., Moreno, F., Schwartz, M. 1980. *J. Bacteriol.* 143: 1374–83
130. Wanner, B. L., Sarthy, A., Beckwith, J. 1979. *J. Bacteriol.* 140:229–39
131. Watson, D. H. 1980. *Biochem. J.* 185:463–71
132. Watts, C., Silver, P., Wickner, W. 1981. *Cell* 25:347–53
133. Waxman, D. J., Strominger, J. L. 1981. *J. Biol. Chem.* 256:2059–66
134. Waxman, D. J., Strominger, J. L. 1981. *J. Biol. Chem.* 256:2067–77
135. Wickner, W. 1979. *Ann. Rev. Biochem.* 48:23–45
136. Wilson, G. Van., Hogg, R. W. 1980. *J. Biol. Chem.* 255:6745–50
137. Zwizinski, C., Date, T., Wickner, W. 1981. *J. Biol. Chem.* 256:3593–97
138. Zwizinski, C., Wickner, W. 1980. *J. Biol. Chem.* 255:7973–77

Ann. Rev. Microbiol. 1982. 36:467–93

THE LABORATORY APPROACH TO THE DETECTION OF BACTEREMIA

Richard C. Tilton

Department of Laboratory Medicine, University of Connecticut Health Center, Farmington, Connecticut 06032

INTRODUCTION

This review on laboratory approaches for the diagnosis of bacteremia critically evaluates traditional methods for the detection of bacteremia and assesses the state of the art. I make no claim for a comprehensive treatise on the clinical aspects of bacteremia, nor is this an attempt to exhaustively review the myriad of organisms that may be isolated from the blood. Rather, this review sumamrizes the primary factors that face both the clinician and the microbiologist in the utilization of the laboratory for the detection of bacteria and some fungi in blood.

Since early times, it has been known that disturbances of the blood cause people to be sick. Blood was one of the five humors described as the basis for disease, and alterations in one or more of these humors was thought to predispose to disease or to health. Bloodletting, or partial exsanguination, has been practiced for centuries, and even in the 1980s some social cultures still believe that removal of whole blood may cure disease. Such practices predate the discovery of microorganisms, however. The modern era of infectious disease and microbiology now understands blood as not only a tissue that can itself become infected, but also as a vehicle that may act as a microbial transport system from one organ to another.

In 1925, Wright (138) reported on endocarditis and bacteremia. His review, minus some technology, bears a startling resemblance to this one. Many of the problem areas extant in the 1980s, such as how much blood to collect, how often to collect it, and the significance of organisms recov-

0066-4227/82/1001-0467$02.00

ered from the blood, both pre- and postmortem, were issues in 1925. Wright discussed the effects of inhibitory substances in blood that affected blood culture and concluded that the primary reasons for culture failure was frequently due to the inhibitory nature of the blood, i.e. phagocytosis. He also thought that the delay in the growth of many of the organisms was due to prolonged lag phases in their growth cycle.

For the purposes of this review, the terms bacteremia and septicemia are used interchangeably, with recognition that bacteremia reflects the presence of bacteria in blood, whereas septicemia may indicate not only microbial presence but also the characteristic symptoms of septicemia, such as fever, chills, tachycardia, shock, and leukocytosis.

With the exception of intravascular infections, such as bacterial endocarditis, mycotic aneurysm, and thrombophlebitis, bacteria usually enter the circulation through the lymphatic system. When bacteria multiply at a local site of infection such as the lung, the determining factor in bacteremia is a function of the occurrence of local conditions that favored drainage of lymph from the infected area to the thoracic duct and eventually to the venous blood (122). In most individuals, once bacteria enter the blood, they are effectively and rapidly removed by the reticuloendothelial system in the liver, spleen, and bone marrow, and by phagocytosis.

Bacteremia may be transient, intermittent, or continuous. Transient bacteremia may result from such innocuous procedures as brushing the teeth or a difficult bowel movement. It can also result from invasive procedures such as dental manipulation, sigmoidoscopy, prostatic massage, or tonsillectomy or other surgical procedures and venous infusion of fluids. The manipulation of any organ or mucous membrane contaminated with bacteria and contiguous to the blood supply may predispose to transient bacteremia. Fortunately for most of us, the average individual copes well with this shower of microorganisms and at the very most suffers from slight fever or a chill. In transient bacteremia it takes approximately 7 min for organisms to be cleared from the circulatory system (105).

In 1932, Reith & Squier (99) reported that 12% of 99 normal people tested had positive blood cultures. However, the majority of the organisms may have been contaminants. MacGregor & Beatty (81) stated that 152 of 1707 (8.9%) blood cultures were judged to contain contaminants. In a recent study by Wilson (137), approximately 2.1% of patients who had no demonstrable site of infection had positive blood cultures. The isolation of these organisms probably represented contamination, not bacteremia. True bacteremia rarely occurs in the normal individual in the absence of manipulative procedures. The biggest risk to transient bacteremia is in those patients who may have valvular heart disease. Infectious endocarditis may result. The transient appearance of microorganisms in the blood may presage infection. The rigor noted in the early stages of pneumonia is due to transient bacteremia (122).

Continuous bacteremia may be seen in typhoid fever, brucellosis, endocarditis, and endarteritis. In most other infections, such as gram-negative sepsis, bacteremia is intermittent. A temperature spike or chill is preceded by a shower of bacteria into the blood approximately 30–45 min before the symptoms. Periodicity is characteristic in some diseases and random in others. It is difficult to predict when blood cultures should be taken, so for maximum recovery blood should be drawn at intervals, not all at once. This issue is discussed in detail below.

Bacteremia may present in several ways (7):

1. a classic septicemia with invasion of the blood stream from a distinct infection (in gram-negative septicemia, fever, leucocytosis, hypotension, and prostration are present; in many patients, however, particularly the elderly, the initial symptoms may be confusion and disorientation);
2. minimal clinical signs in the immunosuppressed patient or failure to thrive in the newborn;
3. adjunct to meningococcal or gonococcal disease;
4. accompanying meningitis, pneumonia, systemic abscesses, or infections of the biliary tree (usually the presenting signs are those of the primary infection);
5. as a result of intravascular infection;
6. as part of a multisystem infection such as brucellosis, leptospirosis, or typhoid fever [many times, the patient may have a fever of unknown origin];
7. as a result of infusion of contaminated intravenous products or of trauma to a heavily colonized area, such as the abdomen.

The spectrum of etiologic agents of bacteremia has changed over the past 20-30 years. This change has been due to events related both to the patient and to technological advances in microbiology. Incidence studies, unfortunately, are method dependent, and the significant increase in anaerobic isolates from the blood may reflect only new technology and increased awareness, not necessarily an absolute increase in anaerobic infections. Numerous incidence reports have been published. Reporting on blood culture isolates at the Mayo Clinic from 1968 to 1975, Ilstrup (67) shows that the rate of positive cultures has increased for *Corynebacterium, Propionibacterium, Serratia,* and *Staphylococcus epidermidis* and has decreased for *Klebsiella, Enterococcus,* alpha-hemolytic streptococci, *Bacteroides,* and *Pseudomonas aeruginosa.* The other organism groups exhibited no trends that were statistically significant. As part of the same study, 25,000 blood culture specimens from 7000 patients received in 1975 were analyzed. There were 2409 positive specimens from 1211 patients. The most frequently seen organism in 1975 was *Corynebacterium* (30% of positives). The next most frequent organism was *S. epidermidis* (19.5%), followed closely by *Escherichia coli* (16.7%), and *Staphylococcus aureus* (9.8%). *Corynebacterium, Propionibacterium* and *S. epidermidis,* are often contaminants in blood cultures as they are normal commensals of the skin. The incidence of *Corynebacterium* is seen only in hospitalized patients; parallel studies done in the normal population with similar methods of skin asepsis do not show the high proportion of *Corynebacterium.* Many of the patients harboring these isolates, however, showed no signs or symptoms of microbial disease. The isolation of *S. epidermidis* from a blood culture leads to similar diagnostic dilemmas as it also is part of the indigenous skin microflora.

The Subcommittee on Health of the Committee on Labor and Public Welfare in the United States Senate commissioned a study on gram-negative rod bacteremia. Four questions were addressed. The first question concerned the incidence of gram-negative rod bacteremia in the United States. The best estimate in 1974 was 71,000 cases. The second question attempted to place responsibility for the increased incidence of gram-negative rod bacteremia. A number of interrelated factors were implicated, including invasive devices such as intravenous and bladder catheters, inhalation therapy equipment, intentional alterations in the hosts immune system, longer survival times of patients with neoplasia, diabetes, and other debilitating illnesses, and increased use of antibiotics. The third question concerned the number of deaths. Fatality ratios were reported to be 25–32% in gram-negative rod bacteremia. The final question sought information on the current status of medical treatment. The answer was that although the use of an appropriate antibiotic has been helpful in the management of patients with gram-negative shock, mortality remains high. Kreger et al (73) evaluated 612 episodes of gram-negative bacteremia over a 10-year period and demonstrated its progressively increasing frequency. The increase was similarly associated with patients with severe underlying disease, extremes of age, manipulative procedures, antibiotics, steroids, and antimetabolite therapy. In these 612 episodes, 72% of blood cultures were positive. Bacteremia was of low magnitude in 77% of the patients (<10 gram-negative bacilli per ml). *E. coli* was the most frequently isolated bacterium, followed by the *Klebsiella-Enterobacter-Serratia* group, *P. aeruginosa, Proteus, Providencia,* and *Bacteroides.*

Several authors stress the importance of polymicrobic bacteremia. Kiani et al (71) reported that polymicrobic bacteremia increased from 6% in 1970 to 13% in 1975. In 88 patients studied, mortality was 44% compared to 18% in patients with monomicrobial bacteria. Kreger et al (73) also noted that bacteremia caused by multiple species of bacteria was associated with higher fatality rates.

These incidence data have little practical value for the microbiologist as methods must be designed to isolate many different types of organisms, not only gram-negative rods. Under normal circumstances, the isolation of aerobic and facultative bacteria is not technically difficult. The isolation of *Legionella, Mycoplasma,* nutritionally fastidious streptococci, antibiotic-damaged organisms, *Leptospira,* and fungi pose a much greater technological challenge. The challenges that face the microbiologist in the 1980s include those of control of utilization, rapid detection, antibiotic susceptibility tests, and isolation of problem organisms.

Current blood culture methodology has been reviewed recently by Blazevic et al (10). A survey was sent out and 341 laboratories responded. Laboratories represented community hospitals, Veteran's Administration hospitals, and hospitals directly affiliated with medical schools. The majority of the laboratories were in hospitals with 200-500 beds and two thirds of the respondents processed 5000 blood cultures per year or less. In 68% of the laboratories, blood was collected by syringe and was inoculated into a variety of blood culture media. The majority of the respondents used two culture bottles, one vented, the other unvented. Approximately half used Trypticase soy broth and a significant portion of the remainder used ^{14}C-labeled broth for radiometric procedures. A 50-ml culture bottle to which was added 5 ml of blood was the most popular. In 60 out of 341 replies, 100-ml volumes were used with 10 ml added. A great majority of respondents (304) supplemented the media used with sodium polyanethol sulfonate 0.025 to 0.03%. Only four laboratories routinely performed quantitative blood cultures and 315 indicated they never did. Two thirds incubated aerobic and anaerobic cultures 7 days and examined them visually for growth on a daily basis. In the absence of turbidity, growth was detected by either radiometric methods, a blind Gram stain, or a blind subculture. In many cases, respondents used combinations of these three methods. Many (188) performed blind subcultures at the end of the incubation period. Approximately one third of the microbiologists queried attempted to control the numbers of blood cultures taken on a single patient.

In determining how often blood cultures should be drawn, it is important to consider the diseases to be diagnosed. In bacterial endocarditis, the low level bacteremia is continuous, so the spacing of the blood culture is not as critical. Belli & Waisbren (9) found that in 52 of 82 cases of bacterial endocarditis, an organism was isolated on the first culture attempt and that in 6 of the 82 cases more than five cultures were necessary. Bartlett et al (7) recommend four to six blood cultures for the diagnosis of bacterial endocarditis. If the organism is not isolated by that time, it is probably not going to be. Werner et al (134) found that in patients with streptococcal endocarditis the organisms were isolated from the first culture in 96% of the cases. In staphylococcal endocarditis, the first culture yielded positive results in approximately 90% of the instances. If the patient has been treated prior to admission with an antimicrobial agent, it is more difficult to isolate the organism and multiple cultures may be necessary. For most patients, three blood cultures within a 24-hr period is usually sufficient and six is maximum. Washington et al (130) studied 80 patients with bacteremia who did not have endocarditis. In all of these patients at least four separate sets of blood cultures were collected in a 14-hr period. Of these cultures, 80% were positive in the first set of bottles, 89% were positive in the first two, and 99% were positive in the first three. The blood culture should be drawn just prior to chills and fever elevation. If such events are predictable, there will be optimal isolation of the organism. These events do not usually cycle on schedule. The minimum time between blood collection should be 1 hr. In the acutely ill patient, it may be necessary to draw multiple sets of blood cultures prior to the initiation of antimicrobial therapy. Because of the potential for abuse of blood cultures, and the increasing cost, many laboratories limit the numbers of cultures, usually three a day (6). Darbridge (30) doubts the value of blood cultures in the diagnosis of septicemia in adults. He indicates that septicemia is usually recognizable clinically and treatment is often started before the results of the culture are known. In most cases, the cultures confirm the initial suspicions of the physician. Denham & Goodman (35, 36) on the other hand, reported that in geriatric practice the patients often present in a nonspecific atypical manner and blood cultures remain a useful diagnostic tool. Eisenberg et al (45) studied 565 febrile patients in the emergency room, of which 210 had blood cultures. Bacteremia was present in 9 out of 86 patients who were admitted to the hospital, but only 1 of 124 patients who were not admitted. They concluded that the use of blood cultures to screen for bacteremia was unnecessary for febrile adult patients who are not admitted to the hospital because the suspicions of the clinician are sufficient to sort out those bacteremic and nonbacteremic patients. Whether or not such astute clinical judgment is routinely present in emergency rooms is questionable. McGowan et al (87) indicated that children seen in a walk-in clinic were at risk if their temperature was 38.3°C or higher. Bacterial pathogens were identified in the blood from 31 of 708 children. They concluded that the blood culture provided valuable help in establishing the specific diagnosis in febrile children without localizing sites of infection.

It is tempting to relate the volume of the blood cultured to the frequency of culture. That is, if a greater volume of blood is cultured at a specific time, then frequency of culture may be decreased. This is not the case, because in most diseases bacteria are present intermittently in the blood in small numbers. The more blood cultured, the greater the chance of isolating the organism. Washington (128) reports that if 5 ml of blood collected is taken as a 100% relative yield, doubling the sample to 10 ml would increase the yield by 12%. Tenney et al (120) found that more organisms, particularly gram-negative rods, were isolated from cultures when larger volumes of blood were used. They found an average percentage increase in the detection of bacteria of 4.5 per ml of blood cultured. The ratio of the volume of blood cultured to the volume of medium is essential. Dilution of blood should not exceed 1:10 (10–12 ml per 100 ml of medium). The anticoagulant and antiphagocytic effects of additives are minimal if the dilution of blood to broth is less than 1–7 or 1–8. The situation is different, however, for neonates. Mangarten (83) describes a method for obtaining blood for culture by heel stick from high risk babies. The blood was collected directly into a 0.02-ml heparinized micropipette and then was squeezed into a sterile test tube containing 2 ml of Trypticase soy broth. In only 3 of 195 infants studied was the culture of venous blood positive and capillary blood negative. In each of these cases, the positive culture was thought to be a contaminant. We (R. C. Tilton and J. Rowe, unpublished data) studied 100 neonates that represented over 250 sets of blood cultures. Venous blood was collected in parallel with 200 μliters of blood from a heel stick. In no case was an organism isolated from the venous blood that was not isolated from the heel stick. The apparent contamination rate of the heel-stick blood cultures was high as many more *Staphylococcus epidermidis* grew from the heel-stick cultures than from the venous blood cultures. Such data suggests that neonates, when infected, have much higher counts of bacteria in their blood than do adults and children.

There is no significant difference between the culture of venous blood and the culture of arterial blood except for the filamentous fungi. Stone et al (117) reported that *Candida* is more often isolated from arterial blood than from venous blood. Rubenstein et al (107) indicates that for patients with aspergillosis, the diagnosis may be made more easily from arterial than from venous blood.

Careful attention to skin disinfection is essential to reduce the numbers of contaminated blood cultures (6). The majority of reports (7) recommend cleansing initially with 70% isopropyl or ethyl alcohol followed by concentric swabbing with either 2% iodine or an iodophor. Alternative agents include chlorhexidine. Preparations such as benzylkonium chloride should not be used as they may harbor microorganisms. Two outbreaks of pseudobacteremia caused by contaminated blood-drawing equipment (42) and benzylkonium chloride (69) have been reported.

Blood is drawn from the patient's vein with a syringe and needle and then is injected into a set of blood culture bottles at the bedside. Alternatively, a blood transfer set may be used and the blood may be passed directly from the vein through the transfer set to the culture bottles. There are no data substantiating the relative efficacy of these two procedures. Becton Dickinson has a blood culture bottle available in which the sample can be aspirated directly into the culture vial by vacuum, called a Vacutainer.® Blood can also be collected in a Vacutainer tube containing an anticoagulant and transported to the laboratory for subsequent inoculation into appropriate media. This procedure introduces additional steps in specimen handling and may lead to an excessive contamination rate unless extreme care is taken. Hoffman et al (62) reported that pediatric-size vacuum tubes containing ethylenediaminetetraacetic acid (EDTA) were contaminated with *Serratia marcescens.* Katz et al (70) also found *S. marcescens* in nonsterile tubes. Washington (127) reported on a collaborative study involving 20 clinical laboratories, in which 1433 nonsterile Vacutainer tubes were cultured. Of those examined 14% contained microorganisms; 9% of the allegedly sterile tubes contained *Pseudomonas aeruginosa,* enterococci, *Serratia,* and *Acinetobacter calcoaceticus.* Not only is there a risk of false positive blood cultures from back flow when such tubes are used, but there is the potential risk of infection, particularly in the immunocompromised host. Gantz et al (57) reported that the rate of isolation of *Bacteroides* from blood cultures increased and the rate of *Candida* decreased after vacuum bottles were adopted for blood cultures in their hospital. All vacuum culture bottles deliberately inoculated with *Bacteroides fragilis* showed growth, and 87% of those inoculated with *Candida* were negative.

Blood culture additives can be divided into two groups: (*a*) those that prevent coagulation and inhibit the bactericidal action of blood; and (*b*) those that remove harmful components from the blood such as penicillinase and antibiotics. The addition of sucrose, sorbitol, and cysteine is discussed below.

The most widely used blood culture anticoagulant is sodium polyanetholsulfonate (SPS), 0.025–0.05%. The utility of SPS both as a polyanionic anticoagulant and an inhibitor of the microbicidal activity of blood was demonstrated in 1938 by Von Haebler & Miles (125). Since then, a large number of studies have proven its efficacy (8, 50, 54, 103, 106).

SPS is toxic to *Peptrostreptococcus anaerobius.* However, Wilkins & West (136) indicated that the sensitivity of *P. anaerobius* to SPS was a function of the medium. If the medium contained gelatin, protease peptone, or casein the organism was protected. However, discs containing SPS are now used in many laboratories for the identification of *P. anaerobius* based on its susceptibility to this compound. Kocka et al (72) described sodium anethol sulfonate (SAS) to replace SPS. The apparent advantage of SAS was that it did not interfere with the growth of *P. anaerobius.* However, SAS also made the medium slightly cloudy. SAS is not widely used.

SPS is inhibitory for certain antibacterial activities of blood. It is anticomplimentary, it inactivates leukocytes, apparently increasing the rate of glucose oxidation, and it also complexes some aminoglycoside and polypeptide antibiotics (7).

The leukocytic property of SPS is important as Elliott (47) reported that leukocytes retain their phagocytic properties for 24 hr in broth media used for blood culture. According to Belding & Klebanoff (8), SPS exerts a direct effect on the metabolic activity of the leukocyte. It increases glucose oxidation of resting leukocytes but not of phagocytizing leukocytes. It increases formate oxidation by both resting and phagocytizing leukocytes. According to these authors, these metabolic alterations have no apparent relationship to the inhibition of phagocytic function by SPS.

Staneck & Vincent (116) reported that 0.025% SPS inhibited the growth of *Neisseria gonorrhoeae.* Of 50 clinical isolates, 16 failed to grow in subculture at 1, 3, and 10 days after incubation. Recovery was delayed with eight isolates as compared to controls. Gelatin (1%) reversed the adverse effect of SPS. Eng & Iveland (51) reported similar experience with *Neisseria meningitidis.* Pai & Sorger (94) confirmed the protective effect of gelatin on *N. meningitidis.*

Other anticoagulants such as EDTA, oxylates, citrates, and fluorides are unsatisfactory as they inhibit many bacteria.

Penicillinase is added to blood cultures to inactivate any penicillin present in the blood of the patient. Some studies (41) seem to confirm the utility of pencillinase, but there have been many incidents of contamination when penicillinase was added to blood culture media. Sterility controls should always be performed.

A commercial device for the removal of antibiotics from blood has been recently introduced. The antibiotic removal device (ARD) is marketed by Marion Scientific Co., Kansas City, Mo. The ARD consists of a 60-ml rubber-capped vial to which has been added both cationic resin and polymeric adsorbent resin. The resin is fluidized with 5 ml of physiologic saline. Blood is drawn directly from the patient and is added to the ARD. The inoculated ARD is shaken for 15 min, and the blood is removed aseptically and then transferred to standard blood culture bottles. It is claimed that ARD can remove more than 100 μg of most antibiotics. A number of preliminary studies have reported increased isolation rates, particularly for patients receiving antibiotics. McLimans (87a) studied 47 patients with proven *Staphylococcus aureus* bacteremia. Blood samples were drawn 2–4 days after the start of the antibiotic. Blood was split: half went to the ARD, and the rest went into a conventional system. Of the 47 cultures, 22 were positive only after ARD processing, 4 were positive by both systems, and 22 were negative by both systems. Recent data on over 4000 blood cultures from the Mayo Clinic (129) suggest that when used routinely, the isolation rate of the ARD is no greater than conventional blood culture media. Wallace et al (126) studied 51 patients, 31 of whom yielded positive cultures. Subcultures within the first 12 hr yielded 12 positive isolates with the ARD compared to 2 when subcultured without the ARD. Overall, 21 of the 31 microbes were isolated more rapidly after ARD treatment and 4 were positive only after such treatment. In one burn patient, the ARD failed to isolate *S. aureus,* whereas the nontreated culture did. Lindsey & Riley (80) showed that 13 antibiotics, including amikacin, ampicillin, carbenicillin, cefazolin, cephalothin, chloroamphenicol, gentamicin, nafcillin, tetracycline, tobramycin, and vancomycin, were removed and cefoxitin and ticarcillin were reduced to very low levels. The system was challenged with five species of anaerobic bacteria, one yeast species and six species of facultative or aerobic bacteria to determine if the ARD would trap or inhibit the microorganisms. All of the organisms were recovered from the device in the same numbers in which they were inoculated. Additional studies must be done to confirm the utility of this device.

The literature is replete on the relative merits of various broth media for the culture of blood. In many cases, it is difficult to determine the actual effect of growth media on the relative rate of organism recovery because not all variables were held constant. The rate of recovery of an organism from a blood culture system is dependent on many factors, including the volume of blood added, incubation temperature, presence of an anticoagulant, amount of CO_2 present, venting, etc. Any study that attempts to determine the superiority of one medium over another must hold all other variables constant.

Of all the media used, three are the most popular: Trypticase soy broth (TSB) (a casein soybean digest broth); Columbia broth (CB) (92); and brain heart infusion (BHI). Although there may be minor differences between these media, they are not significant for the recovery of aerobic and facultative organisms. Morello & Ellner (92) report that bacteria grow more rapidly in CB than in TSB. Since the advent of the Bactec© radiometric system (34), the propietary Bactec aerobic, anaerobic, and hypertonic broth has been used by virtually all owners of the Bactec system. This broth appears to be similar in recovery rates to the three previously mentioned. Several media exist for the culture of anaerobes. Among them, thioglycollate broth, and prereduced anaerobically sterilized BHI. Once again, the literature claims certain advantages for each of these broth systems, although in one study, Washington (128) claimed that TSB incubated in an unvented system recovered as many anaerobes as did thiol and thioglycollate.

Babu et al (4) evaluated 23 blood culture media. They concluded that most media failed to support the growth of the challenge organisms without the addition of a human red blood cell-serum mixture (RBC-SM). Incorporation of the RBC-SM into blood culture media did not improve the recovery rates of microorganisms from 15 of 23 media tested. A few media proved satisfactory. They included BHI broths from Difco, Gibco, and Pfizer, CB from Gibco and Scott Laboratories, TSB from Difco and Gibco, and dextrose phosphate broth from Gibco. Although this study is important, its practical relevance is doubtful. Clinical blood cultures were not used. Organisms were laboratory isolates. Whole blood was not used to simulate the blood culture, rather a mixture of a normal human serum pool and group 0 erythrocytes was used.

A two-bottle culture system, one aerobic and one anaerobic, is standard practice. Several attempts have been made to introduce a one-bottle system. The literature clearly states that a one-bottle system cannot support the requirements of *Candida albicans, Neisseria,* and *Pseudomonas* on the one hand, and *Bacteroides* and *Fusobacterium* on the other. Gantz et al (57) confirmed that when vacuum blood culture bottles were inoculated with *Bacteroides fragilis,* all culture bottles grew *B. fragilis.* After 10 days, 87% of the bottles given similar inocula of *Candida* failed to grow.

Many studies have attempted to show the value of a hypertonic medium for the growth of cell wall-defective microorganisms (104, 119, 130). No one study has confirmed that a hypertonic medium is essential for the isolation of cell wall-defective forms. Ellner et al (49) collected 18,000 clinical blood specimens and inoculated them into modified CB with and without 10% sucrose. Their conclusion was that when the volume was held constant, sucrose had no demonstrable effect on the recovery of aerobic and facultative organisms in vented bottles. In unvented stationary cultures, a significantly greater recovery of facultative organisms and a marginally greater recovery of anaerobes were obtained with a hypertonic broth. Ellner et al (49) pointed out that many of the studies suggesting the efficacy of hypertonic media failed to control the basal medium formulation, the inoculum, or other variables such as blood volume. The use of a three-bottle system is not recommended. However, an aerobic (vented) isotonic medium coupled with an anaerobic (unvented) hypertonic medium may increase recovery. Some hypertonic blood culture media have been rendered hyperosmotic by the addition of sorbitol instead of sucrose. My own results (unpublished data) indicate that recovery rate of any organism is not increased with sorbitol. However, the presence of sorbitol in the medium caused less turbidity and muddiness than did sucrose. If a hypertonic medium is required, then sorbitol may be efficacious.

Washington et al (130), however, reported that 15% sucrose inhibited the recovery of *Haemophilus, Bacteroides,* and *Staphylococcus aureus.* Chong et al (25) also found that 15% sucrose affected the growth of *Salmonella typhi.* When 10–20% sucrose was added to BHI, the growth of *Escherichia coli, Pseudomonas, Streptococcus viridans,* and *Streptococcus pneumoniae* was diminished. Frankel & Hirsch (56) suggested the addition of cysteine to blood culture media as certain strains of nonhemolytic streptococci require cysteine for growth. Cysteine could be substituted by thioglycollate or reduced glutathione. George (58) isolated another *Streptococcus* that required pyridoxal hydrochloride for growth. Since then, other investigators (86) have isolated vitamin B_6-dependent bacteria from blood cultures. These are also called "satelliting streptococci."

Regardless of media used, the inspection of the bottle by turbidity alone cannot be relied upon to detect all positive blood cultures. Studies report that only 65% of positives will be detected when bottles are only inspected for turbidity (11). Rarely do bottles containing *Pseudomonas aeruginosa* become turbid. Gram stain and subculture must be performed.

The majority of blood cultures can be incubated 7 days and discarded. Ellner (48) reviewed 40,000 blood cultures processed over a 10-year period. In no case did a culture become positive after day 5. The majority of studies report that 80–90% of blood cultures become positive after the first 1 to 4 days of incubation.

Bartlett et al (7) has recommended that blood cultures be routinely subcultured at 24 hr or earlier, 48 hr, and finally, after 5-7 days if the cultures appear negative. The College of American Pathologists Inspection Check List suggests the same. A report by Campbell & Washington (22) evaluated 2780 previously negative blood cultures. All were cultured after 7 days of incubation. Of four bottles positive by subculture, three yielded the same organism as previously isolated from the companion bottle and one yielded an organism thought to be a contaminant. They suggested that the routine 7-day blind subculture was not warranted. However, they also stated that in special instances incubation should be continued for 14 days so that organisms such as *Candida,* fastidious gram-negative rods, *Hemophilus, Cardiobacterium,* and *Actinobacillus* might be isolated. Recently it was found (R. C. Tilton, unpublished data) that nine sets of blood cultures were positive from a patient with a subacute bacterial endocarditis, but only after 21 days of incubation. The isolate was *Cardiobacterium.*

Gill (59) questions the usefulness of final blind subcultures. He evaluated 14,000 blood culture bottles subcultured blindly at 7 days of incubation. Only 12 potentially significant organisms were found, but 11 out of 12 cases, the organism had already been reported from prior positive bottles.

A few laboratories still perform colony counts on blood. Randall (personal communication) reports some prognostic significance related to the numbers of organisms in blood. If done, quantitative blood cultures should be performed in parallel with routine blood cultures. If 1 ml of blood is cultured, then up to 40% of bacteremias will be missed because of the low numbers of organisms present. Conversely, a quantitative blood culture, if positive, yields growth in 24 hr or less from which rapid tests for identification and antimicrobial susceptibility can be performed.

LaScolea et al (76) has reported on the diagnosis of bacteremia in children by quantitative direct plating. He compared this to a traditional system as well as to a radiometric procedure. Of 2123 blood cultures processed, 135 (6.4%) were positive. The organisms most frequently recovered were *Haemophilus influenzae* type B, *Neisseria meningitidis,* and *Streptococcus pneumoniae.* Quantitative direct plating systems detected 89% of the cultures positive for *H. influenzae* and *N. meningitidis,* of which 55% yielded results before either broth procedure. However, only 50% of the cultures positive for *S. pneumoniae* showed growth on quantitative direct plating. This difference appeared to be a function of the number of organisms present in the blood. In only 7% of cases were there more than 100 organisms per ml of *S. pneumoniae.* Studies revealed that although *H. influenzae* grew well in the radiometric broth, a significant growth index was not observed during day 1 of incubation. The quantitative direct plating system was optimal for *H. influenzae* and *N. meningitidis.* It provided information on the magnitude of bacteremia as well as an earlier diagnosis, but it cannot be recommended for *S. pneumoniae* bacteremia.

The techniques of lysis filtration and centrifugation, although distinctly different, have common goals, that is the removal of bacteria from a hostile host environment after the lysis of red blood cells. Pickett & Nelson (95) laked the blood and then centrifuged it to concentrate *Brucella.* After the development of membrane filters in the middle 1950s, several investigators attempted to lyse blood and pass it through membrane filters to trap the bacteria (102). The technique at the time was not successful, as it resulted in unacceptable high rates of contamination. *Haemophilus influenzae* did not survive the lysis filtration technique. In the lysis filtration technique, the anticoagulated blood is subjected to enzymatic and chemical agents. The blood may be either sedimented or lysed. Many lysing solutions have been proposed, including 3% dextran, 6% sodium chloride, sodium citrate, Triton X-100, 0.35% SPS, Rhozyme 41, Varidase, sodium carbonate, and solryth.

Zierdt et al (139) have reported on the routine use of lysis filtration for blood cultures. Blood is received from the wards in a tube containing an anticoagulant. The blood is lysed with a combination of a nonionic detergent and an enzyme. The lysate is filtered through a 0.45-μm membrane and the membrane filter then is added to an impedance monitoring device. According to Zierdt et al (139), the optimal lysing solution is 0.04% Triton X-100, 0.6% Rhozyme, and sodium carbonate buffer (final pH 7.8). The problem of leucocyte clogging was solved with the addition of one of two enzyme preparations to the lysing solutions, either Varidase (Lederle Laboratories, Pearl River, NY) or Rhozyme 41 (Rohm & Haas, Philadelphia, Pa.).

Virtually all of the studies that have used either lysis centrifugation or lysis filtration have reported that organisms are detected more rapidly and more reliably by their systems. The techniques of lysis centrifugation and lysis filtration must be weighed carefully as they may be time-consuming and not applicable to the average laboratory without commercial streamlining. Anhalt (2) reports that these concentration techniques are at least comparable in sensitivity to the conventional methods, but that the advantages provided by these concentration techniques are dependent to a great extent upon the elegance of the conventional system used for comparison.

Lysis-centrifugation differs in that the blood is centrifuged instead of filtered subsequent to lysis. The major developments have been by Dorn and his colleagues (38–40). The centrifugation device consists of a double-stoppered tube to which has been added a sucrose-gelatin density gradient layer. The anticoagulated blood is added to the density gradient layer and is centrifuged. The aqueous lysing solution contains EDTA, polypropylene glycol, saponin, and SPS. The bottom stopper has an angular plane complementary to the angle of the centrifuge used. Following centrifugation, a syringe is inserted into the bottom stopper, the lysed blood is removed, and then trapped microorganisms may be removed and cultured.

The preliminary clinical studies by Dorn et al (39, 40) suggest that the lysis centrifugation tube (the Isolator) is at least as sensitive as a traditional blood culture system.

There are subtle differences between the isolation of microorganisms and their detection. Microorganisms (bacteria, in particular) are usually isolated by traditional cultural means, sometimes subsequent to either lysis filtration or lysis centrifugation. The organisms usually have not been detected previously except in those blood culture bottles that have become turbid. In these, bacteria may be detected by the Gram stain or other stains such as acridine orange (74). The detection of bacteria in blood directly by Gram stain generally is not useful as in few infections are there $>10^5$ CFU/ml of blood.

Nontraditional means of detecting the presence of bacteria may involve systems that measure the microbial synthesis of a metabolic product. This product may be carbon dioxide, mannan, arabinitol, electrical current, heat, bacterial polysaccharide, or other metabolic components characteristic of bacteria or fungi.

These detection systems offer the potential of both sensitivity and specificity.

Radiometry, the detection of radiolabeled metabolites in blood cultures, has had a significant impact on the clinical microbiology laboratory. Radiometric methods for the detection of bacteria were not initially developed for blood cultures. Levin et al (77) described the method to detect bacterial contamination of water by coliforms. This method trapped liberated $^{14}CO_2$ in a solution of barium hydroxide following metabolism of [^{14}C]lactose. It was not until the report of Schrot et al (109) that the technique was applied to blood. The technique was essentially a manometric one in which bacteria were recovered on a membrane filter after the blood was lysed and then the filter was placed in an isotopically labeled growth medium. As in the coliform procedure, the $^{14}CO_2$ was trapped in barium hydroxide solution. Although this predated the Bactec (Johnston Laboratories, Colkeysville, Md.), it appeared that the method was awkward and time-consuming so that it could never be applied routinely in the clinical laboratory. Nonetheless, the advantages of such a technique were evident. DeLand & Wagner (33) first described what can be termed as the current approach to the radiometric detection of bacteria. Both simulated and actual blood cultures were used. The bacteria in blood were incubated in a nutrient medium containing [^{14}C]glucose. After growth of the organism, the head space gas was flushed out and the amount of $^{14}CO_2$ was determined in an ionization chamber. They noted that small quantities of $^{14}CO_2$ could be detected very early (6 hr) after the cultures were inoculated. The early clinical data were encouraging. Bacteria were detected in 30 out of 500 blood cultures by the radiometric procedure and in 26 out of 500 by the routine procedure. There were no false positives or false negatives. Shortly after this report, DeBlanc et al (32) studied the prototype automated radiometric instrument. The Bactec was developed and tested on 3000 blood cultures from 1280 patients. Of these patients, 57 were positive by one or both methods. Conventional techniques were positive in 87% of the cases and by the radiometric method 85% were positive. Seventy percent of the cultures were first detected by the Bactec, over half of these on the initial day of inoculation. If the system failed to detect a positive culture, it was usually a microaerophilic streptococcus. The initial enthusiasm was tempered by a report from Washington & Yu (131) in which 59 simulated and 65 actual blood cultures were tested. They were unable to detect growth in the first 6 hr. In 18 hr, growth was detected in all cultures with the exception of five that contained group D streptococci. They reported false negatives with *Pseudomonas aeruginosa,* and *Bacteroides fragilis* as well.

As a result of these early studies (5, 20, 49, 69, 100, 112, 121), several modifications were recommended for the optimalization of the Bactec system. They included necessity for labeled substrates other than glucose, notably amino acids and alcohols, necessity for agitation of the aerobic cultures, the inclusion of an anaerobic prereduced bottle, the critical nature of the atmosphere of incubation, the ability to detect *Haemophilus,* the role of hypertonicity in Bactec media, thresholds for growth, necessity for routine subculture, frequency of radiometric detection, specificity and sensitivity of radiometric detection, and detection time.

Similar criticisms can be made of the innovative detection systems as have been made of routine methods. That is, many studies have not isolated variables. For example, the volume required in a Bactec blood culture bottle is 3 ml, but the volume may be 10 ml for a conventional system.

Although the Bactec has been utilized for many functions, such as aminoglycoside assays, leukocyte function, mycobacterial susceptibility testing, detection of bacteria in foods, and pharmaceutical sterility testing, only its application for the detection of bacteremia is discussed. Four instruments have been marketed. The Bactec 301 is a manual device. It requires the user to place the bottle underneath the detection needles and manually lower the needles into the blood culture bottle. The Bactec 225 completely automates the manual functions. The bottles are incubated on the machine and are sampled automatically at preset intervals. The Bactec 460 is similar to the 225 except that the bottles are incubated separately and are brought to the machine for automated reading. The newly introduced Bactec 5 provides both aerobic and anaerobic incubation in the instrument for large numbers of bottles, automated reading, and reporting.

Initially, Bactec media contained 1.5 μC of ^{14}C-labeled substrate. This has been recently increased to 2.0 μC. The basal medium (both aerobic and anaerobic) contains a tryptic digest of casein soy broth, hemin, menadione, and SPS, and the anaerobic medium contains L-cysteine and yeast extract. The hypertonic medium is similar to the aerobic medium except that 20% sucrose has been added.

DeLand & Wagner (34) reported that shaking increased the release of $^{14}CO_2$ from a broth suspension. Ellner et al (49) have reported that agitation decreased the detection times of conventional broth cultures without adversely affecting bacterial growth. Agitation had no effect on organisms growing in the prereduced anaerobic medium (141).

The original culture gas used for the Bactec contained only air with ambient amounts of CO_2. Waters & Zwarun (133) noted that air used for the flushing and the incubation gas of aerobic cultures significantly reduced the rate of false positive machine readings. Air also reduced the numbers of organisms isolated, particularly those organisms that require exogenous CO_2, such as the pneumococci or *Haemophilus,* and some of the pseudomonads. If the CO_2 in the air mixture was increased to 10%, the capnophilic organisms grew well. However, there were nearly 20% false positives due to background production of $^{14}CO_2$. Current recommendations are that the air mixture contain 5% CO_2. The user can expect 2–5% false positives. A mixture of 80% N, 10% CO_2, and 10% H has been recommended for anaerobic bottles, although in my experience substitution of the anaerobic gas with 5% H makes little difference.

The necessity for a routine hypertonic culture medium is controversial, as with traditional blood culture systems. Several investigators (27, 140) compared aerobic, anaerobic, and hypertonic Bactec media. Both claim that the use of a hypertonic medium reduced false positive cultures with the aerobic bottle. However, 10% CO_2 was used for the aerobic incubation, not 5%. Coleman et al (27) reported that septicemia was detected in 57 patients with the hypertonic medium compared to 37 patients with the aerobic medium. Hull et al (65) indicated that the use of all three media gave no statistically significant advantage in recovery of organisms compared to combinations of two media alone. No recent data claim efficacy of a three-bottle blood culture. A combination of either an aerobic bottle (6A), an anaerobic bottle (7B), or a hypertonic bottle (8A) are satisfactory.

Most studies using Bactec stress the early detection of bacteremia made possible by this instrument. Randall (97), in a review, reported that approximately 50% of the *Enterobacteriaceae,* 24% of nonfermenting gram-negative rods, 24% of streptococci, 32% of staphylococci, and 6% of miscellaneous organisms were detected by Bactec in 12 hr or less. More than half (51%) of all aerobic organisms were detected in 24 hr. The conventional system usually required 48 hr. Bactec detects bacteremia most rapidly when the agent is either a pseudomonad, a staphylococcus, or one of the Enterobacteriaceae. Others have noted there is nothing inherently faster in the Bactec system. The ease of sampling allows blood cultures to be checked on multiple occasions during the first couple of days of growth. This may result in the faster detection times reported for the Bactec. Sliva & Washington (111) reported that for traditional blood culture systems, the optimal time for routine subculture is between 6 and 17 hr after blood collection. If such data are extrapolated to the radiometric system, the blood cultures should be sampled at least three times on the day the culture arrives in the laboratory and daily thereafter. Just as in traditional systems, Bactec cultures should not be held for more than 7 days unless there is a clinical indication of slow-growing organisms such as *Brucella, Actinobacillus,* or *Cardiobacterium.*

The efficacy of the Bactec for *Haemophilus influenzae* detection has been questioned. Larson et al (75) could not detect *Haemophilus* in the aerobic medium in simulated blood cultures. Bannatyne & Harnett (5) and Smith & Little (112) detected *H. influenzae* in clinical blood cultures using the 6A medium. Rosner (104) found *H. influenzae* in pleural fluid, synovial fluid, and cerebrospinal fluid using the 8A hypertonic medium without added $NADH_2$. A study by LaScolea et al (76) reported that even though *H. influenzae* grew in the Bactec broth, it did not register a significant growth index (GI) within 24 hr in 15 cultures that grew out *H. influenzae.* Of the total of 16 positive cultures of *H. influenzae,* 9 were detected by Gram stain, and 6 were only positive after a 24-hr subculture and resulted in a significant

GI. Of the 15 cultures with a negative GI at 24 hr, 11 out of 15 were positive at 36 hr, 2 were positive at 48 hr, and 2 were positive at 72 hr. The authors also noted that 14 of the 15 cultures did not show a sequential increase in GI before registering a GI of $\geq$30. This failure of the GI reading to reflect growth of the Bactec bottle at 24 hr was not observed with *Neisseria meningitidis* or *Streptococcus pneumoniae.*

The necessity for terminal subculture of Bactec bottles has also been an issue (24). Washington (128) presented data on specificity and sensitivity of the Bactec. He reported a range of from 1.6 to 43% false positive. This false positive rate appears to be a function of the amount of CO_2 in the incubation gas, as well as a GI cut off that is too low. False negative results were seen with alpha-hemolytic streptococci and enterococci. Washington concludes that one blind subculture of Bactec vials is necessary to detect all organisms present, particularly the enterococci. Strauss et al (118) noted false negative cultures could be a function of the strain of bacteria isolated. Of the 17 isolates of group D enterococci not detected by Bactec, 12 were from a single patient. Recent evidence (J. McLaughlin, personal communication) indicates that routine subculturing of Bactec bottles is not necessary.

Cady (21) reported that the growth of microorganisms could be detected by changes in the electrical impedance of the culture. Several investigators (61, 124) have shown that impedance changes parallel the growth of microorganisms in broth. The impedance curve is essentially the same as the viable cell count growth curve. The sensitivity of impedance changes can be reduced by media with a high buffering capacity (124) or a high ionic strength and failure to precisely control incubation temperature. The only commercially available impedance instrument in the US is the Bactometer. The Bactometer was first used by Hadley & Senyk (61) to study impedance changes caused by microbial growth in simulated blood cultures. They found that the Bactometer was a sensitive instrument and could detect from 10^6 to 10^7 CFU of bacteria per ml. Hadley & Kazenka (60) reported on results of 785 actual blood cultures using the Bactometer and a reference culture method. Of the cultures 12.4% were positive. Twelve additional cultures were positive only by the conventional method and 15 were positive only by the impedance method. There were 11 false negatives using the impedance method. Except for *Candida* and some enterococci, impedance detected bacteria more rapidly than did the conventional method. Recently, Zierdt et al (139) have combined lysis filtration and impedance monitoring. Blood is lysed and filtered through a membrane; the membrane is placed in a nutrient medium that contains stainless steel electrodes. They reported that the lysis-filtration-impedance method detected 92% of the positive cultures and the conventional method detected 56%. Impedance monitoring usually detected growth in less than 12 hr, where conventional cultures were recognized at 24 hr. Slight turbidity was noted in those bottles that were positive by impedance monitoring.

The colony count that corresponded to an initial impedance change was 5×10^5 CFU/ml (approximately the colony count at which turbidity appears).

Holland et al (63) described a system in which platinum electrodes were implanted in blood culture bottles and were monitored electronically. This technique for detecting microbial growth by using a platinum measuring electrode and a saturated calomel reference electrode was described by Wilkins & Stoner (135). The changes in voltage across the electrodes are associated with bacterial metabolism and the elaboration of electroactive substances as a function of the growth phase. Matsunoga et al (85) recently described a method for electronically detecting microbial populations by using platinum and silver electrodes in combination. The system of Holland et al (63) has utilized the Abbott MS-2 microprocessor for detecting electrical potentials and recording voltage changes. Of 163 simulated cultures, 90.8% were positive by the electronic detection system. Of 21 false negatives, 12 were caused by *Cryptococcus neoformans*. Of 156 patient blood cultures examined, 13 were positive both by the electronic detection system and by conventional methods.

Either system is promising. The monitoring of electrical potential or impedance in growing cultures can be automated with presently available instrumentation. The primary advantages of these systems over Bactec is that reliance on isotopically labeled growth medium is not necessary.

Forrest (55) reported that the heat given off by bacterial metabolism could be measured and patterns of heat release were similar to the growth curve of the organism. Boling et al (13) determined that by using microcalorimetry, one microbial species could be separated from another on the basis of characteristic heat curves. Such characteristic heat curves were extremely difficult to repeat (W. J. Russell, personal communication). Ripa et al (101) have shown that microcalorimetry could be used to detect bacteria growing in blood cultures. To date, no clinical studies have been done on microcalorimetry.

Levin & Bang (78) reported that soluble lysates from *Limulus polyphemus* amoebocytes form a gel in the presence of endotoxin. The test is highly sensitive. Early studies (79) indicated that the *Limulus* amoebocyte lysate test could be used to detect circulating endotoxin in a patient with endotoxemia. Levin et al (79) reported results from 98 patients in whom gram-negative sepsis was suspected. In 15 patients, gram-negative rods were cultured and the *Limulus* amoebocyte lysate test was positive in 10. Four of the five remaining patients had gram-negative bacteria grown in cultures from other sites but had negative blood cultures. Subsequently, several groups reported that the *Limulus* amoebocyte lysate test was of limited use clinically (46, 53, 84). These adverse reports were based primarily on the inability to detect endotoxin when gram-negative bacteria were grown from the blood. This was apparently due to a lack of sensitivity of the test and the rapid clearance of endotoxin from the circulation.

Gas-liquid chromatography (GLC) is a sensitive and specific method for the microbial components in body fluids. Mitruka et al (89) analyzed serum extracts from experimentally infected mice. He was able to see organism-specific peaks in the serum of the infected mouse. Subsequently, the serum obtained from 84 patients with fever were studied (90). Prior to analysis, the serum specimens were acid hydrolyzed and derivativized and when chromatographed appeared to have characteristic peaks. Organisms giving characteristic peaks included members of the Enterobacteriaceae, pneumococci, streptococci, mycobacteria, and pseudomonads. In 80 additional derivatized samples, it was noted that pneumococcal bacteremia in patients with pneumonia could be detected with greater than 90% accuracy. Some attempts to repeat Mitruka's work have failed (31). Recently, Sondag et al (114) reported on the rapid presumptive identification of anaerobes in blood cultures by GLC. Products of metabolism were detected after blood was added to culture medium and incubated. The spent medium was then analyzed for volatile and nonvolatile acids. Some positive blood cultures (128) were analyzed, including 30 blood cultures inoculated with stock cultures. An additional nine polymicrobial cultures, from which 24 organisms were recovered, were processed. *Bacteroides fragilis, Fusobacterium,* anaerobic gram-positive cocci, and clostridia could be identified from their characteristic fermentation products. The authors suggest that this is an accurate, presumptive method for the identification of most anaerobic blood culture isolates. The identification must still be confirmed by appropriate biochemical tests or standardized GLC analysis of metabolic byproducts. The disadvantage of this is the time of processing. Brooks et al (1, 15–19) have demonstrated that the chemistry of body fluid changes as a function of pathology. These investigators used frequency pulsed electron capture (FPEC) GLC. Brooks has analyzed synovial fluid, serum, cerebrospinal fluid, and pleural fluid. His studies on derivatized extracts of normal diseased and treated body fluids and spent culture media have revealed several interesting observations: (*a*) Chromatograms obtained from all types of body fluids showed that the constituents changed with disease; (*b*) chemical changes differ as a function of the disease, both infectious and noninfectious; (*c*) FPEC-GLC could be used as a diagnostic tool for meningitis, Rocky Mountain Spotted Fever, arthritis, and pleural infusions; and (*d*) in many instances, certain microorganisms could be identified at the species level by using GLC patterns obtained from derivatized extracts and spent culture media.

Once a microorganism has been detected in the blood culture, it must be identified. Several studies have recently shown that it is possible to identify microorganisms directly in blood culture fluids without the necessity for isolation by subculturing on agar medium to obtain isolated colonies. Rapid indentification procedures that use the growth in blood cultures were reviewed recently by Morello & York (93). Wasilauskas & Ellner (132) described the application of some traditional tests to the rapid identification of growth in the blood culture broths. They were able to identify almost 90% of the organisms within hours after growth was detected.

The introduction of rapid biochemical tests such as Micro-ID, API-20E, and API-20E (4 hr) has facilitated the identification of blood culture isolates in 3–4 hr. Several investigators (12, 43) have described a technique in which 5–10 ml of medium from a positive blood culture is centrifuged at low speeds to sediment the red cells, then at high speeds to remove the bacteria. The pellet may be washed and used to inoculate systems for either identification or antimicrobial susceptibility tests. The inoculum may also be adjustd for the instrumental analysis by using Autobac, MS-2, or the Vitek Auto Microbic System. Doerne (personal communication) has recently completed a study in which direct disc diffusion tests were done on the growth obtained from a large number of blood cultures. Agreement with the reference Bauer-Kirby test was approximately 96%. He confirms the work of others (52, 68, 88). Morello & York (93) reported that rapid-method testing of growth obtained from 50 positive blood cultures containing oxidase-negative Gram-negative bacteria, 94% were correctly identified by Micro-ID and 92% were correctly identified by the 4-hr API system. They indicated that system failures were usually due to insufficient inoculum. Edberg et al (43) and others reported the successful 4-hr presumptive identification of Enterobacteriaceae using growth from blood cultures, with the Micro-ID. Cleary et al (26) presented similar data on the Mini-Tek system. Recently, Moore et al (91) described the rapid identification and antimicrobial susceptibility testing of gram-negative bacilli from blood cultures with the Auto Microbic Systems. Blood cultures (196) were seeded and used for comparison with standard techniques. Identification of most cultures was accomplished in 8 hr, and antimicrobial susceptibility results were available in 4.7 hr. Correct identification was obtained with 95% of the cultures. The antimicrobial susceptibility data agreed 87% of the time with 3.8% very major and 1.4% major errors.

Immunology offers a rapid means of identifying bacteria directly from blood cultures. Several methods are available, including counterimmunoelectrophoresis (CIE), latex agglutination (LA), coagglutination (COAG), enzyme-linked immunoabsorbant assay, radioimmunoassay, capillar tube precipitin tests, and direct agglutination. The use of these techniques in microbiology has been recently reviewed (108, 123). The application of these techniques to detection of bacteremia was originally proposed by Dorff et al (37). They detected antigens in the sera of patients with pneumococcal bacteremia by CIE.

Recently, Artman et al (3) described a method for the detection of pneumococcal and *Haemophilus influenzae* type B antigen directly in blood cultures of bacteremic patients. During a 6-month period, all blood cultures received in the laboratory were studied by CIE to detect both pneumococcal and *H. influenzae* antigens. Pneumococcal antigen was detected in 22 blood cultures. Of these cultures 16 showed visible evidence of growth at the time of sampling. Pneumococcal antigen was detected in six additional blood cultures in which there was no visible evidence of growth at the time of sampling. *H. influenzae* type B antigen was detected in seven blood cultures, none of which showed turbidity. No false positive nor false negative reactions were observed in this study. Others have proposed LA as well as COAG for detection of antigen directly from blood cultures. For the past 5 years I have routinely used CIE or LA for detection of antigen in blood cultures when the Gram stain was suggestive of either gram-negative pleomorphic rods, gram-positive cocci, or gram-negative diplococci. Antigen can also be detected directly in the blood of patients with other infectious diseases. *H. influenzae* antigen has been detected in serum and the cerebral spinal fluid of patients with meningitis (28, 110). Antigenemia has been seen in cases of epiglottitis and pericarditis (113). Spencer & Savage (115) reported that the amount of antigen did not correlate with poor prognosis, but the presence of pneumococcal antigen in the blood did. When no antigen was found, 90% of the patients survived. However the mortality rate in patients with pneumococcal antigenemia was 50%. Several other antigens have also been detected directly in the blood by CIE including group B streptococci, *Pseudomonas aeruginosa, Klebsiella pneumoniae, Escherichia coli* K1, and staphylococcal teichoic acid.

Several investigators (44, 82) have reported the isolation of *Legionella pneumophila* from blood cultures. *Legionella* is not routinely isolated from blood probably because of its fastidious growth requirements. For those workers attempting to isolate *Legionella* from blood, the blood must be subcultured on charcoal yeast extract medium.

During the initial stages of leptospirosis, the organism may be seen in blood by dark field examination. Care must be taken not to confuse the erythrocyte membrane organelles with leptospires. Leptospira may be grown directly from anticoagulated blood using Fletcher's medium (76a).

Direct Gram staining of peripheral blood is not usually successful, as in most cases of bacteremia there are considerably less than 10^5 CFU per ml present, which is the limit of detection by Gram stain. In the case of neonatal sepsis, the Gram stain of peripheral blood may be positive, as there may be an increased number of organisms per milliliter present. *Candida* may occasionally be observed in a peripheral blood smear.

In 1944, Humphrey (66) suggested a Wright's stain of the buffy coat layer to detect bacteremia before blood cultures became positive. There has been controversy as to whether the technique is useful or not. Brooks et al (14) described the procedure as useful. Of 135 blood cultures, 14 were positive and 5 of the 14 cultures from three patients showed gram-positive intracellular organisms when the buffy coat was Gram stained. Gram-positive cocci were seen in two of three patients with staphylococcal bacteremia and gram-positive rods and in one patient with *Clostridium perfringens* bacteremia. Carlson & Andersen (23) thought buffy coat smears were unreliable. Recently, Reik & Rubin (98) prepared buffy coat smears from 599 blood samples. Of these, 21 blood cultures from 17 patients were positive; 2 (11.7%) of the 17 patients had positive buffy coat smears. The authors commented that the low rate of detection of positive buffy coat smears was a function of the low concentration of organisms present. Their data show that in both cases of a positive buffy coat smear at least 10^5 to 10^6 CFU of either *Streptococcus pneumoniae* or *Staphylococcus aureus* per ml was present. The authors also commented that acridine orange could be used to great advantage in buffy coat smears as it is 10 times more sensitive than the Gram stain. Buffy coat smears are not recommended as a routine procedure. Reik & Rubin (98) suggested abandoning the requirement that organisms be seen intracellularly, as has been stipulated (14). In their study, the greater proportion of bacteria was present extracellularly. The preselection of patients might also increase the yield. Patients with positive buffy coat smears are likely to be seriously ill at the time of admission. The patients are likely to have a debilitating underlying illness. They are usually infected with gram-positive organisms or *Neisseria meningitidis.* The patients are likely to suffer from pneumonia, a would infection, or cellulitis. Most patients with urinary tract infections and bacteremia have low level bacteremia. The authors (98) also historically assessed clinical data of eight patients with positive buffy coat smears between 1974 and 1980. All of these patients died. Positive buffy coat smears are a poor prognostic sign.

Cooper (29) recently reviewed rapid methods for detecting fungi in blood cultures. He indicated that detecting fungi in blood smears usually is a fortuitous observation and that although *Histoplasma* occasionally may be seen in peripheral blood smears, the examination of a stained buffy coat has not been productive. Traditional blood culture methods have not been an efficient approach to isolating fungi from the blood. Venting the bottle improves the yield of positive cultures, but use of a biphasic media yields more isolates. Cooper (29) reported that lysis filtration and lysis centrifugation appeared to offer some advantage. Hopfer et al (64) suggested that the radiometric method may be used for the rapid detection of yeasts. Prevost & Bannister (96) indicated that radiometry detected yeast about as well as biphasic media.

This review has focused on both traditional and other methods for the detection of bacteremia. The conclusion is that relatively few, if any, methods will detect the presence of bacteria in blood with acceptable levels of sensitivity and specificity and at a reasonable cost within a few hours. Many gains have been made resulting in same-day identification and antimicrobial susceptibility of positive blood cultures. Much more remains to be done in this exciting and dynamic area of investigation.

Literature Cited

1. Alley, C. C., Brooks, J. B., Kellogg, D. S. Jr. 1979. *J. Clin. Microbiol.* 9:97–102
2. Anhalt, J. P. 1978. See Ref. 128, pp. 109–44
3. Artman, M., Weiner, M., Frankl, G. 1980. *J. Clin. Microbiol.* 12:614–16
4. Babu, J. P., Schell, R. F., LeFrock, J. L. 1978. *J. Clin. Microbiol.* 8:288–92
5. Bannatyne, R. M., Harnett, N. 1974. *Appl. Microbiol.* 27:1607–69
6. Bartlett, R. C. 1974. *Medical Microbiology: Quality Cost and Clinical Relevance.* New York: Wiley
7. Bartlett, R. C., Ellner, P. D., Washington, J. A., eds. 1974. *Blood Cultures.* Washington DC: Am. Soc. Microbiol.
8. Belding, M. E., Klebanoff, S. J. 1972. *Appl. Microbiol.* 24:691–98
9. Belli, J., Waisbren, B. A. 1956. *Am. J. Med. Sci.* 232:284
10. Blazevic, D. J., McCarthy, L., Morello, J. 1981. *Clin. Microbiol. Newsletter, Vol. 3, No. 12*
11. Blazevic, D. J., Stemper, J. E., Matsen, J. M. 1974. *Appl. Microbiol.* 27:537–39
12. Blazevic, D. J., Trombley, C. M., Lund, M. E. 1976. *J. Clin. Microbiol.* 4:522–23
13. Boling, E. A., Blanchard, G. C., Russell, W. J. 1973. *Nature* 241:472–73
14. Brooks, G. F., Pribble, A. H., Beaty, H. N. 1973. *Arch. Intern. Med.* 132:673–75
15. Brooks, J. B., Alley, C. C., Liddle, J. A. 1974. *Anal. Chem.* 46:1930–34
16. Brooks, J. B., Craven, R. B., Schlossberg, D., Alley, C. C., Pitts, F. M. 1978. *J. Clin. Microbiol.* 8:203–8
17. Brooks, J. B., Edman, D. C., Alley, C. C., Craven, R. B., Girgis, N. I. 1980. *J. Clin. Microbiol.* 12:208–15
18. Brooks, J. B., Kellogg, D. S., Shepherd, M. E., Alley, C. C. 1980. *J. Clin. Microbiol.* 11:45–51
19. Brooks, J. B., Melton, R. 1978. *J. Clin. Microbiol.* 8:402–9
20. Brooks, K., Sodeman, T. 1974. *Am. J. Clin. Pathol.* 61:859–66
21. Cady, P. 1975. In *New Approaches to the Identification of Microorganisms,* ed. C. G. Heden, T. Illeni. pp. 73–100. New York: Wiley
22. Campbell, J., Washington, J. A. II. 1980. *J. Clin. Microbiol.* 12:576–78
23. Carlson, B. E., Andersen, B. R. 1976. *J. Am. Med. Assoc.* 235:1465–66
24. Caslow, M., Ellner, P. D., Kiehn, T. E. 1974. *Appl. Microbiol.* 28:435–38
25. Chong, Y., Yi, K. N., Lee, S. Y. 1975. *Yonsei Med. J.* 16:99–102
26. Cleary, T. J., Valdes, S., Selem, M. 1979. *J. Clin. Microbiol.* 10:248–50
27. Coleman, R. M., Laslie, W. W., Lambe, D. W. Jr. 1976. *J. Clin. Microbiol.* 3:281–86
28. Coonrod, J. D., Rytel, M. W. 1972. *Lancet* 1:1154–57
29. Cooper, B. H. 1982. In *Proc. 3rd Int. Symp. Rapid Methods and Automation in Microbiology* ed. R. C. Tilton,

pp. 45–48. Washington DC: Am. Soc. Microbiol.
30. Darbridge, T. C. 1977. *Lancet* 1:1206
31. Davis, C. E., McPherson, R. A. 1975. In *Microbiology—1975*, ed. D. Schlessinger pp. 55–63. Washington DC: Am. Soc. Microbiol.
32. DeBlanc, H. J., Deland, F. H., Wagner, H. N. Jr. 1971. *Appl. Microbiol.* 22: 846–49
33. DeLand, F. H., Wagner, H. N. Jr. 1969. *Radiology* 92:154–55
34. DeLand, F. H., Wagner, H. N. Jr. 1970. *J. Lab. Clin. Med.* 75:529–34
35. Denham, M. J. 1977. *Lancet* 1:459
36. Denham, M. J., Goodman, G. S. 1977. *Age Ageing* 6:85–88
37. Dorff, G. J., Coonrod, J. D., Rytel, M. W. 1971. *Lancet* 1:578–79
38. Dorn, G. L. 1978. *J. Clin. Microbiol.* 1:52–54
39. Dorn, G. L., Burson, G. G., Haynes, J. R. 1976. *J. Clin. Microbiol.* 3:258–63
40. Dorn, G. L., Haynes, J. R., Burson, G. G. 1976. *J. Clin. Microbiol.* 3:251–57
41. Dowling, H. F., Hirsh, H. L. 1945. *Am. J. Med. Sci.* 210:756–62
42. Duclos, T. W., Hodges, G. R., Killian, J. E. 1973. *Am. J. Med. Sci.* 266:459–63
43. Edberg, S. C., Clare, D., Moore, M. H., Singer, J. M. 1979. *J. Clin. Microbiol.* 10:693–97
44. Edelstein, P. H., Meyer, R. D., Finegold, S. M. 1979. *Lancet* 1:750–51
45. Eisenberg, J. M., Rose, J. D., Weinstein, A. J. 1976. *J. Am. Med. Soc.* 236:2863–65
46. Elin, R. J., Robinson, R. A., Levine, A. S., Wolff, S. M. 1975. *N. Engl. J. Med.* 293:521–24
47. Elliott, S. D. 1938. *J. Pathol. Bacteriol.* 46:121–31
48. Ellner, P. D. 1968. *Appl. Microbiol.* 16:1892–94
49. Ellner, P. D., Kiehn, T. E., Beebe, J. L., McCarthy, L. R. 1976. *J. Clin. Microbiol.* 4:216–24
50. Eng, J. 1975. *J. Clin. Microbiol.* 1:119–23
51. Eng, J., Iveland, H. 1975. *J. Clin. Microbiol.* 1:444–47
52. Fay, D., Oldfather, J. E. 1979. *J. Clin. Microbiol.* 9:347–50
53. Feldman, S., Pearson, T. A. 1974. *Am. J. Dis. Child.* 128:172–74
54. Finegold, S. M., Ziment, I., White, M. L. 1968. *Antimicrob. Agents Chemother.* 7:692–96
55. Forrest, W. W. 1972. In *Methods in Microbiology*, ed. J. R. Norris, D. W. Ribbons, pp. 285–318. New York: Academic
56. Frankel, A., Hirsch, W. 1961. *Nature* 191:728–30
57. Gantz, N. M., Swain, J. L., Medeiros, A. A., O'Brien, T. F. 1974 *Lancet* 2:1174–76
58. George, R. H. 1974. *J. Med. Microbiol.* 7:77–83
59. Gill, V. J. 1981. *J. Clin. Microbiol.* 14:116–18
60. Hadley, W. K., Kazenka, W. 1976. *Abstr. Ann. Meet. Am. Soc. Microbiol.,* C69
61. Hadley, W. K., Senyk, G. 1975. See Ref. 31 pp. 12–21
62. Hoffman, P. C., Arnow, P. M., Goldmann, D. A., Parrott, P. L., Stamm, W. E., McGowan, J. E. Jr. 1976. *J. Am. Med. Assoc.* 236:2073–75
63. Holland, R. L., Cooper, B. H., Helgeson, N. G. P., McCracken, A. W. 1980. *J. Clin. Microbiol.* 12:180–84
64. Hopfer, R. L., Orengo, A., Chesnut, S. et al. Wenglar, M. 1980. *J. Clin. Microbiol.* 12:329–31
65. Hull, K. H., Silvanic, J. M., Walsh, L. J. 1976. *Abstr. Ann. Meet. Am. Soc. Microbiol.,* C81
66. Humphrey, A. 1944. *Am. J. Clin. Pathol.* 14:358–62
67. Ilstrup, D. M. 1975. See Ref. 128, pp. 23–26
68. Johnson, J. E., Washington, J. A. II. 1976. *Antimicrob. Agents Chemother.* 10:211–14
69. Kaslow, R. A., Mackel, D. C., Mallison, G. F. 1976. *J. Am. Med. Assoc.* 236:2407–9
70. Katz, L., Johnson, D. L., Neufeld, P. D. et al. Gupta, K. G. 1975. *Can. Med. Assoc. J.* 113:208–11
71. Kiani, D., Quinn, E. L., Burch, K. H., Madhavan, T., Saravolatz, L. D., Neblett, T. R. 1979. *J. AM. Med. Assoc.* 242:1044–47
72. Kocka, F. E., Magoc, T., Searcy, R. L. 1972. *Ann. Clin. Lab. Sci.* 2:470–73
73. Kreger, B. E., Craven, D. E., Carling, P. C. 1980. *Am. J. Med.* 68:332–43
74. Kronvall, G., Myhre, E. 1977. *Acta Pathol. Microbiol. Scand. Sect. B* 85:249–54
75. Larson, S. M., Charache, P., Chen, M., Wagner, H. N. 1973. *Appl. Microbiol.* 25:1011–12
76. LaScolea, L. J. Jr., Dryja, D., Sullivan, T. D., Mosovich, L., Ellerstein, N., Neter, E. 1981. *Microbiol.* 13:478–82
76a. Lennette, E. H., Balows, A., Hausler, W. J., Truant, T., eds. 1980. *Manual of Clinical Microbiology.* Washington DC: Am. Soc. Microbiol. 3rd ed.

77. Levin, G. V., Harrison, V. R., Hess, W. C. 1956. *Am. J. Public Health* 46:1405–14
78. Levin, J., Bang, F. B. 1968. *Thromb. Diath. Haemorrh.* 19:186–97
79. Levin, J., Poore, T. E., Zauber, N. P., Oser, R. S. 1970. *N. Engl. J. Med.* 283:1313–16
80. Lindsey, N. J., Riley, P. 1981. *J. Clin. Microbiol.* 13:503–7
81. MacGregor, R. R., Beaty, H. N. 1972. *Arch. Intern. Med.* 130:84–87
82. Macrae, A. D., Greaves, P. W., Platts, P. 1979. *Br. Med. J.* 2:1189–90
83. Mangurten, H. H., LeBeau, L. J. 1977. *J. Pediatr.* 90:990–92
84. Martinez, L. A., Quintiliani, R., Tilton, R. C. 1973. *J. Infect. Dis.* 127:102–5
85. Matsunoga, T., Karube, I., Suzuki, S. 1979. *Appl. Environ. Microbiol.* 34:117–21
86. McCarthy, L. R., Bottone, E. J. 1974. *Am. J. Clin. Pathol.* 61:585–91
87. McGowan, J. E., Bratton, L., Klein, J. O., Finland, M. 1973. *N. Engl. J. Med.* 288:1309–12
87a. McLimans, C. A., Hall, M. M., Thompson, R. L. 1980. *Abstr. Annu. Meet. Soc. Microbiol.,* C259
88. Mirrett, S., Reller, L. B. 1979. *J. Clin. Microbiol.* 10:482–87
89. Mitruka, B. M., Jonas, A. M., Alexander, M. 1970. *Infect. Immun.* 2:474–78
90. Mitruka, B. M., Kundargi, R. S., Jonas, A. M. 1972. *Med. Res. Eng.* 11:7–11
91. Moore, D. F., Hamada, S. S., Marso, E., Martin, W. J. 1981. *J. Clin. Microbiol.* 13:934–39
92. Morello, J. A., Ellner, P. D. 1969. *Appl. Microbiol.* 17:68–70
93. Morello, J. A., York, M. K. 1982. See Ref. 29, pp. 348–52
94. Pai, C. H., Sorger, S. 1981. *J. Clin. Microbiol.* 14:20–23
95. Pickett, M. J., Nelson, E. L. 1951. *J. Bacteriol.* 61:229–37
96. Prevost, E., Bannister, E. 1981. *J. Clin. Microbiol.* 13:655–60
97. Randall, E. L. 1975. See Ref. 31, pp. 000–00
98. Reik, H., Rubin, S. J. 1981. *J. Am. Med. Assoc.* 245:357–59
99. Reith, A. F., Squier, T. K. 1932. *J. Infect. Dis.* 51:336–43
100. Renner, E. D., Gatherdige, L. A., Washington, J. A. II. 1973. *Appl. Microbiol.* 26:368–72
101. Ripa, K. T., Mardh, P. A., Hovelius, B., Ljungholm, K. 1977. *J. Clin. Microbiol.* 5:393–96
102. Rose, R. E., Bradley, W. J. 1969. *Med. Lab.* 3:22–24
103. Rosner, R. A. 1972. *Am. J. Clin. Pathol.* 57:220–27
104. Rosner, R. A. 1975. *Am. J. Clin. Pathol.* 63:149–52
105. Rosner, R. A. 1976. *Current Aspects of Blood Cultures.* Groton, Ct: Pfizer
106. Rosner, R. A. 1975. *J. Clin. Microbiol.* 1:129–31
107. Rubinstein, E., Noriega, E. R., Simberkoff, M. S., Holzman, R., Rahal, J. J. 1975. *Medicine* 54:331–34
108. Rytel, M. 1979. In *Rapid Diagnosis in Infectious Disease,* pp. 7–18. Boca Raton: CRC
109. Schrot, J. R., Hess, W. C., Levin, G. V. 1973. *Appl. Microbiol.* 26:867–73
110. Shackelford, P. G., Campbell, J., Feigin, R. D. 1974. *J. Pediatr.* 85:478–81
111. Sliva, H. S., Washington, J. A. II. 1980. *J. Clin. Microbiol.* 12:445–46
112. Smith, A. G., Little, R. R. 1974. *Ann. Clin. Lab. Sci.* 4:448–55
113. Smith, E. W. P., Ingram, D. L. 1975. *J. Pediatr.* 86:571–73
114. Sondag, J., Ali, M., Murray, P. R. 1980. *J. Clin. Microbiol.* 11:274–77
115. Spencer, R. C., Savage, M. A. 1976. *J. Clin. Pathol.* 29:187–90
116. Staneck, J. L., Vincent, S. 1981. *J. Clin. Microbiol.* 13:463–67
117. Stone, H. H., Kolb, L. D., Currie, C. A., Geheber, C. E., Cuzzell, J. Z. 1974. *Ann. Surg.* 179:697–711
118. Strauss, R. R., Throm, R., Friedman, H. 1977. *J. Clin. Microbiol.* 5:145–48
119. Sullivan, N. M., Sutter, V. L., Carter, W. T., Attebery, H. R., Finegold, S. M. 1972. *Appl. Microbiol.* 23:1101–6
120. Tenney, J. H., Reller, L. B., Stratton, C. W., Wang, W. L. L. 1976. *Abstr. Int. Sci. Conf. Antimicrob. Agents Chemother.,* 309
121. Thiemke, W. A., Wicher, K. 1975. *J. Clin. Microbiol.* 1:302–8
122. Thorn, G. W., Adams, R. D., Braunwald, E. E., Isselbacher, K. J., Petersdorf, R. G. 1977. *Harrison's Principles of Internal Medicine.* New York: McGraw-Hill 8th ed.
123. Tilton, R. C. 1978. In *Critical Reviews of Laboratory Medicine,* pp. 347–65. Boca Raton: CRC
124. Ur, A., Brown, D. F. J. 1975. *J. Med. Microbiol.* 8:19–28
125. Von Haebler, T., Miles, A. A. 1938. *J. Pathol. Bacteriol.* 46:245–52
126. Wallis, C., Melnick, J. L., Wende, R. D., Riely, P. E. 1980. *J. Clin. Microbiol.* 11:462–64
127. Washington, J. A. II. 1977. *Ann. Int. Med.* 86:186–88

128. Washington, J. A. II. 1978. In *The Detection of Septicemia,* ed. J. A. Washington II, pp. 41–88. Boca Raton: CRC
129. Washington, J. A. II. 1982. See Ref. 29. pp. 33–35
130. Washington, J. A. II, Hall, M. M., Warren, E. 1975. *J. Clin. Microbiol.* 1:79–81
131. Washington, J. A. II, Yu, P. K. W. 1971. *Appl. Microbiol.* 22:100–1
132. Wasilauskas, B. L., Ellner, P. D. 1971. *J. Infect. Dis.* 124:499–504
133. Waters, J. R., Zwarun, A. A. 1973. *Dev. Ind. Microbiol.* 14:80
134. Werner, A. S., Cobbs, C. G., Kaye, D., Hook, E. W. 1967. *J. Am. Med. Assoc.* 202:199–203
135. Wilkins, J. R., Stoner, G. E. 1975. *NASA Tech Brief LAE-11525, April*
136. Wilkins, T. D., West, S. E. H. 1976. *J. Clin. Microbiol.* 3:393–96
137. Wilson, W. R. 1978. See Ref. 128, pp. 1–22
138. Wright, H. D., 1925. *J. Pathol. Bacteriol.* 25:541–78
139. Zierdt, C. H., Kagan, R. L., MacLowry, J. D. 1977. *J. Clin. Microbiol.* 5:46–50
140. Zwarun, A. A. 1973. *Appl. Microbiol.* 25:589–91
141. Zwarun, A. A. 1974. *Abstr. Ann. Meet. Am. Soc. Microbiol.,* M274

Ann. Rev. Microbiol. 1982. 36:495–517

THE PLANT PATHOGENIC CORYNEBACTERIA

Anne K. Vidaver

Department of Plant Pathology, University of Nebraska, Lincoln, Nebraska 68583-0722

CONTENTS

INTRODUCTION

The plant pathogenic corynebacteria are a fascinating group of bacteria that produce effects as diverse as any in the microbial world, ranging from wilts to abnormal growths. Several new pathogens have been discovered since this group was reviewed by Jensen (75), Starr (157), and Lelliott (96). Most subsequent reviews have dealt with taxonomy. The areas covered in this review necessarily reflect my biases and limitations. However, this review attempts to be comprehensive in some areas, and it calls attention to other areas that would benefit from more detailed investigation.

This review is limited to corynebacteria whose association with plants is confirmed and to other bacteria for which taxonomic work indicates association or placement with known corynebacteria. Incompletely identified corynebacteria have been reported and are not discussed here, such as the

0066-4227/82/1001-0495$02.00

causal agents of brown-stem disease of beans (185), stunting and other symptoms of soybeans (42, 114), cardamon blight (56), rotting yams (116), and carnation rot [mixed infection with *Pseudomonas caryophylli* (12)]. Two other plant-associated coryneforms are excluded: the cytokinin-producing bacteria isolated from pine seedling roots that might be saprophytic (79), and the specific coryneform that has been found in the leaf cavity of the fern *Azolla caroliniana* in its symbiotic association with the nitrogen-fixing blue-green alga, *Anabaena azollae* (54), a phenomenon of unknown significance.

Some general properties of these bacteria should be noted. These are all Gram-positive, aerobic, pleomorphic rods (at some stage or condition of growth). In contrast to most Gram-positive bacteria, but like other coryneform bacteria [except *Brevibacterium linens* (51) and an unusual coryneform (2)], the plant pathogenic coryneforms lack teichoic acid in their cell walls (38, 51). Most of these bacteria can be seed borne (129) or are transmitted by insects (67), and they have a restricted natural host range. These bacteria are not considered soil, air, or water inhabitants, with the possible exception of *Corynebacterium fascians* (41, 104); their survival depends on association with plant material (141). The majority of the phytopathogenic corynebacteria produce systemic, rather than localized, infections; only *C. fascians* and the newly discovered bacterial wheat mosaic pathogen (14) are in the latter category. These two pathogens also do not require injury or wounding of plants to produce their effects, in contrast to the rest of the group in which wounding is the predominant (if not the only) form of entry into plants. Some of these properties are elaborated on below.

TAXONOMY

The bacteria in this group of plant pathogens present a continuing challenge at all levels of taxa placement. There have been more studies on the taxonomy of these bacteria in recent years than in any other area. Most of the data and the rationale for placement of these bacteria into different taxa have been recently reviewed (15, 40, 42, 77, 83, 155, 181, 187). Therefore, only some general comments are made to orient the reader.

At the highest taxonomic level, Stackebrandt et al (155) propose that most coryneforms (including single strains of two plant pathogenic species) and actinomycetes form at least five unnamed clusters (families) of a single order. They used comparative analysis of oligonucleotides in the ribosomal 16S RNA as a criterion of relatedness because it is generally agreed that rRNA is more conserved than are other nucleic acids. DNA : DNA hybridization, for example, yields little information on genetic relationships of species exhibiting homology values below 20% (154).

Below the family level, the phytopathogens are currently retained within the genus *Corynebacterium* (30). Whether this is appropriate is a matter of considerable disagreement. The plant pathogens are in this genus because historically all aerobic, nonsporeforming, irregularly shaped Gram-positive rods (like *C. diphtheriae*) were assumed to be related (75, 82). However, recent studies show that all the plant-disease-associated coryneform bacteria, with the possible exception of *C. fascians,* are significantly different from *C. diphtheriae,* particularly in chemical composition (84).

The plant pathogenic bacteria have a G+C content of 65–76% compared to 52–60% for *C. diphtheriae* (15, 28, 40, 169). There are different diamino acids in the cell wall peptidoglycan of some species. These are L-lysine in *C. ilicis* and possibly in *C. rathayi;* diaminobutryic acid in the *C. michiganense* group, *C. iranicum,* and *C. tritici;* L-ornithine in the *C. flaccumfaciens* group; and meso-diaminopimelic acid in *C. fascians* and the human pathogen *C. diphtheriae* (24, 84, 125). Three of the plant pathogens were classified in a rare peptidoglycan group designated B2, based on cross linkages between peptide subunits and the presence on an inter-peptide bridge containing a D-diamino acid (138). In addition, examination of vitamin K and its derivatives (menaquinones) showed that only *C. fascians* has a menaquinone system with an 8-isoprene-unit side chain in common with *C. diphtheriae;* the remaining plant pathogens have major menaquinones of 9 or 10 isoprene units (22–24, 190). Phospholipid analyses also showed that only *C. fascians* shared phosphatidylethanolamine with *C. diphtheriae* and its relatives; other plant pathogenic corynebacteria lack this compound (88). There are also differences between *C. diphtheriae* and two phytopathogenic corynebacteria in their 16S rRNA sequences (155). Wall carbohydrate analyses (84) show differences in composition for different strains of the same species; such differences are therefore unlikely to be useful for taxonomic classification, but they may have significance in pathogenicity and other recognition phenomena. Cellular protein analyses by single-dimension polyacrylamide gel electrophoresis also show significant differences among certain plant pathogenic corynebacteria and the animal-associated *C. bovis* (15). The groupings suggested by Carlson & Vidaver (15) are consistent with the most recent and extensive DNA : DNA hybridization studies, including plant pathogens (40). Collectively, then, the data show that the plant pathogens as a group are not closely related to *C. diphtheriae* and its relatives.

Differences among the coryneform bacteria led to several suggestions that the plant pathogenic corynebacteria be transferred into one or more genera by the 1950s (75) and most recently into *Arthrobacter, Curtobacterium, Microbacterium,* or *Rhodococcus* (40, 77, 84, 155). These conclusions are supported by some numerical taxonomy studies based on phenotypic properties [Level 4 of microbial expression (115)] (9, 32, 76, 103, 185) and DNA : DNA hybridization (40). Dye & Kemp (43), however, question the transfer of these bacteria into one or more of the suggested genera because of considerable homology among them in phenotypic properties; they suggest that these bacteria be retained in the genus *Corynebacterium* for the time being.

Thus, despite a growing and strongly supported view that *Corynebacterium* is not an appropriate repository for the plant pathogens, even proponents of change say that "much more information on a large number of strains is required before the creation of [other] genera can be contemplated" (83). Keddie (82) concluded that it is still convenient to use the term coryneform bacteria to define "a broad morphological group, sometimes imperfectly, but [it] does not imply relatedness within it." For purposes of this review, the plant pathogens are retained within the genus *Corynebacterium.*

Below the generic level, recent studies show that some currently described species (147) do not have sufficient differences among them to justify their recognition as distinct taxa at the species level (15, 40, 43). There is disagreement, however, whether the differences found among them warrant taxon placement at the subspecific level (15) or at the infrasubspecific level of pathovars (43). The data of Carlson & Vidaver (15) led them to formally propose the following new combinations: *C. flaccumfaciens* ssp. *flaccumfaciens; C. flaccumfaciens* ssp. *poinsettiae; C. flaccumfaciens* ssp. *betae; C. flaccumfaciens* ssp. *oortii; C. michiganense* ssp. *michiganense; C. michiganense* ssp. *nebraskense; C. michiganense* ssp. *insidiosum; C. michiganense* ssp. *sepedonicum.* A new taxon *C. michiganense* ssp. *tessellarius* was also proposed.[1] Taxa that remained unchanged are *C. fascians, C. ilicis, C. tritici, C. iranicum,* and *C. rathayi.* This classification is generally consistent with the DNA : DNA hybridization data of Döpfer et al (40). Thus, it is my belief that this classification is the most useful at the present time and is likely to be accepted by plant pathologists and microbiologists; hence, it is used here.

There are occasional anomalies in reports of plant pathogenic corynebacterial species. *C. ilicis,* which causes a disease of holly, apparently has only been seen once (99, 100). Some of the corynebacteria classified with plant pathogens either have not been reported as plant pathogens ["*C. mediolanum*," (155)] or were reported once and are no longer available in known culture collections ["*C. agropyri*" and others (30)]. These species are not in the *1980 Approved Lists of Bacterial Names* and consequently have no

[1]Hereafter, these names are abbreviated as *C. f.* ssp. *flaccumfaciens, C. f.* ssp. *poinsettiae, C. f.* ssp. *betae, C. f.* ssp. *oortii, C. m.* ssp. *michiganense, C. m.* ssp. *insidiosum, C. m.* ssp. *sepedonicum, C. m.* ssp. *nebraskense,* and *C. m.* ssp. *tessellarius.*

standing in the literature. "*C. agropyri,*" however, was recently isolated from a 37-year-old herbarium specimen (106a); its relationship to other plant pathogens can now be determined. *C. rathayi* and *C. tritici* were inadvertently left off the *Approved Lists* and have been revived (15) on the basis of differences between them and other species.

The classification of other plant-associated corynebacteria is not clear. Recent information on the sugarcane ratoon stunting bacterium (33, 98) suggests that this coryneform is similar to several known pathogens in having a high G+C (67–68%) and diaminobutryate in the cell wall (M. J. Davis, personal communication). Kao et al (80), however, report 60% G+C and both ornithine and lysine in the cell wall. It is not clear whether or not differences in methods and strains account for these discrepancies. And it will be of interest to know how a newly isolated coryneform associated with leaf scorch of grape (D. C. Gross, personal communication) is related to the other corynebacteria.

All coryneform pathogens except *C. fascians* are generally host specific under natural conditions (see Table 1), a fortuitous situation, as disease diagnosis and pathogen identification are virtually certain if the host is known (see 180, 181). For relatively quick pathogenicity determinations, alternate hosts may be more convenient, e.g. eggplant can be used to test *C. m.* ssp. *sepedonicum* strains (see 29). There is some difficulty, however, in easily distinguishing the plant pathogens in culture independent of the host plant, but motility, pigmentation, colony morphology, and growth on triphenyl tetrazolium agar usually serve to distinguish the pathogens commonly found in the USA (180).

Morphological differences and mode of cell division are questionable criteria in the classification of corynebacteria, including plant pathogens (82, 155). For example, electron micrographs of *C. rathayi* cultivated in vitro can show a simple rod form whereas the same culture from plants shows pleomorphism characteristic of coryneform bacteria (6). It is by no means clear how one should decide which is correct or typical.

Colony pigmentation can be a useful aid in identification of plant pathogenic corynebacteria, but pigment analysis is of unproven value in classification. The majority of these bacteria are pigmented in shades of yellow or orange on complex media, but other colors are sometimes seen from freshly isolated material or after extensive cultivation (52, 160; R. R. Carlson & A. K. Vidaver, unpublished results). The water-insoluble blue pigment indigoidine (89) can assist in identifying *C. m* ssp. *insidiosum* when it is present. A water-soluble extracellular purple pigment can be produced by some strains of *C. f.* ssp. *flaccumfaciens* (143). The lipid-soluble pigments of coryneform bacteria have been recently reviewed (105); the plant pathogens *C. f.* ssp. *poinsettiae, C. m.* ssp. *michiganense,* and *C. fascians* all have carotenoids that appear to differ from one another. However, as no systematic studies of pigments have been reported, the utility of pigments in classification is still an open question.

Rapid and specific identification may be possible by two methods. Serological identification of specifc corynebacteria has produced contrasting results (see 34, 66, 136). It is not clear whether such results are due to the specific immunogen or strain, its age, condition or treatment prior to immunization (e.g. 19), or to variables in the techniques employed. It is clear that sera prepared against one pathogen can cross-react with taxonomically closely related and distantly related corynebacteria (e.g. 29, 34, 136). In view of the high degree of similarity among many of the plant corynebacteria (15, 43), such cross-reactions may be expected. For the ratoon stunting organism, however, specific antisera have been obtained (M. J. Davis, personal communication). The convenient, though less sensitive, latex-agglutination test (149) is currently being used as much as indirect fluorescent antibody techniques (34). There are differences of 10^4 in the sensitivity of the indirect fluorescent antibody technique for rapid detection and presumptive identification of *C. m.* ssp. *sepedonicum* (35, 148), perhaps due partly to differences in sampling or concentration of antigen or antibody. The advent of monoclonal antibody techniques may allow for the preparation of specific antibodies for the corynebacteria. This would enable some of the problems in serological identification and specificity to be resolved in the near future. The second method, bacteriocin production, offers promise of specific and relatively rapid identification of the majority of strains and species of phytopathogenic corynebacteria (64), but confirmation is needed.

HABITATS, ISOLATION, CULTIVATION, AND GROWTH

Since this area was reviewed recently (181), only highlights are mentioned here. In general, the corynebacterial plant pathogens are studied because of their economically important associations with plants, including latent infections by *C. m.* ssp. *sepedonicum* (7, 29, 70, 150) and *C. f.* ssp. *flaccumfaciens* (170). Survival requires association with diseased plants or residue materials (108, 139, 141). With the possible exception of *C. fascians* (104, 139), these bacteria are considered poor survivors in soil. This could be due to the action of other microorganisms, such as rhizobia, *Bacillus,* or *Arthrobacter* sp., which produce antibiotic-like substances in vitro (62, 69). Soil survival studies of corynebacteria can be criticized on several grounds: lack of inoculum quantitation, unspecified growth state of the inoculum, use of genetically unmarked bacteria [except one study of *C. m.* ssp. *insidiosum* (111)], indirect recovery using plants as bait, and ignorance of detection

Table 1 Diseases caused by plant pathogenic corynebacteria

Bacterium[a]	Natural host	Predominant symptoms	References[b]
C. fascians	General: annuals, perennials	Leaf, bulb distortion; bud deformity; proliferation	30, 41
C. flaccumfaciens ssp. *flaccumfaciens*	Field bean	Wilt	41
C. flaccumfaciens ssp. *betae*	Beet	Silvering of leaves; wilt	30
C. flaccumfaciens ssp. *oortii*	Tulip	Leaf, bulb spot; wilt	30
C. flaccumfaciens ssp. *poinsettiae*	Poinsettia	Leaf spot; wilt	30
C. ilicis	American holly	Shoot, branch blight	99
C. iranicum	Wheat	Yellow slime of leaves, inflorescences; leaf spots	30
C. michiganense ssp. *michiganense*	Tomato pepper	Wilt, fruit spot	41, 160
C. michiganense ssp. *insidiosum*	Alfalfa	Wilt, stunting	30, 41
C. michiganense ssp. *nebraskense*	Corn	Wilt, leaf blight	140, 146a
C. michiganense ssp. *sepedonicum*	Potato	Wilt, tuber rot	41, 101, 144
C. michiganense ssp. *tessellarius*	Wheat	Leaf spot	14
C. rathayi	Cocksfoot grass, annual ryegrass	Yellow slime of leaves, inflorescences	10
C. tritici	Wheat	Yellow slime of leaves, inflorescences	5, 11
Ratoon stunting bacterium	Sugarcane	Stunting	33

[a] See text for classification rationale. Only bacteria currently in culture collections are listed.
[b] A current compilation of disease descriptions is not available; these references will enable interested readers to obtain representative descriptions. Commonwealth Mycological Institute descriptions (e.g. ref. 10, 11) are available for many of these pathogens.

limits. All of these bacteria can be isolated from both infected and infested seed, tubers, or bulbs (41, 139, 180; Carlson & Vidaver, unpublished results), particularly if material is obtained from areas of known infection. *C. fascians* can be difficult to isolate from leafy-galls (93) and bulbs (176) and may require enrichment through an intermediate host plant (139, 176). Epiphytic survival on, and natural infection of, weeds has been reported for *C. m.* spp. *nebraskense* (140), *C. f.* ssp. *flaccumfaciens* (see 141), and *C. m.* ssp. *michiganense* (160; 171); whether or not weeds are hosts for the remaining pathogens is not known. At least some of these pathogens can survive in untreated water for many hours [*C. rathayi,* reported in Kuznetsov et al (91)] or even weeks [*C. m.* spp. *nebraskense* (158)], suggesting that this may be a source of inoculum. Secondary sources of inoculum are less well studied. *C. m.* ssp. *sepedonicum* survives on the surfaces of potato handling equipment; this plays an important role in its epidemiology (e.g. 107, 109). No airborne survival studies have been published, to my knowledge. Thus, for all practical purposes, infected seed, tubers, bulbs, and plants are the primary sources of inoculum and also are source material for isolation of the bacteria.

Growth of the phytopathogenic corynebacteria on complex media is relatively slow, ranging from about 3 days for visible colony formation of *C. f.* ssp. *flaccumfaciens* to 10–12 days for microscopic colony formation of the fastidious ratoon stunting bacterium (15, 33, 43, 96, 98) at optimal growth temperatures ranging from 23 to 28°C. Such slow growth and the relative lack of selective media have hampered ecological studies (181). The selective media devised for *C. m.* ssp. *nebraskense* (63), *C. m.* ssp. *sepedonicum* (151), and *C. m.* ssp. *michiganense* (78) delay colony formation 1 or more days. An improved selective medium for *C. m.* ssp. *sepedonicum* and selective media for other corynebacteria are in the process of development (M. Sasser, personal communication). These are needed and should prove useful in determining the survival and spread of these bacteria in various habitats. Soil, air, and water sampling for pathogens should then be feasible; these potential avenues of spread have been virtually ignored by investigators because of the limitations of current isolation procedures in which the majority of saprophytic bacteria outnumber and outgrow the pathogens.

Nutritionally, not many advances have been made since Starr's (156) pioneering study of these bacteria. He found *C. fascians* strains to be least fastidious, requiring only thiamine as a growth factor, whereas thiamine, biotin, and pantothenate were required for growth of *C. f.* ssp. *flaccumfaciens* and *C. f.* ssp. *poinsettiae* in a minimal salts, amino-acid medium. Thus, it is interesting that Keddie et al (85) found the same requirements needed by herbage isolates but not soil coryneform bacteria. For the related pathogens, *C. m.* ssp. *michiganense, C. m.* ssp. *insidiosum,* and *C. m.* ssp. *sepedonicum,* amino acids, thiamine, biotin, and nicotinic acid (niacin) were required (156). Lachance (94) confirmed these requirements for *C. m.* ssp. *sepedonicum.* In view of the close taxonomic relationship of these bacteria to *C. m.* ssp. *nebraskense* and *C. m.* ssp. *tessellarius* (15), it is not surprising that the same three growth factors were highly stimulatory (R. R. Carlson & A. K. Vidaver, unpublished results) in a purified agar minimal salts medium (178).

Amino acid requirements for growth have been examined even more sparsely (see 30). L-Methoinine is required by *C. m.* ssp. *sepedonicum* (73),

C. m. ssp. *nebraskense,* and *C. m.* ssp. *tessellarius* (R. R. Carlson & A. K. Vidaver, unpublished results) and it is highly stimulatory for *C. f.* ssp. *flaccumfaciens* (130). In this regard, Owens & Keddie (119) reported that the amino acid requirement of 23 out of 38 herbage coryneforms could be satisfied by L-methionine alone in a basal salts medium with vitamins and inorganic nitrogen. These results suggest a close relationship between these bacteria and the plant pathogens. It is also interesting that 0.3 mM cystine and cysteine inhibited a strain of *C. m.* ssp. *sepedonicum* (73); unfortunately, potato tubers have not been found to contain these amino acids in such high concentration (73). L-Cysteine was also inhibitory for *C. f.* ssp. *flaccumfaciens* (130).

The only biochemical studies of metabolism in this group of bacteria deal with the examination of carbohydrate dissimilation. These are all aerobic organisms and oxidize glucose in conventional ways. Zagallo & Wang (193), using radiorespirometric methods, showed that a single strain of *C. tritici* was metabolically similar to a strain of *Arthrobacter globiformis* in primarily using the Embden-Meyerhof-Parnas pathway, and to a lesser extent the hexose monophosphate pathway, whereas in a strain of *C. m.* ssp. *sepedonicum* both pathways appeared to be equally important in glucose dissimilation. There are also reports of a pentose cycle in *C. m.* ssp. *michiganense, C. f.* ssp. *flaccumfaciens,* and *C. fascians* (81) and in *C. m.* ssp. *insidiosum* (194). More data are needed on major metabolic pathways, both anabolic and catabolic, for other strains and species.

PATHOGENICITY FACTORS

Tbe phytopathogenic corynebacteria probably cause all or part of their pathogenic effects by the production of various metabolites, including "toxins," polysaccharides, hormones, and possibly enzymes (31, 122, 163, 181). More is known about the chemistry of such compounds than of their biosynthesis and modes of action.

The first reports of extracellular polysaccharides in phytopathogenic corynebacteria was by Gorin & Spencer (59), who examined named strains (NCPPB) of *C. m.* ssp. *michiganense, C. m.* ssp. *sepedonicum, C. m.* ssp. *insidiosum, C. fascians, C. f.* ssp. *flaccumfaciens, C. tritici,* and *C. rathayi.* The strains produced polysaccharides differing markedly in sugar content and composition. Subsequently, some of these compounds were shown to be large peptidyl glycans of about 2×10^4 to 5×10^6 daltons; those isolated from strains of the taxonomically related bacteria *C. m.* ssp. *insidiosum* (59, 131, 132), *C. m.* ssp. *sepedonicum* (59, 161), and *C. m.* ssp. *michiganense* (127, 128) could cause wilting of plants. According to Gorin & Spencer (59), these compounds contained high concentrations (29–46%) of fucose, although Strobel and associates, using unidentified strains, found little or no fucose in *C. m.* ssp. *sepedonicum* (162, 165) and differing quantities in *C. m.* ssp. *insidiosum* strains (131, 132). Fucosyltransferase activity, possibly involved in biosynthesis of the polysaccharide material, was reported in a culture of *C. m.* ssp. *insidiosum* (134).

The polysaccharide structure from a single strain of *C. m.* ssp. *insidiosum* appears to be a polymer of D-glucose, D-galactose, L-fucose, and pyruvic acid residues (60), whereas a *C. m.* ssp. *sepedonicum* polysaccharide was reported to consist largely of mannose, glucose, galactose, and 2-keto-3-deoxygluconic acid (165). The latter results differ from those of Gorin & Spencer (59), who found a preponderance of galactose, glucose, and fucose in NCPPB 9850. These nonspecific wilt-inducing macromolecules were all isolated from stationary phase cultures; compounds that produced similar wilting activity were isolated from plants infected by *C. m.* ssp. *sepedonicum* (153, 162) and *C. m.* ssp. *insidiosum* (132, 153). Ultrastructural membrane and cell wall damage was reported by Strobel & Hess (164) for cut tomato stems treated with 10 mg of *C. m.* ssp. *sepedonicum* "toxin" per ml for 2 hr and for potato stems sampled 4 to 5 weeks after inoculation with the same organism (71). Other workers have not detected any physiological (175) or ultrastructural (184) evidence of membrane damage in either alfalfa cuttings treated with low concentrations of *C. m.* ssp. *insidiosum* "toxin" (175) or in infected alfalfa plants (37) or infected tomato plants (184). In the Van Alfen & Turner (175) study, as little as 2 μg of toxin forced into the stem resulted in a marked (19%) decrease in water conductance, and 200 μg/ml was effective in visible wilting. No evidence of mechanical plugging was detected either by conductivity studies (37) or by electron microscopy (184). There is rather convincing evidence that the only requirement for effectiveness in wilting is large size (173, 174). The claim by Strobel and associates (159) that the wilt toxin of *C. m.* spp. *insidiosum* can be used to differentiate susceptible and resistant alfalfa varieties has not been confirmed (55, 174). All the above considerations, and the numerous internal qualitative and quantitative inconsistencies detailed by Daly (31), make it difficult to have full confidence in the work of Strobel and associates (71, 127, 128, 131, 132, 134, 159, 161–164).

Electron microscopy showed that membrane damage to chloroplasts was the earliest detectable site of damage (D. W. Fulbright, personal communication) with the newly described, presumably nonvascular, wheat pathogen, *C. m.* ssp. *tessellarius* (14, 15). This observation is consistent with the appearance of irregular chlorotic spots characteristic of this leaf disease. Whether or not toxin is involved in this phenomenon remains to be determined.

Three corynebacterial pathogens, *C. rathayi, C. tritici,* and *C. iranicum,* secrete copious quantites of gums (apparently polysaccharides) onto infected leaves and developing seeds; *C. rathayi*-infected plants can have toxicity to grazing animals, especially sheep (5).

The *C. rathayi* toxic factor(s) can be efficiently isolated from annual ryegrass (*Lolium rigidum*) galls that are induced by nematodes (see below) and colonized by bacteria (182). The toxic factor can be less efficiently isolated from inoculated ryegrass (*Lolium multiflorum*) endosperm tissue cultures (166). Stynes & Petterson (166) suggested that the low concentration of toxin produced may be due to either the use of a nonoptimal grass species or a nonoptimal physical and chemical environment for the tissue culture:bacterium interaction. It is also possible that different strains of the bacterium could produce differing quantities of toxin. No toxin production was detected by the bacteria in vitro with any of 10 culture media, nor was toxin isolated from the culture media or uninoculated plant cell cultures (167). Thus, bacteria are required for toxin production. Whether the bacterium induces the plant to produce a toxic substance or vice versa is not yet clear. These studies show the possibilities not only of obtaining toxin in quantity and studying its biosynthesis under controlled conditions, but they also offer the potential of an in vitro assay for testing plant breeding material for resistance to the bacteria or toxin production.

Partially purified glycoplipid toxin fractions have been isolated from *C. rathayi*-infected plants (182); a minimum lethal dose for nursling rats was less than 5 μg for six of eight fractions tested. The toxic gall material was purified by reverse-phase high performance liquid chromatography (HPLC) to yield a group of eight compounds; control extracts of normal rye grass seeds proved to be nontoxic and did not give any high performance liquid chromatography traces typical of the toxic mixture (182). (However, neither controls of nematode galls alone or toxin produced in tissue culture was reported.) All eight compounds were principally glycolipids. The two main toxin fractions, designated corynetoxin 3 and 4, contained 3-hydroxyheptadecanoic acid and heptadec-2-enoic acid, respectively, as principal components. Each also contained an amino sugar, suggested as glucosamine, and all eight compounds contained residues presumed to be either uracil or cytosine. In this regard it is of interest to note that the wilting material isolated by Pearson (124) for *C. m.* ssp. *sepedonicum* contained an estimated 10–30% nucleic acid and was inactivated by DNase and RNase in combination (not tested separately). The role of the *C. rathayi* toxin in rye grass toxicity of animals seems firm; whether or not any of the toxic fractions have any effects on plants has not yet been reported.

Many questions remain as to the role of toxic antimetabolites in pathogenicity. It needs to be pointed out (192) that the majority of criteria used to evaluate the pathological significance of toxins, including symptom production, presence in plants, kinetics of production during disease development, and correlation of biosynthetic rate with virulence, are logical but inconclusive as primary lines of evidence. Genetic analyses are needed for definition of the role of toxins in disease (192). The same, of course, can be said for other pathogenicity factors.

The abnormal growth characteristics of leafy galls or fasciation (distortion of flowers or stems) associated with *C. fascians* infections are undoubtedly due to one or more of the plant hormones it can produce; this topic was recently reviewed (181). The relationship of hormone production to the presence of a "large plasmid" is tenous (106); its loss in one of four strains was correlated with loss of virulence. No association between pathogenicity and plasmid presence was detected in other strains (E. N. Lawson & M. P. Starr, manuscript in preparation).

Evidence for enzymatic activity as a pathogenicity factor is indirect. Ultrastructural studies of tomato plants infected with *C. m.* ssp. *michiganense* showed plant cell wall degradation occurring before any visible symptoms (184). Both cellulytic and pectolytic enzymes were implicated in progressive breakdown of cell walls and middle lamella, respectively. Histochemical and fluorescence microscopy of naturally infected tomatoes showed middle lamella, rather than cell wall, breakdown (102). As the breakdown occurred in the apparent absence of bacteria in both studies, enzymes would have to be transported in the conducting tissue to target sites. Pectolytic enzyme production has been reported for strains of *C. m.* ssp. *michiganense* in vivo (123) and in vitro (72, 123). In the latter case, pectolytic activity was detected at low, but not high, pH. The negative report of pectolytic activity by *C. m.* ssp. *sepedonicum* (121) may be due to high pH in the test medium. Patiño-Méndez (123) also detected cellulase activity from infected plants and in vitro, whereas Goto & Okabe (61) detected cellulolytic activity in vitro by *C. m.* ssp. *sepedonicum.* Electron microscopy of beans infected with *C. f.* ssp. *flaccumfaciens* also showed damage to the xylem and decomposition of the middle lamella prior to wilting (39). In the cases cited, however, localized membrane damage cannot be excluded. Yet, in *C. m.* spp. *insidiosum*-infected alfalfa plants, membrane damage was not detected physiologically (see above), and, as with the other cases cited, there was no relationship between the site containing the largest concentrations of bacteria and site of damage (37). An unidentified macerating agent has also been reported from *C. betae* (86). Earlier histological work on the vascular pathogens *C. m.* ssp. *sepedonicum, C. m.* ssp. *michiganense, C. m.* ssp. *insidiosum, C. f.* ssp. *flaccumfaciens,* and *C. f.* ssp. *poinsettiae,* as summarized by Nelson & Dickey (113), is generally consistent with the interpretation of enzymatic activity preceding bacterial appearance. The complexities of investigating the degradation of plant cell walls and membranes has been reviewed by Bateman & Basham (4).

Another possible pathogenicity factor may be the production of surfactants to promote adhesion to plant surfaces and possibly to enhance entry into plant cells. Thus, it is intriguing that Akit et al (1) found all six *C. fascians* strains tested produced biosurfactants from hexadecane supplemented media, since this bacterium is considered to be principally a surface inhabitant of plants (41, 139). However, a strain of *C. m.* ssp. *insidiosum* produced the most surfactant of any bacteria tested. Strains of *C. m.* ssp. *michiganense, C. f.* ssp. *poinsettiae, C. rathayi,* and *C. tritici* did not produce any detectable surfactant. Nonpathogens were also tested and produced varying amounts of surfactant, showing that there was no correlation between this property and pathogenicity. Surfactants were detected by reduction of both surface tension and interfacial tension of liquid cultures compared to controls, as well as by critical micelle concentration. Preliminary work, based on pentane extraction, suggested that the surfactants were neutral lipids (1).

The presence of pili has not been reported for phytopathogenic corynebacteria, probably because no one has looked for them. In human and animal corynebacteria, the 11 species examined had pili, with variations in the number per cell and percentage of piliated cells (191). Pili are associated with virulence in some animal pathogens (152). Thus, it would be of interest to learn of their presence or absence and biological significance in the plant pathogens.

Environmental factors sometimes crucially affect disease development. Temperature and humidity can have either synergistic or antagonistic effects on symptom expression, as can pollution. Surprisingly, sulfur dioxide partially inhibited lesion development by *C. m.* ssp. *nebraskense* in corn without affecting healthy plants (95).

THE ROLE OF NEMATODES IN PLANT DISEASES ASSOCIATED WITH CORYNEBACTERIA

Certain phytopathogenic corynebacteria are unusual in that nematode transmission is required for characteristic disease development under natural conditions. In these diseases, neither the bacterium nor the nematode alone account for disease, although some growth distortions can occur with either the nematode or sometimes with the bacterium alone. The most dramatic example of such interactions is that of the nematode *Anguina agrostis* and *C. rathayi,* the only complex that continues to be a serious problem. The association results in both a plant disease, yellow slime, and a neurological animal disease, rye-grass toxicity (5). The disease is characterized by a yellow slime or gumming symptom in the upper parts of plants, especially the inflorescences. Dwarfing, distortion, and yellow seed galls are also seen (5, 10). At maturity the galls will contain a predominance of either nematodes or bacteria, even though all galls are induced by nematodes (167). Vessels and parenchyma are invaded by the bacteria (10) and several genera of grasses are susceptible (5). Animals feeding on infected galls or grass stubble can experience neurotoxicity; in sheep this includes staggering, collapse, and periods of violent convulsions until death (see 5). The toxic agent(s) can be isolated from infected galls as well as from callus tissue (see preceding discussion).

Other examples of nematode : bacterium interactions are less well studied. Yellow slime or "tundu" disease of wheat requires the presence of *Anguina tritici* (11, 168). As above, this nematode forms a characteristic seed gall, as well as producing distinctive curling leaves with gummy exudates. The nematode acts as a vector for the bacterium, which produces slime. In addition to gumming of leaves, seed production is impaired. Despite its similarity with the annual rye-grass syndrome, there is no evidence of mammalian toxicity. The "cauliflower" syndrome of strawberry (126) also requires the presence of a nematode, *Aphelenchoides ritzemabosi,* and *C. fascians* for characteristic disease; neither agent alone is sufficient. Unlike *C. rathayi* and *C. tritici,* however, there is ample evidence for *C. fascians* as a pathogen per se (41, 104). The ecology of the former species is largely unknown.

Not surprisingly, disease may be more severe in the presence of nematodes that weaken the plant physically or physiologically. Tomato canker, caused by *C. m.* ssp. *michiganense,* is more severe in some plant cultivars in the presence of *Meloidogyne incognita* (36). Similarly, in a wilt-susceptible cultivar, the number of *C. m.* ssp. *insidiosum*-infected alfalfa plants may increase in the presence of *Ditylenchus dipsaci* (68).

GENETICS

The genetics of plant pathogenic corynebacteria are largely unknown; the same can be said for closely related bacteria. The topic is singled out for discussion because of optimism for the future, since no gene transfer systems have been reported, much less any substantive data.

Plasmids have been found in many species and subspecies (65, 106; E. N. Lawson & M. P. Starr, manuscript in preparation), but there is, as yet, no clear association between their presence and phenotypic properties of pathogenicity, bacteriocin production, drug resistance, or ability to grow on unusual hydrocarbons.

Chromosomal genetic studies are also very limited. There are reports of auxotrophic mutants of *C. fascians* (74) and *C. m.* ssp. *michiganense* (49). The former group obtained mutants with single requirements for glycine, aspartic acid, arginine, methionine, creatine, or adenine, whereas Ercolani (49) obtained mutants requiring valine, methionine, or alanine. In each case, the mutants were unable to infect and grow in plants or induce typical symptoms unless the appropriate amino acid or base was supplied exogenously. None of the amino acids had any effect on the infectivity of mutants if these were supplied to the plants 6 days after inoculation, corresponding with a decrease in auxotroph survival. Thus, the absence of bacterial growth in plants could be ascribed to inadequate nutrition, as with other plant pathogenic bacteria (e.g. see 49). It would therefore be of interest to know whether avirulent prototrophic mutants could still attach or whether subsequent steps in the infection process were affected. It would not be surprising to find that, as with many other bacteria, the capsule or cell wall contained necessary virulence factors. However, in one case, extracellular gums seemed to be produced as readily by both a virulent parent and avirulent derivative of *C. m.* ssp. *insidiosum* (19). Chemical characterization was not done and would be required to detect any differences in gum composition or structure between the strains.

Natural or induced colony morphology or pigment changes have been correlated with loss of virulence in some cases but not in others (3, 17, 52, 74, 92, 135, 142, 160, 179; R. R. Carlson & A. K. Vidaver, unpublished results). A colorless derivative of *C. f.* ssp. *poinsettiae* (90) was more sensitive to ultraviolet light than was the pigmented parent; similar results have been reported for other bacterial genera.

Freshly isolated bacteria vary in virulence when tested for pathogenicity (e.g. 104, 160, 179). The data suggest that differences in regulation of virulence factors occurs or that more than one gene or pathogenicity factor is involved in disease. It may be that a limited number of nonallelic genes are involved in pathogenicity since no race variation, or specificity associated with plant cultivars, has so far been reported, unlike the case for several Gram-negative pathogens. A decline in virulence of cultures maintained on agar is frequently reported for the *C. michiganense* subspecies (16, 47, 110, 142, 160).

There is still the challenge of finding any means of genetic transfer in these bacteria. The recent report of a plasmid marker in the animal pathogen *C. diphtheriae* (137) is a hopeful sign that successful genetic studies with these bacteria will be as forthcoming as they have been for other Gram-positive bacteria.

BACTERIOPHAGES AND BACTERIOCINS

Bacterial viruses have been isolated for some of the phytopathogenic corynebacteria. Most of the phages have been reported for members of the taxonomic group belonging to *C. michiganense: C. m.* ssp. *michiganense*

(46, 183); *C. m.* ssp. *insidiosum* (25); and *C. m.* ssp. *nebraskense* (146; Y. Shirako & A. K. Vidaver, manuscript in preparation). There are also reports of phages for *C. f.* ssp. *flaccumfaciens* (87) and *C. rathayi* (cited in 10). The characterized phages had relatively long latent periods, low burst sizes, and morphologies similar to the flexous-tailed types reported for *C. diphtheriae.*

It is interesting that no virulent phages have been found for *C. m.* ssp. *sepedonicum,* with the possible exception of an uncharacterized "transmissible lysin" (169a). This may be due to a type of resistance (immunity), as suggested for some phytopathogenic *Erwinia* sp. (21), or the particular enriching strain employed; e.g. samples apparently devoid of phage for one enriching strain yielded several phages for another strain of *Xanthomonas campestris* pv. *phaseoli* (A. K. Vidaver, unpublished results). It is not known whether phage-like particles associated with *C. m.* ssp. *sepedonicum* are temperate or defective phage, since no biological activity was reported (172).

No evidence of lysogeny has been reported, yet this is a matter of intriguing interest in view of the toxigenic properties carried by temperate phage of *C. diphtheriae* (120).

Bacteriocins are prevalent among this group of corynebacteria (45, 64, 112), but they are not well characterized except for some specificity data and general physical and biochemical properties.

CONTROL

Broadly speaking, control of plant pathogens can be categorized into chemical control, biological control, use of resistant varieties, cultural practices, and physical control (26). The principal forms of control for phytopathogenic corynebacteria have been sanitation, cultural practices, seed and tuber certification programs (145) that test for the presence of the organism, and plant breeding for resistance. Chemical control has been effectively used only to interrupt the chain of transmission of the potato ring rot bacterium, *C. m.* ssp. *sepedonicum,* by disinfecting equipment, tools, and sacks with formaldehyde or quaternary ammonium compounds (97, 117, 144). Heat treatment is used commonly to reduce populations of the ratoon stunting bacterium in vegetatively propagated sugarcane (M. J. Davis, personal communication). Biological control is in preliminary stages of development. Echandi (44) found that a nonpathogenic mutant of *C. m.* ssp. *michiganense* protected tomato plants significantly against subsequent challenge by a pathogenic strain, whereas Oruinbaev (118) reported a 2–8% reduction of ring rot, with increased germination, plant height, and yield by treatment of tubers prior to planting with an unidentified actinomycete. In vitro antagonists of *C. m.* ssp. *insidiosum* (69) and *C. m.* ssp. *nebraskense* (62) have also been reported, but their significance under natural conditions is unknown.

For microbiologists, there are some curious features to breeding for resistance that should be pointed out. To breed for alfalfa wilt resistance, for example, an otherwise undescribed mixture of wild-type strains of *C. m.* ssp. *insidiosum* has been used as the inoculum. This mixture is preserved from season to season by freezing roots of infected plants, presumably because pure cultures lose virulence (47). Nevertheless, enduring resistance ($>$ 25 years) has been found for this disease. This may be because of the multigenic nature of resistance and the heterogeneity of alfalfa cultivars. This analysis may be valid because resistance is defined as satisfactory if 46% of the variety is resistant to wilting (see 27). The difficulty of breeding for resistance to alfalfa wilt is compounded by the discovery that the agronomically desirable property of nitrogen fixation can actually decrease in some resistant breeding material (177). In addition, the mechanism of plant resistance is not known in any of the corynebacterial diseases. One possible base from which to begin such analyses is suggested by the finding that sap from resistant tomato (53) is inhibitory to *C. m.* ssp. *michiganense.*

In analysing resistance, plant breeders have not used quantal responses (i.e. healthy or diseased) as a criterion of resistance. Yet infectivity titrations analyzed this way in tomatoes have potential usefulness in determining not only varieties resistant to the tomato canker bacterium *C. m.* ssp. *michiganense* (8, 48), but also the heterogeneity of resistance (48). Resistance to canker in tomatoes seems to be controlled by multiple genes in complex arrangements (133).

Cultivars resistant to other diseases are not in use for several reasons. Disease-resistant cultivars have been developed for the potentially devastating disease, ring-rot, of potatoes, but have not achieved economic prominence (101). There is concern that such cultivars, if used commercially, may serve as disease carriers because of possible latent infections with the causal bacterium, *C. m.* ssp. *sepedonicum* (29, 70, 150). Very high populations (10^6 CFU/g) can be reached without showing symptoms even in susceptible cultivars (7). Field beans resistant to the wilt-inducing *C. f.* ssp. *flaccumfaciens* also have been developed (see 27), but they are not used commercially because the disease does not currently limit production. Germplasm resistant to bacterial mosaic of wheat also has been identified (J. H. McBeath, personal communication); whether or not it will be used in plant breeding depends on disease incidence and severity. In field corn, resistance to Goss's wilt and blight (caused by *C. m.* ssp. *nebraskense*) was identified and incorporated into commerical varieties soon after the disease was reported (13); the genetic basis of resistance is not yet known. However, as resistant cultivars do not yet yield as well, susceptible cultivars are still widely grown. In contrast, popcorn cultivars have shown little resistance (186). In all the cases cited, strains of bacteria that might overcome resistance have not yet been detected. This durability of resistance may be due to the complex nature of resistance, a complex of environmental conditions, or a mutation(s) in the bacteria that has not yet occurred.

There are two reports of induced (as opposed to constitutive) resistance of plants to corynebacteria. Carroll & Lukezic (18) found that alfalfa leaflets infiltrated with cell suspensions of avirulent strains protected plants inoculated later with pathogenic strains. Even gnotobiotically grown plants preinoculated in the roots with avirulent strains of *C. m.* ssp. *insidiosum* protected against challenge pathogenic strains for up to 5 weeks (18). The protective effect in leaflets and roots was localized at the site of interaction between host cells and avirulent cells of *C. m.* ssp. *insidiosum.* The results were specific for the avirulent inducer: Cell-free filtrates, other bacteria, and killed or sonicated virulent cells would not protect. The obvious question is whether specific cell components or living cells are required for this effect, the chemical basis of which is unknown. Ercolani (50), on the other hand, found that a heat-killed wild-type strain of *C. m.* ssp. *michiganense,* as well as Gram-negative bacteria, protected tomato against subsequent challenge better than an auxotrophic derivative. He suggested that some step after attachment to a multiplication site was necessary for induction of the protective response. Whether or not compounds such as phytoalexins (a class of nonspecific antibiotics produced by plants as a response to infection) play any role in resistance needs to be determined. Four different purified phytoalexins were inhibitory at low concentrations to seven Gram-positive bacteria, including *C. fascians,* but not Gram-negative bacteria (57). With Gram-negative bacteria, either localized or systemic-induced resistance can occur (e.g. 58). Obviously this is another area where more research is needed because of the possibility for eventual practical application of induced resistance.

CONCLUDING REMARKS AND PROSPECTS FOR THE FUTURE

The plant pathogenic corynebacteria are an intriguing group of microorganisms. Many general questions remain unanswered, such as their taxonomic relationship to nonpathogenic coryneforms, their interactions with the environment, their metabolism, and their genetics. Other, more specific questions were addressed in the text. There have been a number of important developments in the last quarter century; these should serve as a basis for future work.

Finally, it should be mentioned that these bacteria may have useful applications. The copious production of polysaccharides by the plant pathogenic corynebacteria might be exploited, as these may be as useful as xanthan gum, produced by the plant pathogen *Xanthomonas campestris* pv. *campestris*. Industrial production of D-alanine by *C. fascians* has been proposed (188, 189), as has the production of cytokinins by the same bacterium (181). *C. f.* ssp. *flaccumfaciens* might be useful in steroid transformation, as it can hydrolyze steroids (20). The potential usefulness of these bacteria should be evaluated.

ACKNOWLEDGMENTS

I am indebted to the following for preprints, helpful discussions, or advice: M. J. Davis, S. H. De Boer, K. E. Damann, Jr., D. W. Fulbright, D. C. Gross, J. H. McBeach, M. Sasser, S. A. Slack, E. Stackebrandt, and M. P. Starr. The editorial assistance of R. R. Carlson is gratefully appreciated.

Literature Cited

1. Akit, J., Cooper, D. G., Manninen, K. I., Zajic, J. E. 1981. *Curr. Microbiol.* 6:145–50
2. Anderton, W. J., Wilkinson, S. G. 1980. *J. Gen. Microbiol.* 118:343–52
3. Ark, P. 1951. *J. Bacteriol.* 61:293–97
4. Bateman, D. R., Basham, H. G. 1976. In *Physiological Plant Pathology*, ed. R. Heitefuss, P. H. Williams, pp. 316–55. New York: Springer-Verlag
5. Bird, A. F. 1981. In *Plant Parasitic Nematodes*, ed. B. M. Zuckerman, R. A. Rohde, 3:303–23. New York: Academic. 508 pp.
6. Bird, A. F., Stynes, B. A. 1977. *Phytopathology* 67:828–30
7. Bishop, A., Slack, S. A. 1981. *Phytopathology* 71:861 (Abstr.)
8. Boelema, B. H. 1977. *Phytophylactica* 9:5–10
9. Bousfield, I. J. 1972. *J. Gen. Microbiol.* 71:441–55
10. Bradbury, J. F. 1973. In *CMI Descriptions of Pathogenic Fungi and Bacteria*, No. 376. Kew, England: Commonwealth Mycological Inst. 2 pp.
11. Bradbury, J. F. See Ref. 10, No. 377
12. Brathwaite, C. W. D., Dickey, R. S., 1971. *Phytopathology* 61:476–83
13. Calub, S. G., Compton, W. A., Gardner, C. O., Schuster, M. L. 1974. *Plant Dis. Rep.* 58:956–60
14. Carlson, R. R., Vidaver, A. K. 1982. *Plant Dis.* 66:76–79
15. Carlson, R. R., Vidaver, A. K. 1982. *Int. J. Syst. Bacteriol.* In press
16. Carroll, R. B., Lukezic, F. L. 1971. *Phytopathology* 61:688–90
17. Carroll, R. B., Lukezic, F. L. 1971. *Phytopathology* 61:1423–25
18. Carroll, R. B., Lukezic, F. L. 1972. *Phytopathology* 62:555–64
19. Carroll, R. B., Lukezic, F. L., Levine, R. G. 1972. *Phytopathology* 62:1351–60
20. Charney, W. 1966. *J. Appl. Bacteriol.* 29:93–106
21. Chatterjee, A. K. , Starr, M. P. 1980. *Ann. Rev. Microbiol.* 34:645–76
22. Collins, M. D., Goodfellow, M., Minnikin, D. E. 1979. *J. Gen. Microbiol.* 110:127–36
23. Collins, M. D., Goodfellow, M., Minnikin, D. E. 1980. *J. Gen. Microbiol.* 118:29–38
24. Collins, M. D., Jones, D. 1980. *J. Appl. Bacteriol.* 48:459–70
25. Cook. F. D., Katznelson, H. 1960. *Can. J. Microbiol.* 6:121–25
26. Corbett, J. R. 1978. In *Applied Biology*, ed. T. H. Coaker, 3:230–330. New York: Academic. 418 pp.
27. Coyne, D. P., Schuster, M. L. 1980. In *Advances in Legume Science*, ed. R. J. Summerfield, A. H. Bunting, 1:225–33. Kew, England: Royal Botanical Gardens. 667 pp.
28. Crombach, W. H. J. 1978. In *Coryneform Bacteria*, ed. I. J. Bousfield, A. G. Calley, pp. 161–79. London: Academic. 315 pp.
29. Crowley, C. F., De Boer, S. H. 1982. *Am. Potato J.* 59:1–8

30. Cummins, C. S., Lelliott, R. A., Rogosa, M. 1974. In *Bergey's Manual of Determinative Bacteriology,* pp. 602–17. Baltimore: Williams & Wilkins. 8th ed.
31. Daly, J. M. 1981. In *Toxins in Plant Disease,* ed. R. D. Durbin, pp. 331–94. New York: Academic. 515 pp.
32. Da Silva, G. A. N., Holt, J. G. 1965. *J. Bacteriol.* 90:921–27
33. Davis, M. J. 1980. *Science* 210:1365–66
34. De Boer, S. H. 1982. *Phytopathology.* In press
35. De Boer, S. H., Copeman, R. J. 1980. *Am. Potato J.* 57:457–65
36. De Moura, R. M., Echandi, E., Powell, N. T. 1975. *Phytopathology* 65:1332–35
37. Dey, R., Van Alfen, N. K. 1979. *Phytopathology* 69:942–46
38. Diaz-Mauriño, T., Perkins, H. R. 1974. *J. Gen. Microbiol.* 80:533–39
39. Dinesen, G. 1978. *Proc. Int. Conf. Plant Pathogenic Bacteria, 4th Angers,* ed. Station de Pathologie Vegetale et Phytobacteriologie, pp. 929–33. Beaucouze: Inst. Natl. Recherche Agr.
40. Döpfer, H., Stackebrandt, E., Fiedler, F. 1982. *J. Gen. Microbiol.* In press
41. Dowson, W. J. 1957. *Plant Diseases Due to Bacteria.* Cambridge: Cambridge Univ. 2nd ed.
42. Dunleavy, J. 1962. *Phytopathology* 52:8 (Abstr.)
43. Dye, D. W., Kemp, W. J. 1977. *NZ J. Agric. Res.* 20:563–82
44. Echandi, E. 1975. *Proc. Am Phytopath. Soc.* 2:56 (Abstr.)
45. Echandi, E. 1976. *Phytopathology* 66: 430–32
46. Echandi, E., Sun, M. 1973. *Phytopathology* 63:1398–401
47. Elgin, J. H. Jr., Barnes, D. K., Ratcliffe, R. H., Frosheiser, F. I., Nielson, M. W., Leath, K. T., Sorensen, E. L., Lehman, W. H., Ostazeski, S. A., Stuteville, D. L., Kehr, W. R., Peaden, R. N., Rumbaugh, M. D., Manglitz, G. R., McMurtry, J. E. III, Hill, R. R. Jr., Thyr, B. D., Hartman, B. J. 1982. *Standard Tests to Characterize Pest Resistance in Alfalfa Cultivars,* USDA-ARS-NC-19, Revised. In press
48. Ercolani, G. L. 1967. *Phytopathol. Mediterr.* 6:19–29
49. Ercolani, G. L. 1970. *Phytopathol. Mediterr.* 9:145–50
50. Ercolani, G. L. 1970. *Phytopathol. Mediterr.* 9:151–59
51. Fiedler, F., Schäffler, M. J., Stackebrandt, E. 1981. *Arch. Microbiol.* 129: 85–93
52. Fulkerson, J. F. 1960. *Phytopathology* 50:377–80
53. Galach'yan, R. M. 1961. In *Voprosy Mikrobiologii,* ed. A. K. Panosyan, 1:21–40. Erevan: Akad. Nauk Armyan. SSR (In Russian)
54. Gates, J. E., Fisher, R. W., Candler, R. A. 1980. *Arch. Microbiol.* 127:163–65
55. Gaunt, R. E. See Ref. 39, pp. 973–74
56. George, M., Joseph, T., Potty, V. P., Jayasankar, N. P. 1976. *J. Plantation Crops* 4:23–24
57. Gnanamickam, S. S., Smith, D. A. 1980. *Phytopathology* 70:894–96
58. Goodman, R. N. 1980. In *Plant Disease,* ed. J. G. Horsfall, E. B. Cowling, 5:305–17. New York: Academic
59. Gorin, P. A. J., Spencer, J. F. T. 1961. *Can. J. Chem.* 39:2274–81
60. Gorin, P. A. J., Spencer, J. F. T., Lindberg, B., Lindh, F. 1980. *Carbohydr. Res.* 313–15
61. Goto, M., Okabe, N. 1958. *Nature* 182:1516
62. Gross, D. C., Vidaver, A. K. 1978. *Appl. Environ. Microbiol.* 36:936–43
63. Gross, D. C., Vidaver, A. K. 1979. *Phytopathology* 69:82–87
64. Gross, D. C., Vidaver, A. K. 1979. *Can. J. Microbiol.* 25:367–74
65. Gross, D. C., Vidaver, A. K., Keralis, M. B. 1979. *J. Gen. Microbiol.* 115: 479–90
66. Hale, C. N. 1972. *NZ J. Agric. Res.* 15:149–54
67. Harrison, M. D., Brewer, J. W., Merrill, L. D. 1980. In *Vectors of Plant Pathogens,* ed. K. F. Harris, K. Maramorosch. New York: Academic. 467 pp.
68. Hawn, E. J. 1963. *Nematologica* 9: 65–68
69. Hawn, E. J., Lebeau, J. B. 1962. *Phytopathology* 52:266–68
70. Hayward, A. C. 1974. *Ann. Rev. Phytopathol.* 12:87–97
71. Hess, W. M., Strobel, G. A. 1970. *Phytopathology* 60:1428–31
72. Hildebrand, D. C. 1971. *Phytopathology* 61:1430–36
73. Ikin, G. J., Hope, H. J., LaChance, R. A. 1978. *Can. J. Microbiol.* 24:1087–92
74. Jacobs, S. E., Habish, H. A., Dadd, A. H. 1965. *Ann. Appl. Biol.* 56:161–70
75. Jensen, H. L. 1952. *Ann. Rev. Microbiol.* 6:77–90
76. Jones, D. 1975. *J. Gen. Microbiol.* 87: 52–96
77. Jones, D. See Ref. 28, pp. 13–46
78. Kado, C. I., Heskett, M. G. 1970. *Phytopathology* 60:969–70
79. Kampert, M., Strzelczyk, E. 1980. *Acta Microbiol. Polonica* 29:117–24
80. Kao, J., Blakeney, E. W., Gerencser, M.

A., Damann, K. E. Jr. 1980. *Phytopathology* 70:568 (Abstr.)
81. Katznelson, H. 1958. *J. Bacteriol.* 75: 540–43
82. Keddie, R. M. See Ref. 28, pp. 1–12
83. Keddie, R. M., Bousfield, I. J. 1980. In *Microbiological Classification and Identification,* ed. M. Goodfellow, R. G. Board. London: Academic. 408 pp.
84. Keddie, R. M., Cure, G. L. See Ref. 28, pp. 47–84
85. Keddie, R. M., Leask, B. G. S., Grainger, J. M. 1966. *J. Appl. Bacteriol.* 29:17–43
86. Kern, H., Naef-Roth, S. 1971. *Phytopathol. Z.* 71:231–46
87. Klement, Z., Lovas, B. 1959. *Phytopathology* 49:107–12
88. Komura, I., Yamada, K., Otsuka, S., Komagata, K. 1975. *J. Gen. Appl. Microbiol.* 21:251–61
89. Kuhn, R., Starr, M. P., Kuhn, D. A., Bauer, H., Knackmuss, H. J. 1965. *Arch. Mikrobiol.* 51:71–84
90. Kuniswawa, R., Stanier, R. Y. 1958. *Arch. Mikrobiol.* 31:146–56
91. Kuznetsov, S. I., Dubinina, G. A., Lapteva, N. A. 1979. *Ann. Rev. Microbiol.* 33:377–81
92. Lacey, M. S. 1939. *Ann. Appl. Biol.* 26:262–78
93. Lacey, M. S. 1961. *Ann. Appl. Biol.* 49:634–44
94. Lachance, R. -A. 1960. *Can. J. Microbiol.* 6:171–74
95. Laurence, J. A., Aluisio, A. L. 1981. *Phytopathology* 71:445–48
96. Lelliott, R. A. 1966. *J. Appl. Bacteriol.* 29:114–18
97. Letal, J. R. 1977. *Am. Potato J.* 54:405–10
98. Liao, C. H., Chen, T. A. 1981. *Phytopathology* 71:1303–06
99. Mandel, M., Guba, E. F., 1962. *Phytopathology* 52:925 (Abstr.)
100. Mandel, M., Guba, E. F., Litsky, W. 1961. *Bacteriol. Proc.* 1961:61 (Abstr.)
101. Manzer, F., Genereux, H. 1981. In *Compendium of Potato Diseases,* ed. W. J. Hooker, pp. 31–32. St. Paul, Minn: Am. Phytopathol. Soc.
102. Marte, M. 1980. *Phytopathol. Z.* 97:252–71
103. Masuo, E., Nakagama, T. 1969. *Ann. Rep. Shionogi Res. Lab.* 19:121–33
104. Miller, H. J., Janse, J. D., Kamerman, W., Muller, P. J. 1980. *Netherlands J. Plant Pathol.* 86:55–68
105. Minnikin, D. E., Goodfellow, M., Collins, M. D. See Ref. 28, pp. 85–160
106. Murai, N., Skoog, F., Doyle, M. E., Hanson, R. S. 1980. *Proc. Natl. Acad. Sci. USA* 77:619–23
106a. Murray, T. D. 1982. *Phytopathology.* In press (Abstr.)
107. Nelson, G. A. 1978. *Am. Potato J.* 55:449–52
108. Nelson, G. A. 1979. *Am. Potato J.* 56:71–78
109. Nelson, G. A. 1980. *Am. Potato J.* 57:595–600
110. Nelson, G. A., Harper, F. R. 1973. *Am. Potato J.* 50:365–70
111. Nelson, G. A.,Neal, J. L. Jr. 1974. *Plant Soil* 40:581–88
112. Nelson, G. A., Semeniuk, G. 1964. *Phytopathology* 54:330–35
113. Nelson, P. E., Dickey, R. S. 1970. *Ann. Rev. Phytopathol.* 8:259–80
114. Nikitina, K. V., Korsakov, N. I. 1978. *Tr. Prikl. Botan. Genet. Sel.* 62:13–18 (In Russian)
115. Norris, J. R. See Ref. 83, pp. 1–10
116. Obi, S. K. C. 1981. *Appl. Environ. Microbiol.* 41:563–67
117. O'Brien, M. J., Rich, A. E. 1979. *Potato Diseases.* Washington DC: USDA-ARS Agric. Handb. No. 474. 79 pp.
118. Oruinbaev, S. 1958. *Tr. Inst. Microbiol. Virusol. Acad. Nauk Kaz. SSR* 2:51–60 (In Russian)
119. Owens, J. D., Keddie, R. M. 1969. *J. Appl. Bacteriol.* 32:338–47
120. Pappenheimer, A. M. 1977. *Ann. Rev. Biochem.* 46:69–94
121. Paquin, R., Lachance, R. -A., Coulombe, L. J. 1960. *Can. J. Microbiol.* 6:435–38
122. Patil, S. S. 1974. *Ann. Rev. Phytopathol.* 12:259–79
123. Patiño-Méndez, G. 1967. *Studies on the pathogenicity of Corynebacterium michiganense (E.F. Sm) Jensen and its transmission in tomato seed.* PhD thesis. Univ. Calif. Davis. 53 pp.
124. Pearson, D. 1971. *The biological activity of phytotoxic extracts from Corynebacterium sepedonicum.* MS thesis. Univ. Neb. Lincoln. 51 pp.
125. Perkins, H. R., Cummins, C. S. 1964. *Nature* 201:1105–07
126. Pitcher, R. S., Crosse, J. E. 1958. *Nematologica* 3:244–56
127. Rai, P. V., Strobel, G. A. 1969. *Phytopathology* 59:47–52
128. Rai, P. V., Strobel, G. A. 1969. *Phytopathology* 59:53–57
129. Richardson, M. J. 1979. *An Annotated List of Seed-Borne Diseases.* London: Commonwealth Agric. Bureaux. 320 pp. 3rd ed.
130. Rikard, S. F., Walker, J. C. 1965. *Phytopathol. Z.* 52:131–44

131. Ries, S. M., Strobel, G. A. 1972. *Plant Physiol.* 49:67–84
132. Ries, S. M., Strobel, G. A. 1972. *Physiol. Plant Pathol.* 2:133–42
133. Russell, G. E. 1978. *Plant Breeding for Pest and Disease Resistance.* London: Butterworths. 485 pp.
134. Sadowski, P. L., Strobel, G. A. 1973. *J. Bacteriol.* 115:668–72
135. Saperstein, S., Starr, M. P., Filfus, J. A. 1954. *J. Gen. Microbiol.* 10:85–92
136. Schaad, N. W. 1979. *Ann. Rev. Phytopathol.* 17:123–47
137. Schiller, J., Groman, N., Coyle, M. 1980. *Antimicrob. Ag. Chemother.* 18: 814–21
138. Schleifer, K. H., Kandler, O. 1972. *Bacteriol. Rev.* 36:407–77
139. Schroth, M. N., Thomson, S. V., Weinhold, A. R. 1979. In *Ecology of Root Pathogens,* ed. S. V. Krupa, Y. R. Dommergues, pp. 105–56. Amsterdam: Elsevier. 281 pp.
140. Schuster, M. L. 1975. *Univ. Neb. Agric. Exp. Sta. Res. Bull. 270.* 40 pp.
141. Schuster, M. L., Coyne, D. P. 1974. *Ann. Rev. Phytopathol.* 12:199–221
142. Schuster, M. L., Hoff, B., Compton, W. A. 1975. *Plant Dis. Rep.* 59:101–04
143. Schuster, M. L., Vidaver, A. K., Mandel, M. 1968. *Can. J. Microbiol.* 14: 423–27
144. Secor, G. A., Lamey, H. A. 1981. *Cooperative Extension Service, No. Dakota State Univ., Circular PP-507.* 2 pp.
145. Shepard, J. F. Claflin, L. E. 1975. *Ann. Rev. Phytopathol.* 13:271–93
146. Shirako, Y., Vidaver, A. K. 1981. *Phytopathology* 71:903 (Abstr.)
146a. Shurtleff, M. C., ed. 1980. *Compendium of Corn Diseases,* pp. 7–8. St. Paul, Minn: Am. Phytopathol. Soc. 2nd ed.
147. Skerman, V. B. D., McGowan, V., Sneath, P. H. A., eds. 1980. *Int. J. Syst. Bacteriol.* 30:225–420
148. Slack, S. A., Kelman, A., Perry, J. B. 1979. *Phytopathology* 69:186–89
149. Slack, S. A., Sanford, H. A., Manzer, F. E. 1979. *Am. Potato J.* 56:441–46
150. Sletten, A. 1980. *Potato Res.* 23:111–13
151. Snieszko, S. F., Bonde, R. 1943. *Phytopathology* 33:1032–44
152. Sokatch, J. R. 1979. In *The Bacteria,* ed. J. F. Sokatch, L. N. Ornston, 3:229–289. New York: Academic
153. Spencer, J. F. T., Gorin, P. A. J. 1961. *Can. J. Microbiol.* 7:185–88
154. Stackebrandt, E., Fiedler, F. 1979. *Arch. Microbiol.* 120:289–95
155. Stackebrandt, E., Lewis, B. J., Woese, C. R. 1980. *Zentralbl. Bakteriol. Parasitenkd. Infektionskr. Hyg. Abt. 1 Orig. Reihe C.* 1:137–49
156. Starr, M. P. 1949. *J. Bacteriol.* 57: 253–58
157. Starr, M. P. 1959. *Ann. Rev. Microbiol.* 13:211–38
158. Steadman, J. R., Bay, R. W., Hammer, M. J. 1979. *Proc. Water Reuse Symp., Washington,* pp. 2038–45
159. Straley, C. S., Straley, M. L., Strobel, G. A. 1974. *Phytopathology* 64:194–96
160. Strider, D. L. 1969. *No. Carolina Agric. Exp. Sta. Tech. Bull. 193.* 110 pp.
161. Strobel, G. A. 1967. *Plant Physiol.* 42: 1433–41
162. Strobel, G. A. 1970. *J. Biol. Chem.* 245:32–38
163. Strobel, G. A. 1977. *Ann. Rev. Microbiol.* 31:205–24
164. Strobel, G., Hess, W. M. 1968. *Plant Physiol.* 43:1673–88
165. Strobel, G. A., Talmadge, K. W., Albersheim, P. 1972. *Biochim. Biophys. Acta* 261:365–74
166. Stynes, B. A., Petterson, D. S. 1980. *Physiol. Plant Pathol.* 16:163–68
167. Stynes, B. A., Petterson, D. S., Lloyd, J., Payne, A. L., Lanigan, G. W. 1979. *Aust. J. Agric. Res.* 30:201–10
168. Suryanarayana, D., Mukhopadhaya, M. C. 1971. *Indian J. Agric. Sci.* 41: 407–13
169. Suzuki, K., Kaneko, T., Komagata, K. 1981. *Int. J. Syst. Bacteriol.* 31:131–38
169a. Thomas, R. C. 1948. *Ohio J. Sci.* 3: 102–06
170. Thomas, W. D., Graham, R. W. 1952. *Phytopathology* 42:214
171. Thyr, B. D. 1971. *Plant Dis. Rep.* 55:336–37
172. Trofimets, L. N., Shneider, Y. I. 1969. *Biol. Nauki* 12:96–100
173. Van Alfen, N. K., Allard-Turner, V. 1979. *Plant Physiol.* 63:1072–75
174. Van Alfen, N. K., McMillan, B. D. 1982. *Phytopathology* 72:132–35
175. Van Alfen, N. K., Turner, N. C. 1975. *Plant Physiol.* 55:559–61
176. Van Hoof, H. A., Huttinga, H., Knaap, A., Mass Geesteranus, H. P., Mosch, W. H. M., de Raay-Wieringa, D. G. J. 1979. *Netherlands J. Plant Pathol.* 85:87–98
177. Viands, D. R., Barnes, D. K., Forsheiser, F. I. 1980. *Crop Sci.* 20:-699–702
178. Vidaver, A. K. 1967. *Appl. Microbiol.* 15:1523–24
179. Vidaver, A. K. 1977. *Phytopathology* 67:825–27
180. Vidaver, A. K. 1980. In *Laboratory Guide for Identification of Plant Patho-*

genic Bacteria, ed. N. W. Schaad, pp. 12–16. St. Paul, Minn: Am. Phytopathol. Soc.

181. Vidaver, A. K., Starr, M. P. 1981. In *The Prokaryotes: A Handbook on Habitats, Isolation, and Identification of Bacteria,* ed. M. P. Starr, H. Stolp, H. G. Trüper, A. Balows, H. G. Schlegel, 2:1879–87. Berlin: Springer-Verlag
182. Vogel, P., Petterson, D. S., Berry, P. H., Frahn, J. L., Anderton, N., Cockrum, P. A., Edgar, J. A., Jago, M. V., Lanigan, G. W., Payne, A. L., Culvenor, C. C. J. 1981. *Aust. J. Exp. Biol. Med. Sci.* 59:455–68
183. Wakimoto, S., Uematsu, T., Mizukami, T. 1969. *Ann. Phytopathol. Soc. Jpn.* 35:168–73
184. Wallis, F. M. 1977. *Physiol. Plant Pathol.* 11:333–342
185. Wilson, V. E., Dunleavy, J. M. 1964. *Plant Dis. Rep.* 48:453–55
186. Wysong, D., Doupnik, B., Lane, L. 1982. *Proc. Ann. Corn Sorghum Ind. Conf., 36th, 1981,* In press
187. Yamada, K., Komagata, K. 1972. *J. Gen. Appl. Microbiol.* 18:417–31
188. Yamada, S., Maeshima, H., Wada, M., Chibata, I. 1973. *Appl. Microbiol.* 25:636–40
189. Yamada, S., Wada, M., Izuo, N., Chibata, I. 1976. *Appl. Environ. Microbiol.* 32:1–6
190. Yamada, Y., Inouye, G., Tahara, Y., Kondo, K. 1976. *J. Gen. Appl. Microbiol.* 22:203–14
191. Yanagawa, R., Honda, E., 1976. *Infect. Immun.* 13:1293–95
192. Yoder, O. C. 1980. *Ann. Rev. Phytopathol.* 18:103–29
193. Zagallo, A. C., Wang, C. H. 1967. *J. Gen. Microbiol.* 47:347–57
194. Zajic, J. E., DeLey, J., Starr, M. P. 1956. *Bacteriol. Proc.* 1956:116 (Abstr.)

AUTHOR INDEX

(Names appearing in capital letters indicate authors of chapters in this volume.)

C

D

E

K

M

Q

R

S

Y

SUBJECT INDEX

B

D

T

U

V

W

X

Y

Z

CUMULATIVE INDEXES

CONTRIBUTING AUTHORS, VOLUMES 32–36

CHAPTER TITLES, VOLUMES 32–36